Microwave Handbook

VOLUME 3

BANDS AND EQUIPMENT

Edited by M. W. Dixon, G3PFR

RSGB

Radio Society of Great Britain

Published by the Radio Society of Great Britain, Cranborne Road, Potters Bar, Herts EN6 3JE.

© Radio Society of Great Britain, 1992. All rights reserved. No part of this publication may be reproduced, stored in a retrieval system, or transmitted, in any form or by any means, electronic, mechanical, photocopying, recording or otherwise, without the prior written permission of the Radio Society of Great Britain.

Reprinted 1995

ISBN   0 900612 88 6 (set of three volumes)
        1 872309 12 7 (volume three)

Cover design by Linda Penny, Radio Society of Great Britain.
Printed in Great Britain by Bath Press Ltd, Lower Bristol Road, Bath BA2 3BL.

Acknowledgements

The principal contributors to Volume 3 were as follows:

Chapter 14, The 1.3GHz ("23cm") band – C W Suckling, G3WDG.

Chapter 15, The 2.3GHz ("13cm") band – D J Robinson, G4FRE.

Chapter 16, The 3.4GHz ("9cm") band – D J Robinson, G4FRE.

Chapter 17, The 5.7GHz ("6cm") band – D J Robinson, G4FRE.

Chapter 18, The 10GHz ("3cm") band – M W Dixon, G3PFR and J N Gannaway, G3YGF.

Chapter 19, The 24GHz ("12mm") band – S J Davies, G4KNZ.

Chapter 20, The bands above 24GHz – B Chambers, G8AGN, S J Davies, G4KNZ and H W Rees, G3HWR.

Acknowledgement is also made to other members of the RSGB Microwave Committee and many others who have helped in the writing and compilation of this book, in particular to:

Committee or past-committee members: G3HWR, G3JVL, G3PHO, G3RPE, G3WDG, G3YGF, G4CNV, G4DDK, G4FRE, G4FSG, G4KGC, G4KNZ and G8AGN.

Non-committee members: G3BNL, G3SEK, G4DDN, G4JNT, G4MBS, G4PMK, G8APP, G8DEK, HB9MIN, OZ9CR and Mr S Page.

We would also like to acknowledge material and assistance from the following journals and companies: *CQ-TV*, *DUBUS-Info*, *Microwave Journal*, *Microwave System News*, *Microwaves*, *QST*, *Radio Communication*, *VHF Communications*, Microwave Associates, Mitsubishi Corporation and Plessey Semiconductors. Other sources are acknowledged individually where appropriate.

The typesetting was undertaken by S J Davies, G4KNZ.

Contents

Preface x

14 The 1.3GHz ("23cm") band 14.1
    14.1 Introduction 14.1
    14.1.1 Television 14.1
    14.1.2 Repeaters 14.1
    14.1.3 Moonbounce 14.1
    14.1.4 Satellites 14.1
    14.1.5 Allocation to the amateur services 14.2
    14.1.6 Bandplanning 14.2
    14.1.7 Beacons 14.2
    14.1.8 Repeaters 14.3
    14.1.9 Propagation and equipment performance 14.3
    14.2 Receivers 14.4
    14.2.1 An interdigital converter for 1.3GHz 14.4
    14.2.2 A balanced receive mixer for 1.3GHz 14.6
    14.2.3 A simple preamplifier for 1.3GHz 14.7
    14.2.4 A high performance GaAs fet preamplifier for 1.3GHz . 14.8
    14.3 Transmitters 14.11
    14.3.1 An alternative solidstate tripler for 1,152/1,296MHz .. 14.11
    14.3.2 A linear transmit mixer/amplifier for 28 or 144MHz to 1,296MHz 14.12
    14.3.3 A high-level linear transmit mixer for 1.3GHz 14.16
    14.3.4 A high gain single valve 1.3GHz power amplifier 14.18
    14.3.5 Using 2C39 amplifiers 14.21
    14.3.6 Heater supply circuits for 2C39 amplifiers 14.22
    14.3.7 Alternative bias circuits for 2C39 amplifiers 14.22
    14.4 Antennas 14.22
    14.4.1 A loop Yagi antenna for 1.3GHz 14.22
    14.4.2 Stacking 1.3GHz Yagis 14.23
    14.4.3 Dish antennas at 1.3GHz 14.24
    14.4.4 A dipole/reflector feed for 1.3GHz 14.25
    14.4.5 A high-efficiency dish feed for 1.3GHz 14.25
    14.4.6 A high efficiency circularly polarised feed horn for 1.3GHz eme 14.26
    14.4.7 Helix antennas for 1.3GHz 14.27
    14.4.8 A horizontally-polarised omnidirectional antenna for 1.3GHz 14.27
    14.5 Ancilliaries and test equipment 14.29
    14.5.1 A simple in-line power indicator for 1.3GHz 14.29
    14.5.2 A simple general purpose filter 14.29
    14.5.3 An interdigital bandpass filter for 1.3GHz 14.30
    14.5.4 A simple signal source for 1.3GHz 14.30
    14.5.5 A milliwattmeter for use at 1.3GHz 14.31
    14.5.6 A double-slug tuner for 1.3GHz matching 14.32
    14.5.7 A high power in-line power indicator for 1.3GHz 14.32

	14.5.8	A local oscillator for 1.3GHz converters and transverters	14.32
	14.6	Integrated equipment	14.38
	14.6.1	Local oscillator sources	14.38
	14.6.2	Low power transverters	14.38
	14.6.3	A high powered transverter	14.38
	14.7	Conclusions	14.39
	14.8	References	14.39

15 The 2.3GHz ("13cm") band ... 15.1

	15.1	Introduction	15.1
	15.1.1	Frequency allocation to amateurs	15.1
	15.1.2	Beacons	15.1
	15.1.3	Propagation and equipment performance	15.1
	15.2	Receivers	15.2
	15.2.1	Interdigital converter for 2,320MHz	15.2
	15.2.2	Alternative lo source for 2,320MHz	15.5
	15.2.3	GaAs fet preamplifier for 2,320MHz	15.9
	15.3	Transmitters	15.11
	15.3.1	Varactor multiplier for 2.3GHz	15.11
	15.3.2	Varactor mixer for 2,320MHz	15.13
	15.4	Transmit amplifiers	15.14
	15.4.1	Single valve amplifier for 2,320MHz	15.14
	15.4.2	Single valve mixer for 2,320MHz	15.18
	15.4.3	Two valve amplifier for 2,320MHz	15.18
	15.5	Antennas	15.23
	15.5.2	Loop quad antenna	15.24
	15.5.3	Alford slot antenna	15.25
	15.5.4	Feed horn for 2.3GHz	15.25
	15.5.5	Dipole/splashplate feed	15.27
	15.5.6	Multiway combiners for 2.3GHz antennas	15.27
	15.6	Accessories	15.28
	15.6.1	Interdigital bandpass filter for 2.3GHz	15.28
	15.6.2	Slug tuner for 2.3GHz	15.28
	15.7	Acknowledgements	15.29
	15.8	References	15.29

16 The 3.4GHz ("9cm") band ... 16.1

	16.1	Introduction	16.1
	16.1.1	Allocation to the amateur service	16.1
	16.1.2	Beacons	16.1
	16.1.3	Propagation and equipment performance	16.1
	16.2	Receivers	16.2
	16.2.1	An interdigital converter for 3,456MHz	16.2
	16.2.2	A transistor preamplifier for 3.4GHz	16.5

16.3	Transmitters	16.6
16.3.1	A varactor multiplier for 3.4GHz.	16.7
16.3.2	A varactor mixer for 3.4GHz	16.10
16.4	Antennas	16.10
16.4.1	The loop quad antenna	16.11
16.4.2	Feed horn for 3.4GHz	16.12
16.4.3	Dipole/splashplate feed	16.12
16.5	Accessories	16.13
16.5.1	A 3.4GHz interdigital bandpass filter	16.13
16.6	References	16.14

17 The 5.7GHz ("6cm") band 17.1

17.1	Introduction	17.1
17.1.1	Frequency allocation to amateur services	17.1
17.1.2	Beacons	17.1
17.1.3	Propagation	17.1
17.2	Receivers	17.2
17.2.1	Receive mixer for 5.7GHz	17.2
17.3	Transmit/receive converter	17.3
17.4	Transmitters	17.7
17.4.1	A varactor multiplier for 5.7GHz	17.7
17.5	Antennas and accessories	17.8
17.5.1	A simple waveguide feed for short focal length dishes	17.8
17.5.2	Feed horn for 5.7GHz	17.9
17.5.3	Coaxial N-type to WG14 transition	17.9
17.6	References	17.10

18 The 10GHz ("3cm") band 18.1

18.1	Introduction	18.1
18.1.1	History	18.1
18.1.2	Nature of the band	18.2
18.1.3	Modes	18.2
18.1.4	Power	18.3
18.2	Waveguide components	18.4
18.2.1	Waveguides and flanges	18.4
18.2.2	Tuning and matching waveguide components	18.6
18.2.3	Waveguide diode mounts – mixers/detectors	18.8
18.2.4	Matched loads and attenuators	18.10
18.2.5	Directional couplers	18.10
18.2.6	Adjustable waveguide short circuits	18.10
18.2.7	Wavemeters	18.11
18.2.8	Coaxial to waveguide transitions	18.11
18.2.9	Waveguide switches	18.13
18.2.10	Waveguide filters	18.16

18.2.11	Miscellaneous components	18.18
18.3	Oscillators	18.20
18.3.1	Gunn oscillators and associated mixers	18.21
18.3.3	Dielectric resonator stabilised oscillators	18.29
18.3.4	Power supplies and modulators	18.31
18.3.5	Narrowband sources for 10GHz	18.36
18.3.6	Phase locked narrowband sources	18.44
18.3.7	Injection locked narrowband sources	18.44
18.4	General receiver/transmitter considerations at 10GHz	18.45
18.4.1	Receivers	18.45
18.4.2	Transmitters	18.48
18.4.3	Transmitter/receivers and transceivers	18.48
18.5	Wideband systems	18.48
18.5.1	10GHz down-converters	18.49
18.5.2	Intermediate frequency preamplifiers	18.55
18.5.3	A 30MHz to 10.7MHz crystal controlled down-converter	18.58
18.5.4	A complete 10.7MHz ("baseband") receiver and wideband transmit board	18.59
18.5.5	"Going to the races": a summary of wideband systems	18.62
18.5.6	Fast-scan atv	18.63
18.6	10GHz narrowband systems	18.67
18.6.1	Introduction	18.67
18.6.2	The relative performance of am/ssb and wbfm	18.67
18.6.3	The G3JVL image-recovery mixer/transmit converter for 10GHz ("Mark I")	18.68
18.6.4	Construction of the mixer	18.69
18.6.5	Improved versions of the G3JVL mixer	18.71
18.6.6	Preliminary alignment of the 'JVL transverter	18.73
18.6.7	A low-noise i.f preamplifier and changeover systems for 'JVL transverter	18.74
18.6.8	Final alignment of the 'JVL transverter	18.76
18.6.9	Some measurements of the performance of the 'JVL transverter	18.76
18.7	10GHz antennas and feeds	18.80
18.7.1	Horn antennas	18.80
18.7.2	Construction of 10GHz horns	18.80
18.7.3	Dish antennas and feeds at 10GHz	18.83
18.8	Test "benches" and methods	18.88
18.9	Recent developments in microwave technology	18.93
18.10	Practical 10GHz equipment using the new technology	18.100
18.10.1	The G3WDG–001 2.5 to 10GHz multiplier/amplifier	18.100
18.10.2	The G3WDG–002 10GHz to 144MHz receive converter	18.107
18.10.3	The G3WDG–003 144MHz to 10GHz transmit converter	18.112
18.10.4	Switching microwave transverters	18.116
18.11	Travelling wave tube amplifiers	18.118
18.11.1	Introduction	18.118
18.11.2	Using "surplus" twts	18.119
18.11.3	A practical twt power supply	18.119
18.12	Conclusions	18.128
18.13	Acknowledgements	18.129
18.14	References	18.129

	A1	Performance checking using natural noise sources	18.130

19 The 24GHz ("12mm") band ... 19.1

	19.1	Introduction	19.1
	19.2	Propagation	19.1
	19.3	Equipment capability	19.2
	19.4	24GHz components	19.3
	19.4.1	Waveguide and flanges	19.3
	19.4.2	Coaxial cable	19.4
	19.4.3	Waveguide to coax adapter	19.4
	19.4.4	RF sources and detectors	19.4
	19.5	Wideband transciever	19.5
	19.5.1	GDO33 and GDHO33 oscillators	19.6
	19.5.2	Gunn power supply	19.6
	19.5.3	1N26 diode mount	19.8
	19.5.4	Directional coupler	19.9
	19.5.5	Matched load	19.10
	19.5.6	Setting up the equipment	19.11
	19.6	Attenuators and loads	19.11
	19.7	Wavemeters	19.12
	19.7.1	Self calibrating wavemeter	19.12
	19.7.2	High Q wavemeter	19.13
	19.8	Antennas	19.13
	19.8.1	Horns	19.13
	19.8.2	Dish gains	19.14
	19.8.3	Dish feeds	19.14
	19.9	Narrowband transceiver	19.15
	19.9.1	Power generation	19.15
	19.9.2	Phase locked loop	19.15
	19.9.3	Local oscillator source	19.16
	19.9.4	Waveguide multiplier	19.16
	19.9.5	Bandpass filter	19.17
	19.9.6	Mixer	19.18
	19.9.7	Gunn oscillator	19.18
	19.9.8	Cross-coupler	19.19
	19.9.9	PLL circuits	19.19
	19.9.10	Operation	19.21
	19.9.11	Summary	19.23
	19.10	Filters	19.23
	19.11	Conclusion	19.24

20 The bands above 24GHz ... 20.1

	20.1	Introduction	20.1
	20.2	The bands	20.1

20.3	Propagation	20.2
20.3.1	Free space loss	20.2
20.3.2	Water vapour and oxygen attenuation	20.2
20.3.3	Rain attenuation	20.5
20.4	Band capabilities and uses	20.5
20.4.1	The 47 and 76GHz bands	20.5
20.4.2	The higher mm bands	20.7
20.4.3	System performance analysis program	20.7
20.5	Transmission lines	20.7
20.5.1	Waveguide and flanges	20.7
20.5.2	Overmoded waveguide	20.8
20.5.3	Other transmission lines	20.10
20.5.4	Dielectric waveguide	20.10
20.5.5	Coaxial lines	20.10
20.5.6	"Optical transmission lines"	20.10
20.6	Solid state devices	20.11
20.6.1	Oscillators	20.11
20.6.2	Mixers and detectors	20.11
20.6.3	Multipliers	20.11
20.7	Techniques	20.11
20.7.1	Mounting components	20.11
20.7.2	Self-oscillating mixers	20.12
20.8	Antennas	20.12
20.8.1	Horns	20.12
20.8.2	Dishes	20.12
20.8.3	Lenses	20.13
20.9	Examples of experimental equipment for 47GHz	20.14
20.9.1	Availability of components	20.14
20.9.2	A mixer/detector for 47GHz	20.14
20.9.3	An experimental Gunn oscillator for 47GHz	20.16
20.9.4	Alternative approaches to 47GHz	20.18
20.10	Acknowledgements	20.20
20.11	References	20.20

Index	xi

Preface

Volume 3 completes the *RSGB Microwave Handbook*, rounding it off with band-by-band accounts of the characteristics and potential of the amateur microwave bands and constructional descriptions of suitable, practical equipment for most of them. Many designs may be regarded as "old technology" – that is, well proven designs! Right at the outset this was justified on the basis that it would be very unwise to concentrate on "impossibly high technology" at the expense of what we know to be attainable by amateurs of modest means and skills.

There have been big changes in technology (for instance microstrip pcbs and "consumer" GaAs devices), some of which offer realistic, attainable, high performance goals for amateurs. Albeit needing some special skills, we have included some advanced but entirely practical examples and made suggestions for experimentation. We hope that these ideas will be exploited now that many of the devices and techniques are available at prices which most amateurs can afford, and that individuals will develop their own interpretations of the ideas. Above all, we hope these interpretations (and experiences) will be written up for the benefit of others – we welcome positive feedback!

We also hope that readers will continue to follow – and use – amateur developments published from time to time in *Radio Communication* and the *Microwave Newsletter* or, indeed, in any other amateur publication.

Since Volumes 1 and 2 were published, running of the Microwave Committee Components Service has been transferred to G4KGC, whose address at the time of writing (1992) is 314A Newton Road, Rushden, Northants, NN10 0SY.

Mike Dixon, G3PFR
Chairman (1991–92) Microwave Committee

CHAPTER 14

The 1.3GHz ("23cm") band

14.1 INTRODUCTION

The 1.3GHz band is of particular interest because it represents a transition between vhf/uhf and microwaves, both in terms of equipment design and operating practices. At this frequency, many of the critical components are becoming a significant fraction of a wavelength in size, which means that the techniques involved tend to be those more typical of microwave practice. On the other hand most stages of the equipment, other than those at the final frequency, are at vhf or lower frequencies. In one sense, therefore, the techniques employed are only one step in advance of what many would regard as "conventional" radio; this forms part of the appeal of experimentation in this region of the radio spectrum.

The relatively short wavelength also has a powerful influence on antenna design, in that it is easy to make antennas that have high gain but which are physically much smaller than lower frequency antennas. It is also the lowest frequency band where dishes are used in ordinary operation.

Amateurs first made regular use of the 1.3GHz band in the late 1950s. However, the equipment in use at that time was not particularly powerful or effective. A typical system might have employed low gain antennas such as corner reflectors or small dishes. The receiver could well have had a noise figure as high as 15–20dB and the transmitter output power was generally limited to a few watts. Unless operated from particularly favourable sites, the overall performance of this kind of equipment was such that it limited the range under normal conditions to tens of kilometres only, which gave the false impression that the band was suitable only for "local" contacts.

In the 1970s, a number of technical developments transformed attitudes to operating at this frequency. In particular, receivers improved as better converter designs became available, followed by the widespread adoption of preamplifiers. Transmitter powers increased and some good Yagi designs were developed, making high gain antennas practicable for all stations.

An important feature of these developments was that reasonable results could be obtained even from ordinary locations. As a consequence, operation on this band is now very well established. The ranges under normal propagation conditions are typically up to several hundreds of kilometres; during enhanced conditions, contacts over 1,000km are now quite common on this band.

More recently, there have been a number of interesting developments which will no doubt result in yet more activity on this band. These include:

14.1.1 Television

The 1.3GHz band is now the lowest frequency where there is sufficient spectrum space allocated to amateurs to permit full definition fast scan tv operation, without in-band interference problems. In particular, the wider bandwidth of fm tv can easily be accommodated. As a consequence, this technique is becoming more popular as it simplifies the construction of equipment to some extent.

14.1.2 Repeaters

These are currently being established for single channel speech. TV repeaters are also in use in some countries, including the UK. Although the design of the rf sections of microwave repeaters is different to that of their lower-frequency counterparts, the general design principles are not greatly different, as has already been outlined in chapter 9, "Beacons and repeaters". Considerable information on antenna and filter design respectively has been given in chapter 4, "Microwave antennas" and chapter 12, "Filters". Suitable rf modules will be found in this and subsequent chapters. A further recent use of specified frequencies within the band is high speed packet radio links between user-access nodes operating in other (often lower) frequency bands, effectively forming cross-band digital repeaters.

14.1.3 Moonbounce

With the improved techniques now available, it is possible to communicate worldwide on 1.3GHz via moonbounce (eme), using antennas which are small enough to be fitted into most back gardens. The equipment required is within the capabilities of many stations.

14.1.4 Satellites

With the advent of AMSAT Phase III satellites, a new type of transponder has become available, known as Mode L. The uplink and downlink frequencies for these transponders are 1,269MHz and 436MHz respectively and the

Fig 14.1. UK bandplan and usage of 1.3GHz band

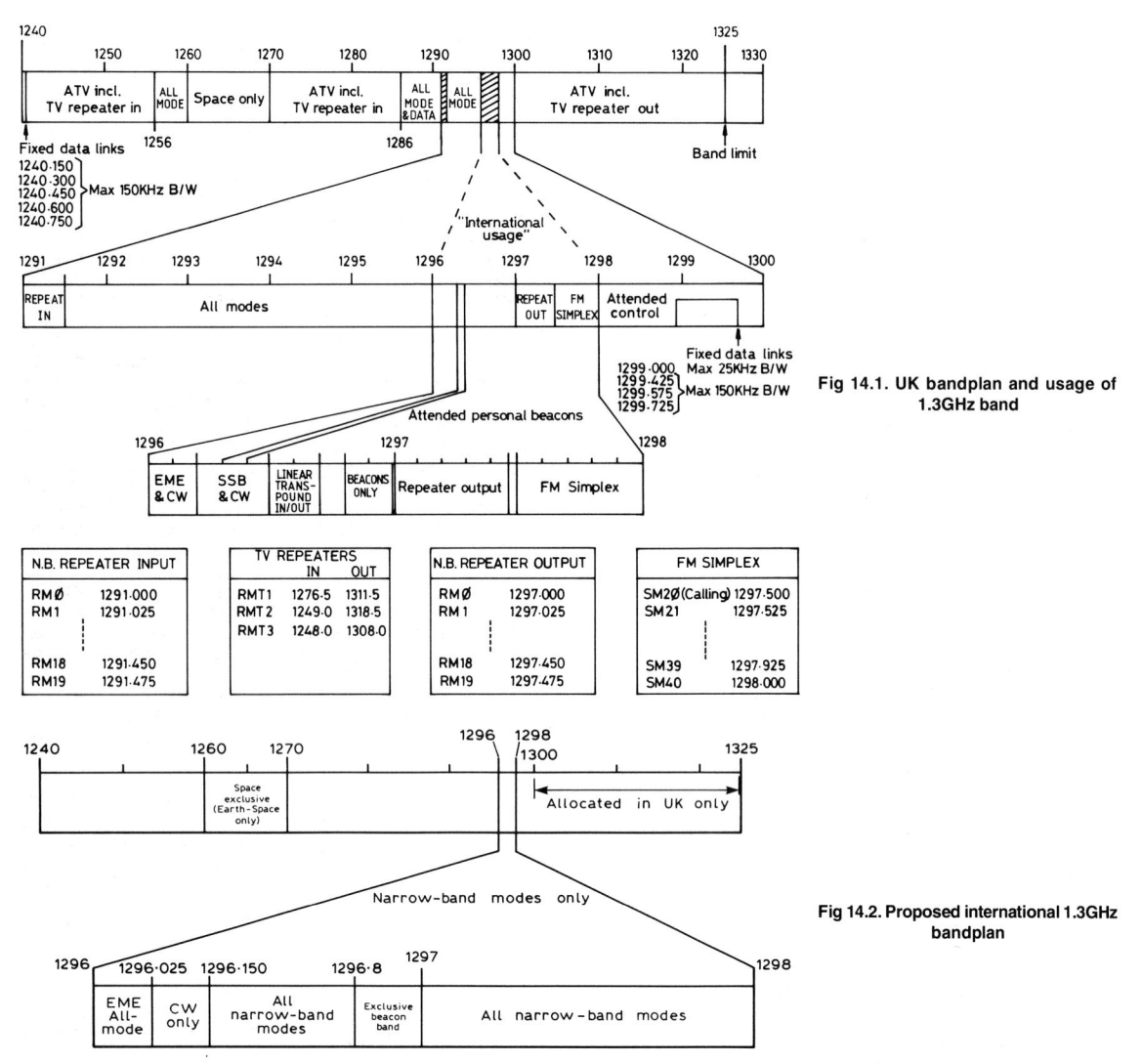

Fig 14.2. Proposed international 1.3GHz bandplan

transponder bandwidth is 800kHz. The wider bandwidth provided by this type of transponder (approximately five times that of lower frequency transponders) can support much higher activity levels and it is likely that Mode L transponders will be included in future AMSAT satellites.

14.1.5 Allocation to the amateur services

Most countries, worldwide, have an allocation in the range 1,240–1,300MHz. The primary use is for radiolocation and radio-navigation-satellite, with the amateur service designated as a secondary user. In the UK, the frequencies 1,300–1,325MHz are also available for use by amateurs on a secondary basis. Despite the use of relatively powerful equipment by both primary and secondary users, in general there has, until recently, been little, if any, mutual interference in most people's experience, although the introduction of high power radars is starting to cause some problems to the amateur service at the present time in the UK and nearer continent.

14.1.6 Bandplanning

The current UK bandplan is shown in Fig 14.1, whilst the less detailed proposed IARU Region 1 (international) bandplan is shown in Fig 14.2.

14.1.7 Beacons

Fig 14.3 shows the locations and nominal frequencies of the current UK beacons which are particularly useful for propagation monitoring. Table 14.1 is a list of callsigns, locations and frequencies of the more important European beacons currently licensed. Many of these can be heard in the UK, under favourable conditions.

THE 1.3GHZ ("23CM") BAND 14.3

Fig 14.3. UK microwave beacons, 1.3GHz

Table 14.1. List of 1.3GHz beacons in Europe, many of which are audible in the UK under tropo-lift conditions

Frequency	Callsign	Locator	Frequency	Callsign	Locator
1,295.995	LA8UHG	JO59JW	1,296.890	IC8C	JN70CT
1,296.025	DF5EO/A	JO31DV	1,296.895	DB0JC	JO40RV
1,296.050	HB9BBD/P	JN47FA	1,296.900	DB0AN	JO31SX
1,296.180	DB0AJ	JN57VX	1,296.900	OK0EA	JO70VP
1,296.280	SP9VHB	JN99TS	1,296.905	DB0AD	JO30XS
1,296.285	DK1WY	JO42MG	1,296.910	DB0JB	JN48FX
1,296.340	OE1XCS	JN88DD	1,296.917	PA0QHN	JO22FH
1,296.400	OE3XMB	JN77BX	1,296.920	DB0VC	JO54IF
1,296.800	DB0JS	JN59GB	1,296.920	SK7UHG	JO77BQ
1,296.800	SK6UHI	JO66LJ	1,296.925	SK6UHG	JO57TQ
1,296.805	DB0GP	JN48VQ	1,296.930	OZ7IGY/A	JO65HP
1,296.815	DB0VI	JN39MF	1,296.935	DB0YI	JO42XC
1,296.823	FX4UHZ	JN06WD	1,296.940	DL0UH	JO41RD
1,296.830	I1I	JN35SH	1,296.945	HB9F	JN36SW
1,296.835	SK0UHG	JO89VI	1,296.945	DB0OS	JO40CW
1,296.844	DB0KI	JO50SF	1,296.950	OZ5UHF	JO55VO
1,296.845	OZ4UHF	JO75JE	1,296.955	OZ1UHF	JO47WB
1,296.850	I5C	JN53DV	1,296.956	DF0ANN	JN59PJ
1,296.850	DL0UB	JO62QL	1,296.960	SK4UHG	JP60VA
1,296.854	DB0JO	JO31RL	1,296.975	DB0JU	JO31BU
1,296.855	I5I	JN53LL	1,296.980	PA0ZM	JO32JF
1,296.860	LA1UHG	JO59FE	1,296.985	OZ3ALS	JO45VB
1,296.865	OZ2UHF	JO46JD	1,296.990	DB0FB	JN47AU
1,296.870	FX1SHF	JN18JU	1,296.990	DB0JN	JO31WP
1,296.875	PA0EHG/A	JO32KE	1,296.995	DB0JQ	JN68TU
1,296.880	ON5UHF	JO10UN	1,297.010	DB0JW	JO30DU
1,296.880	LA3UHG	JO38WD	1,297.040	DB0LB	JN48NV
1,296.886	FX4UHY	JN06BX	1,297.153	OE3XPA	JN78SB

14.1.8 Repeaters

Fig 14.4 shows the locations and nominal input/output frequencies of a number of 1.3GHz repeater stations in the UK. A feature of the repeaters is that when they are not relaying signals they function as beacons, to allow potential users to gauge whether they will be able to access the repeater.

14.1.9 Propagation and equipment performance

Most of the propagation modes encountered on the vhf/uhf bands are also present at 1.3GHz, with the exception of those which require ionised media, such as aurora, meteor scatter and sporadic E. The most commonly encountered propagation modes at 1.3GHz are line of sight, diffraction, tropospheric scatter, ducting, aircraft scatter and moonbounce. Further information on some of these modes can be found in chapters 2 and 3.

Using the techniques described in these chapters, it is possible to predict whether communication is possible over a given path, when the performance parameters of the equipment are known. Fig 14.5 shows the path loss capability of five different sets of equipment, varying in size from a relatively modest station, such as a commercial transverter feeding a single Yagi antenna, to a station with full moonbounce capability. Also shown are the path losses of three types of propagation, line of sight (in general the lowest loss mode of propagation), tropospheric scatter (a reliable terrestrial mode of propagation) and moonbounce. It can be seen that even the modest equipment is capable of working very long line of sight paths easily (the signal to noise ratio can be estimated by subtracting the path loss from the path loss capability of the equipment). In practice, paths are not line of sight and propagation losses are higher. The tropospheric scatter loss is a reasonable upper limit for a terrestrial path and can be used to estimate the maximum range of the equipment under "normal" propagation conditions. It can be seen from the graphs that the modest equipment might be expected to work paths up to about 150km in length.

More powerful equipment will cover longer paths and equipment C, typical of that used by well-equipped stations,

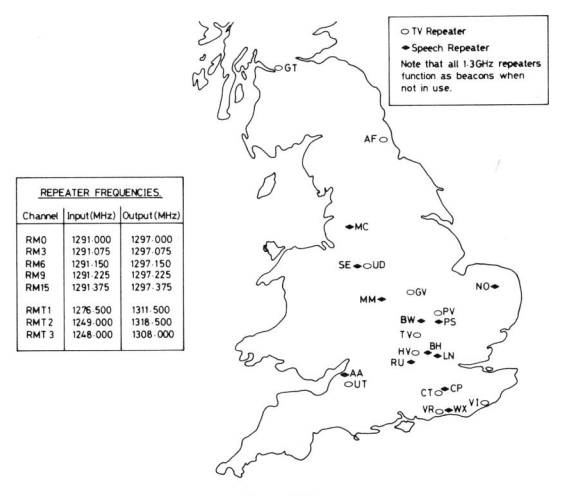

Fig 14.4. UK 1.3GHz repeaters

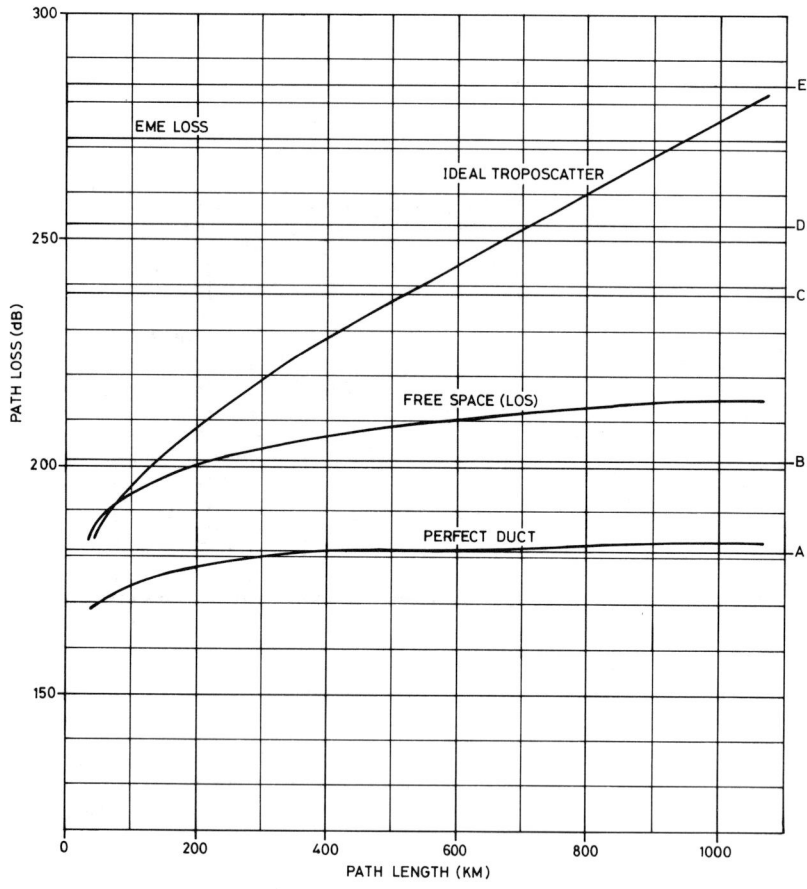

Fig 14.5. Path loss capability graph: the approximate 1.3GHz equipment requirements (normal conditions), to give the plc indicated on graph, calculated using the system analysis program given in chapter 2, section 2.7.6:
(a) 1W tx, cw, 7dB rx nf, 3kHz B/W, 15dB antenna gain, 3dB feeder loss
(b) 1W tx, cw, 3dB rx nf, 3kHz B/W, 21dB antenna gain, 3dB feeder loss
(c) 100W tx, cw, 1.5dB rx nf, 1kHz B/W, 27dB antenna gain, 2dB feeder loss
(d) 400W tx, cw, 1.5dB rx nf, 1kHz B/W, 27dB antenna gain, 1dB feeder loss
(e) Full eme capability

could be expected to work paths up to 500km on ssb, or 700km on cw. The limit for terrestrial communication under normal conditions is about 1,000km, beyond which it becomes easier to use moonbounce! Of course, under "lift" conditions path losses are considerably reduced, often to values near to line of sight and even modest equipment is then capable of working very long paths.

14.2 RECEIVERS

All 1.3GHz receivers now employ crystal-controlled converters to mix incoming signals to a convenient intermediate frequency, usually 144MHz. In this section, two types of converter will be described. The first design uses an interdigital mixer and is recommended as a simple, general purpose converter. The second design employs microstrip techniques and requires somewhat less mechanical engineering in its construction.

For most applications the noise figure of a converter on its own is too high and preamplifiers are used to reduce the noise figure. Two types of preamplifier will be described, a bipolar transistor preamplifier suitable for general applications, and a high-performance GaAs fet preamplifier.

14.2.1 An interdigital converter for 1.3GHz

The converter described below is based on a design originally published in QST [1]. It can be constructed and aligned without the use of any special facilities and has a noise figure of approximately 8dB.

The main advantage of this design is that all the necessary microwave functions, i.e. final local oscillator multiplication, filtering, mixing and signal filtering, are all performed by one assembly which is based on an interdigital filter. This means that a separate final local oscillator multiplier does not have to be constructed, nor is any additional signal filtering necessary to suppress the image response. The interdigital design has been found very easy to duplicate and get working and is in use by a large number of stations. Although a certain amount of metalwork is involved, the design is especially recommended for beginners.

The heart of the unit is the interdigital network. Referring to Fig 14.6, this consists of five rod elements L1–L5. L1, L3 and L5 are low-Q coupling elements, while L2 and L4 are resonant at the signal and local oscillator frequencies respectively and perform the filtering functions. D2 acts as a ×4 multiplier producing a few mW at the local oscillator frequency of 1,152MHz for a drive power of

THE 1.3GHZ ("23CM") BAND   14.5

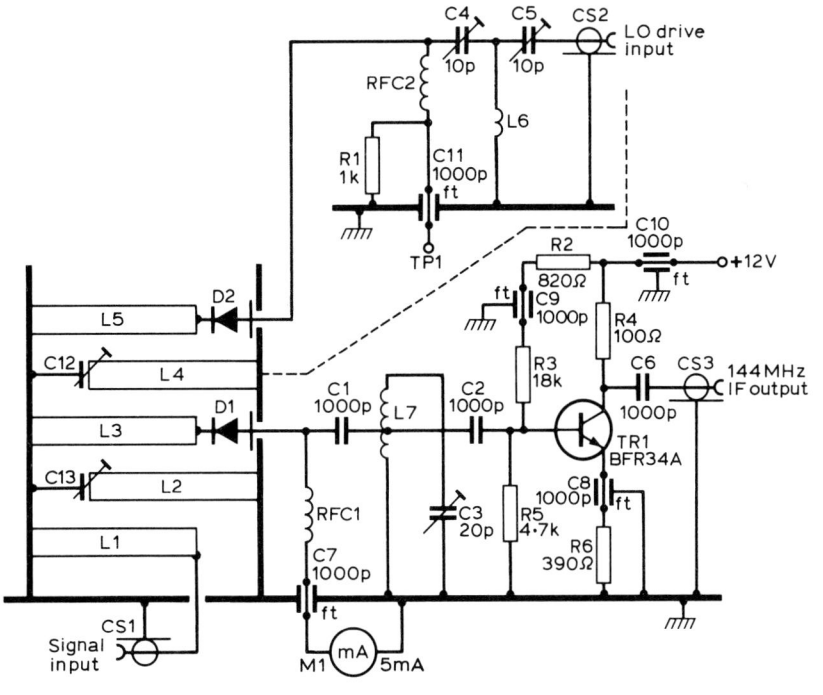

Fig 14.6. Layout of the interdigital converter

50–150mW at 384MHz. The local oscillator design to provide this drive power is described in chapter 8, "Common equipment". D1 is the mixer diode and the 144MHz i.f output is fed to a low-noise amplifier stage using a BFR34A transistor.

Constructional details

The unit is built on a 191 × 115 × 1.6mm piece of double-sided copper laminate board, which is also used as the lid for the diecast box which houses the converter. Constructional

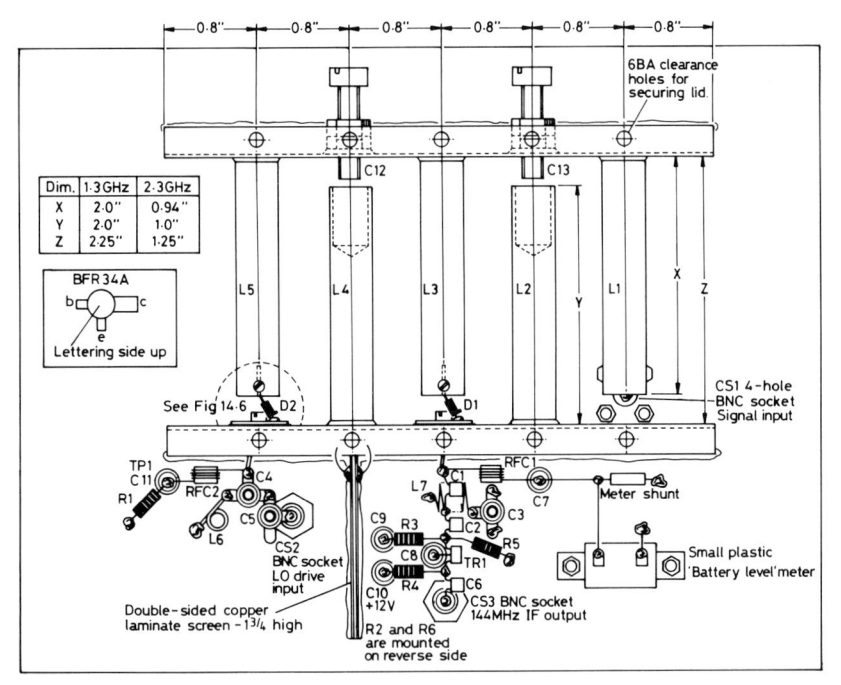

Fig 14.7. Layout of mixer assembly

Dim.	1.3GHz	2.3GHz
X	2.0"	0.94"
Y	2.0"	1.0"
Z	2.25"	1.25"

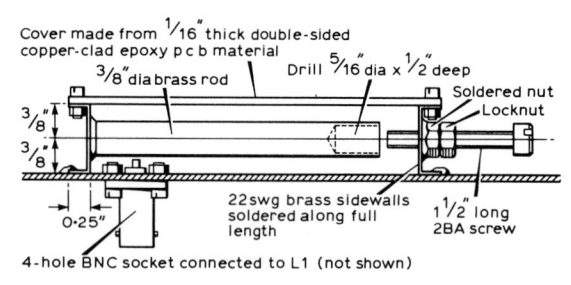

Fig 14.8. Side view of mixer assembly

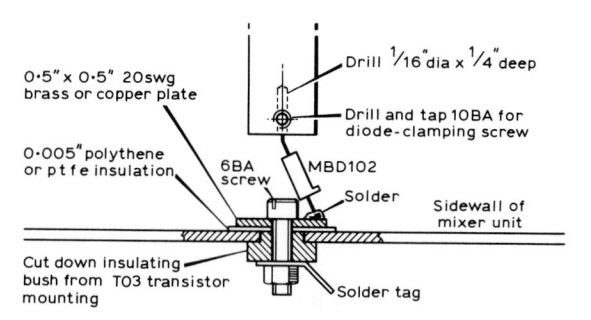

Fig 14.9. Method of fitting D1 and D2

points should be clear from Figs 14.7, 8 and 9. The side walls of the mixer assembly may be bent up from sheet brass or copper, or it can be made from rectangular brass bar. The interdigital elements can be made from brass or copper rod or tube. The method used to fix the elements to the sidewalls depends on whether rod or tube elements are used. Rod elements can be fixed by means of a tapped hole in one end, or by soldering while tube elements can be soldered, or held in place by means of a filed nut, soldered into one end of the element. A cover made from double-sided copper laminate is fitted to the mixer assembly as the final stage of construction. No screening is required at the ends of the interdigital unit. If the local oscillator source is built into the same box as the mixer assembly, the i.f preamplifier will have to be screened by means of an enclosed box which can be fabricated using double-sided copper laminate.

Alignment

The converter can be aligned without any special equipment in the following way:

1. Apply 50–150mW drive at 384MHz to the local oscillator drive input socket, and connect a voltmeter between TP1 and earth. Carefully adjust the 10pF trimmers using an insulated trimming tool until a maximum reading of at least 1.5V is obtained.

2. Monitor mixer current on the meter fitted for the purpose, insert the tuning screw S1 and screw in until maximum mixer current is obtained. This should be at least 1mA. With the specified dimensions the first peak observed will correspond to the correct tuning point.

3. Connect a 144MHz receiver to the i.f output socket and adjust the 20pF trimmer for maximum noise. If no peak is found, it will be necessary to experiment with the tapping points of the 1,000pF capacitors on L7.

4. Connect a 1,296MHz preamplifier or noise source to the antenna socket, insert tuning screw S2 and screw-in until a noise peak is heard. Alternatively, an antenna can be connected, and S2 screwed in until maximum signal strength is obtained from a local station or beacon.

5. Carefully readjust tuning screws S1, S2 and the 20pF trimmer for best signal-to-noise ratio on a weak signal. Alternatively, an automatic noise figure optimisation aid may be used. A description of available amateur equipment (receiver alignment aid) and its method of use was given in chapter 10, "Test equipment".

14.2.2 A balanced receive mixer for 1.3GHz

In contrast to the converter described above, this mixer uses two diodes in a balanced configuration and microstrip circuitry. It requires a local oscillator drive of 5–10mW at 1,152MHz from an external source, such as that described in section 14.5.7.

Construction

The mixer is built on standard 1.6mm thick glass fibre double-clad copper printed circuit board. A schematic diagram of the mixer with an untuned preamplifier is given in Fig 14.10. Dimensions of the microstrip circuitry are given in Fig 14.11. The pcb can be fabricated either by conventional etching techniques, or by using the alternative method described in chapter 7, "Constructional techniques".

If the noise figure of the subsequent receiver is greater than 2–3dB, it will be beneficial to use a low-noise i.f

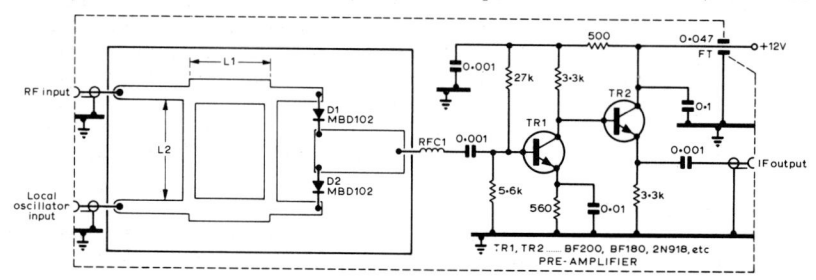

Fig 14.10. Hybrid balanced mixer and i.f. preamplifier

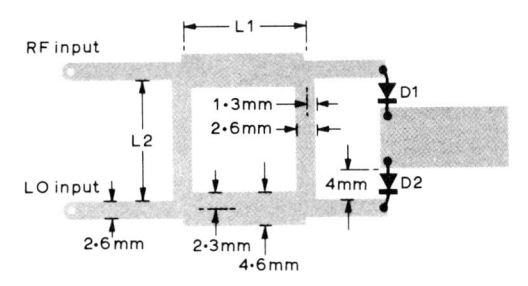

Fig 14.11. Branched-arm 3dB hybrid balanced mixer dimensions

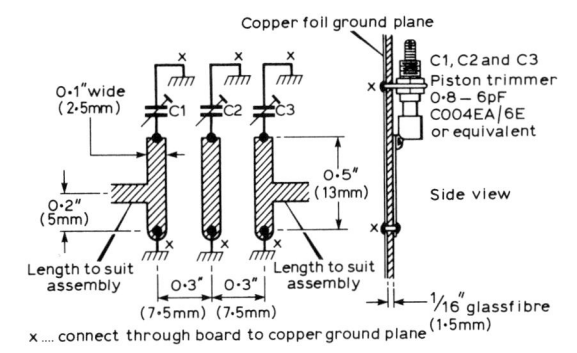

Fig 14.12. A microstrip filter. Suitable capacitors: type C004EA/6E or equivalent

amplifier after the mixer. A suitable alternative, a tuned 144MHz i.f amplifier has already been given in Figs 14.6 and 14.7. This may be built on the same board as the mixer for convenience and efficiency.

Alignment

This converter requires no alignment, except for the adjustment of the i.f amplifier by tuning it for maximum gain.

Using the converter with a preamplifier

Unlike the interdigital converter design described above, this mixer possesses no front end selectivity. If a preamplifier is used to improve the receiver performance, the overall sensitivity may be improved to a lesser extent than might be expected due to image noise. This is because the preamplifier will almost certainly generate some noise at the image frequency, which is mixed down to the i.f by the converter. This will add to the background noise, thus degrading the signal-to-noise ratio. This problem can be overcome by placing a suitable filter between the preamplifier and the mixer. The filter needs to have a reasonably low loss, otherwise the resulting reduction in pre-mixer gain may increase the mixers' contribution to the overall noise figure by an unacceptable degree.

The design and construction of a suitable filter having adequate rejection at the image frequency (1,008MHz) is described in section 14.5.2, or a simple microstrip filter (such as that shown in Fig 14.12) may be used and it could be fabricated on the same pcb as the balanced mixer.

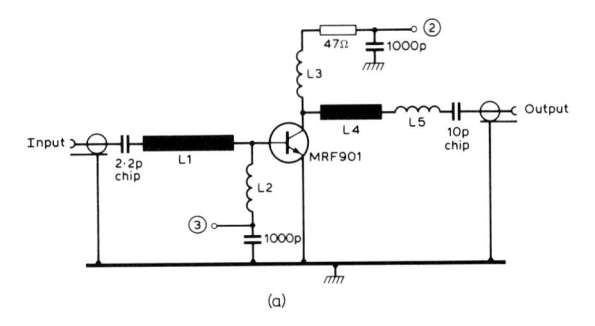

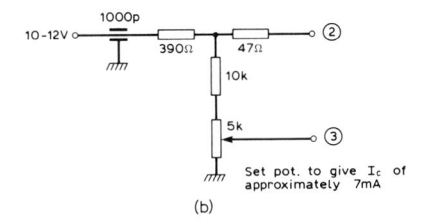

L1: 16.5mm by 2.5mm wide microstrip line
L2: 6t, 2mm diameter (0.25mm copper wire)
L3: 5t, 2mm diameter (0.25mm copper wire)
L4: 4mm long x 2.5mm wide microstrip line
L5: 6.5mm long wire flat on board (0.5mm enamelled copper)

Frequency (MHz)	Gain	NF (dB)
1240	10.7	2.9
1260	10.4	2.7
1280	10.6	2.7
1300	10.5	2.7
1320	10.0	2.8

Fig 14.13. (a) Circuit of microstrip preamplifier using MRF901. (b) Bias circuit for microstrip preamplifier. (c) Inductor values for microstrip preamplifier. (d) Performance of microstrip preamplifier

14.2.3 A simple preamplifier for 1.3GHz

Inexpensive bipolar transistors such as the MRF901, BFR90/91 and BFR34a are capable of giving adequate performance for many purposes. Preamplifiers using transistors of this type can be used on their own, and will certainly improve the performance of most converters by a considerable amount. In receivers where higher performance preamplifiers are used, bipolar amplifiers can be used to follow the first low-noise stage as this will probably not have sufficient gain to overcome the converters' noise figure alone.

The circuit of the preamplifier is shown in Fig 14.13. The input circuitry uses a series capacitor and a length of 50Ω microstrip to provide the required impedance transformation. A simple output matching network is used to provide a low output vswr, so that subsequent stages can operate at optimum performance. The gain is approximately 10dB across the band, with a noise figure of about 3dB.

Construction

The preamplifier is constructed on 1.6mm thick glass fibre double copper-clad printed circuit board. The layout is

Fig 14.14. Layout of the microstrip preamplifier

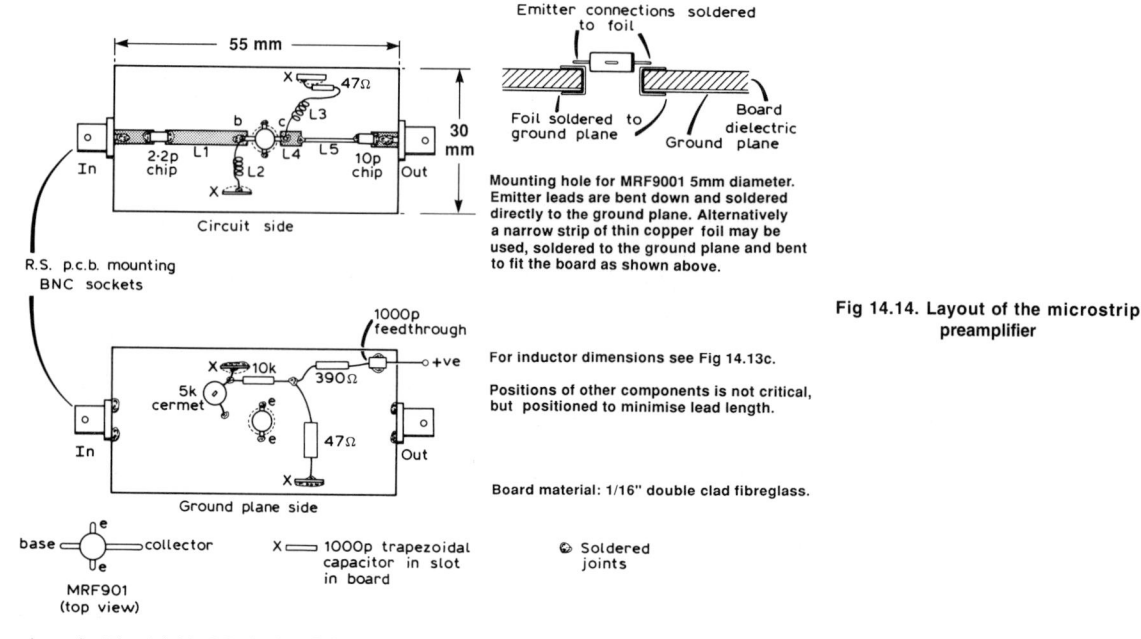

given in Fig 14.14. Methods of fabrication have already been mentioned.

Alignment

The only adjustment needed is to apply power and set the collector current to the required value (5–10mA) using the potentiometer. Provided that the preamplifier has been carefully constructed, it should then function correctly.

14.2.4 A high performance GaAs fet preamplifier for 1.3GHz

Modern GaAs fets are capable of producing very low noise figures at 1.3GHz, and preamplifiers using these transistors can give excellent performance. Indeed, the pick-up of noise from the earth can almost be said to be the major factor limiting receiver performance when GaAs fet preamplifiers are used. However, in order to realise the full potential of these devices it is essential that the input circuit of a preamplifier has a very low loss, or the added noise may degrade the performance unacceptably. Also, any practical design must allow for the fact that these devices are potentially unstable at 1.3GHz and steps should be taken to ensure stability. The preamplifier to be described was designed to take these features into account while being relatively easy to construct.

The circuit diagram of the preamplifier is shown in Fig 14.15. The low-loss input circuit consists of L1, C1 and C2. Source bias is used so that the preamplifier can be run off a single positive rail supply. The two source decoupling capacitors are leadless types, to ensure that the source of the GaAs fet is well-grounded at rf. The output circuit is untuned and consists of R2 and C5. This configuration ensures that the amplifier is stable and has a low output vswr. The value of C5 was chosen so that the capacitor is series resonant, i.e. it has a very low series impedance at 1.3GHz. A three-terminal voltage regulator is included within the preamplifier housing. This not only provides the 5V supply required, but also affords some degree of protection from voltage "spikes" on the power supply line, which could otherwise damage the GaAs fet.

Construction

Constructional details of the preamplifier are shown in Figs 14.16(a) to 14.16(g). Firstly cut out a sheet of copper for the main box, mark out the positions for the holes, make scribe lines for the corners and to indicate the position of the centre screen. Cut out the corner pieces (preferably by sawing) and drill all holes with the exception of holes E, and tap hole D. Clean the sheet using a printed circuit rubber or 'Brasso' taking care to remove any residue of the cleaning agent at the end. Bend up all four walls and adjust the corners using pliers to ensure that the walls touch. The centre screen should be fabricated next. It is best to make this piece slightly oversize initially and then to file it to be a tight fit in the main box. The screen should be cleaned after drilling and deburring the hole. The leadless capacitors are then soldered to the screen, in the positions shown in Fig 14.16(c). The soldering is best done by clamping the sheet in a horizontal plane and applying a small flame from underneath. Tin each capacitor on one side using a soldering iron and small amount of solder and tin the screen in three places, again using only a small amount of solder. Place the capacitors on the screen, tinned sides down, and reheat the screen under each capacitor in turn.

As soon as the solder melts, press down on the capacitor until it is flat on the screen. Using an ohm meter, check that

THE 1.3GHZ ("23CM") BAND

Fig 14.15. Circuit of high performance GaAs fet amplifier

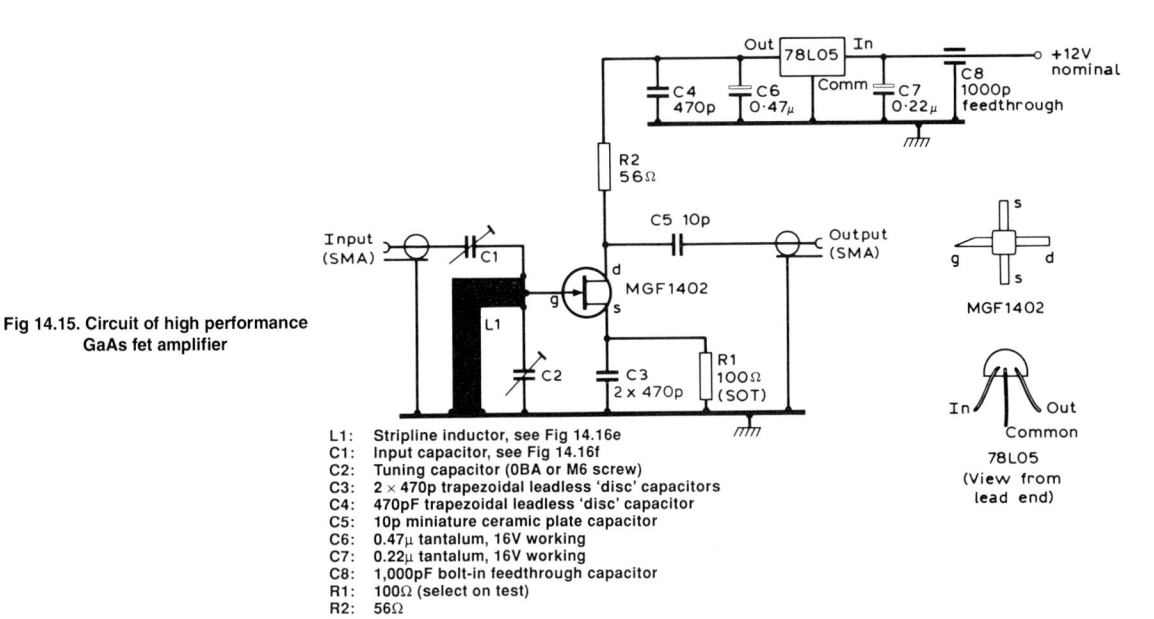

- L1: Stripline inductor, see Fig 14.16e
- C1: Input capacitor, see Fig 14.16f
- C2: Tuning capacitor (0BA or M6 screw)
- C3: 2 × 470p trapezoidal leadless 'disc' capacitors
- C4: 470pF trapezoidal leadless 'disc' capacitor
- C5: 10p miniature ceramic plate capacitor
- C6: 0.47μ tantalum, 16V working
- C7: 0.22μ tantalum, 16V working
- C8: 1,000pF bolt-in feedthrough capacitor
- R1: 100Ω (select on test)
- R2: 56Ω

the capacitors are not shorted to the screen. Remove the excess solder by filing if shorts are found. Fabricate the retaining plate as shown in Fig 14.16(d) and place it on to the screen so that the bumps align with the capacitors. Fix it in position using an M2 screw and nut.

Locate the screen in the box using the scribe marks as a guide to correct alignment. Measure the distance between the screen and the inside of the end wall of the larger compartment and check that this agrees with the dimension given in Fig 14.16(a). Move the screen if necessary. Clamp the screen in position using a toolmaker's clamp applied to the sidewalls of the main box around the screen. Jig a clean brass nut in position at hole D using a stainless steel or rusty screw to hold it in place. Tighten the nut slightly against the wall of the box. Mount the box in a vice so that the junctions of the sidewalls are horizontal and solder along the junctions using a small flame from underneath each corner in turn, re-positioning the box each time. It is easiest to preheat the corner using the flame and to make the joints with a soldering iron. This reduces the chance of re-melting the joints already made. The screen should be soldered next, along all three sides using the same technique, with the box mounted so that the open side is uppermost. Finally, invert the box and solder the nut in position.

When the assembly is cool, remove the jigging screw and run a tap through the nut and sidewall. Remove the tap and check that the tuning screw runs freely in and out. If necessary, file away any excess solder around the nut so that the output connector fits into position correctly. Drill the fixing holes for the input and output connectors. The lid is made next. Cut out the material for the lid and place the box symmetrically on the lid. Scribe around the outside of the box on the lid, cut out the corners and bend the lid over the box to form a tight fit. Using a 1.6mm bit, drill through the sidewalls of the lid and box (holes E). Tap the holes in the box and open out the holes in the lid to 2.1mm. The next stage in the construction is to make the input line, details of which are given in Fig 14.16(e). The line should be cleaned before bending up the tabs. The overall length of the line is quite critical and since errors can occur during bending, it is best to make the line slightly longer initially and then file off excess material at the "input" end of the line. Re-check all dimensions of the line using vernier calipers before proceeding. The line can then be fitted into the box, using M2 fixings.

Measure the distance between the top surface of the line and the bottom of the box, at the fixing end. This should be 4mm (±0.1mm). If necessary, remove the line and file the holes in the box so that this dimension can be achieved. Similarly, measure the height of the input end and bend the line up or down as necessary. The input tab is made next and should be bent so that when located on the input connector, the sides and top of the tab align with the tab on the input line. The tab can then be soldered to the input connector, leaving a 0.5–1mm gap between the tabs. The assembly is completed by fitting the components. The layout is shown in Fig 14.16(g). After mounting the grounding tag, solder R1 into position. When making the joint to the source decoupling capacitor, do not allow solder to run along the capacitor or it will be difficult to mount the GaAs fet correctly. Cut the leads of C5 to 2mm length and solder one end to the output connector. Bend the leads of C5 so that the free lead is in line with the hole in the screen.

Mount the feedthrough capacitor and connect up the voltage regulator circuitry. Apply power and check that +5V is available at the output of the regulator (connect a 1kΩ resistor between the output of the regulator and

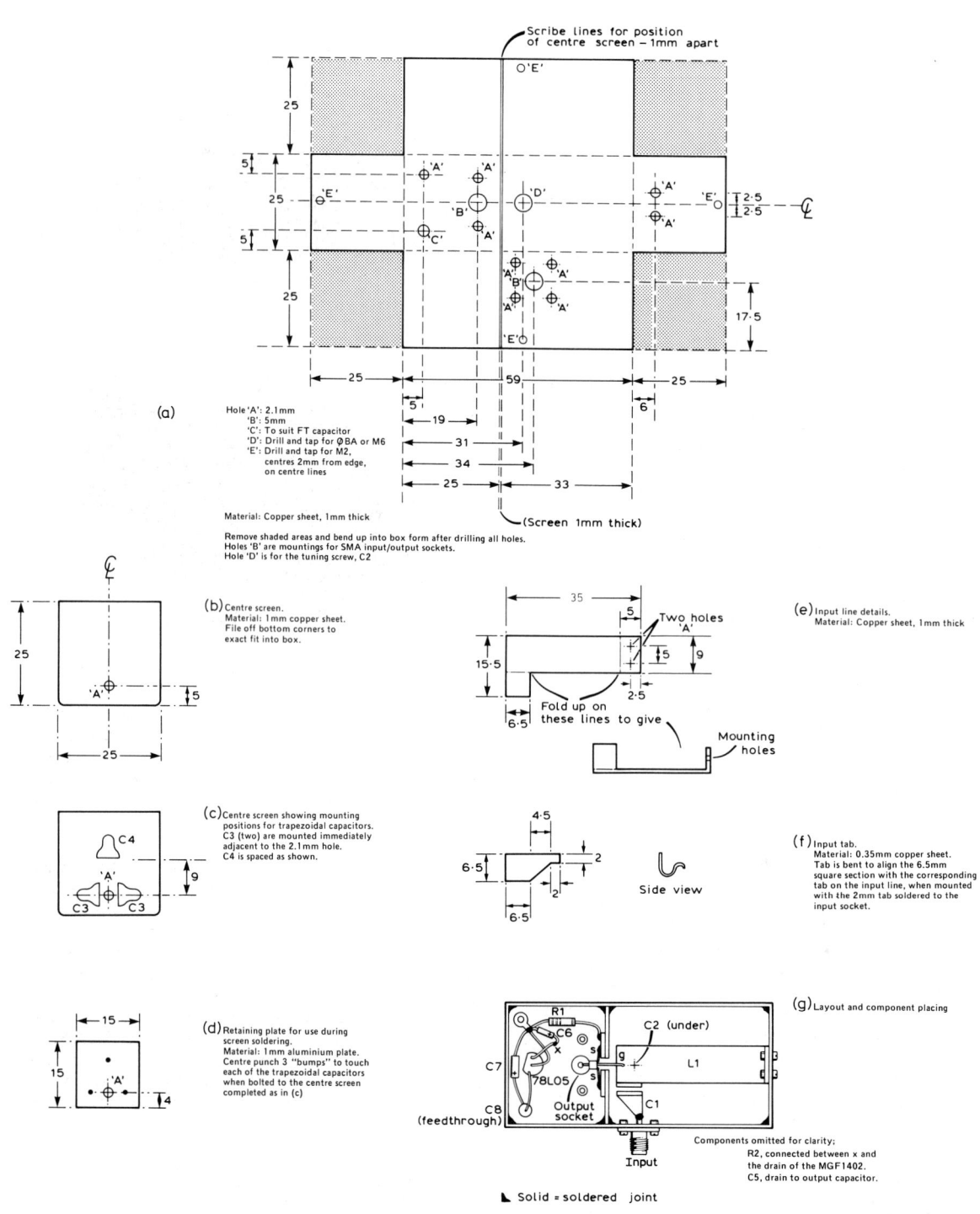

Fig 14.16. (a) Details of the main box for the GaAs fet amplifier. (b) Centre screen. (c) Centre screen showing mounting positions for trapezoidal capacitors. (d) Retaining plate for use during screen soldering. (e) Input line details. (f) Input tab: material, 0.35mm copper sheet. (g) Layout and component placing

Table 14.2. GaAs fet preamplifier performance

Frequency (MHz)	Gain (dB)	Noise Figure (dB)
1,235	15.6	1.04
1,245	16.4	0.90
1,255	16.6	0.76
1,265	16.4	0.65
1,275	16.1	0.58
1,285	15.6	0.56
1,295	14.8	0.55
1,305	13.9	0.56
1,315	12.9	0.58
1,325	12.2	0.62

ground during this measurement). Cut the source leads of the GaAs fet to 2.5mm length. Using tweezers, twist the drain lead through 90° and, holding the end of the drain lead, mount the GaAs fet into position, with its source leads touching the bypass capacitors. Unplug the soldering iron and solder the drain lead to the free end of C5. Check that the gate lead passes centrally through the hole in the screen and adjust the position of the device by careful bending if necessary. The source leads of the GaAs fet are soldered next: again, unplug the iron just before making the joints. Soldering the gate lead requires some care.

Firstly, slightly loosen the fixings of the input line, but not enough to allow the line to move. Using a minimum of a 45W iron, heat the end of the line at its edge until the solder flows. The body of the iron should be connected via a flying lead to the box during this operation. Move the iron to the centre of the line next to the gate lead. Then solder the gate lead to the line. Re-tighten the fixings of the input line. The final operation is to fit R2. Cut one lead to 2mm length and tin. With the iron unplugged, solder the other end of R2 to C4 such that the free end is touching the drain lead of the GaAs fet about 1mm from the package. Then solder R2 to the drain of the GaAs fet.

Adjustment

With the preamplifier connected in circuit, apply power and check the current drawn. This should lie somewhere between 10 and 15mA. If not, the value of R1 will need to be altered. Next, fit the tuning screw for maximum noise. Tune in a weak signal and adjust the position of the input tab and the tuning screw for optimum signal to noise ratio. Alternatively, an automatic noise figure alignment aid can be used (see chapter 10, "Test equipment"). The performance of the preamplifier is shown in Table 14.2.

For applications where noise figure is at a premium, the MGF1402 device specified can be replaced by the MGF1412–09. This device will yield a lower noise figure, by approximately 0.2dB.

14.3 TRANSMITTERS

Transmitters for 1.3GHz fall into two basic categories: frequency multipliers and linear transverters. The latter type are capable of all-mode operation and are in current

Fig 14.17. Photographs of completed GaAs fet amplifier

use by most stations. In recent times, however, simpler transmitters using frequency multiplication have reappeared on the scene, mainly for fm-only operation (e.g. for repeaters) or for television. Linear power amplifiers capable of power outputs up to about 60W will be described.

Valve and solid state triplers for 1.3GHz

The simplest way of generating rf power in the 1.3GHz band is to triple from 432MHz using triode valves such as the 2C39A. However, such triplers are seldom used these days, it being easier and more cost-effective to use solid state devices to generate up to a couple of watts or so by mixing and filtering. Those still interested in using valve triplers should read [2].

Any of the 1,152MHz multipliers described in chapter 8, "Common equipment", will readily re-tune to 1,296MHz and can then be used with existing station 432MHz equipment. As noted elsewhere in this handbook, the setting up and use of solid state multipliers needs considerable care and a minimum of test equipment to assure clean, stable output.

14.3.1 An alternative solidstate tripler for 1,152/1,296MHz

An additional design for a solid state tripler is given here, particularly suitable for multiplication from 384MHz to

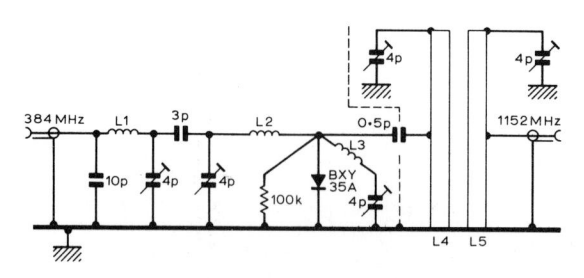

Fig 14.18. Circuit of 384/1,152MHz (or 432/1,296MHz) varactor tripler

1,152MHz, as a driver source for a high-level valve transmit mixer. It will re-tune to triple from 432MHz to 1,296MHz if this obsolescent but simple method is chosen to generate nbfm directly in the band.

The circuit of the design is given in Fig 14.18 and the leading dimensions of mechanical construction in Fig 14.19. The original design used a BXY35A varactor which is capable of a maximum input of 30W and, as a tripler, should generate 5 to 10W output; for an input of 4W at 384MHz, an output of just over 1W should be obtained at 1,152MHz. The original design was built into an RS Components diecast box, type 993. The 4pF tuning capacitors used were Mullard (Philips) type 82025–4E.

14.3.2 A linear transmit mixer/amplifier for 28 or 144MHz to 1,296MHz

The transmit converter to be described was originally published in VHF Communications [3]. It consists of a push-pull mixer and three-stage linear amplifier, which produces a power output of 300mW when operated from a 12V supply. A special feature of the circuit is the push-pull

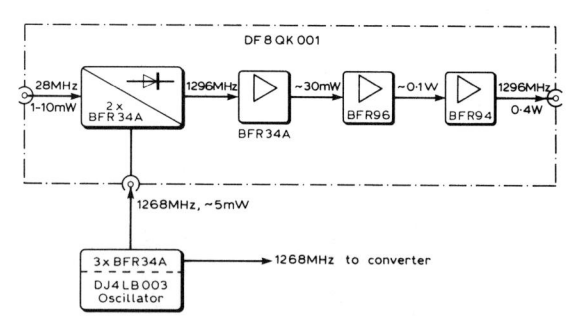

Fig 14.20. Block diagram of the solid state transmit converter (DF8QK, *VHF Communications*)

mixer which is able to suppress the local oscillator signal by at least 40dB.

A block diagram of the transverter is shown in Fig 14.20. A drive power of approximately 5mW at 28 or 144MHz is required, together with a 5mW local oscillator signal at 1,268 or 1,152MHz. The local oscillator signal should be of high purity and stable. A suitable local oscillator could consist of the uhf oscillator board described in chapter 8, "Common equipment", followed by the 384–1,152MHz tripler shown in section 8.5.7 of the same chapter, with suitable attenuation (8–10dB). Alternatively, the G4DDK–001 1,152MHz local oscillator source, described later, in section 14.5.8, could also be used and would eliminate the need for the varactor multiplier.

The circuit diagram of the transmit converter is shown in Fig 14.21. The push-pull mixer is the heart of the unit. The input signal is fed in push-pull, whereas the local oscillator signal is fed in push-push. In order to ensure that the 28MHz inductor L1 is balanced, it is wound in a bifilar

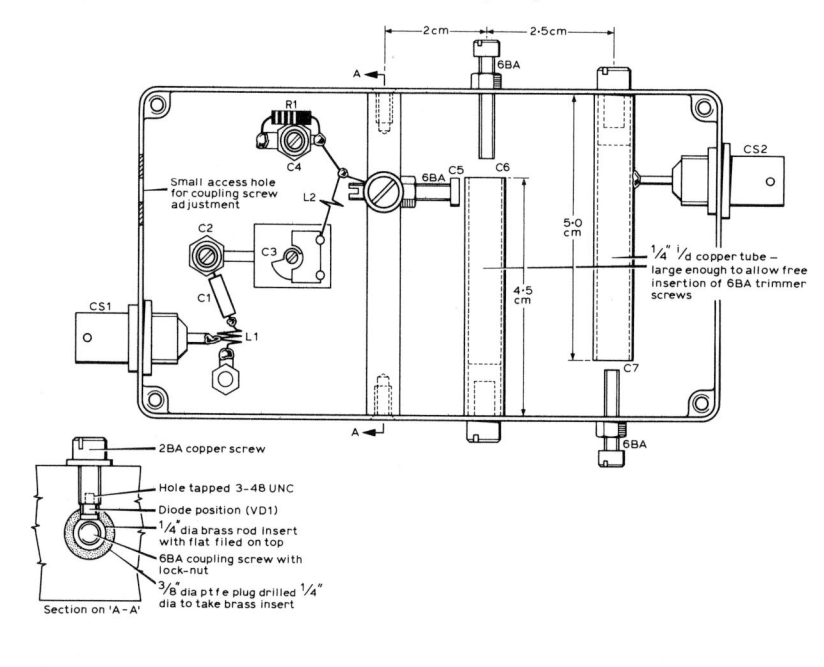

Fig 14.19. Layout of varactor tripler

THE 1.3GHZ ("23CM") BAND 14.13

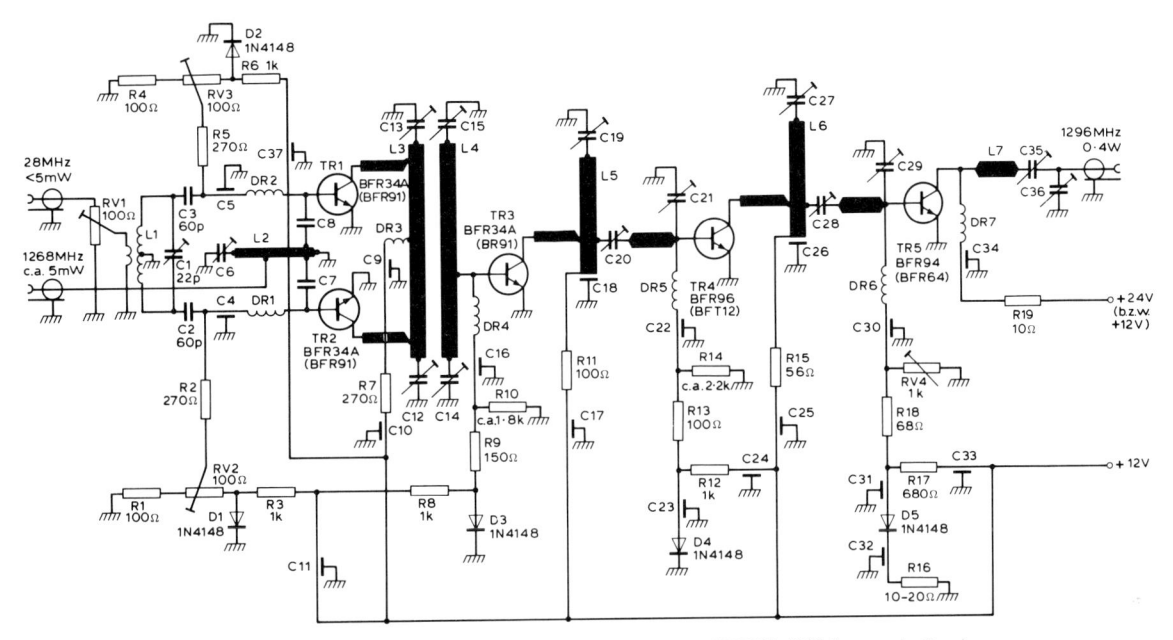

Fig 14.21. Circuit diagram of the solidstate transmit converter (DF8QK, *VHF Communications*)

manner. Since more than enough drive power is usually available, RV1 is provided to select the optimum drive level. The oscillator signal is fed in approximately 8mm from the "cold" end of the shortened quarter-wave circuit comprising L2/C6, which is then tuned to 1,268 or 1,152MHz as appropriate. The dc operating points of the two transistors are adjusted to the same values using the trimmer resistors RV2 and RV3. The microstrip inductor L3 is tuned to the output frequency of 1,296MHz by trimmer capacitors C12 and C13. Any difference in the output capacitance of the two transistors in the mixer is compensated for by suitable adjustment of the trimmers. This means that exact balancing of the output circuit can be achieved, which enables maximum suppression of the local oscillator signal to be obtained.

The loading of the output circuit of the mixer is also balanced. The first linear amplifier transistor T3 is coupled in the vicinity of the "cold" position (centre) of L4. Inductors L3 and L4 form a bandpass filter. The collector of transistor T3 is coupled via a short matching inductor to a capacitively shortened quarterwave circuit, the cold end of which is bypassed to ground by a disc capacitor. The base of the following transistor, T4, is coupled via a matching network comprising C20, C21 and a printed inductor. The collector of T4 also works into a capacitively shortened quarterwave circuit, L6.

The output transistor, T5, is matched in a similar manner. The output is transformed to 50Ω using inductor L7 and trimmer capacitors C35 and C36. All inductors, with the exception of L1, are printed striplines to ensure a high degree of reproducibility. Base bias voltages are fed via printed quarterwave chokes which are bypassed at the ends using printed capacitors.

In the design of the transmitter, special attention was paid to the base voltage dividers and the quiescent current adjustment in order to achieve minimum variation of the quiescent current with change of operating voltage, good stability with respect to ambient temperature fluctuations, non-critical alignment and constancy of the quiescent currents during continuous operation.

To achieve the latter, it is necessary for diodes D4 and D5 to be mounted in thermal contact with the associated transistor. D4 is placed on to the plastic case of transistor T4 with heat-conductive paste: diode D5 is glued to the threaded bolt of T5 also using heat-conductive paste. It is necessary for T5 to be provided with a heat sink. A strip of aluminium, approximately 50 × 10 × 2mm, is adequate; this has the ends bent up approximately 15mm from each end to form a U shape and is screwed to the tapped bolt of T5. It is to be noted that the supply voltage connection to the output stage is provided separately so that it may, if required, be connected to a 24V supply. In this case, the output power obtained will then be approximately 400mW.

Construction

The construction of the transmit converter should not cause any difficulty since there is only a minimum of mechanical work to be done. The layout of the printed circuit board is shown in Fig 14.22. The dimensions of the board are 170 × 75 × 1.6mm, and it is made from epoxy material. After drilling the holes for the components, it is necessary for slots to be sawn in the pcb with the aid of a fret saw for

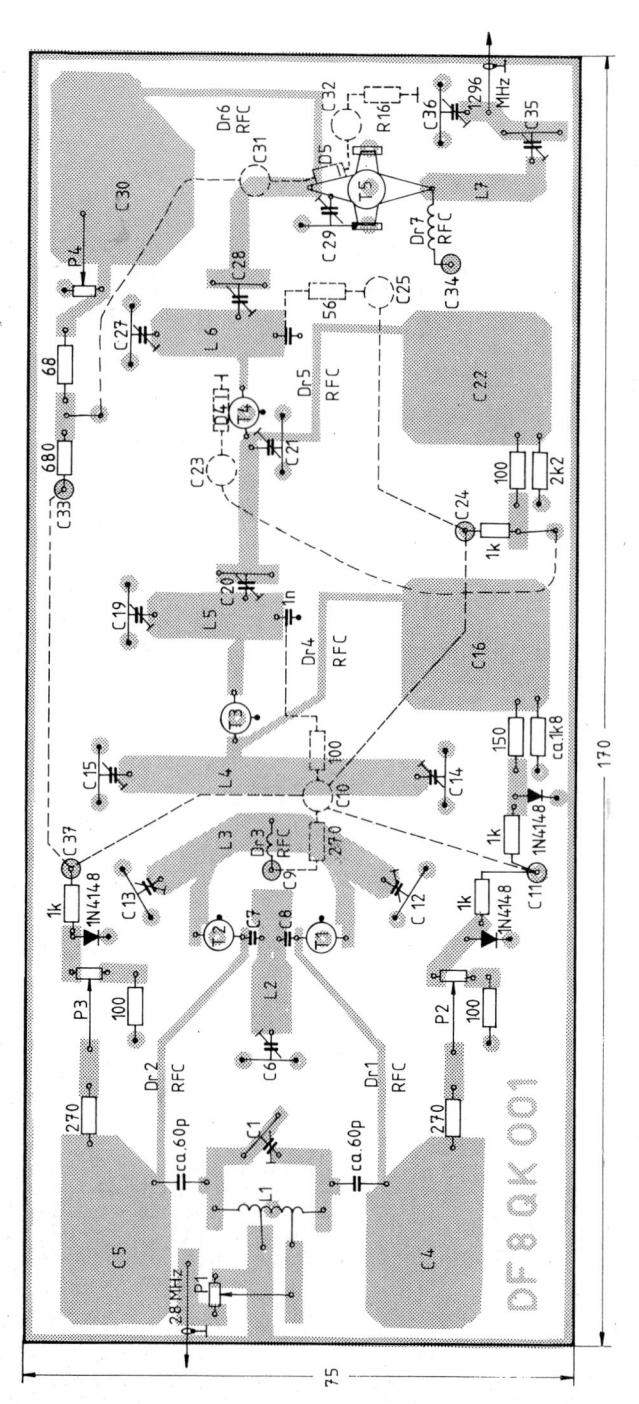

Fig 14.22. Board layout (full size) for solidstate transmit converter (DF8QK, *VHF Communications*)

capacitors C18 and C26. A further slot is necessary at the grounded end of the oscillator circuit, comprising L2, in order to be able to provide a through-contact to ground with the aid of a strip of copper foil. Two further slots of approximately 5mm in length are also required for the emitter connections of transistor T5. After this, the components can be mounted on the board. Transistors T1 to T4 are mounted into 5mm holes. The emitter lead of each transistor is bent down by 90° (bent up in the case of T1) and is placed into the hole for the transistor. This means that the hole must be slightly widened at one side for this. A hole of 7mm diameter is required for transistor T5.

One of the moving plate connections of C29 is pushed through to the ground side; the second and the stationary plate connections are bent down by 90°. Disc capacitors C10, C17, C23, C25, C31, C32, C38 are mounted on the lower side of the board.

The local oscillator signal and the output coupling from the transmit converter can be connected by soldering the connection cable directly to the board from the lower side. The required hole is made by removing about 1mm around the hole on the lower (ground plane) surface of the pcb. The inner conductor is placed through the hole, soldered to the appropriate conductor track, and the outer conductor soldered to the ground plane. It is also possible to use a BNC socket, by removing the protruding insulation so that it directly touches the ground and conductor surfaces.

The module can be enclosed in a case constructed from single-sided pcb. The height of the side panels should be approximately 40mm and the spacing of the board to base plate 15mm. Fig 14.23 shows a photograph of the completed board and Table 14.3 gives the component values.

Adjustment

Alignment of the unit requires a multimeter, an absorption wavemeter for use up to 1.3GHz and a power meter with a 50Ω load capable of measuring 500mW (see chapter 10, "Test equipment", or section 14.5.5).

1. The quiescent currents of the transistors should be set to the following values using the appropriate presets:

    T3:   2mA (0.2V across R11)
    T4:   2mA (0.11V across R15)
    T5:   16mA (0.6V across R19)

Depending on the types of diodes used in the base bias circuitry, it may not be possible to obtain the above currents. If this happens, it will be necessary to experiment with different diodes.

2. Set RV1 fully anticlockwise, RV2 and RV3 to a central position (approximately 0.52V base voltage); set C1 and C20 to approximately half mesh, C29 and C36 to one quarter mesh and all other trimmers to minimum capacitance.

3. Connect the local oscillator signal and set the level so that approximately 7V can be measured across R7. Switch off the oscillator.

4. Feed in the 28 or 144MHz signal. Align the i.f circuit using C1 and adjust the level with RV1 so that approximately 0.1V can be measured across R7.

5. Switch on the local oscillator again. Observing the voltage drop across R11, align C6 and C12–C16 for maximum

Table 14.3. Components values for transmit converter

Capacitors

C1	22pF
C6,12–15,19–21, 27–29,35,36	6pF
C2,3	56pF ceramic disc capacitor
C7,8	Approx. 10pF ceramic disc, <3mm diam.
C9,11,24,33,34,37	1nF feedthrough capacitor 3mm dia.
C10,17,23,25,31,32	Leadless disc capacitor 220pF–1nF
C18,26	1nF leadless disc
C4,5,16,22,30	Printed capacitors

All trimmers are plastic foil types of 7mm diameter

Resistors

RV1–3	100R
RV4	1k0

All potentiometers are horizontal mounting with 10/5mm spacing

Inductors

L1 (28MHz)	2 × 9 turns bifilar wound on a 6mm dia coil former without core; coupling: 1-2 turns at the centre
L1 (144MHz)	4 turns 1mm diameter copper wire (preferably silver plated) wound on a 5mm former, with centre tap and 1 turn coupling
CH3	1.5 turns enamelled copper wire wound through a ferrite bead
CH7	2 turns copper wire (preferably silver plated) wound on a 3mm former, self-supporting

Semiconductors

T1,2,3	BFR34A (Siemens)
T4	BFR96 (Philips), or BFT12 for lower output power
T5	BFR94 (Philips), or BFR64 for lower power levels
D1–5	1N4148

Fig 14.23. Photograph of completed solid state transmit converter (DF8QK, *VHF Communications*)

reading. Switch off the 28/144MHz signal after each adjustment of the trimmers to check that the voltage falls, otherwise the alignment will have been made at 1,268 or 1,152MHz.

6. Measure the voltage drop across R15 and align capacitors C19–C21 for maximum. The 1,296MHz signal at the output should now be clearly seen. Align capacitors C27–C29, C35 and C36 for maximum output power. The coarse alignment is now complete.

7. Realign all trimmers carefully for maximum output power. After this has been completed, it should be checked once again that the output frequency is 1,296MHz and not 1,268 or 1,152MHz.

8. Next, to balance the mixer, switch off the i.f signal and place an absorption wavemeter set to 1,268 or 1,152MHz as appropriate in the vicinity of the output circuit. At this point, the oscillator signal will no doubt produce a clear indication, especially in the case of 28/1,268MHz version. Adjust RV2 or RV3, followed by C12 or C13, respectively, for minimum output; the trimmers should be virtually at minimum capacitance. Switch on the 28/144MHz signal once more and align C13, or C12, as well as C14 and C15 for maximum output power at 1,296MHz. These two alignment processes should be repeated several times, at the same time increasing the coupling of the absorption wavemeter until no further improvement is possible. A spectrum analyser could, of course, be used to monitor the levels of the spurious signals during these adjustments.

After the above alignment procedure, the output spectrum of the transmit converter should be relatively clean. Additional filtering may be required, however, in the case of the 28/1,268MHz version.

The voltages given in Table 14.4 should be present across the given resistors after completing the alignment, with a supply voltage of 13V.

The transmit converter has also been used successfully for television on 1,252.5MHz when fed with an input frequency of 62MHz. An input voltage of 0.2V (approximately 1mW) is sufficient drive. In this case, L1 is 2 × 5 turns and the required oscillator frequency is 1,190MHz.

Table 14.4. Test voltages, transmit converter

28/144MHz	1,268/1,152MHz	V(R7)	V(R11)	V(R15)	V(R19)
–	–	0V	0.2V	0.1V	0.6V
–	X	7V	0.2V	0.1V	0.6V
X	–	0.1V	0.2V	0.1V	0.6V
X	X	6.5V	3V	2.7V	0.6V

14.3.3 A high-level linear transmit mixer for 1.3GHz

As an alternative to the above design, a high-level transmit mixer capable of several watts output will be described. This uses a 2C39A valve in grounded-grid. The circuit diagram of the mixer is shown in Fig 14.24 and the component values in Table 14.5. The 144MHz drive and 1,152MHz local oscillator signals are combined in the cathode circuitry and fed to the cathode of the valve. The circuit is designed to provide high isolation between the two input ports. Since the valve is operated in grounded-grid, an isolated heater supply must be used.

Construction

The cathode circuitry, together with the component values, is shown in Fig 14.25. It is built in a trough-section box, mechanical details of which are shown in Fig 14.26 and 14.27. A tight fitting lid is recommended, to prevent radiation of the 144 and 1,152MHz signals. A screen/cathode bypass capacitor fabricated from 1.6mm double copper-clad epoxy pcb material is soldered into the trough. The cathode and heater connections are made from thin brass sheet. The anode cavity recommended is the same as that used in the single valve power amplifier described below in section 14.3.4. The input circuit can be mounted on the anode cavity by a number of small L-shaped pieces of brass, soldered to the trough and screwed to the anode cavity (Fig 14.27).

Adjustment

Apply 5.5–6V to the heater and +400V to the anode and ensure that there is adequate forced-air cooling applied to the latter. Set the cathode current to approximately 20mA

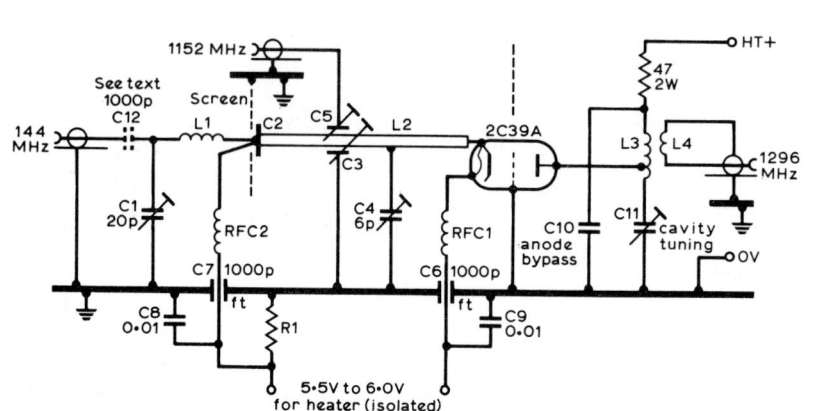

Fig 14.24. Circuit for a high-level valve mixer for 1.3GHz

Table 14.5. Component values for 1.3GHz high-level valve mixer

C1	20pF Mullard ceramic tubular trimmer
C2	20pF homemade capacitor
C3	3/8" diameter disc carries by 2BA screw
C4	6pF Mullard ceramic tubular trimmer
C5	Brass tab
C6, 7	1,000pF feedthrough
C8, 9	10nF ceramic disc with very short leads
C10	Anode bypass
C11	Cavity tuning
C12	1,000pF disc ceramic
L1	3 turns 18 swg copper wire, 8mm i.d.
L2	1,152MHz input line
L3	Anode cavity
L4	Output loop
RFC1,2	8 turns 22 swg e.c.w close wound on 1/8" mandrel
R1	Chosen to give 20-30mA standing current (typically 100 to 270 Ω with 400V ht.

by varying R1. A value of 270Ω should be found suitable for initial tests.

Approximately 2–3W of 1,152MHz drive should then be applied and the 6pF trimmer adjusted for maximum anode current. Fine tuning may be accomplished by adjusting the 2BA screw. The input matching can be optimised by bending the input tab to vary its spacing from the line and then retuning. When the anode current has been maximised it should be in the range 50–60mA. Leaving the 1,152MHz drive connected, apply about 5W of 144MHz drive and adjust the 20pF capacitor for lowest input vswr. The anode current should then be around 80–100mA. The anode tuning and loading should then be optimised for maximum power output.

Performance

When the unit is operating correctly, the anode voltage may be increased to 800–1,000V. Since the mixer is only about 10% efficient, adequate forced-air cooling is essential to prolong the life of the valve. Power output at 1,000V is about 10W and 3–4W if a 400V supply is used.

If it is intended to use the mixer alone as the transmitter, a filter should be used to reject the 1,152MHz feedthrough

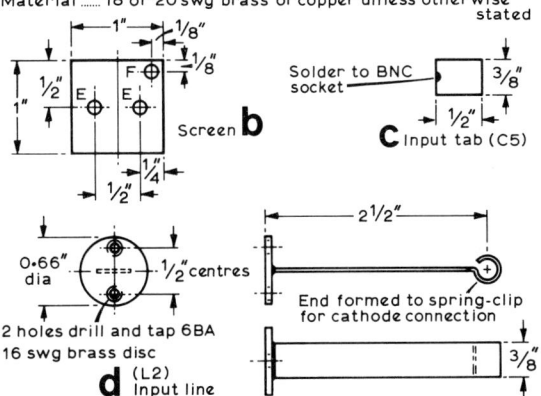

Fig 14.26. High level mixer: details of construction of (a) the trough, (b) the screen, (c) the input tab, (d) the input line

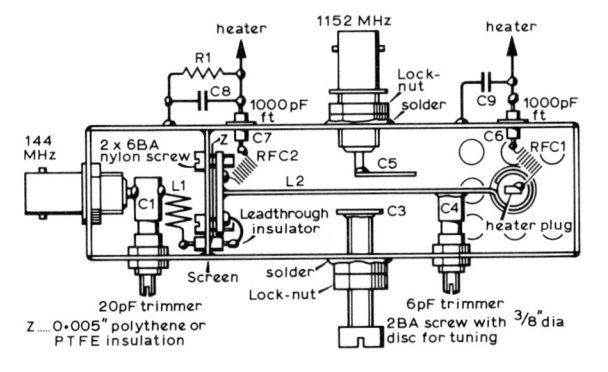

Fig 14.25. Layout of the cathode circuit, high level mixer

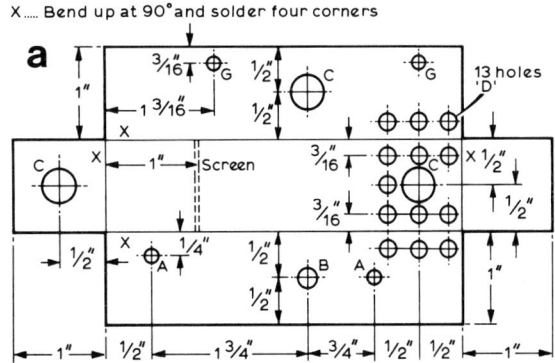

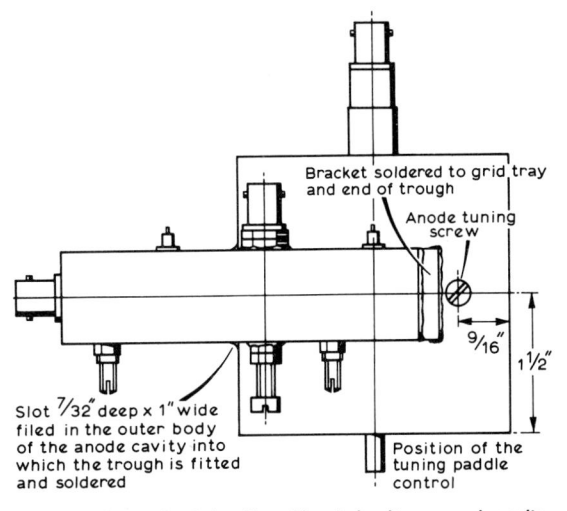

Fig 14.27. Approximate location of input circuitry on anode cavity

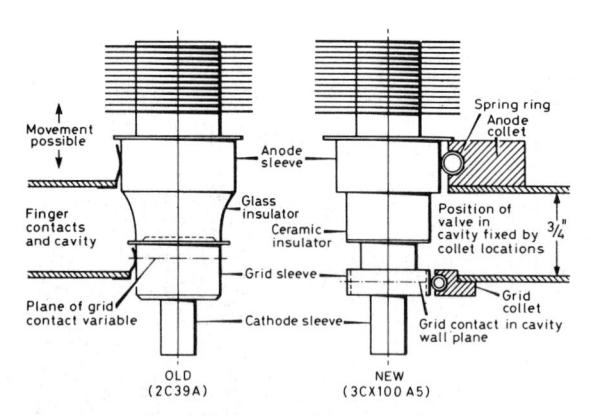

Fig 14.28. High gain 1.3GHz amplifier: old and new methods of mounting valves in cavity (omitting anode supply details for clarity)

and 1,008MHz image. A suitable design is described in section 14.5.3.

14.3.4 A high gain single valve 1.3GHz power amplifier

One of the problems confronting the 1.3GHz operator is how to raise the one to two watts available from a solid state transverter *economically* to significant output levels of tens of watts. Solid state devices for this task are available but are very expensive and, due to low stage-gain available from most transistors, several stages may be needed to reach, say, 25W, unless hybrid "power gain blocks", such as those by Mitsubishi, are used.

Most 1.3GHz power amplifiers are still designed around valves of the 2C39 family and one of the problems which beset early attempts at efficient amplifier design was achieving high gain. Most of the current amplifier designs have evolved from the box-cavity tripler first published by G2RD [4]. Although the use of power triplers is virtually obsolete in this day of transverters, the design has found new life as the basis for amplifier design.

Conversions of the G2RD tripler to both high level mixers and power amplifiers were described by G3LTF and G3WDG [5]. These designs yielded a gain of no more than 8.5dB. References to gains of 10–12dB were reported, [6] and [7], from multivalve amplifiers using the more recent ceramic-insulated versions of the 2C39. A brief reference, [8], was made in *QST* magazine to a design by N6CA in which a single valve amplifier, modified for water cooling,

yielded a gain of 13dB at "normal" drive levels and up to 15dB at lower drive levels. The keys to success in seeking high stage gain lay in the use of the highest possible eht (e.g. 1.3kV) and paying scrupulous attention to the rf grounding of the grid. Tests by N6CA showed that deliberately increasing the grid-to-ground inductance by raising only one in four of the contact fingers on the grid ring led to some 3dB loss in gain, due to the introduction of negative feedback. The design was more fully described in [9].

The present design (by G4PMK and G3SEK, [10]), based on these principles, allows gains of 15dB or more to be realised. Note that all dimensions to be given in this section are in inches, since it is still easier for the UK constructor to obtain brass strip and tubing in Imperial sizes rather than metric. Examination of valve types and the professionally designed UPX4 amplifier (converted to amateur use by W2IMU, [9], enabled the depth of the new version of the cavity to be fixed at 0.75in and a coarse tuning screw to be provided, as described by G3LTF and G3WDG. The differences in the valve mounting positions for the old and new designs is shown in Fig 14.28.

The greatest problem in amateur designs, using the 2C39 series of valves, has always been the contact rings for the grid and anode. To insist on an extremely low-inductance grid contact in the plane of the baseplate makes matters worse than ever!

"Straight" finger-stock (Fig 14.28) is ruled out because it projects either into or out of the anode cavity. Folded-over finger-stock is used in commercial preformed grid rings, but is not readily available; N6CA's experiments suggest that the inductance of the resulting contact is barely low enough.

The solution is to use a ring of spiral spring to contact the valve, the spring-ring itself being held in a collet (Figs 14.29 and 30). In effect, the valve is contacted by several quarter turns of the spring, all of which are electrically in parallel, combining to make a contact of extremely low

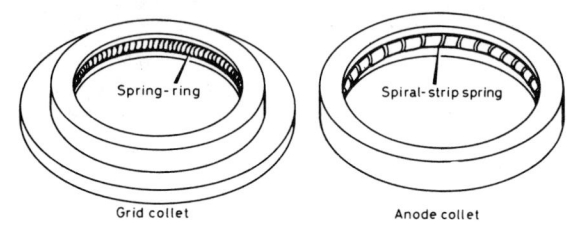

Fig 14.29. Grid and anode collets, showing two alternative types of spring material

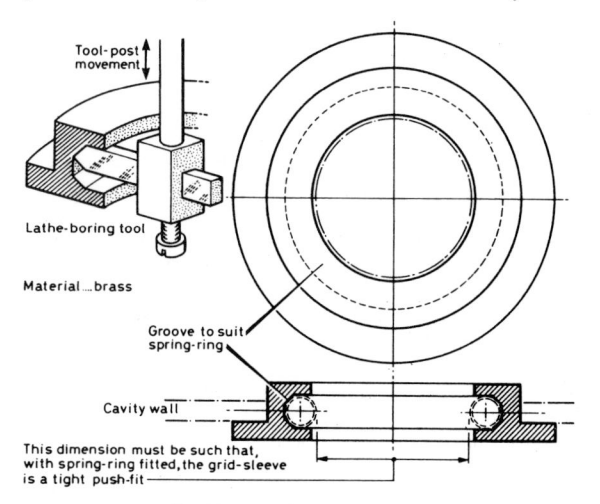

Fig 14.30. Details of grid collet

Fig 14.31. General view of the G4PMK/G3SEK single valve 1.3GHz power amplifier

inductance. The collet can be let into the base of the cavity so that the contact is made in the correct plane.

The ideal spring-ring material is a loosely-wound, silver-plated spring of about 0.25in diameter. A perfectly acceptable homemade substitute is a spiral wound from narrow (e.g. 0.1in wide) phosphor-bronze strip, such as draught-excluder; this gives fewer contacts to the valve but each turn of the strip has lower individual inductance. Fig 14.29 shows the two alternative types of spring-ring in their collets.

Precise dimensions of the collet depend, to a large extent, on the available spring-ring material and the prototypes were turned by "cut-and-try" out of old brass vacuum fittings. The first step is to bore out the blank to just clear the grid sleeve of the valve. Then the internal groove is formed using a small boring tool (inset, Fig 14.30), repeatedly trying first the spring-ring alone for size and, in the later stages, both the spring-ring and valve. The fit of the valve can also be adjusted by pulling or squeezing the spring-ring. When all is well, the valve will be gripped gently but uniformly as it is twisted into place. Owing to the "lay" of the turns of the spring-ring, the valve can only be twisted in one direction – the same for insertion and removal – so if spring-rings are used for both the grid and anode connectors they *must* be wound in the same sense.

The anode is much more forgiving of stray inductance than the grid contact, so a ring of ordinary finger-stock would probably suffice. The prototypes used spring-ring anode connectors.

Many other features of the G2RD/G3LTF/G3WDG designs were retained. The cathode input circuit closely follows the original, the coarse tuning screw has already been mentioned and the fine tuning paddle and coupling loop are also as before. Coupling with the magnetic field in the cavity is strongest when the loop is almost fully withdrawn to the cavity wall and at right-angles to the baseplate. Coarse loading adjustment is by sliding the loop in and out, and fine adjustment by rotating it.

A general view of the amplifier is shown in the photograph (Fig 14.31) and the leading dimensions are given in Figs 14.32 to 14.36. Non-critical dimension are not given,

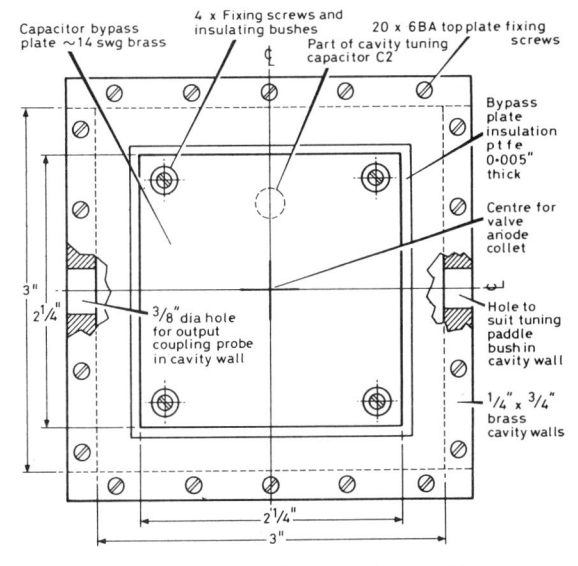

Fig 14.32. Top view of anode cavity assembly

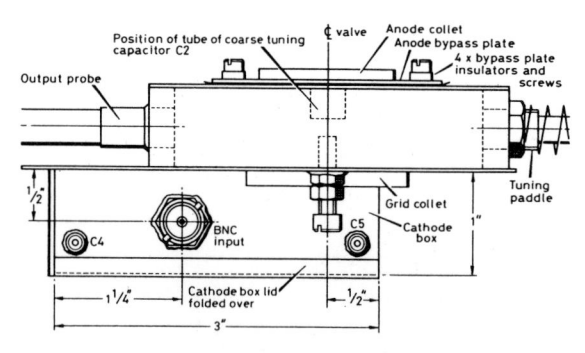

Fig 14.33. Side view of anode cavity assembly

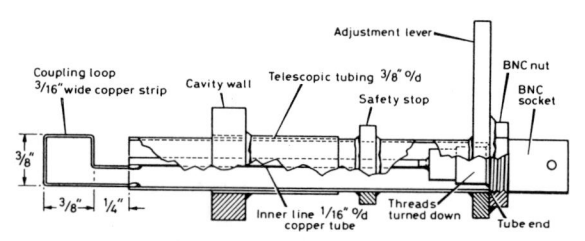

Fig 14.35. Output coupling probe

being at the discretion of the constructor. As noted earlier, the dimensions of the anode and grid connectors are only critical in that they must be adjusted to provide a good fit to the valve. However, the two collets *must* be coaxial in order to avoid shear forces on the valve and detailed assembly instructions are given at the end of this section.

The cathode circuitry below the baseplate (Figs 14.33 and 14.34) is assembled after the grid collet has been soldered into place. Rather than fabricating the rf bypass capacitor for the "cold" end of the cathode stripline, the present design uses the entire end-wall to act as the capacitor by making it from double-sided glass-fibre pcb (Fig 14.34), chamfering the copper from the inside edges to prevent a dc short-circuit.

Contrary to popular belief, grounded grid amplifiers are not unconditionally stable and this high-gain design requires some attention to the possibility of stray feedback paths. Some stability problems were encountered when one of the prototypes was operated very close to the transverter driving it, the system gain at 1.3GHz being of the order of 40dB. The top of the cathode compartment was therefore covered with a close-fitting lid of perforated copper sheet. This, together with careful bypassing of the heater supplies, solved the problem completely.

The sliding loop coupling probe (Fig 14.35) is made using telescoping brass tubing available from good model shops. A safety stop *must* be provided to prevent the loop from touching the anode sleeve of the valve. The tuning paddle (Fig 14.36) needs to be well grounded to rf. This can be ensured by a strong compression spring over the shaft, which maintains a firm contact between the paddle and shaft bushing. It is helpful if the external controls indicate the true orientations of the loop and paddle within the cavity. As an optional refinement, all components can be silver-plated. The brass parts of the prototypes were given a thin but tenacious coating of silver by the method given in the constructional chapter.

The circuit diagram of the amplifier is very simple (Fig 14.37). For maximum gain, a fairly high standing current of the order of 50mA is required, i.e. dc efficiency has to be sacrificed. At low drive levels the amplifier will operate at virtually constant anode current, so simple cathode-resistor biasing will suffice. During development of the amplifier a 250Ω wire-wound potentiometer proved quite satisfactory, and a 22kΩ resistor connected from the cathode bypass to ground allows the valve to cut off safely during receive periods or if the bias resistor fails. At higher drive levels, constant voltage biasing must be used in order to maintain linearity on ssb, and an arrangement in which a single transistor acts as both bias regulator and t/r switch is shown in Fig 14.37. The zener diode sets the cutoff bias on receive and limits the transistors' collector voltage to below Vceo.

The usual precautions regarding heater voltage should be observed when using the amplifier. At no time must the heater voltage exceed 6V. It is important that the cathode of the valve be allowed to reach full operating temperature

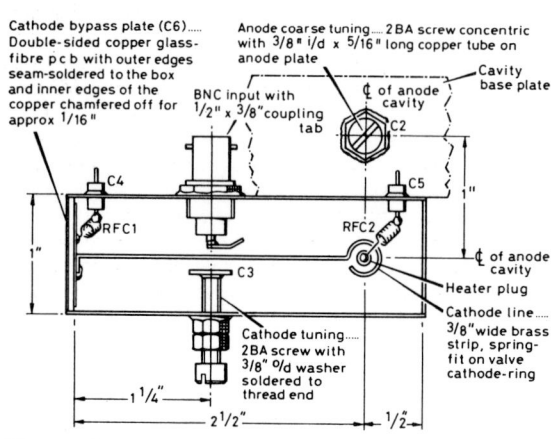

Fig 14.34. Details of cathode box and anode coarse tuning capacitor

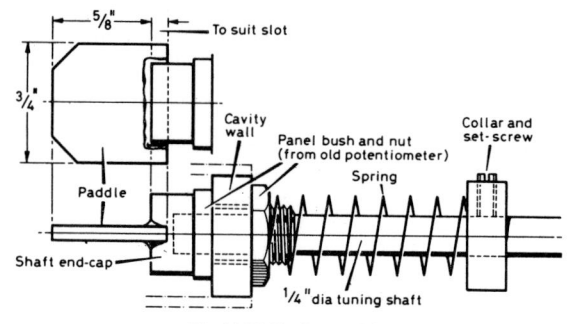

Fig 14.36. Tuning paddle

before the anode voltage is applied. A delay of 60–90s is adequate.

The power gain achievable will depend on the type of valve and on its operating history, if it is second-hand. One of the prototype amplifiers, using a good but not remarkable 7289, gave the following measured performance with an eht supply of 1kV, when the input and output matching were optimised to suit the available level of drive power.

Available drive power (W)	0.35	0.5	1.0
Output (W)	27	32	40
Power gain (dB)	19	18	16

The prototypes were developed using eht supplies of 1.0 to 1.1kV. Some reduction in gain was found at 800V and, in a brief test using 1.5kV, one of the prototypes gave 60W rf output for 1W of drive.

Assembly of the cavity

1. Mark out the locations of the side walls and the centre of the cavity on the baseplate.

2. Solder the sidewalls in position. If the ends of the bars can be faced-off square (not impossible by hand or with a three-jaw lathe chuck) they may be pre-assembled into a square frame before soldering. After soldering, hone the top face of the sidewalls flat.

3. Mark out and pilot drill the 20 fixing holes in the cavity top-plate. Do *not* drill the centre hole yet. Tape the top plate accurately into position on the side walls and, on a drill press, drill two holes in diagonally-opposite positions through the top plate and into the side walls. Tap these two holes and secure the top plate more firmly before drilling and tapping the rest of the fixing holes.

4. Again on a drill-press, drill square through the pilot hole in the base plate and through the top plate.

5. Use the pilot hole in the cavity top plate to locate the centre of the anode bypass plate when marking and drilling through the latter for the four retaining screws.

6. Open out all pilot holes to full size. Be careful to retain concentricity.

7. Drill the four holes in the anode bypass plate slightly oversize for the shoulders of the available insulating bushes. Leave the retaining screws slack until the valve has been fitted squarely into place for the first time, then tighten them.

Adjustment

Apply about 800V ht, and set the cathode current to about 50mA with no drive. Apply about 0.5W of drive and adjust the cathode tuning to maximise total cathode current. Then tune the anode cavity and adjust the output loop for maximum power output. The anode voltage can now be increased up to 1.1kV and the bias should be reset to give 50mA cathode current with no drive. All tuning adjustments should be re-optimised. Drive power can be increased until the cathode current reaches 150mA. Higher drive levels than this will produce more power, but valve life will be shortened. At high drive levels, do not run the amplifier at full output continuously for more than a few seconds.

Adequate cooling must be provided, preferably directing the air-stream over the anode cooling fins by means of a duct made from Perspex or Formica. The spring-ring provides a good thermal contact and helps to keep the grid cool. Overheating of the grid can cause electron emission, leading to dc instability and shortened valve life. Efficient anode cooling can aid in keeping the whole valve cool.

All amplifiers of this type can suffer from differential thermal effects during transmit and receive periods. This difficulty can largely be overcome by cooling the valve adequately on transmit and reducing or removing the air-flow on receive, so that its temperature remains more nearly constant. A further improvement could be expected from the use of one of the modern temperature-compensated derivatives of the 2C39, e.g. the 7855.

14.3.5 Using 2C39 amplifiers

In the tuning instructions given above for the power amplifiers, the methods used are suitable for initial testing only. At the higher power levels, it will be noticed that the

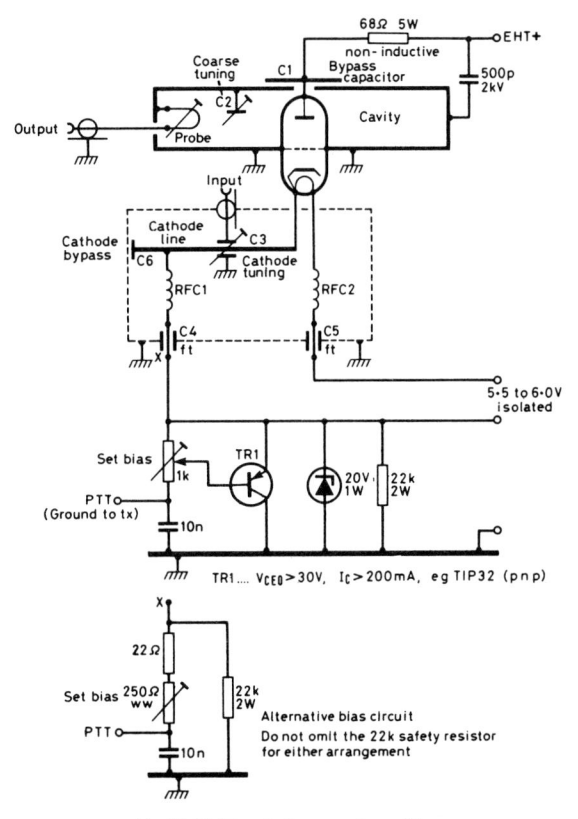

Fig 14.37. Circuit diagram of amplifier

amplifiers tend to drift off tune. This is due to internal heating of the valve(s), which causes the internal capacitances to change due to expansion. The most noticeable effect is detuning of the anode circuit.

A special procedure for tuning the anode circuit is required to minimise the tuning drift during on-air operation. Assuming that the amplifier has already been tuned up approximately, run up the amplifier with full carrier drive for about 10 seconds. Then, change over to the mode in use (cw or ssb) and transmit normally for about 30 seconds, keeping the anode circuit on tune if any drift occurs. Final tuning is accomplished by a short period (no more than a few seconds) of full drive (key down, or whistle) and a quick retune. The amplifier should then stay on tune for long periods, with no more than an occasional "warm tune", as described, being necessary.

After a period of receiving, the amplifier will not give full power immediately on transmit. Resist the temptation to retune but keep transmitting, possibly at full carrier to accelerate the process, until full power is reached. The time taken to reach full power can be reduced if blowers are switched off during receive periods.

14.3.6 Heater supply circuits for 2C39 amplifiers

Since the 2C39 family of valves is normally operated in grounded-grid and the cathode and heater are connected together, this means that an isolated heater supply is required, or the valve cannot be biased correctly. When a mains supply is available this is not a problem since a transformer can be used, with a separate winding for each heater if more than one valve is used. However, when a single 12V dc supply is all that is available (as might be the case when operating portable), providing isolated supplies is rather more difficult. Two circuits are given to overcome this problem.

The first, due to G3ZUD, is suitable for use when one 2C39 is to be powered from a 12V source and is shown in Fig 14.38. TR1, 2 and 3 form a stabilised supply for the

Fig 14.39. Circuit diagram of the GW8AAP 2C39 heater inverter power supply. L1 and L2 wound on LA1 or LA3 pot-core. L2 (1mm enamelled copper wire) wound first, L1 (0.5mm enamelled copper wire) wound second

heater; R1 should be selected on test to give the desired output voltage (5.7 to 6.0V). TR4 and 5 form a simple constant voltage reference for the cathode bias. The voltage at output 2 can be varied from 1.5 to 6V and the corresponding output at output 1 varies between 7.5 and 12V. Thus, connecting the cathode to either output allows a wide range of cathode bias voltage to be used.

The second circuit was developed by GW8AAP with some assistance from G3AVJ and is shown in Fig 14.39. This uses a dc-ac inverter to produce the heater supply; the dc isolation between the windings of the transformer provides the necessary isolation of the heater/cathode from ground. One inverter circuit is required for each valve used. Efficiency is good and the inverter will start reliably from "cold" under load. The values may need some adjustment, however, if different transistors or pot-core are used. The transistors should be mounted on an adequate heatsink.

14.3.7 Alternative bias circuits for 2C39 amplifiers

Many bias circuits have been published for this series of valves. A few selected circuits are illustrated in Figs 14.40 and 14.41. All are uncritical in construction and all are equally effective. The choice is left to the constructor.

14.4 ANTENNAS

Yagi antennas are now used in most 1.3GHz stations. The main reason for this is that they offer relatively high gain with minimum weight and windage. Dishes less than about 1.5m in diameter are rather inefficient at this frequency and, even at this size, may be inconvenient in many home installations. However, larger dishes are becoming more popular with contest and expedition stations since often they do not need to be mounted very high and normal masts can be used to support them.

14.4.1 A loop Yagi antenna for 1.3GHz

This antenna was first designed in 1975 by G3JVL and since then he has made a number of improvements, including a longer version with higher gain. The gain of the basic version is approximately 18dBi.

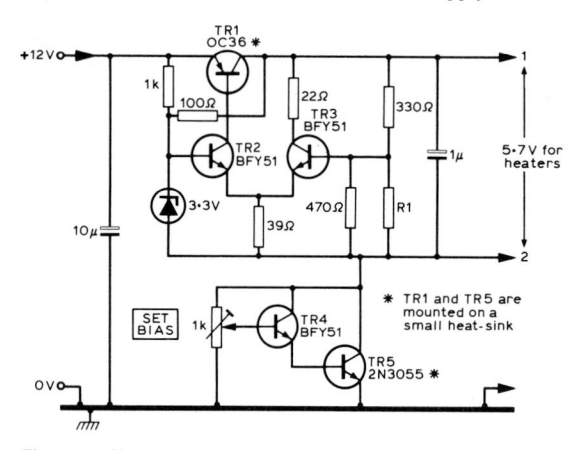

Fig 14.38. Circuit diagram of the G3ZUD 2C39 heater/cathode bias supply. The zener diode is rated at 5W

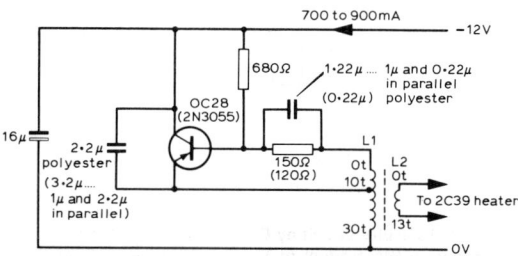

THE 1.3GHZ ("23CM") BAND

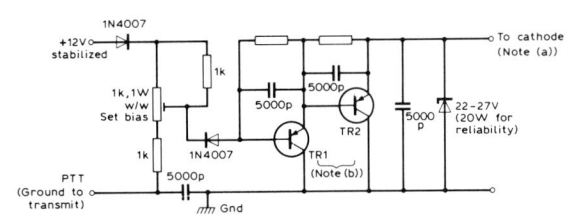

Fig 14.40. 2C39 bias circuit by G6CMS. Note: (a) maximum current dependent on Hfe product of TR1 and TR2 and maximum power dissipation limit of TR2. (b) TR1 Vceo 50, Ic. max 1A; TR2 Vceo 50, Ic.max 10A (use high Hfe device for TR1)

Construction

The design of this antenna is shown in Figs 14.42 and 14.43. Construction is quite straightforward, but the dimensions given must be closely adhered to. In drilling the boom, for example, measurements of the positions of the elements should be made from a single point by adding the appropriate lengths. If the individual gaps are marked out, then errors may accumulate to an excessive degree, resulting in poor performance.

With the exception of the copper driven element, the elements are made from aluminium strips, the two holes in which are drilled before bending. The centre-to-centre spacing of these holes is as follows:

Reflector R	9.92 inches (252mm)
Directors D1–12	8.40 inches (213mm)
Directors D13–20	8.10 inches (206mm)
Directors D21–25	7.80 inches (198mm)

All elements, screws and soldered joints should be protected with polyurethane varnish after assembly, followed by a coat of paint on all surfaces. If inadequate attention is paid to this protection, then the performance of the antenna will deteriorate over a period of time due to corrosion.

If the specified materials are not available, or it is wished to use thicker or wider elements to increase the strength of the antenna, it is necessary to alter the lengths of all the elements to compensate. The necessary data and a program is given in chapter 4, "Microwave antennas". As an example, a version could be made using a 19mm diameter boom and 6.35 by 1.6mm loops. The correction factor would be 0.9 – 0.3 + 0.6 = +1.2 per cent. Thus, all elements (i.e. the reflector, driven element and the directors) should be made 1.2 per cent longer. The element spacing is not altered.

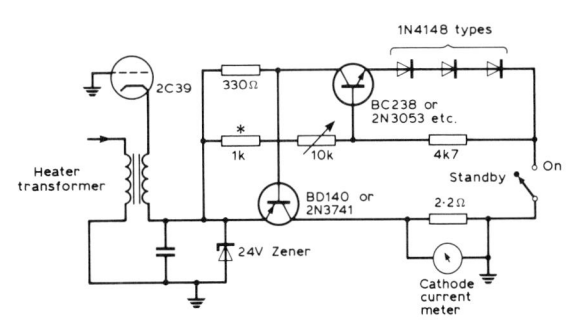

Fig 14.41. Bias circuit by LA8AK. Note: the value of this resistor sets the lower bias voltage limit

Provided that the antenna is constructed carefully, its feed impedance will be close to 50Ω. If a vswr or impedance bridge is available, the matching can be optimised by bending the reflector loop toward or away from the driven element.

An antenna with 1.7dB more gain can be constructed by adding 11 more directors at 90.42mm spacing. The circumference of the new directors is 195.6mm; directors D19–25 are also changed to this size.

The antenna can be mounted using an element clamp from an old antenna. It is essential that the antenna be mounted on a vertical support, as any horizontal metalwork in the vicinity of the antenna can cause severe degradation in its performance. It is usually best to mount the antenna with the loops pointing downwards, to reduce the likelihood of damage caused by perching birds.

14.4.2 Stacking 1.3GHz Yagis

Yagi antennas are often combined together in an array to achieve higher gain. The importance of correct phasing, impedance matching, etc. is discussed in chapter 4, "Microwave antennas". The antennas must be mounted a specific distance apart, known as the stacking distance. This is usually quoted by the manufacturer and may be different for the vertical and horizontal distances. Again, this is discussed in detail in chapter 4. The optimum stacking distance for the 1.3GHz G3JVL loop Yagi is 690mm, in both planes.

At 1.3GHz it is common practice to use a matched power-splitter to provide the feeds to the antennas in an

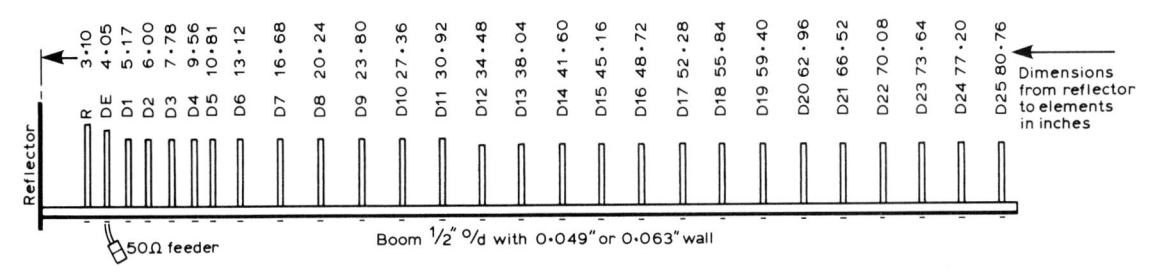

Fig 14.42. The modified G3JVL loop-quad Yagi antenna: element spacing

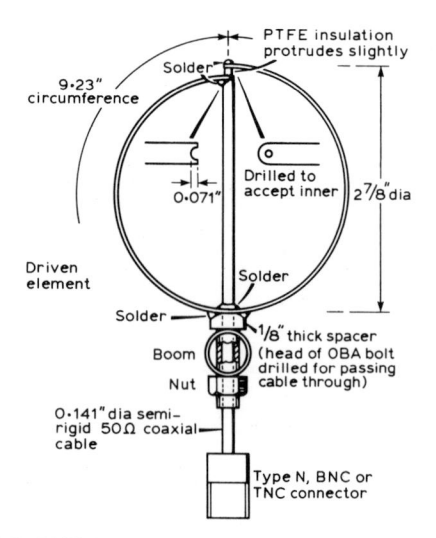

Fig 14.43. G3JVL loop-quad Yagi antenna: detail of driven element

Fig 14.45. Ratios of d/D for other useful coaxial configurations

		50Ω system	
		2 - way $Z_o = 72Ω$	4 - way $Z_o = 50Ω$
d □○□ D		2·82	1·96
d ◎ D		3·32	2·31
d □▮□ D		1·54	—
d ◉ D		1·66	—

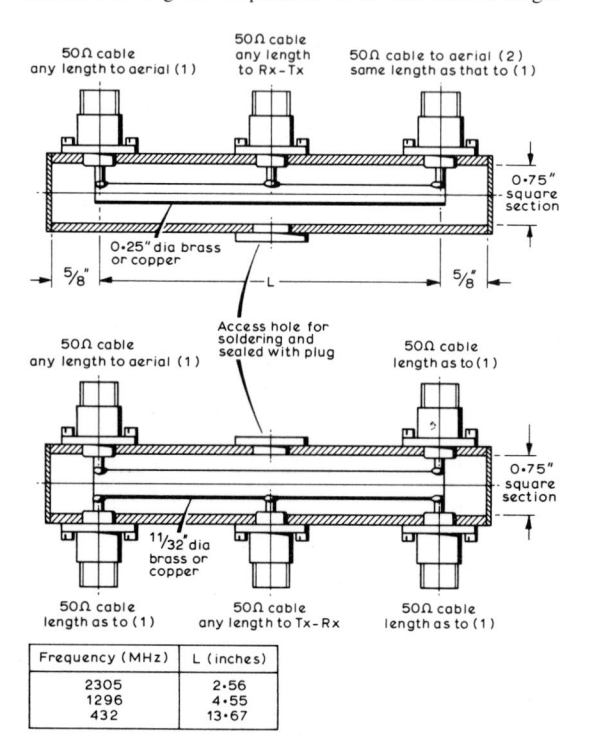

Fig 14.44. Power splitter/combiner for connecting two or four antennas to a common feeder

Frequency (MHz)	L (inches)
2305	2·56
1296	4·55
432	13·67

array, instead of using specific lengths of mis-matched cables to feed the antennas from a common point. The power splitters shown in Fig 14.44 enable two or four antennas of a given impedance to be fed from a single coaxial cable of the same impedance. They consist of a length of fabricated coaxial line which performs the appropriate impedance transformations. The inner is made exactly one half-wavelength long between the centres of the connectors, and the outer is made approximately 1.25in (37.15mm) longer. The outer can be made from square section aluminium tubing, the ends of which and the access hole (for soldering the centre connector) being sealed with aluminium plates bonded with an epoxy adhesive.

Alternatively brass or copper tubing may be used, in which case the plates can be soldered. Other forms of materials can be configured to give the necessary impedance transformations and information to enable other designs is given in Fig 14.45.

14.4.3 Dish antennas at 1.3GHz

The gain of a dish antenna at 1.3GHz as a function of its diameter is shown in Fig 14.46. It can be seen that a dish of 1.5m diameter has similar gain to an array of four Yagis (around 24dBi). Yagi arrays of eight or more antennas are very expensive if bought commercially and are also difficult to feed efficiently; a 2.1m diameter dish would have about the same gain. Dishes of this size are not too difficult to construct (see chapter 4, "Microwave antennas"). A typical dish might have 10–20 ribs, with 13mm chicken wire as the reflecting surface. Constructional tolerances are not particularly severe at 1.3GHz, as deviations from a true parabola of up to 10mm would cause a maximum loss in gain of only 0.5dB. When designing a dish for 1.3GHz, a focal length to diameter ratio of 0.5 to 0.6 is recommended, as this gives a dish which is relatively easy to feed with high efficiency. Of course, a dish may already be to hand with a different f/D ratio and the feed should be chosen to match this. A number of different feed designs are described below.

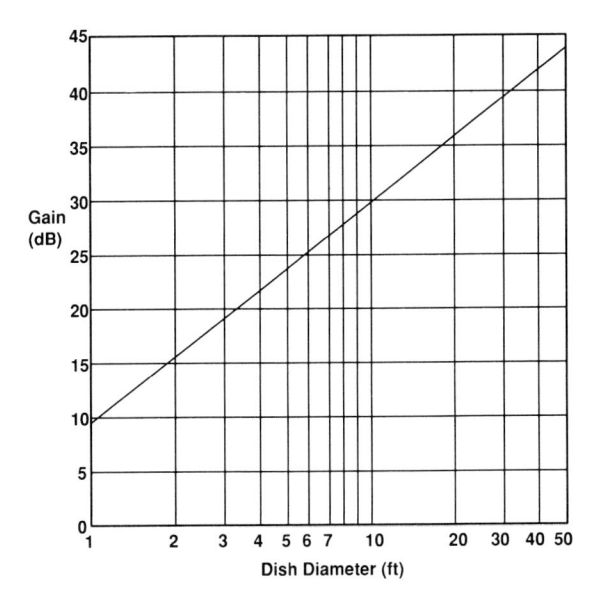

Fig 14.46. Gain of a dish at 1.3GHz as a function of diameter

14.4.4 A dipole/reflector feed for 1.3GHz

The basic arrangement for this feed is given in Fig 14.47. The radiation pattern makes it suitable for use with dishes with f/D ratios in the range 0.25 to 0.35. The feed is built around a length of fabricated rigid coaxial line, the inside dimensions of the outer and the diameter of the inner being chosen to produce a characteristic impedance of 50Ω.

The feed should be positioned in the dish so that the point halfway between the dipole and the reflector is located at the focus of the dish. The feed may then be moved inwards or outwards slightly to optimise the gain.

14.4.5 A high-efficiency dish feed for 1.3GHz

The feed to be described is capable of feeding dish antennas with f/D ratios in the range 0.5 to 0.6 with high efficiency. The construction of the feed is shown in Fig 14.48. The design uses two folded dipoles one half wavelength apart, driven in phase. The dipoles are positioned one quarter-wavelength above a square reflector. The dipoles and the quarter-wavelength open-wire transmission lines which join them to the feedpoint are made from one length of 1.5mm diameter tinned copper wire. The dimensions are fairly critical and should be adhered-to as closely as possible.

The method of feeding the dipoles is also shown in Fig 14.48. The prototype used a length of 50Ω, 6.35mm semi-rigid cable. Alternatively, this could be replaced by a home-made 50Ω rigid cable, using a 6.35mm copper or brass rod, drilled just large enough to take the inner from a piece of standard cable, e.g. UR43. The cable and a strip of copper-clad epoxy board form a 1:1 balun. One side of the open-wire feeder is connected to the junction of the inner conductor of the cable and the top of the epoxy board.

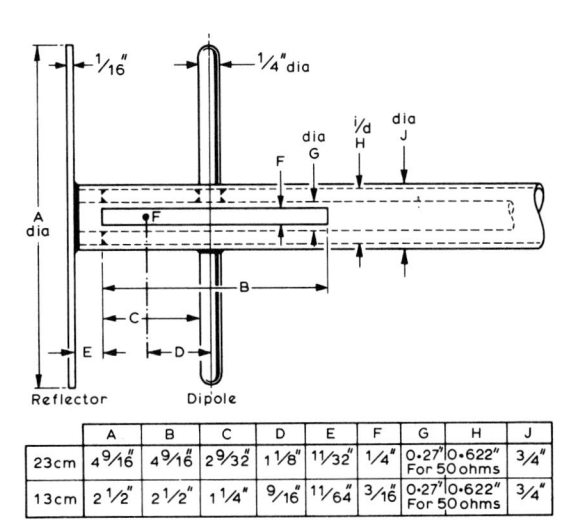

	A	B	C	D	E	F	G	H	J
23cm	4 9/16"	4 9/16"	2 9/32"	1 1/8"	1 1/32"	1/4"	0·27" For 50 ohms	0·622"	3/4"
13cm	2 1/2"	2 1/2"	1 1/4"	9/16"	1 1/64"	3/16"	0·27" For 50 ohms	0·622"	3/4"

Fig 14.47. A dipole/reflector feed for 1.3GHz

Provided that the feed has been carefully made, it should exhibit a low vswr. The vswr can be minimised by squeezing together or pulling apart the open-wire feeder sections and the dipole elements. Try to aim for a symmetrical antenna after doing this, or the two dipoles may not share the power equally. This would distort the radiation pattern and hence reduce the illumination efficiency. If significant power levels, i.e. greater than about 1W, have to be used to

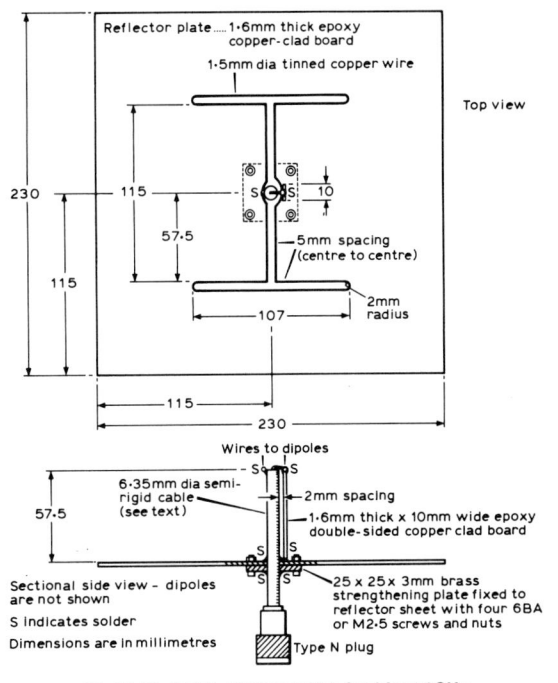

Fig 14.48. A high efficiency dish feed for 1.3GHz

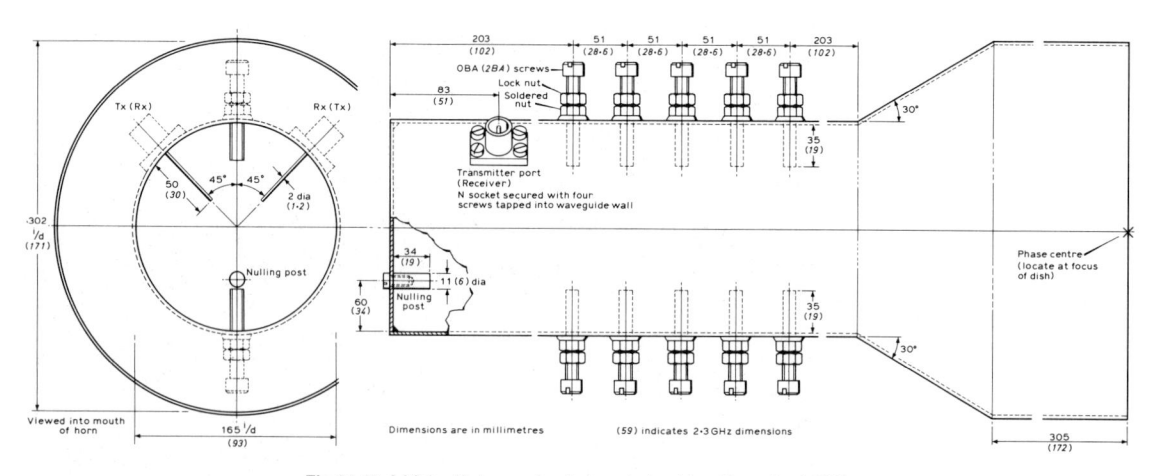

Fig 14.49. A high efficiency circularly polarised feed horn for 1.3GHz

measure the vswr, do not make any adjustments with the power on!

In use, the feed should be located in the dish so that the focal point of the dish is midway between the plane of the dipoles and the reflector. The feed may then be moved inwards or outwards by a small amount to maximise the gain of the dish. The best way to mount the feed is to use a tripod or quadrupod support with the feed fixed between support members just in front of the apex of the support. Horizontal polarisation is obtained when the dipole elements are horizontal.

14.4.6 A high efficiency circularly polarised feed horn for 1.3GHz eme

This dish feed was developed from an original design by W2IMU. It is primarily intended for optimum operation with dishes having f/D ratios in the range 0.5 to 0.6 but can be used, albeit with reduced efficiency, with deeper dishes, e.g. f/D ratio 0.4.

This feed horn differs from more conventional designs in that the waveguide operates in two modes, TE11 and TM11. The horn is termed "dual-mode" for this reason. The TE11 mode is launched by the probe in the smaller diameter section. The tapered section converts some of the dominant TE11 energy into the TM11 mode in the larger diameter section. When both modes reach the front of the horn, their relative phase and amplitudes give zero fields at the periphery of the aperture, resulting in very low rear and side radiation and, thus, minimum noise pick-up.

The generation of circular polarisation is performed by the 10 screws in the smaller diameter section of the horn. These load the waveguide in the direction parallel to the axis of the screws and cause a delay of 90°, but have no effect in the perpendicular direction. The feed probes are orientated at 45° to the polarising screws and thus generate equal components parallel and perpendicular to the screws. The parallel component is delayed in phase by 90° compared to the perpendicular component and thus circular polarisation results. One port produces right-hand circular polarisation while the other produces left-hand. This is exactly what is required for eme operation since the sense of polarisation is reversed on reflection from the Moon. In use, the transmitter is permanently connected to one port (thus no high-power relay is required), while the receiver is connected to the other port via a small isolating relay (which is required since the two probes are not totally isolated in practice). The degree of coupling between the probes depends on a number of factors, such as the exact orientation of the probes with respect to the polarising screws, the efficiency of the polariser and the focal length of the dish. A small post, mounted on the short-circuit back plate is used to minimise the coupling between the probes.

Construction

Constructional details of the feed horn are given in Fig 14.49. The prototype version was constructed out of 0.8mm brass sheet, with the short circuit plate cut from 3mm brass plate. The cone to cylinder joints were secured by means of many small tabs which overlapped the sections to form a butt joint. The rear plate was soldered in position. Aluminium has also been used, with argon-arc welded joints. In this case, the polarising screws require a mounting bush. A suitable bush can be made from the screwed bushing and nut salvaged from an old panel mounting variable resistor. The moulded body of the resistor is broken away from the bushing using a hammer and the shaft is replaced by a piece of brass rod, drilled and tapped to take the polarising screws, soldered into the bush. The polarising screw mounting holes are enlarged to take the bush which is held in place by the salvaged nut.

Adjustment

The adjustment of this feed is rather complicated and requires some care if optimum performance is to be attained. The recommended procedure is to optimise the feed probes individually for best vswr, set up the polariser and then

Fig 14.50. Test range setup of OZ9CR for the adjustment of the circularly polarised feed

adjust the nulling post and polariser for best overall performance. The circularity of the radiation is sampled using two helix antennas of opposite sense.

The full procedure is as follows:

1. Insert one feed probe, with the nulling post not fitted, and adjust the length of the probe for best vswr (<1.2:1). The probe can also be bent slightly towards or away from the back of the horn to improve the match. When no further improvement can be obtained, remove the probe and repeat the procedure with the other probe.

2. Set up a test range as shown in Fig 14.50, making sure there are no obstructions nearby which could cause stray reflections. With one probe inserted and the nulling post not fitted, set the polariser screws to the nominal positions. Check that the probe is at 45° to the polariser screws. Apply a signal to the probe and, using one helix antenna at a time, measure the received signal level. A considerable difference in the signal level received by the two helix antennas should be noted. Select the helix with the lower signal and adjust the polariser screws (move each the same amount) for minimum signal, which corresponds to best circularity.

3. Insert the other probe (make sure it is exactly 90° to the first one) and fit the nulling post. Connect the detector to the probe just fitted. Then adjust the polariser screws, the position of the nulling post (and possibly the probe length/position) for minimum received signal, minimum reflected power and minimum coupling between the probes. It will be found that the adjustments interact to some extent, so the adjustments are a repetitive process! Also, it may happen that one of the polariser screws (e.g. the centre ones) can be used to optimise the vswr without affecting the circularity too much.

4. Stop adjusting when the following specifications have been met: vswr <1.2:1, difference in signal received by the two helix antennas >20dB, coupling between probes <−20dB.

5. Finally, check that the vswr and polarisation discrimination is acceptable for the other feed probe. Continue adjusting if this is not the case.

In use, the feed horn should be positioned with its mouth located at the focus of the dish. It may then be moved inwards or outwards by a small amount to optimise the gain of the dish.

14.4.7 Helix antennas for 1.3GHz

Helix antennas are useful both as test antennas for setting up circularly polarised antennas as described in section 14.4.6, and also as general purpose medium gain antennas of wide bandwidth. An eight-turn antenna will have about 10dB gain. Circular polarisation of either sense is obtained by winding the helix with a right-handed or left-handed thread. A simple matching transformer is included and the vswr can be optimised after construction by slightly shortening the final turn. No details of construction are given since there are now several inexpensive commercial sources of 1.3GHz helix antennas in the UK.

14.4.8 A horizontally-polarised omnidirectional antenna for 1.3GHz

An omnidirectional horizontally polarised Alford slot antenna for 1.3GHz has been developed by G3JVL to serve as a beacon or repeater antenna. Mechanical details of the antenna are shown diagrammatically in Fig 14.51(a) and 14.51(b). It consists of a length of slotted tubing which can be manufactured either by cutting a slot in a piece of

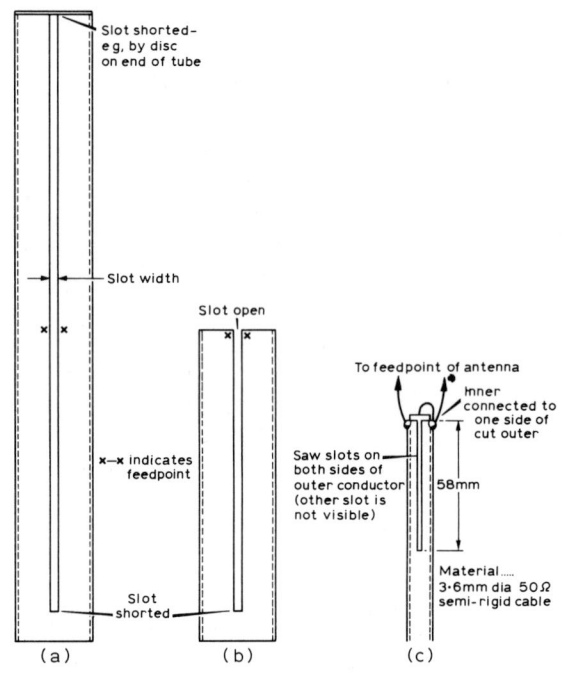

Fig 14.51. The G3JVL Alford slot antenna for 1.3GHz, showing the mechanical details for (a) centre and (b) end-fed versions plus (c) construction of the balun. The dimensions are given in Table 14.6.

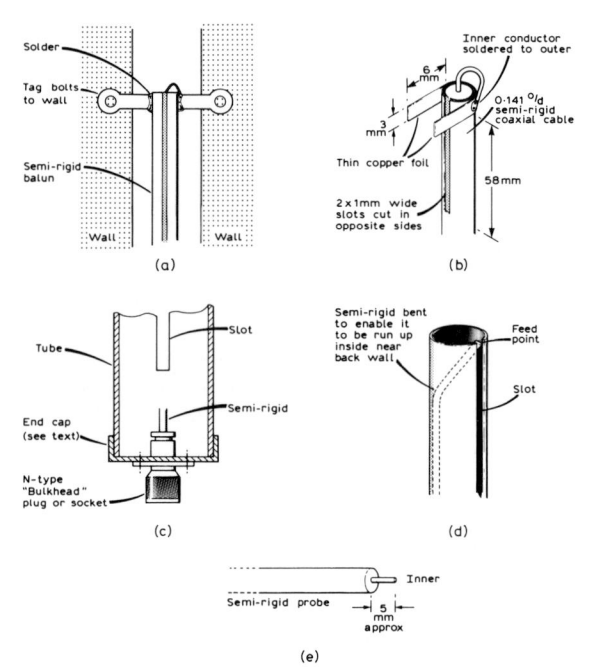

Fig 14.52. (a), (b), (c), (d) and (e) Further detail of construction of feed point and adjustment of antenna

suitable tubing, or by removing metal from a larger diameter piece of tubing and reforming the material left around a suitable former. The feed point can be made either in the centre of the slot or at the end, by suitable design. The width and length of the slot, the wall thickness and the diameter of the tubing are all related and the design data reproduced below was the result of much experimental work. Dimensions for end-fed and centre-fed antennas using different tubing sizes are given in Table 14.6.

The rf is fed to the antenna by a length of 3.6mm (0.141in) semi-rigid cable run inside the antenna to the centre of the slot or the end of the slot, as appropriate (Fig 14.51(a) and (b)), via a 4:1 balun, which is constructed in the end of the cable. Details of this are given in Fig 14.51(c). It should be noted that the outer of the semi-rigid is slotted on both sides. The slots can be made, with care, using a broken blade from a small hacksaw.

Connection is made between the balun and the slot by two tabs made from thin copper foil (Fig 14.52(a) or by the use of two solder tags (Fig 14.52(b)). These may be fixed to the slot by soldering or using small screws. The base of the antenna can be terminated in an N-type bulkhead plug or socket as shown in Fig 14.52(c); the bulkhead fitting may need to be modified to allow correct fitting of the semi-rigid feeder. The feeder should be shaped so that it runs near the inside back face of the tubing as shown in Fig 14.52(d). Standard compression pipe-fittings and blanking plates can be used to close both ends of the centre-fed version of the antenna. A plastic cap can be used to close the open end of the end-fed version, if so desired.

When built, the antenna should exhibit a low vswr if the dimensions given have been followed closely. If suitable test equipment is available, the match may be optimised by adjusting the width of the slot, either by squeezing the antenna in a vice, or by prising the slot apart. These operations should be done carefully! A small probe (Fig 14.52(e)) and sensitive diode detector/meter can be used to explore the voltage distribution along the slot. The probe should be held close to the tube, but not directly in front of the slot (hold it 20 or 30° round from the edge of the slot) and moved along its length. Patterns such as those in Fig 14.53 should be seen.

In use, the antenna should be mounted vertically as accurately as possible, since the vertical beamwidth is quite narrow. The length of the tubing below the slot is uncritical, so the same tube can be used for both mast and antenna. For weather-proofing the antenna can be mounted inside a radome made from a length of 63mm diameter

Table 14.6. Dimensions for Alford slot antennas

Antenna type	Tube dimensions	Slot width	Slot length
End fed	38.10mm o.d. 16 swg wall	11mm	254mm
Centre fed	38.10mm o.d. 16 swg wall	11mm	509mm
End fed	31.75mm o.d. 20 swg wall	4mm	254mm
Centre fed	31.75mm o.d. 20swg wall	4mm	509mm

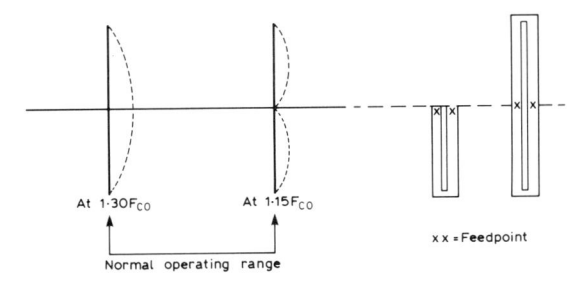

Fig 14.53. Typical field pattern using detector probe

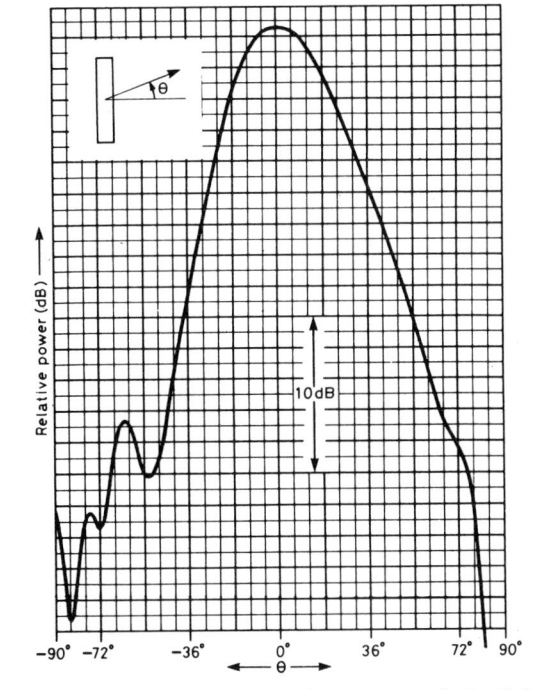

Fig 14.54. Performance of an Alford slot antenna - vertical radiation pattern (Note: measurements made on 2.3GHz antenna; patterns given by a 1.3GHz version will be very similar)

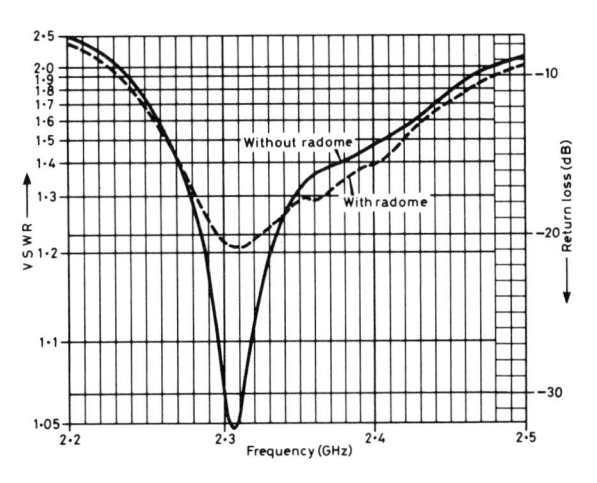

Fig 14.55. Performance of an Alford slot antenna - vswr

plastic drain pipe with 2mm wall thickness. If available, polypropylene tubing can be used with advantage, since it offers slightly lower loss than conventional pvc drainpipe.

The performance of the slot antenna is illustrated in Figs 14.54, 14.55 and 14.56. It should be noted that these measurements, due to G3TQF, were made on an optimised 2.3GHz version of the slot antenna, but the performance of a correctly constructed and optimised 1.3GHz version should be very similar.

14.5 ANCILLIARIES AND TEST EQUIPMENT

14.5.1 A simple in-line power indicator for 1.3GHz

A simple and reliable rf power indicator for insertion in the output line of a 1.3GHz transmitter, of up to 50W output power, can readily be constructed taking advantage of microstrip techniques as shown in Fig 14.57. For this purpose, good quality glass fibre double-clad board is needed, one side being the ground (earth) plane and a section of line on the other, together with the coupling loop for the indicator. A meter is connected between the feedthrough capacitor and ground, and indicates relative power. The insertion loss of this type of indicator is of the order of only 0.5dB, and it may therefore be left permanently in circuit. The unit can be built either by conventional etching, or by using the alternative method described in chapter 7, "Constructional techniques".

The spacing between the main line and the coupled line will need to be decided on the basis of the power expected to be used normally. The whole assembly should be enclosed in a suitable metal box. An alternative design, using a fabricated coaxial line in a diecast box was given in chapter 10, "Test equipment".

14.5.2 A simple general purpose filter

A simple filter for low power applications is shown in Fig 14.58. It can be tuned to operate at 1,152MHz and

Fig 14.56. Performance of an Alford slot antenna - horizontal radiation pattern

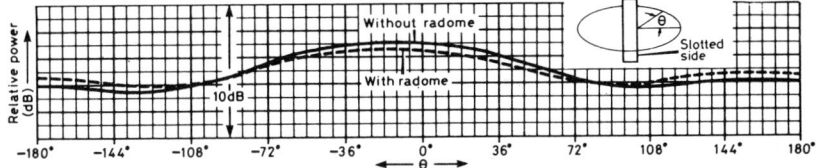

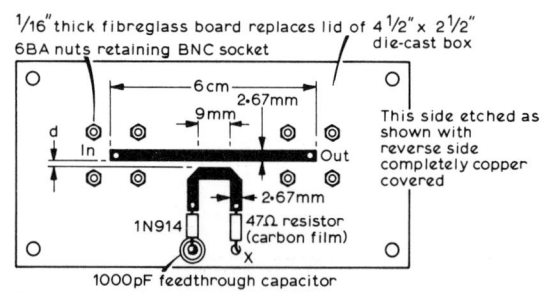

Fig 14.57. Simple microstrip in-line power indicator for 1.3GHz

1,296MHz. The filter consists of a shortened quarter-wave line, tuned by a 2BA screw, with capacitive input/output coupling using two BNC sockets. The outer body can be easily fabricated from copper sheet. The filter is tuned to the desired frequency by means of the 2BA screw and the insertion loss is minimised by screwing the BNC sockets in or out.

14.5.3 An interdigital bandpass filter for 1.3GHz

Most methods of generating rf at 1.3GHz give rise to some unwanted or spurious outputs, e.g. harmonics and mixer products. Varactors and ssb mixers will produce the highest spurious levels and it is often necessary to connect an effective filter between such transmitters and the antenna.

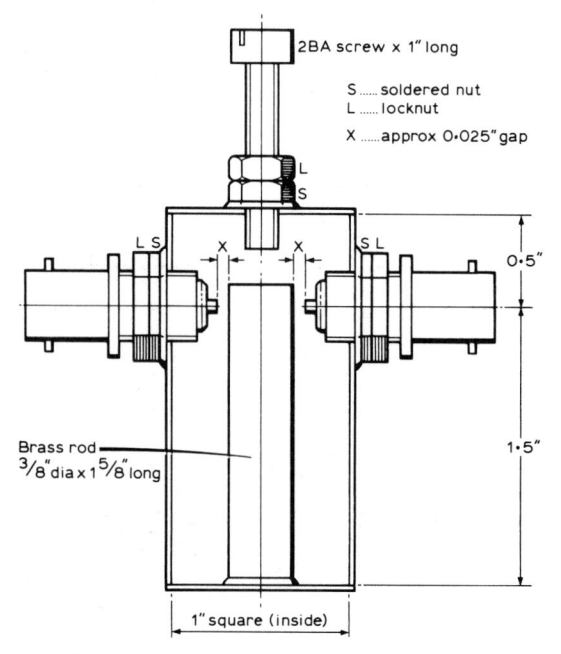

A suitable three pole filter is shown in Fig 14.59. It consists of three shortened, tuned quarter-wave lines, with the input and output connections tapped on to the outer lines. Construction should be evident from the figure, the only critical point being that for minimum insertion loss all surfaces that are screwed together should be a good fit. Ideally, they should be hard soldered together, with the top and bottom plates screwed to the frame by four or five screws along each side to ensure good rf contact.

Adjustment of the filter may be carried out by connecting it to the input of the receiver tuned to a small signal source set up at a suitable distance from the antenna. Once the filter has been tuned up for maximum signal strength with the weakest signal available, further fine tuning may be carried out by inserting a suitable attenuator between the filter and the receiver. A long piece, say 5 or 10m long, of thin (lossy) coaxial cable is suitable for this purpose.

The insertion loss is small (approximately 0.5dB) and so virtually no loss of output power should be seen when the filter is put directly in the output of the transmitter.

A final retune of the transmitter output load and tuning adjustments is recommended after the filter is installed to recover any power lost from small loading changes. The filter could also be used as an image rejection filter in a receiver.

14.5.4 A simple signal source for 1.3GHz

The unit shown in Figs 14.60 and 14.61 is intended to act as a portable "mini-beacon" for testing receivers and for aligning and comparing antennas. Even though the output power is probably only at the microwatt level, it is more than adequate for most purposes. Indeed, when connected

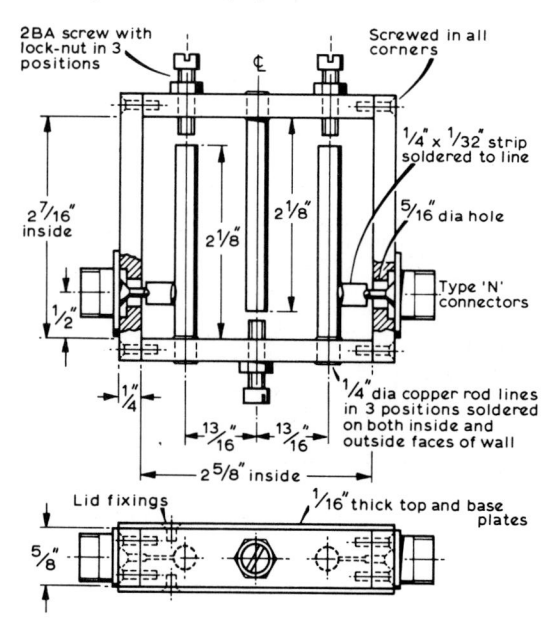

Fig 14.58. A simple general purpose filter for 1.3GHz

Fig 14.59. An interdigital filter for 1.3GHz

Fig 14.60. Circuit of a simple "mini-beacon" signal source for 1.3GHz. Circuit values: L1, 12t 28swg e.c.w., on 0.25" former, tapped 1t from 'cold' end. L2, 4t 18swg t.c.w., 0.25" i.d., 7/16" long, centre tapped RFC1, RFC3 4t 28swg e.c.w. on FX1115 ferrite bead, RFC2 ditto, 2.5t

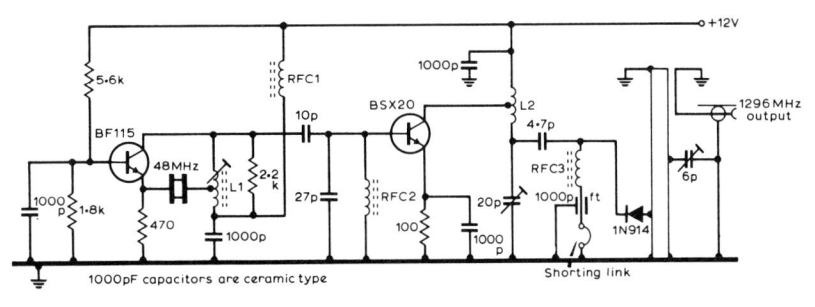

to an antenna of 20dB gain, it has been received at good strength over a distance of 20km.

The unit is built on a piece of double-clad printed circuit board which forms the lid of a standard diecast box. The crystal oscillator at 48MHz is followed by a BSX20 tripler to 144MHz, the output of which is fed to a 1N914 diode used as a ×9 multiplier. The harmonic at 1,296MHz is tuned by the half-wave output line. The shorting link is removed to check the diode current, which should be about 10mA. It may be possible to obtain more output if a varicap diode, type BB405, is substituted for the general purpose 1N914 of the original design.

If a receiver with a 144MHz i.f is being used in conjunction with the signal source, ensure that the signal being received is actually on 1,296MHz and is not 144MHz feed-through. This can be tested by temporarily removing the shorting link, when the 1,296MHz signal should decrease markedly in strength.

14.5.5 A milliwattmeter for use at 1.3GHz

Details of a rf milliwattmeter capable of measuring absolute power at 1.3GHz are given below. It is very useful when aligning local oscillators and transmitters. It is capable of measuring low power levels with a reasonable accuracy, unlike more typical "throughline" power meters which are useful only for measuring higher power levels.

The circuit of the power meter, due to G4ERP, is shown in Fig 14.62 and 14.63. In this design stray pick-up and "hand" effects are eliminated by good high frequency decoupling – two bypass capacitors are used: a conventional 2,200pF component and a fabricated "low inductance" capacitor. Screened cable was used to connect the detector head to the voltmeter; in the prototype RG174 miniature coax was used, but any type of coaxial cable could be used.

Mechanical details of the milliwattmeter are shown in Fig 14.63. The body of the unit consists of two 47mm lengths of 19.5 by 9.5mm aluminium bar held together by M3 screws, to give a piece with a 19mm cross-section. A 7.8mm hole is drilled in the centre of one end to a depth of 40mm. This hole is than continued through the remainder of the bar at 2.6mm diameter. When the braiding of the RG174 is bent back over the outer insulation, a hole of this diameter forms a good cable clamp. The 7.8mm diameter hole provides sufficient clearance for the BNC socket to be mounted flush with the end face. The socket must be modified by removing the ptfe insulation at the pin end and then shortening the pin to 2mm length. The resistor and the 2,200pF capacitor are grounded by solder tags which are mounted as close as possible to the components. The brass rod which forms the inner of the coaxial capacitor is insulated

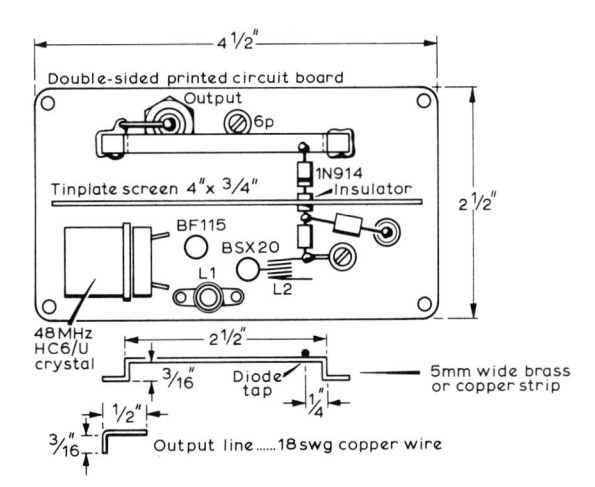

Fig 14.61. Construction of 1.3GHz simple "mini-beacon" signal source

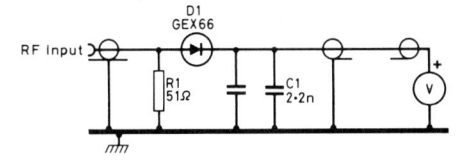

Fig 14.62. Circuit of a simple milliwatt power meter for 1.3GHz

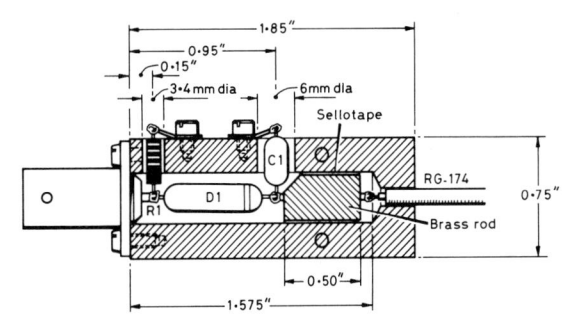

Fig 14.63. Construction of 1.3GHz simple milliwatt power meter

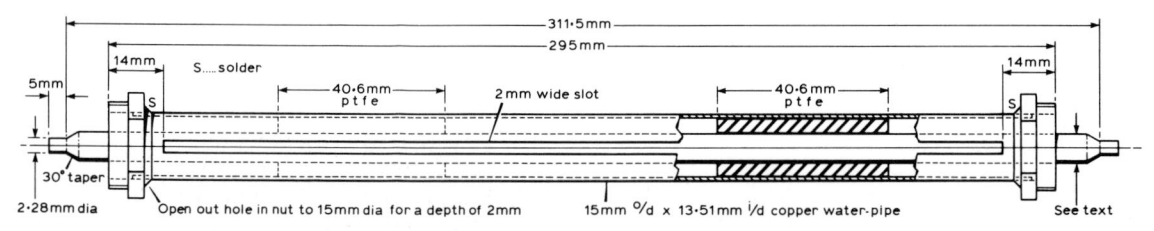

Fig 14.64. Design for a double-slug tuner for 1.3GHz

from the body by a single layer of Sellotape. A small spigot is provided at either end of the "barrel" to assist soldering. Depending on what type of 2,200pF capacitor is used, it may be necessary to file away part of one end of the barrel to give adequate clearance.

The following equation is used to determine the power level being measured:

$$P \text{ (watts)} = 0.01 (V + 0.25)^2$$

where V is the measured voltage. For quick reference, a graph of this function can be drawn.

For measuring power levels in excess of 250mW, a power sampler and dummy load can be used in conjunction with the milliwattmeter. This can give quite accurate results when measuring higher power levels, if the coupling attenuation is known.

14.5.6 A double-slug tuner for 1.3GHz matching

Very frequently it is useful to have available a device for impedance matching. For example, it may be desirable to be able to "tune-out" mismatches arising from poorly matched antennas or changeover relays if vswr-sensitive devices (e.g. low-noise preamplifiers) are used in the station. Also some designs of power amplifier will give higher output power when the impedance presented to them is not exactly 50Ω. The double-slug tuner using ptfe slugs is capable of matching a load with a vswr of up to 4:1 with extremely low loss and can operate with power levels of many hundreds of watts.

A design for a double-slug tuner is shown in Fig 14.64, and is an adaptation of an original design by W2IMU. The diameter of the inner conductor should be such as to give the desired characteristic impedance. For 50Ω, using 15mm OD water pipe as the outer conductor, 6mm diameter rod is near optimum; 6.3mm rod (0.25in) can be used, but the resulting characteristic impedance (45Ω) will reduce (slightly) the range of adjustment of the tuner. Alternatively, a 50Ω line can be made from 4.7mm diameter rod and 11mm i.d. tubing. Both these sizes correspond to standard imperial metal stock.

The construction of the tuner is straightforward, the most important point being to select an undamaged piece of tubing for the outer conductor. The ptfe slugs should be machined to be a sliding fit with both the inner and outer conductors; any dents in the outer conductor will make the slugs difficult to adjust. The slot in the outer conductor can be made either by milling or by drilling a large number of holes next to each other and then filing out. In either case, the tube may tend to open out due to residual manufacturing stresses and it will be necessary to restore its shape. This can be done by squeezing the tube (carefully) back to shape in a vice and soldering a small tab across the slot, halfway along the tuner.

The lengths of the inner and outer conductors are correct for standard UR67-compatible N-type connectors. If different types are used it may be necessary to change the length of the inner conductor slightly, or the centre pins of the connector may not be in the correct positions.

In use, the slugs should be moved using an insulated tool. On no account should a metal tool be employed if significant rf power is present, to avoid the danger of the tool slipping and making contact with the inner conductor.

14.5.7 A high power in-line power indicator for 1.3GHz

An in-line power indicator suitable for operation at power levels above about 50W is shown in Figs 14.65 and 14.66. The design is due to W2IMU. If monitoring of reflected power is not required, only one coupling section is required. Construction should be evident from the figure.

14.5.8 A local oscillator for 1.3GHz converters and transverters

This local oscillator, due to G4DDK, was designed specifically for use with receive and transmit converters such as those described in sections 14.2.2 and 14.3.2. It could also form the basis for a compact low-powered personal beacon, especially when coupled with the keyer/callsign generator described in chapter 9, "Beacons and repeaters". It uses the elements of the two crystal multiplier designs in chapter 8, "Common equipment".

The Butler oscillator stage of the high quality uhf source has been combined with the multiplier stages of the 1,152MHz source to provide a compact unit of very high spectral purity. The good phase noise performance of the oscillator stage ensures minimum problems from reciprocal mixing, close to the carrier. At greater separation from the carrier, low levels of spurious output minimize response in mixer stages from out of band signals. In transmit converters this latter quality is also important to reduce the danger of radiating out of band signals. Versions of this oscillator have been built for various frequencies in the range 1,136MHz (10.368GHz – 144MHz i.f) to 1,557MHz (Meteosat – 137MHz i.f), although it should be noted that

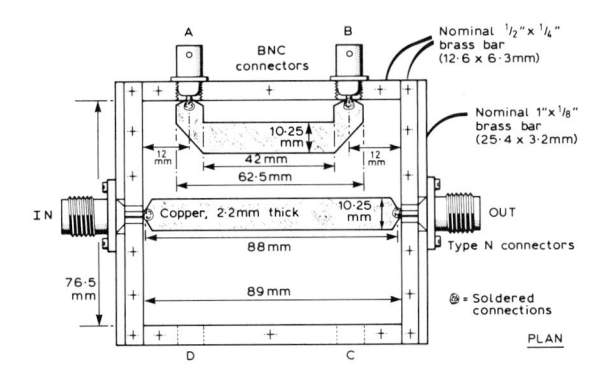

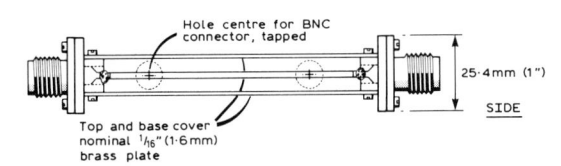

THE 1.3GHZ ("23CM") BAND

Fig 14.65. Design for a high power in-line power indicator for 1.3GHz

Fig 14.66. Photograph of in-line power indicator showing: input socket left, output right. Left rear BNC socket is diode mount, rear right is BNC terminating load. The front BNC sockets are not used in this unit but could carry a second coupling line, load and detector for reverse power measurement. Coupler: G3WDG, Photo: G6WWM

Experience has shown that this circuit is able to provide more reliable operation than the more widely used junction fet circuit, using devices such as the J310 or P8000. Frequently it is found that such circuits lack drive to the following multiplier stage, the maximum output not being co-incident with the wanted frequency. This often leads to frequency offsets of 15 or 20kHz at the final frequency in order to obtain sufficient local oscillator drive to the mixer stages. Also peaking the output in this way can lead to unreliable starting characteristics.

In the unit described, the crystal operates at 96MHz, with the output from TR2 taken at 288MHz. It is possible to use crystals anywhere in the range 90 to 100MHz, with corresponding outputs between 270 and 300MHz. Other frequencies may be possible, but these have not been checked.

The following stage (TR3) is biased for optimum frequency doubling. The trapezoidal capacitor in the emitter circuit provides effective decoupling at both drive and output frequency (576MHz). The collector of the doubler is tuned to the required frequency, using a printed stripline inductor and miniature trimmer capacitor. The collector is tapped onto the inductor at the high impedance end. Transistor output capacitance adds significantly to the total value required to resonate the circuit. This results in a low Q for this circuit, requiring a second tuned circuit to achieve adequate suppression of the doubler drive frequency. It should be noted that the coupling between the two circuits is almost entirely due to capacitance between the two trimming capacitors. The type of miniature trimmer chosen for the circuit ensures that the right coupling is achieved. If physically larger capacitors are used there may be too much coupling, leading to difficulty in resonating the circuits and much reduced suppression of unwanted sub-harmonics.

The second tuned circuit is capacitively coupled to the final doubler stage (TR4). This stage operates in much the same manner as the previous stage, except that its output is tuned to 1,152MHz. Three tuned circuits ensure excellent suppression of unwanted sub-harmonics.

some changes are required to component values and it is important to retain the same multiplication ratios in each of the multiplier stages, i.e. the first multiplier stage after the oscillator (TR3) must be used as a doubler, otherwise it may not be possible to achieve sufficient drive to the following stage to ensure reliable operation. In the case of the Meteosat unit the Butler oscillator output was used as a quadrupler with a slight reduction in output to +10dBm.

The unit is able to provide either a single output at approximately +13dBm or two outputs at +10dBm. When the two output option is chosen, one of the outputs will have slightly inferior spectral purity, but provided this output is used to feed the receive converter this should not be too important. The specification for output may not be met when the unit is used at other than the design frequency. The local oscillator was originally designed for use in a 1,296MHz transverter with a 144MHz i.f.

Circuit description

The circuit of the unit is shown in Fig 14.67. The crystal oscillator circuit used in the first stage of this unit is the well known Butler circuit. This circuit provides a first class phase noise performance due to the low loading on the crystal, allowing it to operate with increased circuit Q. An additional advantage is the ability to extract power from the circuit at a harmonic of the crystal frequency, in this case at three times the crystal frequency.

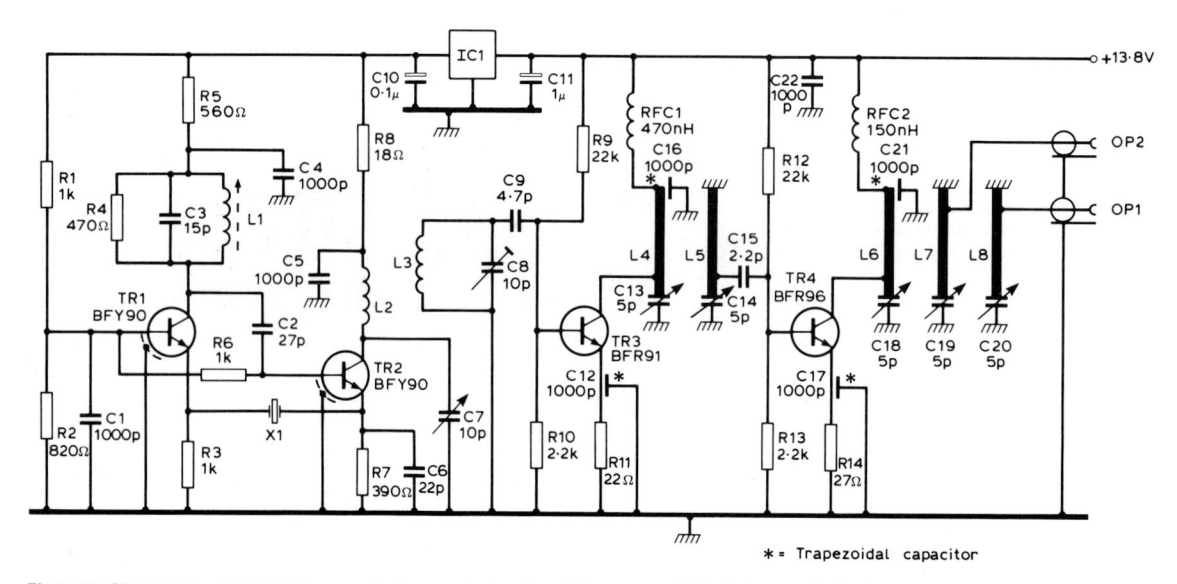

Fig 14.67. Circuit of the G4DDK–001 converter/transverter local oscillator source. This design could also be used as a low power transmitter or beacon source: the tuning range is approximately 1,100 to 1,300MHz.

Transverters often require two separate, well isolated local oscillator outputs. These can be obtained from a single output in a variety of ways, for example by the use of a Wilkinson divider to give two equal outputs, or by directional coupler to obtain one high level and one low level output. In this design a different technique has been used. Two equal outputs are obtained at the expense of some spectral purity at the output taken from the middle tuned line. This approach has several advantages. It is relatively wide band, such that virtually identical output levels are available anywhere in the range 1,136 to 1,300MHz. If the second output is not required it is simple to cut the track where it leaves the middle tuned line. The operating voltage to the oscillator stage is stabilized by an integrated circuit 8V regulator.

Construction

The unit is built on 1.6mm epoxy glass double sided printed circuit board, one side of which is used as a groundplane. The complete board can be mounted in a 127 × 78 × 45mm diecast box such as the Eddystone Radio 27134P. The pcb track layout is given in Fig.14.68, drilling and cutting details in Fig 14.69, and the component placing in Fig 14.70. The component values are given in Table 14.7. A

Table 14.7. Component values for local oscillator

Capacitors
C1,4,5,22	1,000pF ceramic plate
C2	27pF ceramic plate
C3	15pF ceramic plate
C6	22pF ceramic plate
C7,8	10pF Sky or Oxley CD5/10
C9	4.7pF ceramic plate
C10	0.1µF 16V tantalum bead
C11	1µF 16V tantalum bead
C12,16,17,21	1,000pF trapezoidal disc capacitor
C13,14	5pF Sky or Oxley CD5/6
C15	2.2pF ceramic plate
C18,19,20	5pF Sky or Oxley CD5/2
C23	1,000pF feedthrough capacitor

Resistors
R1,3,6	1k0	R8	18R
R2	820R	R9,12	22k
R4	470R	R10,13	2k2
R5	560R	R11	22R
R7	390R	R14	27R

All resistors 0.25W carbon film

Inductors
RFC1	0.47µH moulded axial choke, value not critical
RFC2	0.15µH moulded axial choke, value not critical
RFC3	Two turns through two hole bead, value not critical
L1	5.5 Turn Toko S18 Green Coil, aluminium core
L2,3	3 Turns 22 swg (0.9mm) tinned copper wire 3mm inside diameter; turns spaced one wire diameter. Spacing between coils 5mm. Height of coils 2.5mm above groundplane
L4,5,6,7,8	Printed lines on pcb

Semiconductors
TR1, 2	BFY90
TR3	BFR91
TR4	BFR96
D1	1N4001
IC1	78L08 regulator, 8 to 8.2V

Miscellaneous
X1	96MHz 5th or 7th overtone crystal in HC18/U

THE 1.3GHZ ("23CM") BAND   14.35

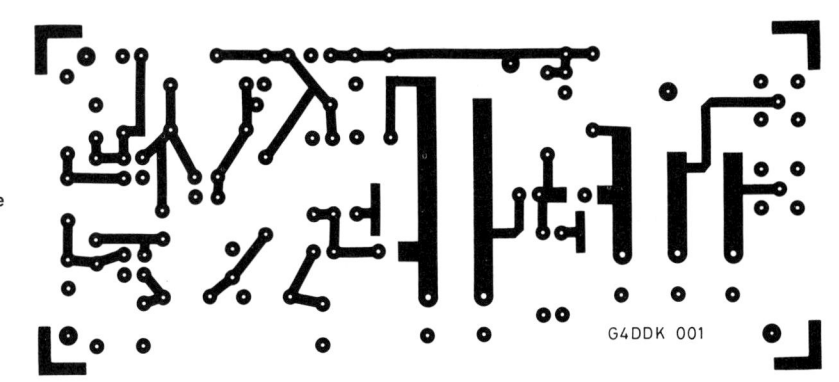

Fig 14.68. PCB track layout for the G4DDK–001 local oscillator source

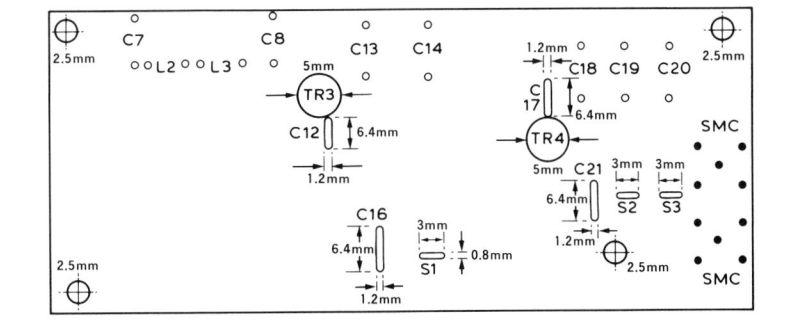

Fig 14.69. Drilling and cutting details for the G4DDK–001 pcb

○ 1mm diameter ALL others 0.8mm diameter, although 1mm is permissable if more convenient.

● 1.2mm diameter holes for two SMC/SMB sockets

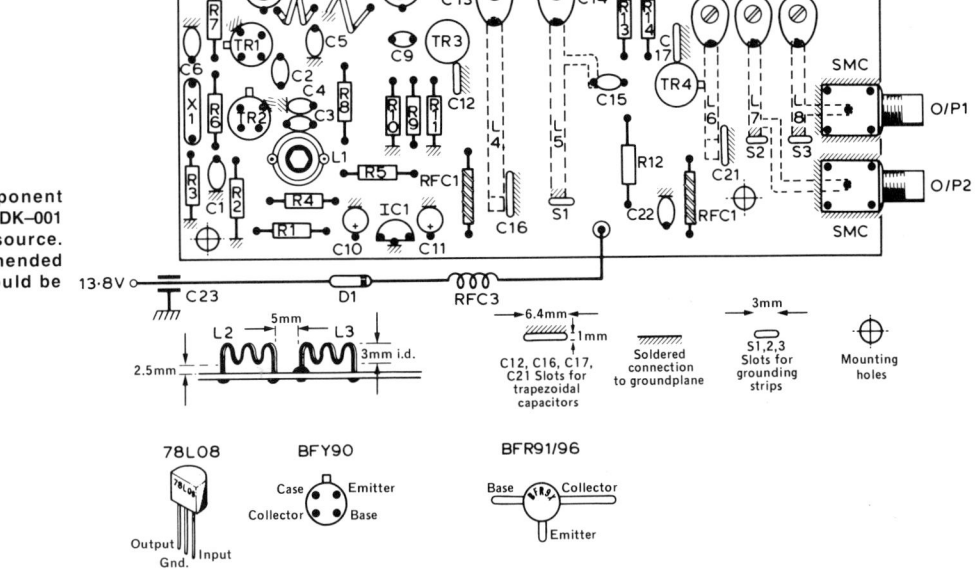

Fig 14.70. Component placing for the G4DDK–001 local oscillator source. Only the recommended components should be used

78L08
BFY90
BFR91/96

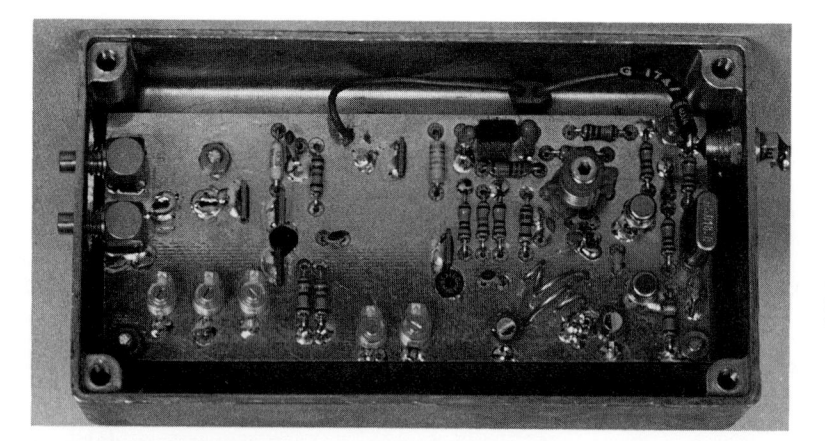

Fig 14.71. Photograph of the completed, housed unit. Note the use of a diecast box to provide screening and thermal stability. Ingoing supplies are well decoupled to avoid unwanted noise or rf pickup. Output(s) via pcb mounting SMC connectors to avoid unscreened leads and possible mismatch. A matching linear power amplifier with 20dB gain (1W output for 10mW input) was described in chapter 8, "Common equipment"

photograph of the appearance of the finished unit is given in Fig 14.71. A ready drilled pcb is available from the Microwave Committee Components Service.

The first stage in assembly is to place the board in the bottom of the diecast box, groundplane uppermost. The end of the board, where the output filter and connectors are located, should be placed against the end wall of the box and the position of the mounting holes marked through to the box. It is extremely difficult to do this with any accuracy once components have been fitted to the board.

If you have made your own board it will be found easier to drill the component mounting holes from the track side of the board and then to countersink the groundplane side with a 2.5mm drill bit. Countersinking should only be done for those component leads that pass through the board and are not soldered to the groundplane. The holes for TR3 and 4 are drilled 5mm diameter. The four mounting holes are drilled 2.5mm. 1.2mm diameter holes are required for the pins of the SMC connectors. All component lead holes should be drilled 0.8mm except those for L2, L3, C7, 8, 13, 14, 18, 19 and 20, which need to be drilled 1mm diameter.

A series of 1mm holes are drilled close together to form a short slot for the trapezoidal decoupling capacitors. In all cases these are drilled alongside the line to be decoupled and not at the end of the lines as is more usual. A series of 0.8mm holes also need to be drilled for the grounding strip slots. These are drilled at the ends of the tuned lines. It is advisable to use either a short section of broken junior hacksaw blade or a very fine needle file to properly join the holes into a slot. The slots for the trapezoidal capacitors must have the copper removed for about 0.5mm on the same side of the groundplane as the track being decoupled to prevent unwanted short circuits to ground.

Fitting of components to the board begins with the grounding strips (copper foil) at the end of each of the printed lines, where shown. The resistors and capacitors are fitted next, taking care to ensure that all grounded leads are soldered on top as well as underneath the pcb. If preferred, the ground leads may be bent out at 90° and soldered to the top groundplane only, however this never looks as good as the previous method.

Coils L2 and L3 are wound as shown and soldered into place, ensuring that the axis of these coils are in line. The two small side projections on L1 are removed with a sharp blade before the coil is soldered into place. The chokes can now be fitted, taking care not to bend the leads too close to the choke body (so that the wire terminations break).

TR1 and TR2 are now soldered into place, taking care to ensure correct lead orientation (see diagram). The body of the transistor should be no more than 2.5mm above the groundplane. The screen lead must be soldered to the groundplane. The regulator may now be fitted, again taking care to solder the centre lead to the groundplane.

Carefully solder the trapezoidal capacitors into place, taking care not to overheat them or they may fracture. Fitting of these capacitors is made easier if they are mounted with the shortest edge downwards, as shown. Solder the capacitor to the printed line or emitter track as appropriate and then the *opposite* side of the capacitor to the top groundplane. Ensure the other side of the capacitor does not short to the groundplane.

Once the decoupling discs are fitted, transistors TR3 and TR4 may be soldered into place. Each transistor should be placed into the hole provided and excess lead length cut off. The three leads may now be soldered to their respective tracks, taking care not to overheat the decoupling capacitors or the transistors. The trimmer capacitors should be fitted such that the earth lead is bent out at 90° to the capacitor body, and then soldered down to the groundplane of the pcb. When SKY trimmers are used it is not obvious which end should be grounded, since neither connection is directly connected to the adjustment screw slot, therefore the trimmer should be mounted as shown. The Oxley trimmer ground lead can easily be identified as the one extending below the trimmer body. If in doubt, check with an ohm meter to establish the lead connected to the rotor (top) of the trimmer. This is the ground lead of the trimmer.

Solder the crystal into place, once again taking care not to overheat this component. The position of the crystal at the end of the board is deliberate so that a heater resistor and thermistor may be attached to it to allow a proportional temperature control circuit to be used.

Miniature SMB or SMC angled connectors are soldered to the board where shown. These connectors must have their bodies soldered to the top groundplane. If suitable sockets are not available, solder short lengths of miniature 50Ω coaxial lead, such as RG174, to the outputs, terminating these on panel mounted sockets on the end wall of the box. Alternatively the coaxial leads may be taken direct to their respective mixers. When the board has been assembled it is advisable to test it out of its box (see the following section). Once initial alignment has been performed, the unit can be fitted into the recommended diecast box.

The box should be drilled 2.5mm for M2.5 screws where previously marked.; the holes in the box should be countersunk on the outside. The pcb is mounted in the box using M2.5 × 10mm countersunk head screws; three nuts are used as spacers between the board and the bottom of the box. The output connectors are not attached to the box, but are allowed to protrude through 10mm holes drilled in the box where shown. By taking the outputs direct from board mounted sockets the problem of maintaining screening integrity from oscillator to mixer is made easier.

Alignment

To align the unit properly does require access to certain basic items of test equipment. An absorption wavemeter and multimeter are the minimum items needed. Initially set the trimmers and coil core to the positions indicated below:

L1	Core level with the top of the former	
C7	50% meshed (assumes 10pF trimmer)	
C8	75% meshed (assumes 10pF trimmer)	
C13,14	90% meshed (assumes 5pF trimmer)	
C18,19,20	10% meshed (assumes 5pF trimmer)	

These positions depend strongly on the type of trimmers used, and to a lesser extent on the dielectric constant of the board. When using Sky trimmers it is important to note that although a higher maximum capacitance is specified, the minimum values obtainable are lower than for the Oxley types. This explains the apparently odd values specified for C13, 14, 18, 19 and 20 in the component table.

Connect the multimeter in the supply lead to the unit and check that the current taken from a 12V source does not exceed about 180mA. If it does, switch off immediately and check for short circuits or incorrectly placed components. When satisfied all is well proceed to align the crystal oscillator stage. This is best done by tuning the wavemeter to 96MHz and placing it close to L1. The core of L1 is tuned until a response is observed on the wavemeter. Peak the response by turning the core. Check that the oscillator restarts after switching off then back on. If it does not then turn the core slightly and repeat until it does. Retune the wavemeter to 288MHz and place it close to L2 and L3. Peak the reading by adjusting the trimmers C7 and 8. Place the probe leads of the multimeter between ground and the emitter of TR3. The meter should be on a sensitive range of no higher than 2 volts fsd. It is best to use a moving coil meter rather than a digital meter, since changes in the reading can be more easily seen. Peak the reading by adjustment of the two trimmers. The wavemeter should be used to confirm the circuits are still tuned to 288MHz.

Transfer the meter probe to the emitter of TR4 and tune C13 and 14 for maximum reading. Again use the wavemeter to confirm these circuits are tuned to 576MHz. Go back to C7 and 8 and repeak the reading at the emitter of TR4.

Connect the power meter or 50Ω diode probe to output 2. Output 1 *must* be terminated in 50Ω. Tune C18 and C19 for maximum indicated power on the meter. Transfer the power meter to output 1 and terminate output 2 in 50Ω. Tune C20 for maximum indicated power. This should be close to 10mW. Repeak all tuned circuits, making sure not to retune them to some other harmonic of the drive frequency. If only one output is required use output 1, cutting the tapped output from L7 where it leaves the stripline.

Exact frequency setting is best done by measuring the oscillator frequency with an accurate frequency counter. If this is done at 96MHz remember that the error is multiplied by 24, hence a 1kHz error becomes 24kHz at 1,152MHz. This would be regarded as too much by most operators and a final offset of no more than 5kHz would be appropriate.

Additional notes

The pcb was designed to accept trimmer capacitors of about 5mm diameter. It is important not to use physically larger capacitors since this may lead to tuning problems as explained above. The recommended trimmers are those made by SKY or Oxley; other trimmers of similar size may be used, provided their capacitance range is similar. The circuit diagram shows an additional decoupling choke and protection diode. These components ensure transients and noise on the power supply do not cause problems with oscillator stability or purity. The diode serves the additional purpose of protecting the unit against reversed power supply.

Performance

Figs 14.72 and 14.73 are spectrum analyser traces of the two outputs. Fig 14.72 is output 1, with output 2 disconnected

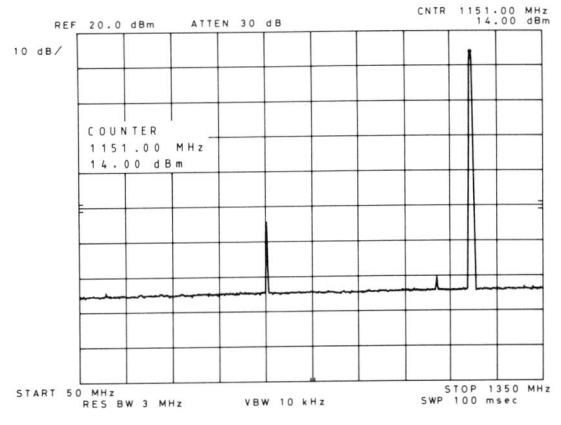

Fig 14.72. Spectrum of output 1, output 2 terminated in 50Ω

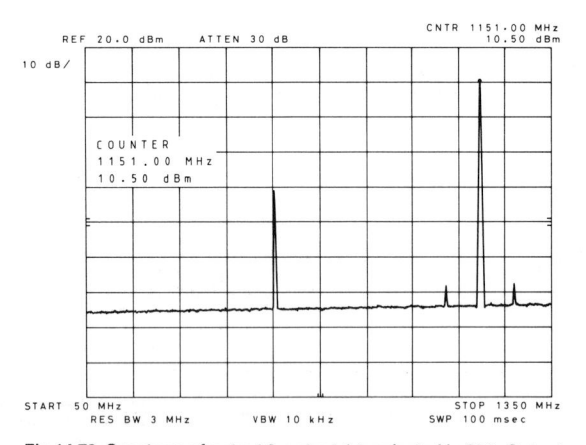

Fig 14.73. Spectrum of output 2, output 1 terminated in 50Ω. Output 1 should be used to drive the transmit converter and the slightly less pure output 2 used to drive the receive mixer

(track to the output socket cut) and, in the prototype measured was at +14dBm. Fig 14.73 shows the slightly less pure output available from output 2, with output 1 terminated in a matched 50Ω load; in this case the two outputs were at +10.5dBm. Output 1 should be used for driving the transmit converter. The slightly less well filtered output 2 is quite clean enough to drive the receive converter.

14.6 INTEGRATED EQUIPMENT

The purpose of this section is to briefly describe how a number of receiver and transmitter "building blocks", described above, can be integrated together to form complete transverters. It is easier to achieve adequate unwanted product rejection if a high i.f, such as 144MHz, is used and it is assumed that this will be the case.

14.6.1 Local oscillator sources

In a transverter, a common local oscillator is used for the receive and transmit converters. A splitter in the oscillator's output is the easiest way to achieve this. Depending on the design of the equipment the split may be at either 384MHz or at 1,152MHz, although the latter frequency is usually more convenient. If it is at 1,152MHz, then a printed Wilkinson splitter can be used.

The local oscillator source described in section 14.5.8 already has a suitable splitter incorporated on the board and this is very convenient for low power transverters.

14.6.2 Low power transverters

Fig 14.74 shows two alternative configurations for a low-powered transverter based on building blocks already described. Where needed, simple resistive attenuators may be designed using the information given in chapter 10, "Test equipment", or in chapter 13, "Data". Each module should be well screened, with particular attention being paid to preventing feedback and noise from reaching the local oscillator.

Inter-wiring should be screened, using miniature coaxial cable, for both the signal connections and the power connections; for the latter, the outer sheath can be removed and the braid frequently grounded by soldering (carefully) to the equipment case and pcbs, as appropriate. The ingoing power and control lines should similarly be well decoupled to prevent unwanted feedback which can be troublesome especially when the transverter is used in conjunction with a high gain, high power linear amplifier. Detailed attention to screening and decoupling of the low powered stages will pay dividends later as the power output and complexity of the station is increased.

14.6.3 A high powered transverter

The block diagram of a high powered transverter is given in Fig 14.75. It uses a valve transmit converter which

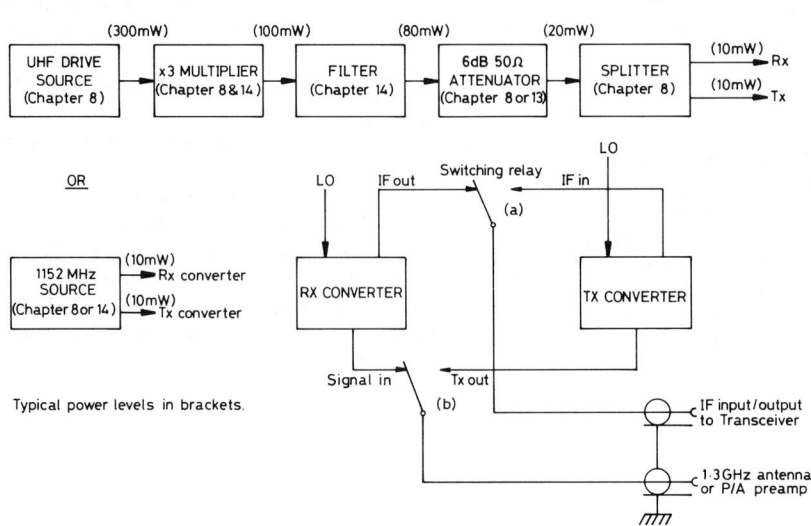

Fig 14.74. Two alternative configurations for a low power transverter based on some of the modules described in the text

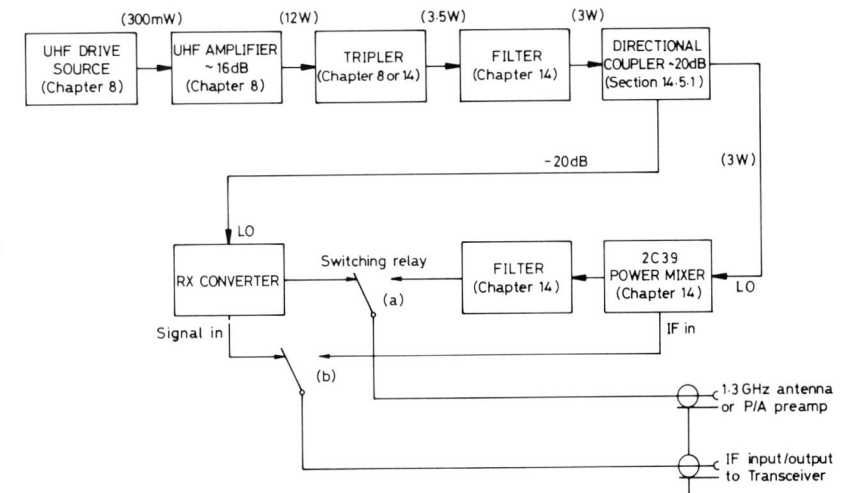

Fig 14.75. Block diagram of a high powered transverter using a valve transmit converter

requires several watts of 1,152MHz drive. This is derived from the high quality uhf source described in chapter 8, "Common equipment", and the output of this board, at 384MHz, will need to be amplified to a level of about 12W before being applied to the varactor multiplier, in order to produce sufficient drive for the valve mixer. This output should be well filtered before feeding it to either the transmit or receive converter. Several of the higher powered varactor multipliers described elsewhere (this chapter and chapter 8) may be used. Having developed considerable power at 1,152MHz, the simplest way to tap off the few mW needed for the receive converter is to use the simple microstrip coupler described as an in-line power indicator. For this use, the power detecting diode is simply removed and the coupled output taken to the receive converter. If the coupling is too tight, i.e. too much power is sampled into the coupled line, either the coupling factor can be reduced by redesigning the coupler, or further 50Ω resistive attenuation can be inserted into the output port of the coupler.

14.7 CONCLUSIONS

This chapter has reviewed the 1.3GHz amateur band and presented several practical designs for simpler equipment, both receivers and transmitters, which can easily be constructed at home without recourse, for the most part, to elaborate workshop facilities. The intention has been to indicate how the average amateur can assemble quite an effective station without going to the extremes necessary to set up a really top class station suitable for eme communication, although this mode has been mentioned as being entirely practical from the average back-garden. For these reasons descriptions of multi-valve amplifiers and large dish antennas have been omitted, although there is a detailed account of large dish construction in chapter 4, "Microwave antennas".

14.8 REFERENCES

[1] "Interdigital converters for 1,296 and 2,304MHz", R E Fisher, W2CQH, *QST*, January 1974, ARRL.
[2] *VHF/UHF Manual*, 4th Edition, RSGB, pp9.10–9.15. ISBN 0 900612 63 0.
[3] "A linear transverter for 28MHz–1,296MHz with push-pull mixer", U Beckmann, DF8QK, *VHF Communications*, Volume 9, No.4 (Winter), 1977.
[4] R Dabbs, G2RD, *RSGB Bulletin*, October 1965, p650. Also in the 3rd and 4th editions of the *VHF/UHF Manual*, RSGB.
[5] C W Suckling, G3WDG and P Blair, G3LTF, *Radio Communication*, January 1976, p24.
[6] Crawford Hills VHF Club (USA), *Technical Report No. 6*, July 1971.
[7] Ibid, *Technical Report No. 13*, December 1972.
[8] Chip Angle, N6CA, *QST*, June 1981, ARRL.
[9] "A quarter kilowatt 23cm Amplifier", Chip Angle, N6CA. Part 1, *QST*, March 1985; Part 2, *QST*, April 1985, ARRL.
[10] "More gain from 1.3GHz Amplifiers", Ian White, G3SEK, and Roger Blackwell, G4PMK, *Radio Communication*, June 1983.

CHAPTER 15

The 2.3GHz ("13cm") band

15.1 INTRODUCTION

The recent development of amateur use of the 2.3GHz band can largely be attributed to the ready availability of commercial equipment on the amateur market. This has allowed the testing of a greater number of paths, due to the ownership of advanced equipment. As the equipment is highly transportable this has allowed easier operation from elevated portable sites.

A number of interesting facets of operation on the band have started to appear, which have increased spectrum usage and improved equipment performance. These include:

Satellites: the provision of beacons operating around 2.4GHz on recent amateur satellites has encouraged development of receiving equipment and techniques for the investigation of space to earth propagation at this frequency.

Moonbounce: the aim of some amateurs to extend moonbounce operation up in frequency from 1.3GHz has led to the appearance of designs for advanced eme equipment that will enable the investigation of much more difficult terrestrial propagation paths.

15.1.1 Frequency allocation to amateurs

A majority of countries have an allocation in the 2.3GHz to 2.45GHz range. In the UK the amateur service is allocated the range 2,310MHz to 2,450MHz on a secondary basis to the fixed service. The secondary status is shared with the mobile and radiolocation services. The amateur satellite service has the allocation of 2,400MHz to 2,450MHz. Both services must accept interference from the industrial, scientific and medical allocations within their spectrum. Fig 15.1 shows the frequency allocation diagrammatically.

15.1.2 Beacons

A map showing the approximate locations and callsigns of the 2.3GHz beacons located in Western Europe is given in Fig 15.2. For exact locations and frequencies, see chapter 9, "Microwave beacons and repeaters". Some of these may be heard in the UK under favourable conditions.

15.1.3 Propagation and equipment performance

Most of the propagation modes encountered on the vhf/uhf bands are also present on the 2.3GHz band, with the exception of those which require ionised media such as aurora, meteor scatter and sporadic E. The most commonly encountered propagation modes on 2.3GHz are line of sight, diffraction, tropospheric scatter, ducting and aircraft scatter. Moonbounce (eme) is also a possibility at this frequency.

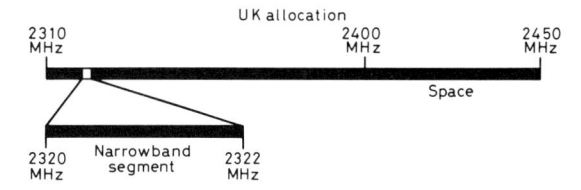

Fig 15.1. 2.3GHz ("13cm") allocation in the UK

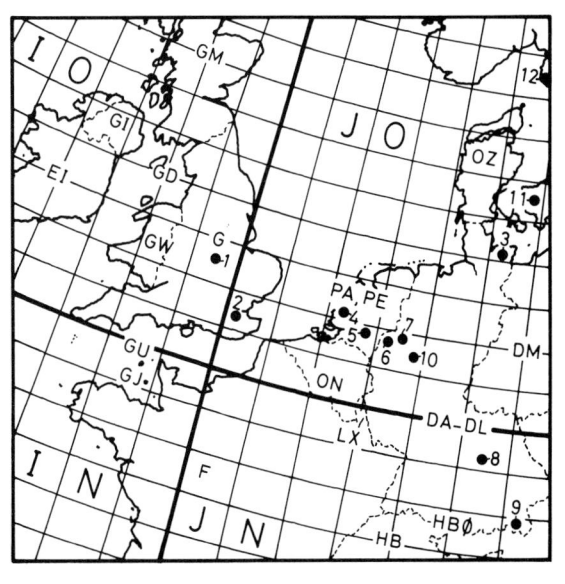

Fig 15.2. Approximate positions of some European 2.3GHz beacons. Key: 1, GB3LES; 2, GB3NWK; 3, DB0VC; 4, PA0QHN; 5, PA0TGA; 6, DB0JU 7, DL0QQ; 8, DF0ANN; 9, DB0AS; 10, DC6MR; 11, OZ7IGY; 12, SK6UHJ. For frequencies and locators, see chapter 9, "Microwave beacons and repeaters"

15.1

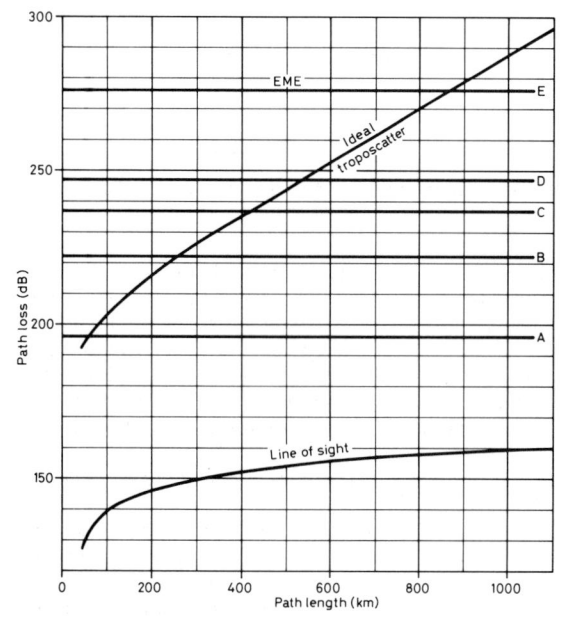

Fig 15.3. Graph of path loss versus distance for line of sight, tropospheric scatter and eme propagation, related to the performance of five sets of equipment:

  (A) 20dBi antenna, 1W transmitter, 15dB nf, 2.5kHz bandwidth
  (B) 1.2m dish, 20W transmitter, 15 dB nf, 2.5kHz bandwidth
  (C) 1.2m dish, 20W transmitter, 2dB nf, 2.5kHz bandwidth
  (D) 1.2m dish, 20W transmitter, 2dB nf, 250Hz bandwidth
  (E) 4.0m dish, 25W transmitter, 2dB nf, 250Hz bandwidth

Using the techniques of chapter 3, "System analysis and propagation", it is possible to predict the possibilities for communication over a given path, when the performance parameters of the equipment are known. Fig 15.3 shows the path loss capability of five different sets of equipment ranging from a basic system of varactor doubler, interdigital converter and single Yagi antenna to a system with moonbounce capability. Also shown on the graph are the path losses associated with three different modes of propagation, line of sight (the least lossy mode of propagation), tropospheric scatter (a reliable, commercially used mode of propagation) and moonbounce.

It can be seen that relatively simple equipment is capable of working very long line of sight paths easily. However, very few paths are purely line of sight and propagation losses are noticeably higher. The tropospheric scatter loss is a reasonable upper limit for a terrestrial path and can be used to estimate the maximum range of equipment under normal propagation conditions.

From the graph it can be seen that the simplest equipment (A) might be expected to work paths of up to 70km long. By adding an amplifier and changing the antenna to a 1.2m dish (B) the coverage distance increases to 260km. By adding a receive preamplifier (E) the coverage distance increases to 425km. Line (D) shows the effect of the equipment of (C) but using a narrow bandwidth i.f filter, as one could when receiving cw, the coverage distance increasing to 535km.

At distances over about 860km, it can be seen that the terrestrial path loss is normally greater than the route via the moon. Line (E) shows the parameters of equipment capable of moonbounce on 2.3GHz. Under anomalous propagation conditions the path losses can be considerably reduced, often approaching the line of sight loss, and even simple equipment is then capable of working very long paths.

15.2 RECEIVERS

There have been many designs for 13cm receive mixers published [1, 2, 3, 4], but two designs are in common use at the present time.

The first type is based on a 3dB hybrid coupler, usually etched on a printed circuit board [1], sometimes with an integrated rf preamplifier [2]. This is the easiest type of mixer to construct, once a pcb has been etched or obtained. For optimum results the pcb should use a low dielectric loss material, usually ptfe. This material can be both expensive and difficult to obtain. However, ready etched boards are available from the publishers of the design in [2].

The second type of mixer is an interdigital design [3, 4]. This involves more mechanical work than the first design but the materials are readily available and constructional techniques relatively simple, making it suitable for beginners. The design has the advantage of inherent front-end selectivity which makes the design less prone to out of band signals and also rejects noise power at the receiver image frequency, which can degrade overall receiver sensitivity. Only the second type of receive mixer will be described here.

To improve receiver sensitivity a preamplifier may be added to the system at a later date. Some suitable designs using bipolar devices have been published [5, 6] but designs using GaAs devices have started to appear [7, 8]. These latter devices have a relatively lower noise figure and therefore greater sensitivity than bipolar devices. However, the devices need careful handling to avoid damage by static. Only the design of [8] will be described in detail here.

15.2.1 Interdigital converter for 2,320MHz

The converter to be described is based on a design published in [4]. It can be constructed and aligned without the use of any special facilities and has a noise figure of around 15dB.

The heart of the unit is an interdigital network. Referring to the circuitry of Fig 15.4, this consists of five rod elements L1 to L5. L1, L3 and L5 are low Q coupling elements. L2 is resonant at the signal frequency of 2,320MHz. L4 is resonant at the local oscillator frequency of 2,176MHz and selects the desired harmonic generated by multiplier diode D2. This diode produces a

THE 2.3GHZ ("13CM") BAND

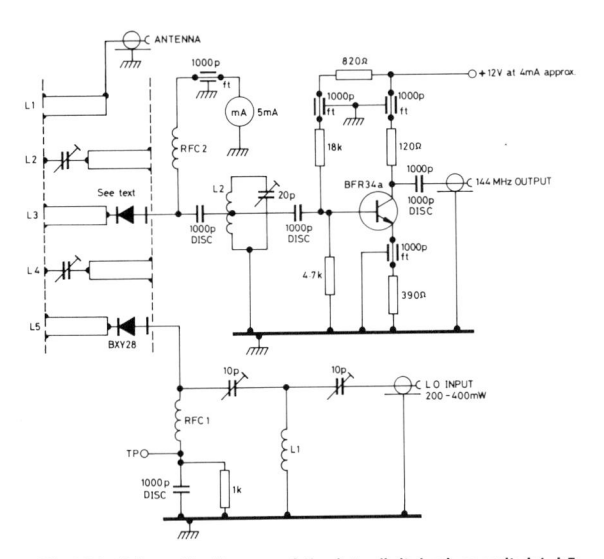

Fig 15.4. Schematic diagram of the interdigital mixer unit. L1–L5: interdigital rods. L6: 45mm 20swg tinned copper wire (t.c.w.) bent to form one turn loop. L7: 5t 20swg t.c.w. 6mm i.d. 20mm long, tapped 1.5 turns from earthy end. RFC1: 3.5t 30swg e.c.w. on FX1115 ferrite bead. RFC2: 20t 30swg e.c.w. 3mm i.d. closewound. FT: 1,000pF feedthrough capacitor. All trimmers: plastic foil types. All 1,000pF capacitors other than feedthroughs are miniature ceramic plate

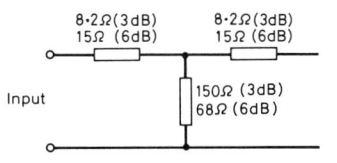

Fig 15.5. A suitable attenuator for fitting between the uhf source and local oscillator multiplier

few milliwatts at 2,176MHz for around 100mW drive at 368MHz. A suitable uhf source designed to generate this signal is described in chapter 8, "Common equipment". It should be fitted with a 90.666MHz crystal for this application.

The network consisting of L6, C4 and C5 matches the impedance of the diode to around 50Ω. To dissipate excess drive power and improve multiplier stability an attenuator should be fitted between uhf source and the local oscillator drive input socket. This attenuator should be chosen to produce around 100mW at the attenuator output. A suitable T-type attenuator to give either 3 or 6dB attenuation is shown in Fig 15.5. D1 is the mixer diode; its 144MHz output signal is fed to a low noise preamplifier stage using a BFR34a which is also described in chapter 8 and elsewhere, being a "universal" design.

The preamplifier should be well screened in a small box made from either copper sheet or copper clad printed circuit board material. Using a HP5802-2817 mixer diode the converter has a noise figure of 16dB. Using an MBD102 mixer diode the converter has a noise figure of 13dB. The multiplier diode is a Hewlett Packard HP5082-2800. The major mechanical details of the converter are shown in Fig 15.6. The baseplate is made from 1.6mm double sided pcb.

Drilling details of the interdigital unit sidewalls are shown in Fig 15.7. The top cover drilling details are shown in Fig 15.8, the material being 1mm copper or brass sheet, or alternatively, double clad pcb material. Fig 15.9 shows the construction of capacitors C1 and C2 in more detail and Fig 15.10 shows the method of mounting L1 to 5.

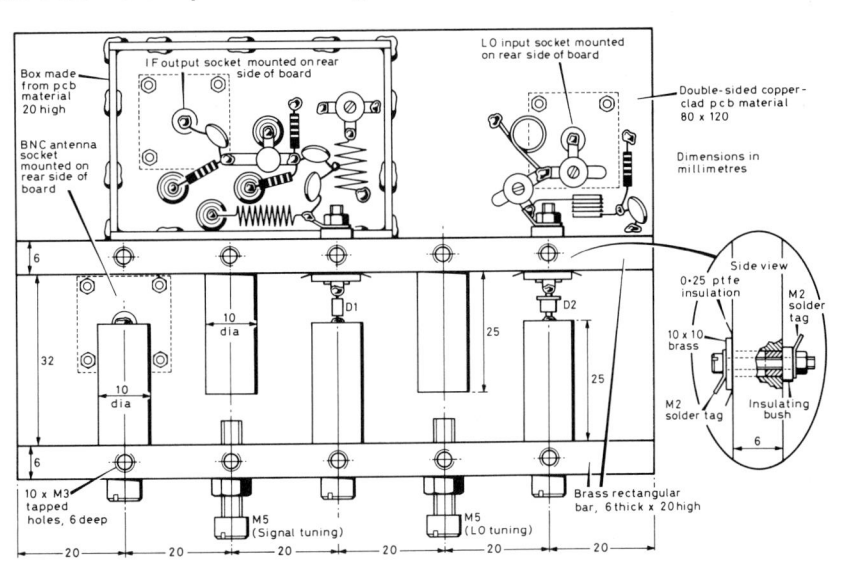

Fig 15.6. Layout of the interdigital converter

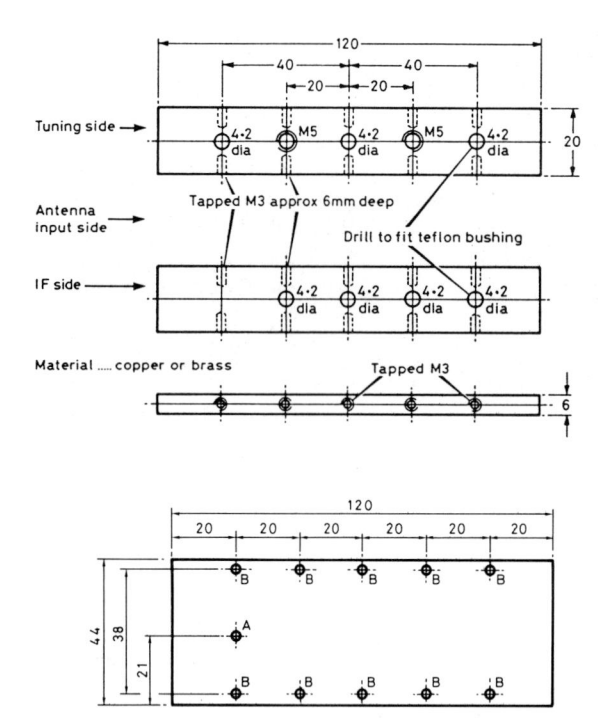

Fig 15.7. (a) Drilling details of the interdigital unit sidewalls. Material: brass bar. (b) Drilling detail of the interdigital unit coverplate. Hole A: 4mm diameter. Holes B: 3mm diameter. Note mounting holes in the baseplate have similar disposition

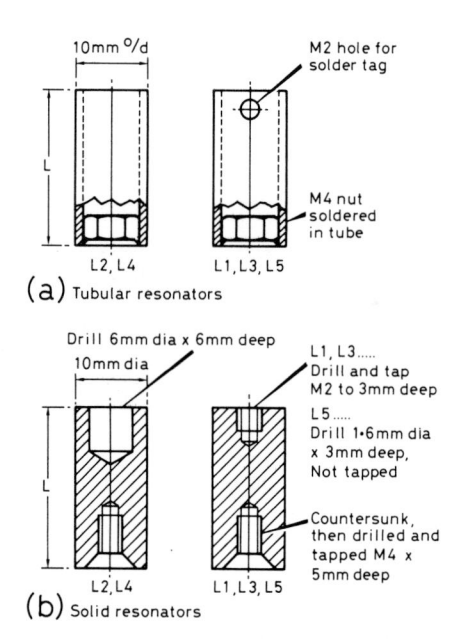

Fig 15.9. Construction of the interdigital rods. Material: L = 25mm × 10mm o.d. × 8mm i.d. brass tube as shown in (a), or suitably drilled and tapped 10mm brass rod as shown in (b)

Alignment

Connect the source of 378.666MHz to the local oscillator input of the converter. Connect a voltmeter between the test point and ground. Adjust the two 10pF trimmers for maximum voltage at the test point (around 1.25 volts). The tuning is fairly critical, and some instability may be encountered if the settings are not quite correct. Remove the voltmeter. Next, fully unscrew the signal tuning screw and adjust the local oscillator tuning screw for maximum current observed on the 5mA meter (1–4mA should be seen).

It is possible to select the wrong harmonic and a simple Lecher line may be used to check the local oscillator frequency if no other test equipment is available. The Lecher line may conveniently take the form shown in Fig 15.11, its usage being as follows: with the mixer current meter still indicating its nominal 4mA, connect the Lecher line to the antenna socket and minimise the indicated current by adjusting the signal tuning screw. It may be of assistance to use a more sensitive mixer current meter at this stage.

Next, short circuit the two Lecher lines together using a suitable metal object such as a screwdriver, and slide this up and down the line until a small peak in the mixer current is indicated. Note the position of the short circuit, and slide it away from the point until the next peak is found. The frequency in MHz of the signal being measured is then equal to 15,000/d where 'd' is the distance between the two points in centimetres.

Having ensured the local oscillator is at the correct frequency a 144MHz receiver is connected to the i.f output socket and the 20pF trimmer adjusted for maximum noise. Correct operation of the mixer can be checked by disconnecting the 378.666MHz drive to the mixer, when the receiver noise level should fall *slightly*.

A large change in receiver noise level is indicative of unstable multiplier operation and the 10pF trimmers in the

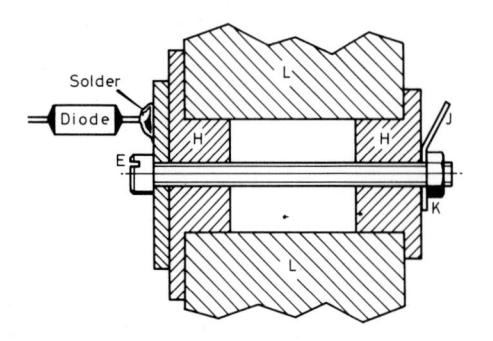

Fig 15.8. Construction of capacitors C1 and C2. E: M2 × 12 brass screw. F: 10mm × 10mm × 1mm copper plate. G: 12mm × 12mm × 0.25mm ptfe sheet. H: Two 4.2mm o.d. × 2mm i.d. × 2mm thick ptfe spacers. J: M2 solder tag. K: M2 nut. L: Interdigital unit sidewall with 4.2mm hole

matching network should be readjusted to eliminate the instability. Finally the signal tuning screw should be adjusted for maximum signal when injecting a weak signal, eg. from a distant beacon in the absence of a signal generator or signal source such as that described in chapter 10, "Test equipment".

15.2.2 Alternative lo source for 2,320MHz

This local oscillator source, by G4DDK, is capable of operating between approximately 2.0 and 2.6GHz and will provide an output of at least 7mW at the upper end of this frequency range. With the additional ATC chip decoupling capacitors in the final multiplier circuit, this output is usually increased by 3 to 4dB, depending on the individual device characteristics.

Such a source could be used as the local oscillator in receive and transmit converters operating in the 2.3GHz amateur band, or as a low power personal beacon or control transmitter operating in the same band under the terms of the new licence. Several modern 10GHz narrowband transverters require drive at 2,556MHz and this source could be used for this purpose also (see chapter 18).

Similar circuits have been in use in Europe, but these units have generally utilised a junction fet (jfet) crystal oscillator. A power jfet is required to give the best performance and difficulties in obtaining such devices has led to the adoption of lower powered devices such as the J310 series. The phase noise performance of these devices is inferior to that of their more powerful cousins. This circuit uses a Butler oscillator whose noise performance is similar to that of a power jfet.

Versions of the circuit have been built with outputs in the range 2,176MHz to 2,556MHz. Based on the available capacitance swing left in the tuned circuits, the unit should be usable with outputs between 2.0 and 2.6GHz. It may even be possible to reach 2.8GHz, opening the possibility of using the unit to drive a doubler to 5,760MHz: this has not been tried at the time of writing.

Circuit

The oscillator is the well known Butler circuit and a readily available coil has been specified for use in the circuit. It will operate satisfactorily with crystals anywhere in the range 80 to 120MHz. If the frequency is significantly outside the range 90 to 100MHz, then the parallel capacitor, C3, should be changed, not the coil. Only first grade BFY90 transistors should be used; "surplus" transistors have been known to cause problems.

Sometimes there are problems of oscillator re-start after switching off if the crystal is pulled too far. The cure for this problem is simple and requires only that a 10 to 33pF capacitor is inserted in series with the crystal. Provision has been made for this additional component: if it is not required, then a *short* wire link should bridge the Cx pads, or a ceramic 1nF capacitor used.

The oscillator may be narrow-band frequency modulated using the circuit described with the uhf source in chapter 8,

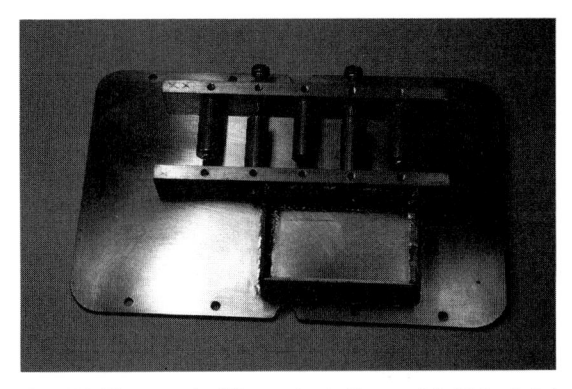

Fig 15.10. Photograph of the mechanically completed interdigital unit, ready for wiring of the i.f pre-amplifier and other circuits

"Common equipment". Either speech or frequency shift keying can be applied using this circuit.

An integrated circuit 9V regulator ensures the crystal oscillator and base bias to the first and second multipliers is stabilised to minimise frequency shift if the supply voltage changes.

The type of trimmer capacitor used is important. SKY (green) trimmers were used for C13 and 14, but any small (5mm diameter) 5pF trimmer could be used in these positions. SKY trimmers should be used for C18, 19 and 20 and the output filter.

Originally the third multiplier stage (TR4) used a BFG91A in an attempt to obtain more output. This resulted in a tendency to instability and replacement by a BFR96 restored stability: the pads for a BFG91 are still present on the artwork to allow experimentation.

The final multiplier uses a BFR91A and operates reliably at these frequencies, easily providing the required output. In additional to decoupling capacitors C24 and C25 (1nF), further decoupling is provided in both the collector and emitter circuits of this stage with 100pF ATC chip capacitors mounted on the *ground plane* side of the pcb. Extra decoupling is provided in the collector supply circuit by using a quarter-wave open-circuit low impedance transmission line. The open circuit at the end of the line is transformed to a short circuit a quarter wavelength away at the junction of R20 and the quarter-wave choke by the track from C29 to R20. Although this arrangement can

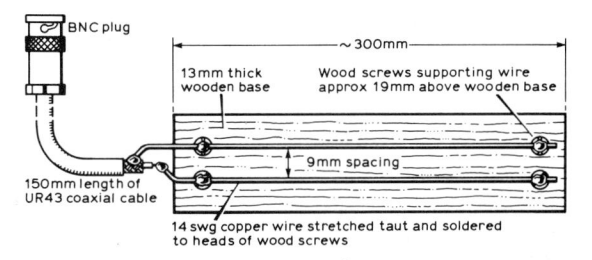

Fig 15.11. Construction of Lecher lines suitable for checking the local oscillator frequency

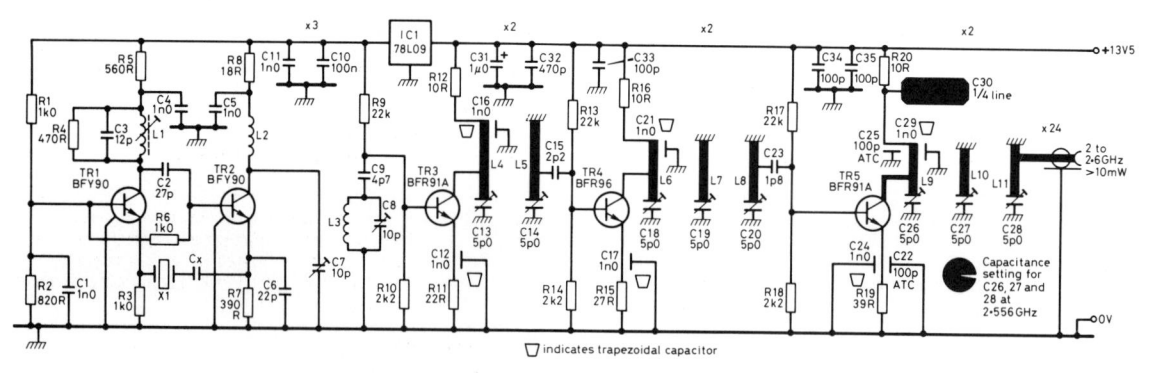

Fig 15.12. Circuit of an alternative 2.3GHz local oscillator source, G4DDK-004

only be optimum at one frequency, in practice the bandwidth is such that it remains effective over the band 2.0 to 2.6GHz at least. The circuit of the complete unit is given in Fig 15.12.

Construction

The unit was designed to fit into a type 45 tin-plate box (7768, Piper Communications), which is 55.5 by 148 by 30mm. A full size pcb layout is given in Fig.15.13 and the component overlay in Fig.15.14. A full components list is given in Table 15.1. The material used is good quality 1.6mm thick epoxy-glass, double clad with 1oz copper. Slight differences in ε_r between board materials may affect performance slightly at the high frequency end of the range. Alternatively, a ready etched, drilled and tinned pcb is available from the Microwave Committee Components Service.

After suitable etching, the board is drilled with 0.8mm holes for all but the transistor mounting holes (5mm) for TR3, 4, and 5. 1.2mm holes are needed for the trimmers and L1 and L3. Slots 1mm wide need to be cut for the grounding strips at the ends of L5, 7, 8, 10 and 11. 1.5mm wide slots are needed for the trapezoidal capacitors C12, 16, 17, 21, 24, 25 and 29.

The corners of the board may need to be filed slightly to clear the two overlapping corners in the tinplate box. The board should be positioned so that the top of L1 comes 5mm below the rim of the box. Mark the position of the output socket, allowing the spill of the connector to lie flush with the output track on the pcb. It is better to use a single hole mounting type and solder it flush to the outside wall of the box with its spill protruding into the box. The socket should be an SMA, SMB or SMC type: N-type connectors are too large and BNC connectors can be unreliable at these frequencies. It may be necessary to remove a small area of the bottom lid of the box to accommodate the socket. Drill the socket mounting hole. Also drill a hole in the same end of the box to take a feedthrough capacitor to bring in power. If a crystal heater is to be used, then also drill holes in the opposite end of the box for the power feed for this.

If the unit is to be used as a low power transmitter, then an additional feedthrough capacitor will be needed for the modulation input. The value of this capacitor will need to be carefully chosen to match the intended modulation.

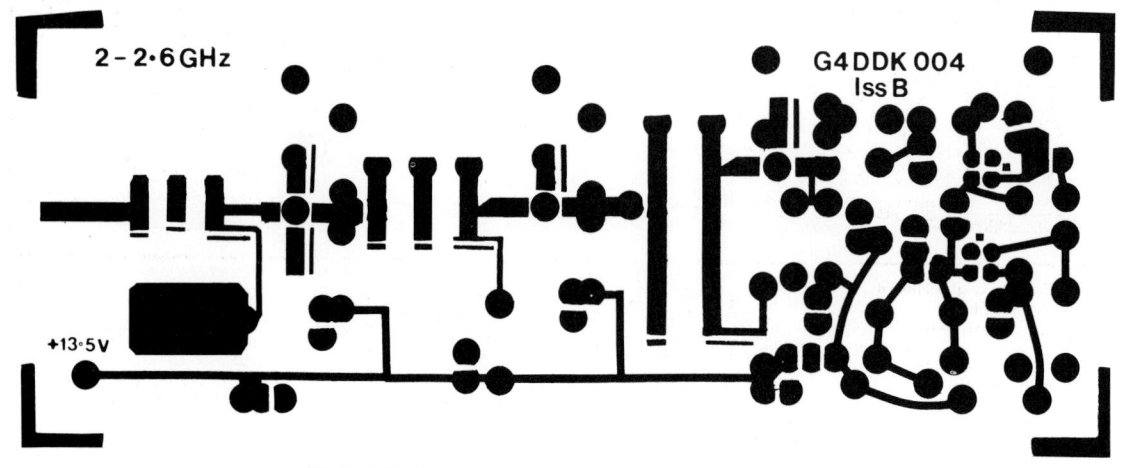

Fig 15.13. Track layout for the alternative local oscillator source

Table 15.1. Component list for alternative local oscillator source for 2.3GHz

Resistors

R1, 3, 6	1k0	R9, 13, 17	22k
R2	820R	R10, 14, 18	2k2
R4	470R	R11	22R
R5	560R	R12, 16, 20	10R
R7	390R	R15	27R
R8	18R	R19	39R

All resistors 0.25W miniature carbon film or metal film

Capacitors

C1, 4, 5, 11	1000pF High K ceramic plate eg. Philips 629 series
C2	27pF low K ceramic plate eg. Philips 632 series
C3	12pF .. ..
C6	22pF .. ..
C9	4p7 .. ..
C15	2p2 .. ..
C23	1p8 .. ..
C31, 32, 33	470pF medium K ceramic plate eg. Philips 630 series
C22, 34, 35	100pF low K ceramic plate eg. Philips 632 series
C10	0.1μF Tantalum bead, 16V working
C12, 16, 17	1nF trapezoidal (RSGB or Cirkit)
C21, 24	470pF trapezoidal ..
C25, 29	22pF trapezoidal ..
C7, 8	10pF miniature trimmer (5mm diameter) eg. Cirkit 06-10008
C13, 14, 18, 19, 20, 26 27, 28	SKY trimmer (green) (Piper Communications)
C30	PCB track
Cx	10 to 33pF type as C3 (see text)

Coils

L1	Toko S18 5.5 turns (green) with aluminium core
L2, 3	2 turns of 1mm diameter tinned copper wire, inside diameter 4mm. Turns spaced to fit hole spacing. Exceptionally 3 turns at low frequency end of range.
L4 to 11	Printed on pcb

Semiconductors

TR1, 2	BFY90
TR3	BFR91A
TR4	BFR96
TR5	BFR91A
IC1	78L09

Miscellaneous

X1	Fifth overtone crystal in HC18/U case. Frequency of crystal = $F_{out}/24$
Tinplate Box	Type 45 (7768) Piper Communications 55.5mm wide by 148mm long by 30mm high.
O/P socket	Single hole SMA, SMB or SMC

Carefully solder the board into one half of the tin box, tack-soldering at several places along the groundplane/box junction. Next solder the other half of the box in place, ensuring a good fit. It may be necessary to file a small amount off the pcb at the ends or sides to get a comfortable fit. Finally when satisfied, solder along the whole of the sides and ends of the box to give a good rf-tight connection.

When these mechanical arrangements are complete, wiring can commence. Cut five short lengths of thin (shim)

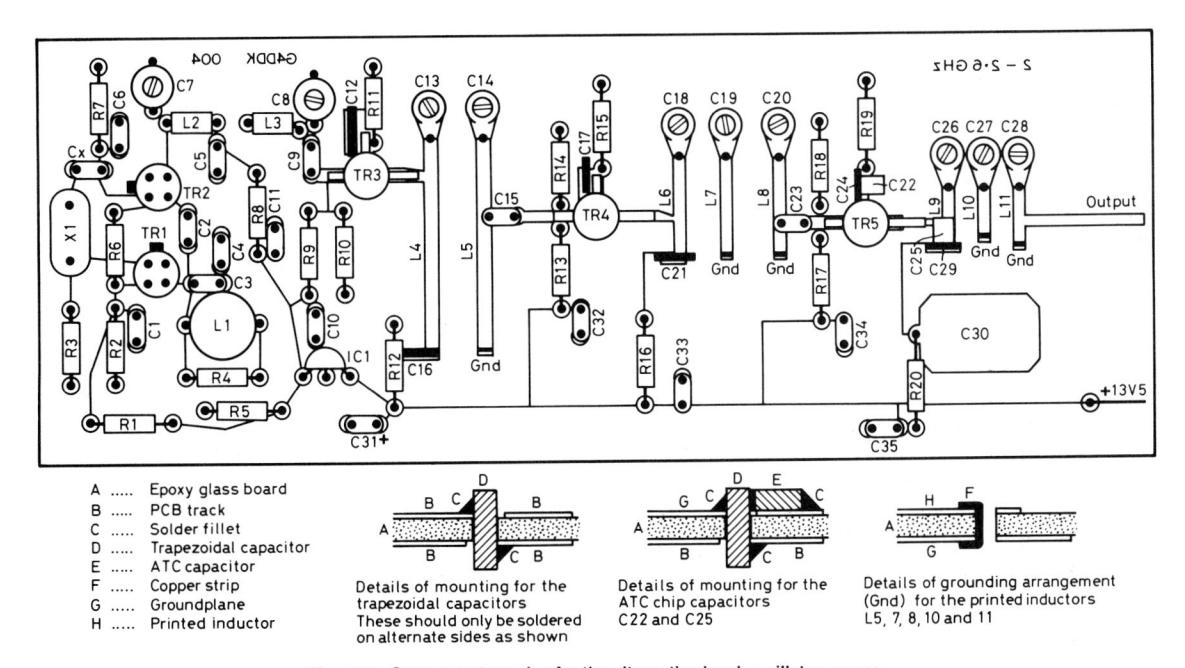

```
A ..... Epoxy glass board
B ..... PCB track
C ..... Solder fillet
D ..... Trapezoidal capacitor
E ..... ATC capacitor
F ..... Copper strip
G ..... Groundplane
H ..... Printed inductor
```

Details of mounting for the trapezoidal capacitors
These should only be soldered on alternate sides as shown

Details of mounting for the ATC chip capacitors C22 and C25

Details of grounding arrangement (Gnd) for the printed inductors L5, 7, 8, 10 and 11

Fig 15.14. Component overlay for the alternative local oscillator source

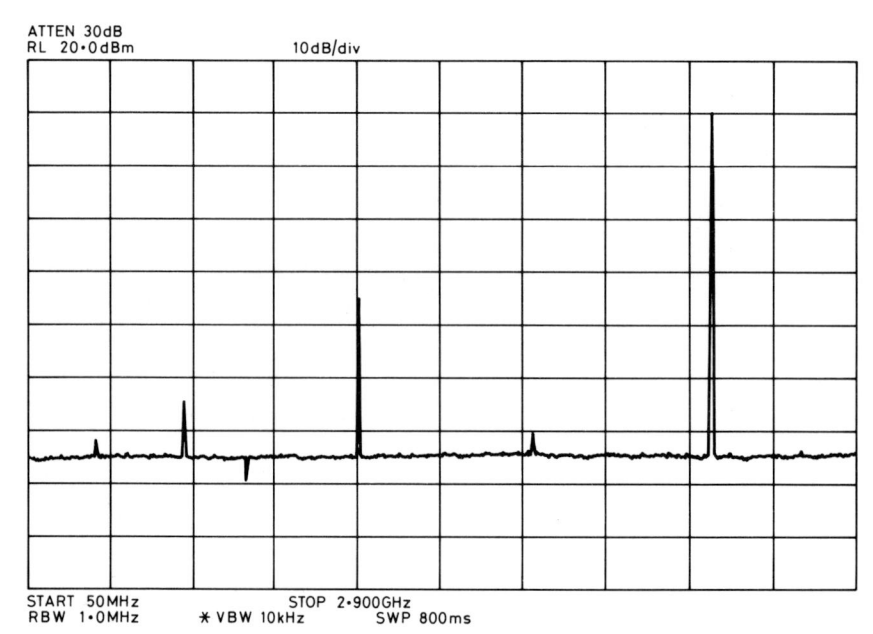

Fig 15.15. Spectrum analyser trace of output from the alternative local oscillator source

copper strip and solder them at the ends of L5, 7, 8, 10 and 11 as shown in the overlay diagram.

Solder all resistors and capacitors into place where shown, taking care to solder grounded leads both top and bottom of the board. Solder the coils in place, remembering to solder the grounded end of L3 both top and bottom.

Solder TR1 and TR2 in place, making sure they are seated well down onto the board but leaving just sufficient room to solder the case lead of both transistors to the groundplane of the board. Solder IC1 into place, remembering to ground the centre lead. Solder the trapezoidal capacitors in place as shown in the overlay diagram.

Carefully bend the flat connection lead of the SKY trimmers out at 90° to the capacitor body. Place the trimmer with the round lead in the hole in the tuned line. Solder this end first and then the flat lead to the groundplane. C7 and C8 can be treated in the same way. If trimmers other than SKY trimmers are used, the mounting arrangements may need to be different – always use the shortest possible leads on the trimmers, whichever type is used. Solder TR3, 4 and 5 into place, taking care to get the leads the right way round. Solder the crystal in place last of all, seating it well down onto the board. It may be advisable to ground the case of the crystal, especially if a heater is to be used. Take care to ensure that the crystal is not of the type where one side is connected to the case, or the bias may be shorted out.

General points on construction are:

1. Solder resistors flat to the board.

2. Keep capacitor leads very short and use only the recommended ceramic types.

3. Place transistors into the mounting holes and seat their leads flat onto the pcb tracks before soldering. Too long leads may cause instability.

4. Trapezoidal capacitors must not be overheated. It is quite possible to solder them into place with a hot iron and not have them fracture – if you are quick!

Alignment

It is possible to align the unit with nothing more than a multimeter. However this will give little information as to what frequency the unit is tuned to or what output level has been achieved. As far as alignment is concerned the most important items of equipment are a good moving-coil (analogue) multimeter and absorption wavemeter. A digital multimeter is worse than useless for tuning up as it can give *very* misleading results. The wavemeter is also preferable to a frequency counter, since not only does it indicate the frequency of the required output but also the presence of unwanted signals and their relative levels. The wavemeter and multimeter are more valuable than even a spectrum analyser: many modern analysers with ranges beyond 2GHz use digital display systems which often do not work in real time. As a result, what is indicated on the screen is not what you are doing, but rather what you have just done!

Connect a 13.5V supply and check that the unit takes no more than about 150mA. If the consumption is much higher than this, switch off and check for faults. When all is well, check the regulated voltage is 9V.

Place a wavemeter close to L1 and tune it to the crystal frequency. Adjust the core of L1 for a strong reading. Switch off then on again to check that the oscillator restarts. If not,

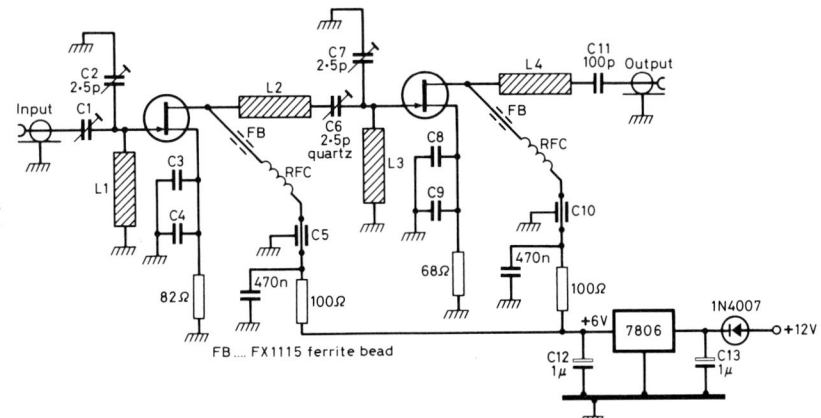

Fig 15.16. Circuit diagram of a two-stage 2.3GHz GaAs fet preamplifier

adjust L1 core and try again. The exact frequency does not matter at this stage.

Set the multimeter to the 2.5V range and connect it across R11. Adjust C7 then C8 for maximum reading. When C7 is adjusted initially, there is only the slightest movement of the meter. This is what you are looking for and a digital meter will probably miss it due to the up-dating method used. Peak the reading with C8. Confirm with the wavemeter that you have set L2 and L3 to the required third harmonic of the crystal frequency.

Transfer the meter to R15 and adjust C13 and C14 for a maximum voltage reading across R15. Confirm frequency with the wavemeter. Transfer the meter to R19 and adjust C18, 19 and 20 for a maximum reading. Because there are three tuned circuit this time, the initial meter movement may be very small. The middle tuned circuit is especially sharply tuned.

Now that there are no more emitters to use, the wavemeter becomes indispensable. Connect it to the output and tune it to the required output frequency. It helps here to have an accurately calibrated scale. Adjust C26, 27 and 28 to obtain maximum reading. This should be close to 10mW if your wavemeter is also calibrated for relative output. Use *only* an insulated trimming tool to adjust the output trimmers. Again the middle tuned circuit will be found to tune very sharply.

It is now worth going back over the previous adjustments, re-peaking to obtain maximum output. If any adjustment requires more than the slightest "tweak", be suspicious of the alignment and recheck it from the beginning.

Now is the time to use access to an expensive spectrum analyser! Check the output frequency is correct and that the output spectrum looks something like that in Fig.15.15. If the output frequency is a little low and pulling it onto frequency results in reduced output or failure to restart after switching off and then back on, it may be necessary to add Cx, previously mentioned. This should cure the problem: if not, change the crystal.

15.2.3 GaAs fet preamplifier for 2,320MHz

This preamplifier was designed by OE9PMJ [7] for use on 2,320MHz eme but the design is equally suitable for 2,320MHz tropospheric working. The original design used two MGF1412-11-08 GaAs fets and produced a noise figure of 0.6dB and a gain of 32dB. The version to be described here is much cheaper to build than the original design. It uses an MGF1412-11-09 and a MGF1402 which will produce a noise figure of 0.7dB and a gain of 29dB. The circuit diagram of the preamplifier is shown in Fig 15.16.

Construction

The major mechanical details are shown in Figs 15.17 and 15.18. The case is folded from a single piece of 0.5mm brass sheet and soldered at the corners. The two holes at the ends of the box are drilled to clear the centre pins of the input and output connectors used, preferably N-type or SMA. The two partitions are made from 0.5mm brass sheet which are firstly cut to fit in the appropriate position, then drilled as shown in Fig 15.19. The dimensions of the

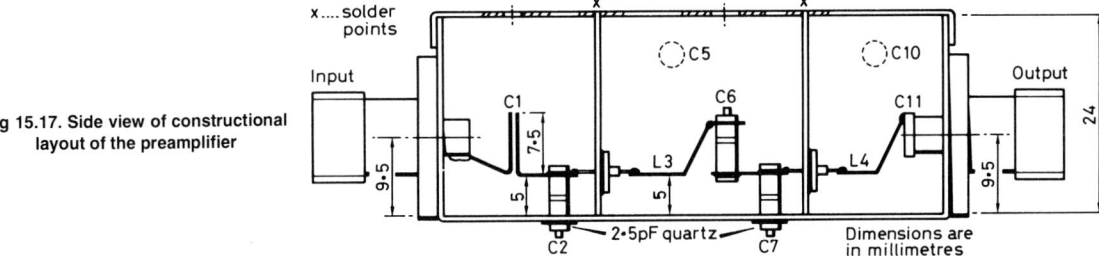

Fig 15.17. Side view of constructional layout of the preamplifier

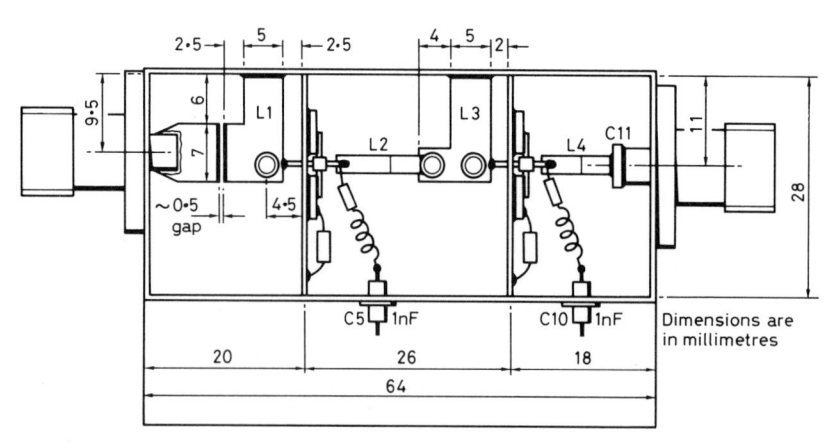

Fig 15.18. Top view of constructional layout of the preamplifier

Dimensions are in millimetres

inductors which are shown in Fig 15.20 are made of 0.5mm copper sheet. The 3mm by 5mm tabs on L1 and L3 are to enable the inductors to be bolted to the case using very short M2 screws. The appropriate sized holes are drilled in the case to suit capacitors C1 and C7 and the 1nF feed-through capacitors C5 and C10. In the authors' version, 0.3 to 3.5pF Johanson trimmers (type 5800) were used which required the holes to be 3mm. Alternatively as shown in Fig 15.21, capacitors C1 and C7 may be replaced by M4 bolts running in M4 nuts soldered to the case. This method is especially useful if good quality trimmers cannot be obtained.

The box, partitions and rf connectors are then soldered together with a large soldering iron. Whilst the assembly is cooling, the 1nF feedthroughs (if solderable) and the leadless disc capacitors (C3, C4, C8 and C9) are carefully soldered into position using a small soldering iron.

The source decoupling capacitors are soldered as close as possible to the hole in the partition to ensure minimum source lead length on the fet. Direct heat should not be applied to the discs but the surface tinned whilst the reverse side of the copper is heated then the disc is gently pressed in position and soldered. If difficulty is experienced in retaining the leadless disc capacitors on the partitions, a pair of aluminium plates made as shown in Fig 15.22 may prove useful. Using a steel M2.5 screw and nut mounted in the centre hole of the two plates, the leadless discs are sandwiched in position, fixing their position during the soldering operation. The tuning trimmers are next bolted or soldered in position as appropriate. Chip capacitor C11 should be soldered between the centre pin of the output connector and L4, shortening the centre pin of the connector as necessary.

The remaining components are added as shown in the circuit diagram, with the regulator and associated components being mounted on the side wall of the box.

The GaAs fets are mounted into position last of all, with their package in the appropriate chamber and source leads soldered to the decoupling capacitors. The gate leads are passed through to the hole in the partition and soldered to the gate inductor. Extreme care should be taken to prevent damaging the fets with static charges.

Alignment

Firstly the dc bias conditions of each fet should be checked by connecting a voltmeter across the appropriate source resistor. For the MGF1402 the current is nominally 10mA and for the MGF1412 the current is 15mA, so the source resistor is adjusted to obtain the appropriate voltage drop across itself. If an automatic noise figure optimisation aid is available (e.g. the G4COM design described in chapter

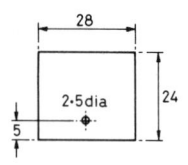

Fig 15.19. Preamplifier partitions, two required. Material: 0.5mm brass

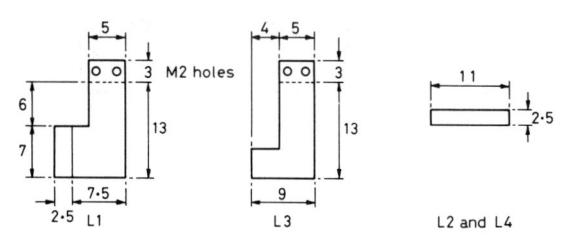

Fig 15.20. Dimensions of the stripline inductors. Material: 0.5mm copper, preferably silver plated

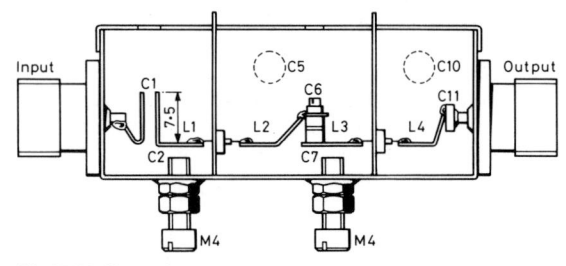

Fig 15.21. Alternative arrangement of C2 and C7, using M4 nuts and bolts

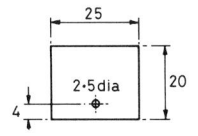

Fig 15.22. Leadless disc capacitor retaining plate. Material: 1.6mm aluminium

10, "Test equipment"), this may be used as described to initially align the preamplifier.

Final adjustments (or the whole alignment process for those without a suitable alignment aid) should be made for maximum signal to noise ratio when monitoring a weak signal source with the preamplifier configured as it is will be used (i.e. the same length of feeder, same antenna, etc.). The gain against frequency and noise figure against frequency measurements made on the OE9PMJ original preamplifier are shown in Figs 15.23 and 15.24.

15.3 TRANSMITTERS

The easiest way to transmit on 2.3GHz is still to use a varactor doubler to generate a few watts of fm or cw. This is driven by an 1,160MHz source which may either be a retuned 1,152MHz high-level mixer chain or a purpose built source. As can be seen from Fig 15.1, it is advantageous to have the crystal controlled source near the centre of activity around 2,320.2MHz, so the crystal frequency should be chosen accordingly.

Once experience of the band has been gained with this arrangement, one can progress to a linear transmit converter to allow all mode operation on the band.

There is a choice between three main types of mixer. The first is a varactor mixer which has the advantage, for portable operation, of being passive and therefore needing no power supply. The second is a valve mixer, usually employing a 2C39A type valve. This has the advantage of generating more power on its own than the first type but does require a high voltage and blower. The local oscillator for either of these can be obtained from retuning the original varactor multiplier transmitter chain to 2,176MHz.

The third type is a transistor (or GaAs fet) mixer. This type of mixer suffers from a low output level, needing many transistor amplifier stages to amplify the output to a reasonable level. However, with the recent increase in availability of suitable devices this method is gaining popularity.

The final stage of development is to increase the output power. Generation of tens of watts at a modest cost at the moment is only possible using valves of the 2C39A family, although there are some expensive transistors available which will generate a couple of watts on the band.

15.3.1 Varactor multiplier for 2.3GHz

A complete transmitter using a varactor doubler may take the form shown in Fig 15.25. The attenuator between the two varactor multipliers is needed in the majority of cases

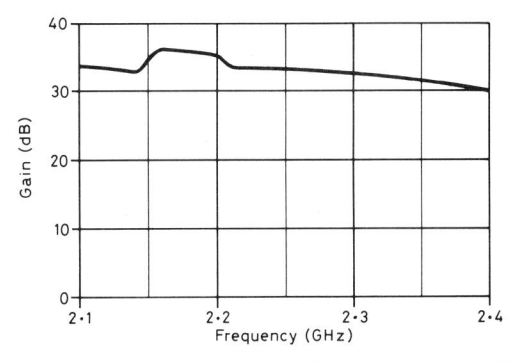

Fig 15.23. Measured gain performance of the prototype preamplifier using two MGF1412-11-08 devices

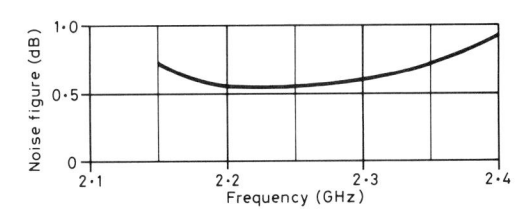

Fig 15.24. Measured noise figure performance of the prototype preamplifier using MGF1412-11-08 devices

to ensure system stability! Due to scarcity of high power and high frequency attenuators, it may conveniently consist of a length of coaxial cable. This method also allows the multiplier to be mounted at the antenna to maximise the radiated power. This position of mounting the multiplier is of advantage, as can be demonstrated by the following example. A five watt 1,160MHz source and a 50% efficient varactor doubler are to be connected to an antenna 20m distant using URM67 cable.

The first method is to locate the doubler adjacent to the source and feed the resulting 2.5W, 2,320MHz signal to the antenna through the cable. At 2,320MHz the cable has a loss of 8dB, therefore the radiated power is 0.4W. The second method is to feed the 1,160MHz rf through the 20m cable to the doubler mounted on the back of the antenna.

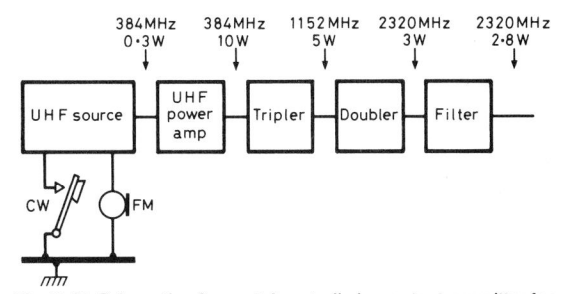

Fig 15.25. Schematic of a crystal controlled varactor transmitter for 2.3GHz

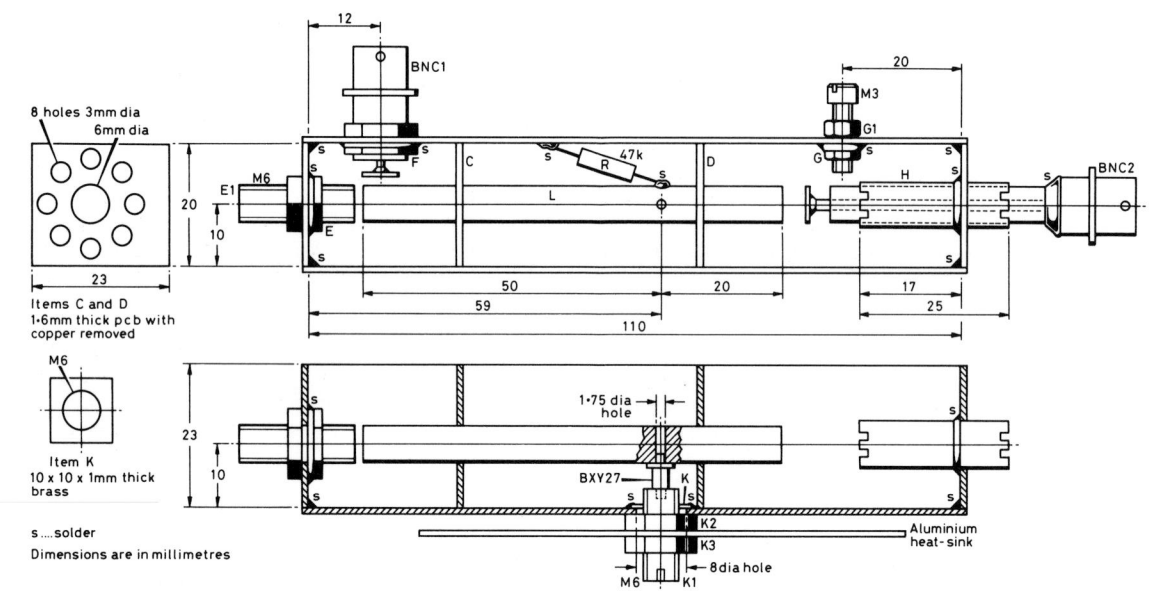

Fig 15.26. Construction of the 2.3GHz doubler

The doubler may be located in the same box as any masthead preamplifier.

At 1,160MHz the cable has a loss of 5dB, therefore 1.6W is available at the input to the doubler. The radiated power is now 0.8W, double the previous level.

Two designs of varactor doubler, both using BXY27e series diodes seem to be in general usage. The first by S Freeman, G3LQR, [9] is made from sheet brass and needs a high power heat source to assemble. The second design, by O Frosinn, DF7QF, [10] to be described here, is made from 1.6mm glass fibre printed circuit board and can easily be assembled using ordinary shack tools.

Construction

The important dimensions of the varactor doubler are shown in Fig 15.26. The pcb pieces are cut to size and the necessary holes drilled. Part H is a 25mm length of 7mm outside diameter tube with two slits in each end, allowing it to be compressed slightly. Nuts E, F, G and tube H are then soldered into position. BNC1 and BNC2 are single hole fixing sockets (type UG1094/U). BNC1 has the exposed ptfe insulation removed (and kept for use later). The centre conductor is then cut to leave only 3mm showing. A 6mm disc is then soldered centrally onto the centre conductor. BNC2 has approximately 40mm of 2mm diameter silver plated copper wire soldered to its centre conductor. A 38mm length of 6mm outside diameter tube is soldered centrally to the outer conductor, the surplus ptfe piece from BNC2 being inserted in the open end to maintain concentricity. The centre is then cut to leave 4mm protruding from the tube and a 5mm brass disc attached centrally.

The whole assembly is then inserted into its guide tube (part H). The input and output connector and tuning screw guide nuts are next soldered to the outer wall of the trough and the connectors and tuning screws with their associated locknuts screwed into place.

Spacers C and D are made from 1.6mm pcb material, from which the copper has been removed. They should be made from an M6 × 12 brass bolt with a 1.8 m hole drilled centrally in its end. The corresponding diode mounting hole is also drilled through the top and bottom of part L also using a 1.8mm drill. Part K is screwed onto part K1 and put into the 6mm hole in the base of the trough, with the plate on the inside. The plate is then soldered into position. Spacers C and D are pushed on to line L and placed into the trough. Using a 1.8mm drill as a jig, both diode mounting holes are aligned. Parts C, D and L are then glued into place using cyanoacrylate adhesive. The bias resistor is then soldered in the position shown using the shortest possible lead length. Alternatively the resistor may be "synthesised" by reducing the resistance of the diode in the reverse direction to approximately 47kΩ. This is done by rubbing a 2B pencil on the ceramic case of the diode whilst monitoring the resistance of the diode in its

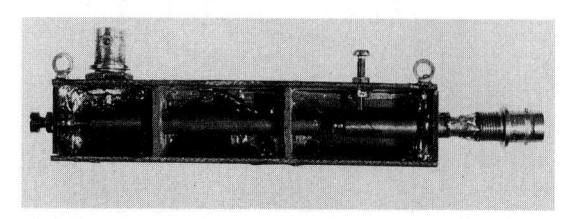

Fig 15.27. Photograph of the completed varactor doubler

Fig 15.28. Equipment for aligning the varactor doubler

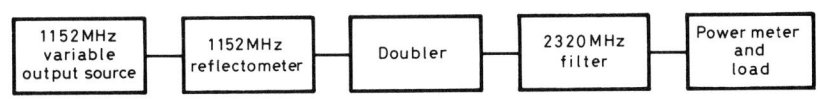

1152MHz variable output source → 1152MHz reflectometer → Doubler → 2320MHz filter → Power meter and load

non conducting direction. The diode is inserted into its mount and carefully screwed into place. When in its final position a locknut is added to the diode mount on the outside of the trough. For efficient heatsinking the whole assembly should be a minimum of 100mm square of 1.2mm thick aluminium. A photograph of the completed unit is shown in Fig 15.27.

Alignment

For testing purposes the varactor input is connected via a reflectometer and 1,160MHz filter to an 1,160MHz power source with adjustable output power (e.g. a fixed power output source with suitable attenuators). The output is connected via a 2,320MHz filter and power detector to a suitable 50Ω load. This may be a long length of 50Ω coaxial cable (e.g. RG174) terminated in a 50Ω resistor. A suitable arrangement is shown in Fig 15.28. Initially apply around 100mW of drive to the varactor and adjust the tuning screws and the position of the output tubes for best input match and maximum output power. The drive power is increased to its final value and the tuning adjustments reset as necessary. Lastly, the multiplier is assembled with its ultimate configuration of antenna and coaxial cables and any final adjustments made including an attenuator, if necessary, on the input to the doubler.

The typical performance of this varactor doubler is a minimum output power of 3W with a drive power of 5W using a BXY27e diode. The maximum drive power for this diode is around 12W.

15.3.2 Varactor mixer for 2,320MHz

This transmit mixer was designed by H Schildt, PA0HEJ, [11] and is capable of 350mW output.

Operation

The layout of the mixer is shown in Fig 15.29. L1 is tuned to 1,088MHz which is half the required local oscillator frequency. Energy at this frequency is coupled to the diode by C1 and loop A. The diode may be either a BXY27e or a BXY28e. The 144MHz lowpass filter consisting of L3, C4 and C5 also performs an impedance match between the 144MHz source and the varactor. C3 prevents significant amounts of 2GHz rf from appearing at the 144MHz terminal. The diode both doubles from 1,088 to 2,176MHz and mixes the 144MHz to produce an output at 2,320MHz. This signal is coupled by tab B to the three section 2,320MHz filter which reduces out of band signals.

Construction

The majority of the mixer is made from brass. The frame is 20mm by 6mm brass bar. The baseplate is a 150mm by 72mm piece of 1.6mm thick brass plate. The cover is also 1.6mm brass sheet in an L shape to enclose the rf circuitry. The assembly is held with M3 tapped holes and M3 × 6 panhead brass screws.

The construction of the mixer is shown in Fig 15.30. Piece A is a 1.5mm by 6mm wide piece of 0.2mm copper foil. Piece B is a 20mm by 8mm wide piece of 0.2mm thick copper foil. Piece C is a piece of M4 studding with a 1.6mm diameter by 2mm deep hole in one end for mounting the diode. It is fitted with a lock nut. Piece D is a 4mm diameter by 35mm long brass rod with a 1.6mm diameter hole drilled radially at its centre. The rod is mounted centrally in two ptfe bushes 6mm outside diameter by 4mm inside diameter by 6mm thick mounted in the walls as shown. L1 and L4 to 7 are cut to the dimensions specified, then an M4 brass nut is soldered into one end, allowing fixing in place with an M4 × 10 brass screw through the sidewall. The completed unit is shown in Fig 15.31.

Alignment

A signal at 1,088MHz is applied via a suitably rated attenuator to produce 1.5W at the local oscillator port of the mixer. A signal at 144MHz is applied via an attenuator to produce around 0.25W at the i.f port of the mixer. A suitable detector is connected to the output of the mixer, after a 2,320MHz filter. Initially connect a voltmeter across the 100kΩ varactor bias resistor and turn on the local oscillator. Tune L1 and C1 for maximum voltage indicated by the meter. Turn off the local oscillator and turn on the i.f source. Align C4 and C5 for maximum reading on the meter. Remove the i.f drive. To align the 2,320MHz filter incorporated in the output of the mixer, feed a 2,320MHz source (e.g. the varactor transmitter described earlier) to the output of the mixer and tune the filter screws for maximum reading on the meter. Remove the meter from the diode. Upon simultaneous application of lo and i.f to the mixer an indication on the output meter should be seen.

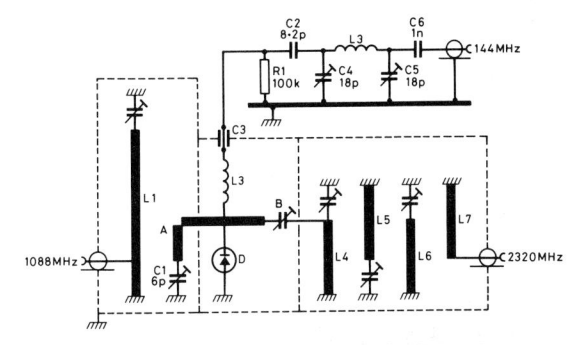

Fig 15.29. Circuit of a 2.3GHz varactor mixer. D1: BXY27. R1: 100kΩ carbon. L1: 54mm long 10mm o.d. 8mm i.d. copper tube. L2: 4 turns 0.4mm diameter t.c.w. 3mm i.d. L3: 6 turns 1mm silver plated copper wire 7mm i.d. L4-L7: 10mm o.d. 8mm i.d. copper tube. C1 6pF tubular ceramic trimmer maximum 3mm diameter. Pieces A-E, see text. RFC1: 3t 26swg enamelled on FX1115 bead

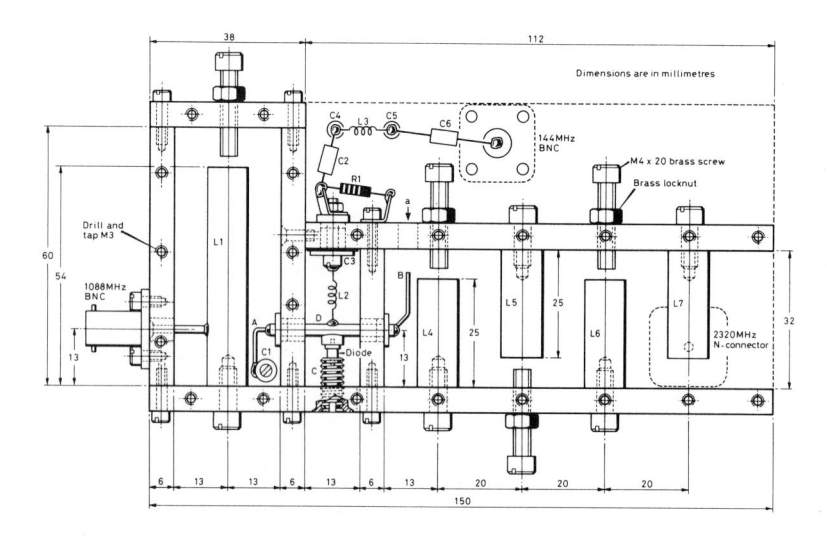

Fig 15.30. Mechanical construction of the 2.3GHz varactor mixer

Fig 15.31. Photograph of the completed varactor mixer

All variables should readjusted for maximum indication on the meter. A check should be made to ensure the mixer has not been tuned up as a 2.3GHz doubler by removing the i.f drive and noting that the output signal disappears.

Linearity should be checked by applying 144MHz ssb to the local oscillator port and listening to the output quality on a 2,320MHz receiver. With the drive levels mentioned above an output power of 0.35W (350mW) at 2,320MHz has been achieved.

15.4 TRANSMIT AMPLIFIERS

15.4.1 Single valve amplifier for 2,320MHz

This amplifier designed by Hans Rasmussen, OZ9CR, uses a single ceramic 2C39A type valve to produce an output of around 30W for a drive level of around 2W. An overall diagram of the amplifier is shown in Fig 15.32.

Partition plate (1), Fig 15.33, separates the anode and cathode cavities. Care should be taken during drilling the 21mm holes that the plate is not bowed as this will make final assembly difficult. The cathode cavity (2), Fig 15.34,

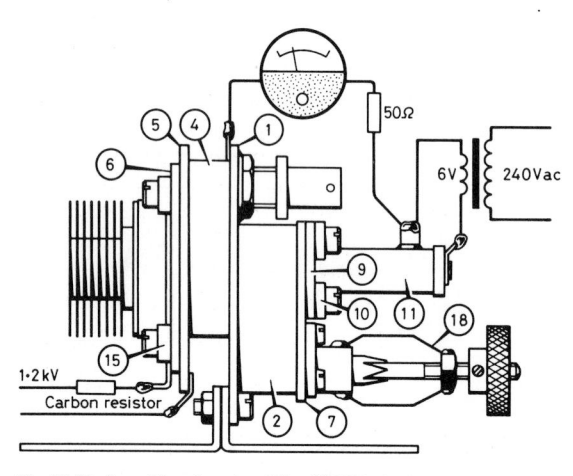

Fig 15.32. Constituent parts of the OZ9CR single valve amplifier. Circled numbers correspond to part numbers in text

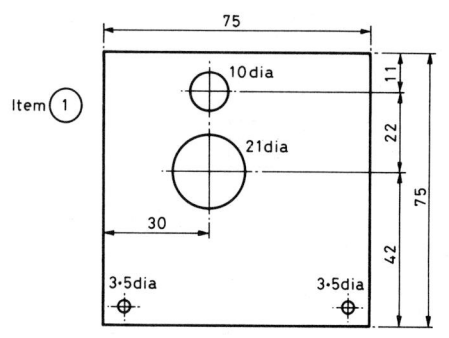

Fig 15.33. Partition plate (part 1). Material: 2.5mm brass sheet

THE 2.3GHZ ("13CM") BAND

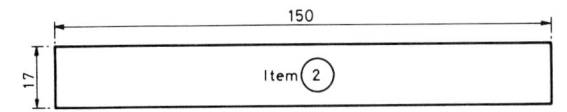

Fig 15.34. Cathode chamber wall (part 2). Material: 1.6mm brass sheet

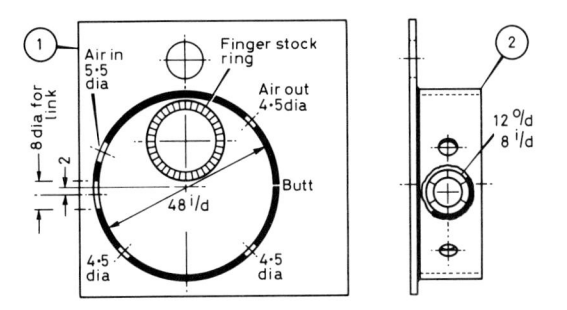

Fig 15.35. Positioning of the cathode chamber on the partition sheet

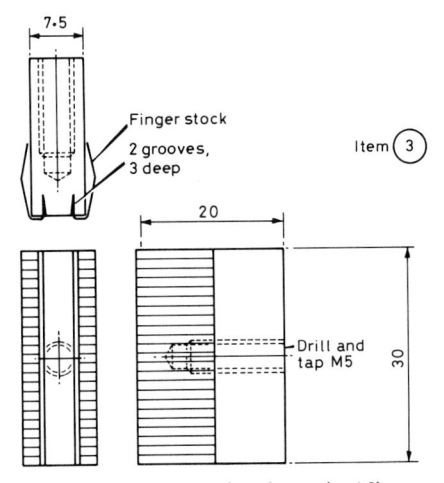

Fig 15.36. Anode tuning plunger (part 3)

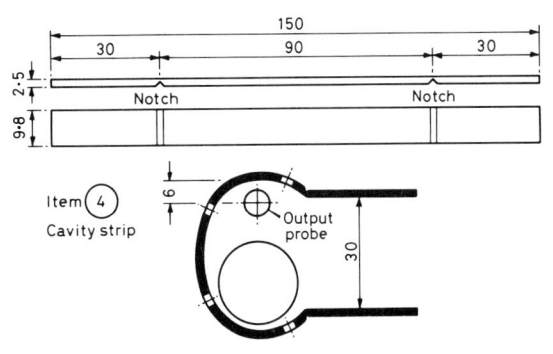

Fig 15.37. Anode cavity wall (part 4)

is made from 1.6mm thick brass strip, formed into a ring. The air holes are drilled after assembling the amplifier. The tuning piston (3), Fig 15.36, is made either from a solid brass block or layers of brass plate riveted together. Two grooves should be cut in the long side of the piston, 3mm deep, using a hacksaw. Finger stock should then be soldered into the groove. The tapped hole is used for connecting the tuning shaft. The anode cavity (4), Fig 15.37, is formed from a brass strip. The two grooves shown on the drawing are needed to assist in bending the ends of the strip to the sharp angle required. Again, the air holes should be drilled after assembly.

After making these components they may be soldered together. Tuning piston (3) is *clamped* in its position in the anode cavity guide. The anode cavity is then soldered to the partition plate (1) along with the grid finger stock. Care should be taken to align (4) with the hole in (3); the cavity wall will be close to the hole as shown in the drawing. The anode cavity top plate (5), Fig 15.38, and the anode capacitor plate (6), Fig 15.39, are cut to the specified size. The centre of the 31mm hole in each plate is marked out, along with the other four holes in the anode capacitor plate. All marked holes are now drilled 2.5mm. The centres of the 31mm holes are the aligned with M2.5 nuts and bolts and assembly clamped together squarely. All holes are then drilled through with a 2.5mm drill. M2.5 nuts and bolts are then put through the outer holes in (5) to clamp it together, then the 31mm hole is opened up to its final diameter. This process enables good hole alignment.

A ring measuring 35mm outside diameter and 8mm high, made from 2.5 mm thick brass strip, is formed around a suitable mandrel and centred on the hole in (6), then soldered on the outside. Finger stock, cut to the appropriate length is then soldered carefully to the inside of the ring with 2mm exposed above the top of the ring. The capacitor plate is then screwed to (5) and the finger stock adjusted to obtain a good fit on a 2C39A valve. The valve is left in place to help alignment during final soldering.

The cathode top plate (7), Fig 15.40, is then cut to size and the holes drilled as shown. The plate can be place on the cathode cavity (2) and aligned with the valve so the latter's cathode connector is centred in the 14.5mm hole. In this position the whole unit should be clamped firmly together, the valve removed and the whole assembly soldered.

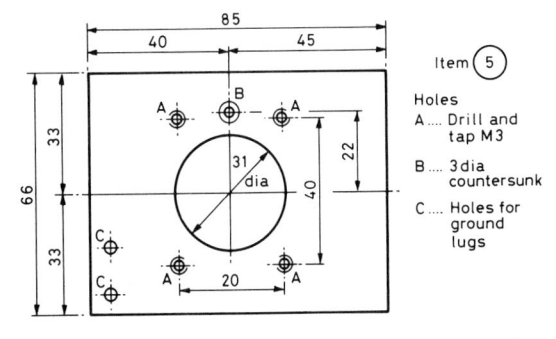

Fig 15.38. Anode cavity cover (part 5). Material: 2.5mm brass sheet

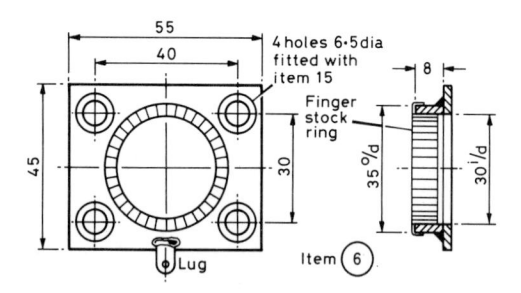

Fig 15.39. Anode capacitor plate (part 6). Material: 2.5mm brass sheet

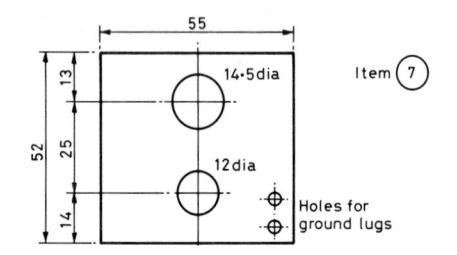

Fig 15.40. Cathode cavity cover (part 7). Material: 2.5mm brass sheet

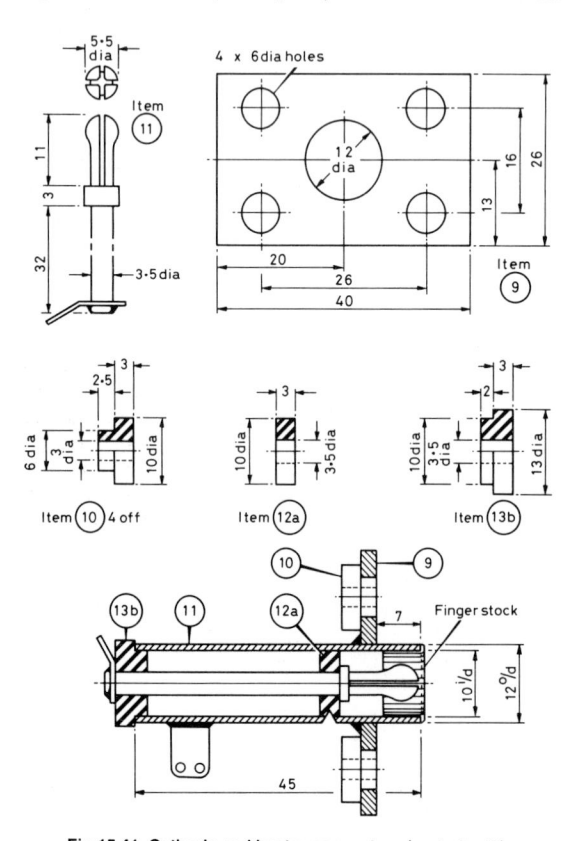

Fig 15.41. Cathode and heater connectors (parts 8 - 13)

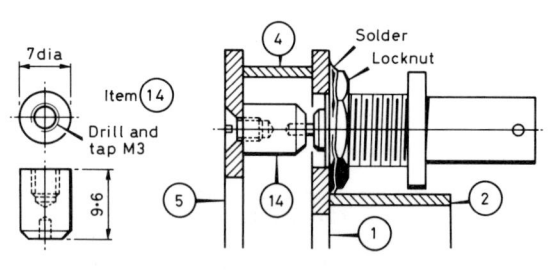

Fig 15.42. Output coupling link (part 14)

The cathode stem (8), Fig 15.41, is made to the dimensions shown. The ends are squared off and the finger stock fitted to one end. The cathode capacitor plate (9) is cut and all holes initially drilled 2.5mm. Care should be taken to ensure that the plate does not warp during either process.

With a 2C39A valve in place, tube (8) is pushed carefully onto the cathode connector and the plate slid onto the tube. This is then used as a template to mark the holes on (7). These holes are drilled 2.5mm and tapped M3. The 2.5mm holes in (9) are then opened out to 3mm. The plate is again slid onto the tube and the plate screwed into position.

With a high wattage soldering iron, pieces (8) and (9) are carefully soldered together, taking great care that the finger stock in (8) does not unsolder. A solder tag should be attached to (8) to enable wires to be attached easily. The four holes in (9) are then opened out to 6mm. Four ptfe bushes (10) are then made as shown. A piece of 0.25mm ptfe sheet is cut to be 4mm larger than the cathode plate. Care should be taken to ensure that there are no gaps left around the holes made in the ptfe for the various fixings.

The heater pin (11) is made from a long piece of 3.5mm brass rod with a suitably shaped brass connector *hard* soldered on the end. Two ptfe bushes (12) are made to maintain concentricity of the line. They are held in position by nicking (eg. with a centre punch) the outside of the tube. A solder tag is soldered to the free end of the lines, after assembly, for the heater connection wire.

The output coupling, Fig 15.42, is next to be constructed. The 3mm hole in (7) is used as a guide to mark the corresponding holes in plate (1). The hole in (1) is then expanded to 10mm. The 3mm hole in (7) is then countersunk to take an M3 countersunk screw. The non-threaded end of the coupling pin (14) is soldered to the centre of a single hole fixing BNC connector (UG1094B/U or similar). An extra mounting nut is soldered to (1) and, when cool, the pin (14) is screwed into place using an M3 × 6 countersunk screw. *It is imperative that the screw head is flush or slightly sub-surface* as the anode insulation sheet may be punctured with subsequent catastrophic effects! The normal mounting nut is used as a lock nut to hold the connector in place. The four holes in the anode capacitor plate should be enlarged to 6.5mm. A piece of ptfe sheet 5mm larger than the plate is then cut to shape with the appropriate holes carefully made (the hole for the valve is 24mm in diameter). The capacitor is then carefully assembled using the four ptfe bushes (15), Fig 15.43, ensuring

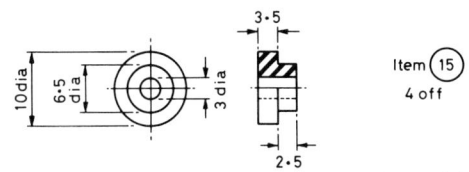

Fig 15.43. Anode capacitor insulating bushes (part 15). Material: ptfe, four required

that no grit or swarf is trapped in the sandwich. The input coupling link (16), Fig 15.44, is made from a length of 1mm wall brass tubing. In one end is filed a 3 × 3mm notch (for subsequent connection of the coupling loop). A piece of 2mm copper wire 65mm long is used as the centre conductor. One of the ends is tapered before soldering to the BNC input connector, type UG1094/U. A ptfe bush, which is 8mm long, a tight fit in the tube, possessing a 2mm hole, is fitted where the notch is filed and the centre conductor fitted. The BNC connector is then soldered to the tube. The coupling strip is made from a 36mm length of copper strip 3mm wide and 0.6mm thick. A 2mm hole is drilled near one end, by which means it is attached to the centre conductor. The free end of the loop is soldered into the 3mm notch previously filed.

The bushing for holding the link tube is made as shown in (17). One end is sawn off at 45° and a wooden dowel used to hold it in place in the cathode chamber whilst soldering. The free end of the bushing is then slit into six segments using a small hacksaw. A clamp ring is bent to shape from a brass strip 1mm thick and 7mm wide.

The cathode tuning capacitor (18), Fig 15.45, is made from a series of parts. The tuning shaft is a length of 5mm diameter brass rod threaded at one end, M5, for a length of 30mm. An 11mm diameter brass disc is then soldered to the opposite end. A 40mm length of 8mm brass tube is split into eight segments for a distance of 6mm at each end and bent to obtain a good fit on the shaft. The flange bushing is made on a lathe. It should be a good fit onto the brass tube. The 12mm diameter flange must fit the 12mm hole in the cathode cover plate (7). The screw holes are marked out, drilled 2.5mm then used as a template to mark the holes in (7). The holes are next opened to 3mm and the holes in (7) tapped M3. The flange should be shaped so as not to interfere with the cathode capacitor plate. These pieces should be assembled as shown. An M5 nut is screwed on the shaft

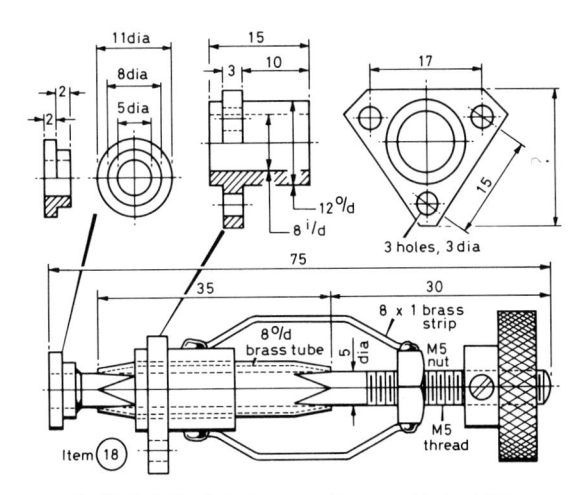

Fig 15.45. Cathode tuning capacitor assembly (part 18)

at a suitable position, to enable soldering to the two support strips which are soldered in place.

Finally the anode tuning plunger (19), Fig 15.46, is constructed from a 75mm length of 5mm brass rod threaded M5 for 30mm at one end. An angle piece is made from 2.5mm brass strip 45mm long and 22mm wide bent 16mm from one end. The hole for the spindle is drilled 7.4mm from the angle. The end play of the spindle is taken up by two nuts which are soldered to the angle piece. The assembly is clamped to (1) in the correct position then 2.5mm holes are drilled through the angle into the partition plate, the holes in the latter being tapped M3. The angle holes are then increased to 3mm. The assembly is attached to the partition plate with two short M6 screws.

A base plate, Fig 15.47, is made to hold the amplifier vertical. This takes the form of a pair of angle brackets as shown in (20). An efficient blower should be provided for the amplifier along with the necessary ducting to direct the air in the right direction. Air from the blower must be directed into both the anode and cathode chambers through the cooling fins of the valve. This latter function may be done using a 'Perspex' scoop to force air across the anodes. A possible arrangement is shown in Fig 15.48. The cavities are cooled by positioning 10mm copper tubes in the air stream and using these to direct some of the flow into the appropriate cavity.

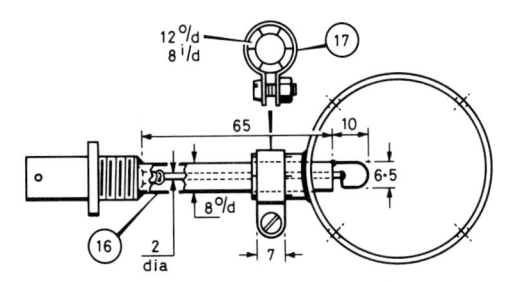

Fig 15.44. Input coupling loop (part 16) and clamp (part 17)

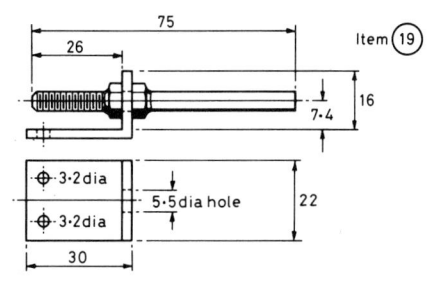

Fig 15.46. Anode tuning plunger (part 19)

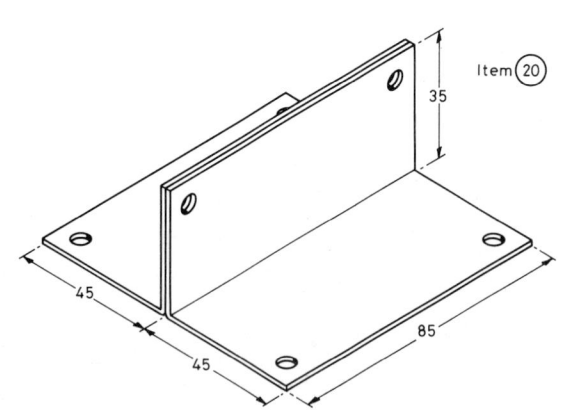

Fig 15.47. Amplifier base plate (part 20). Material: 1.6mm aluminium, two required

Alignment

Initially a 50Ω resistor is connected between the cathode connector and ground to establish the valve standing current. The heater voltage of 5.8V at around 1A is now applied and the heaters allowed to pre-heat for one minute, minimum. A dummy load is now connected to the output of the amplifier. This may consist of a long length of coaxial cable connected through a power meter. Alternatively, a loop may be soldered to the end of the cable where the rf probe may be coupled. A 50Ω *non-inductive* 50W resistor is connected in series with the eht supply, close to the amplifier. This may consist of ten 5W carbon resistors in parallel. The input link is positioned half way into the cavity. A few watts of drive should be applied and an rf probe inserted through one of the cathode chambers' air holes. The meter should show some response which will increase as the cavity is tuned to resonance. Depending on the care taken in construction resonance should occur with the capacitor plate about three turns from its fully in position. The position of the input link can then be adjusted along with the tuning capacitor to maximise the meter reading. The probe is then removed.

Now with an eht of 1,000V applied the valve should draw around 30mA of standing current. By moving the anode plunger the anode cavity should be brought to resonance as indicated by the peak in the output power indication. All adjustments are then re-optimised for maximum output. Different makes of tube may give more output if pulled slightly out of their socket and/or rotated *with the eht removed*! Cooling air should be applied at all times when eht is applied.

With 1,200V on the anode and 3W of drive the author's amplifier produced 40W output.

15.4.2 Single valve mixer for 2,320MHz

This mixer uses the same metalwork as the previously described amplifier. Extra components are incorporated to enable 144MHz to be mixed in the cathode along with the local oscillator. The 2,176MHz signal is supplied through the connector referred to as "rf input" in the amplifier design. The extra components to be added are shown in Fig 15.48. The 144MHz BNC type input connector is mounted on an L-shaped brass bracket attached to the cavity. The new components are connected between the BNC socket and the valve cathode connector.

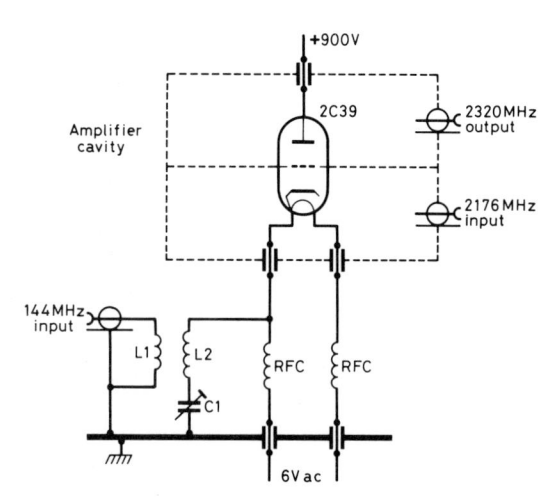

Fig 15.48. The necessary modifications to turn the single valve amplifier into a high level transmit mixer. L1: 2 turns, 1mm silver plated copper wire, 6mm i.d. L2: 6 turns, 1mm silver plated copper wire, close coupled to L1. RFC: 20cm length of 0.5mm enamelled copper wire 6mm i.d. C1: 12pF tubular ceramic preset. C2, C3 1nF feedthrough capacitors

Adjustment

A 144MHz drive source is connected to the 144MHz input and a 2,176MHz source is connected to the 2,176MHz input. The output is connected to a suitable power detector. A 5.5V heater supply is connected to the feedthrough capacitors and the valve allowed to preheat for a minimum of one minute. The 900V eht supply is connected to the anode and the standing current of the mixer is set to 30mA using VR1.

The 144MHz source capable of around 3W output is switched on, and C1 tuned for a maximum in cathode current. The 144MHz drive is then removed. the 2,176MHz local oscillator source, producing around 2W is switched on. The cathode tuning capacitor is adjusted for maximum cathode current. Both 144MHz and 2,176MHz sources are next applied simultaneously and the anode tuning capacitor adjusted for maximum output power. Care should be taken not to tune the cavity up as a 2,176MHz amplifier. By removing the 144MHz drive ensure that the output power disappears. When optimised, the mixer produces an output of around 4W.

15.4.3 Two valve amplifier for 2,320MHz

This design, also by Hans Rasmussen, OZ9CR, uses a pair of ceramic 2C39A type valves to produce an output of around 100W. A drawing of all the constituent parts is

THE 2.3GHZ ("13CM") BAND 15.19

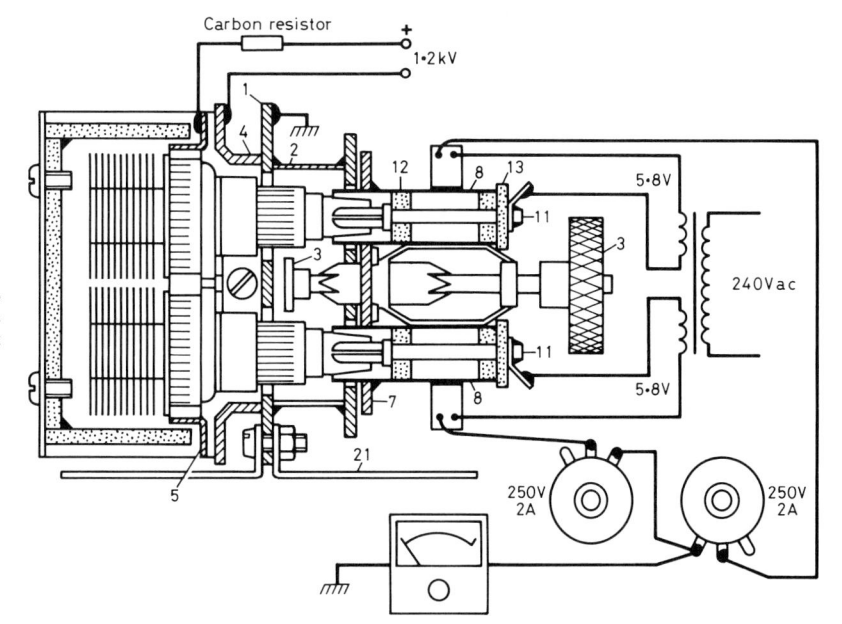

Fig 15.49. Constituent parts of the OZ9CR two valve amplifier. Numbers correspond to part numbers in text

shown in Fig 15.49. The following notes will aid the construction.

Partition plate (1), Fig 15.50, separates the anode and cathode cavities. Care should be taken during drilling the 21mm holes that the plate is not bowed as this will make final assembly difficult. The cathode cavity (2), Fig 15.51, is made from 1.5mm thick brass strip 175mm long, formed as shown. The air holes are drilled after assembling the amplifier. The tuning piston (3), Fig 15.52, is made from three layers of brass plate riveted together.

Two grooves should be cut in the long side of the piston 3mm deep using a hacksaw. Solder may ultimately have to be applied to the block to take up any free play. Finger stock should then be soldered into the groove. The tapped hole is used for connecting the running shaft.

The anode cavity (4), Fig 15.53, is formed from a brass strip 190mm long. The two grooves shown on the drawing are to assist in bending the ends of the strip to the sharp angle required. Again, the air holes should be drilled after assembly. After making these components they may be soldered together. Tuning piston (3) is *clamped* in its position in the anode cavity guide. The anode cavity is then soldered to the partition plate (1) along with the grid finger stock. Care should be taken to align (4) with the holes in (3). The anode cavity top plate (5), Fig 15.54, and the anode capacitor plate (6), Fig 15.55, are cut to the specified size. The centres of the valve holes in each plate are marked out, along with the other four holes in the anode capacitor plate. All marked holes are now drilled 2.5mm. The centres of the valve holes are then aligned with M2.5

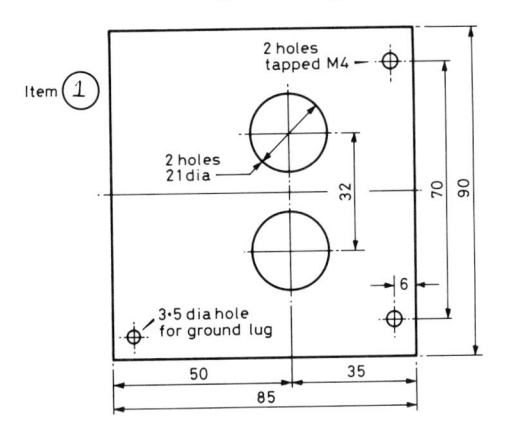

Fig 15.50. Partition plate (part 1). Material: 2.5mm brass sheet

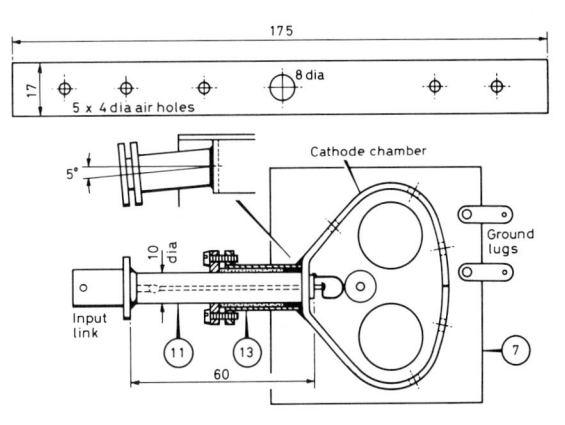

Fig 15.51. Cathode chamber wall (part 2). Material: 1.6mm brass sheet

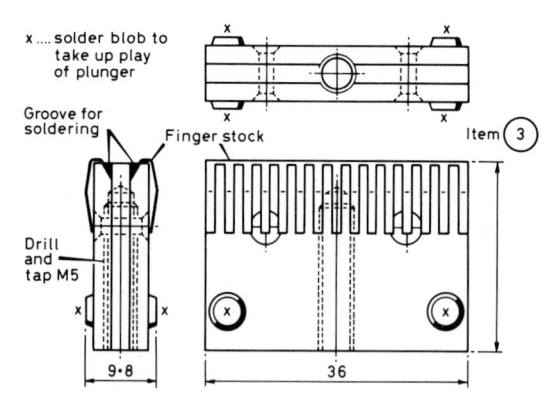

Fig 15.52. Anode tuning plunger (part 3)

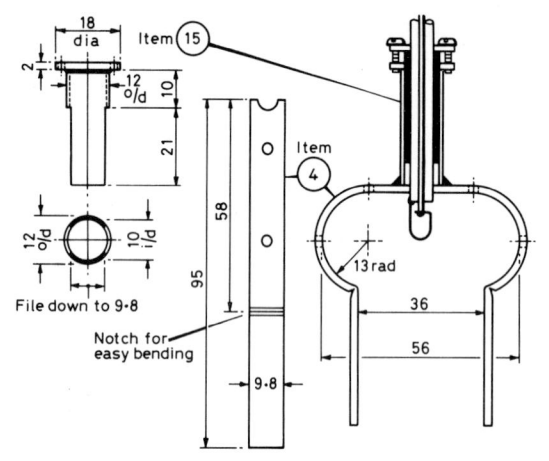

Fig 15.53. Anode cavity wall, Material: 2.5mm brass sheet (part 4) and output coupling guide (part 15)

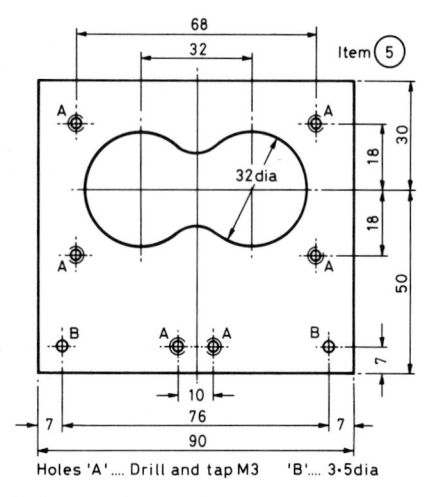

Fig 15.54. Anode cavity cover (part 5). Material 2.5mm brass sheet

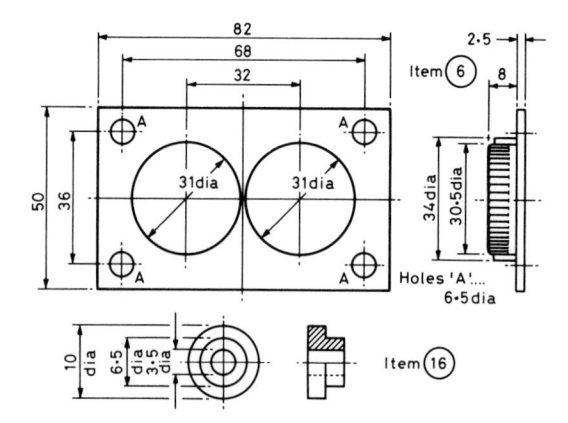

Fig 15.55. Anode capacitor plate (part 6). Material: 2.5mm brass sheet. Anode capacitor bushings (part 16). Material: ptfe

nuts and bolts and the assembly clamped together squarely. All holes are then drilled through with a 2.5mm drill. M2.5 nuts and bolts are then put through the outer holes in (5) to clamp it together then the large holes opened up their respective sizes. This process enables good hole alignment. Two rings measuring 34mm outside diameter 30.5mm inside diameter and 8mm high made from brass strip, formed around a suitable mandrel then centred on the hole in (6), then soldered on the outside. Finger stock, cut to the appropriate length is then soldered carefully to the inside of the ring with 2mm exposed above the top of the ring. The capacitor plate is then screwed to (5) and the finger stock adjusted to obtain a good fit on the 2C39A valves. The valves are left in place to help alignment during final soldering.

The cathode top plate (7), Fig 15.56, is then cut to size and the holes drilled. The plate can then be placed on the cathode cavity (2) and aligned with the valves so the latter's cathode connectors are centred in the 14.5mm holes. In this position the whole unit should be clamped firmly together, the valves removed and the assembly soldered together.

The cathode stems (8), Fig 15.57, are made to the dimensions shown. The ends are squared off and the finger stock

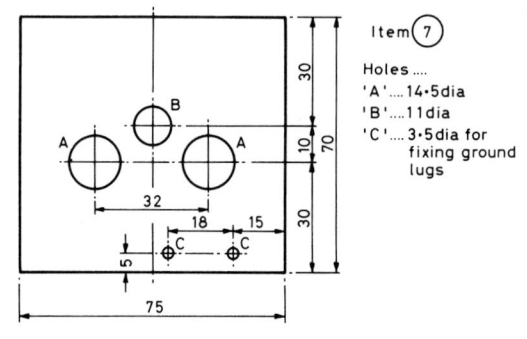

Fig 15.56. Cathode cavity cover (part 7). Material: 2.5mm brass sheet

THE 2.3GHZ ("13CM") BAND   15.21

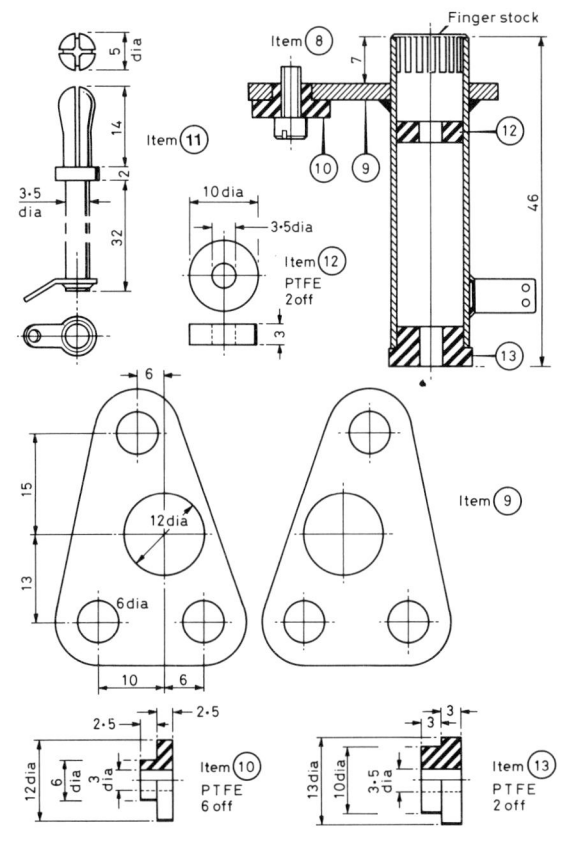

Fig 15.57. Cathode and heater connectors (parts 8-13)

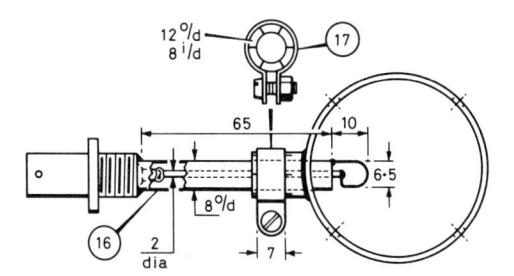

Fig 15.59. Input coupling link (part 17)

fitted in one end. The cathode capacitor plates (9), Fig 15.57, are cut and drilled as shown, ensuring that the plate does not warp during either process.

With a 2C39A valve (or "dummy", such as that shown in chapter 13, "Data") in place in the cavity tubes (8) they are pushed carefully onto the cathode connector and the plates slid onto the tube which are then used as a template to mark the holes on (7). These holes are drilled 2.5mm and tapped M3. The 2.5mm holes in (9) are then opened out to 3mm. The plate is again slid onto the tube and the plate screwed into position. With a high wattage soldering iron pieces (8) and (9) are carefully soldered together taking great care that the finger stock in (8) does not unsolder.

A solder tag should be attached to (8) to enable wires to be attached easily. The four holes in (9) are then opened out to 6mm. Four ptfe bushes (10), Fig 15.57, are then made as shown. A piece of 0.25mm ptfe sheet is then cut to be 4mm larger than the cathode plate. Care should be taken to ensure that there are no gaps left around the holes made in the ptfe for the various fixings.

The heater pin (11), Fig 15.57, is made from a long piece of 3.55mm brass rod with a suitably shaped brass connector hard soldered on the end. Two pairs of ptfe bushes (12) and (13), Fig 15.57, are made to maintain concentricity of the line. They are held in position by nicking the outside of the tube. A solder tag is attached to the free end of the line, after assembly for the heater connection wire.

The output coupling link (14), Fig 15.58, is made from two lengths of brass tube. The connector is an N-type intended for four hole chassis attachment. The guide tube for mounting (14) designated (15), Fig 15.60, is constructed on a lathe from brass rod as shown. The flats are filed after turning whilst the work is held in the chuck. The four holes in anode capacitor plate should be enlarged to 6.5mm. A piece of ptfe sheet 5mm larger that the plate is then cut to shape with the appropriate holes carefully made (the hole for the valve is 24mm in diameter). The capacitor is then carefully assembled using the four ptfe bushes (16), Fig 15.55, ensuring that no grit or swarf is trapped in the sandwich.

Input link (17), Fig 15.59, is made from a length of 1mm wall brass tubing. In one end is filed a 3 × 3mm notch (for connecting the coupling loop later). A piece of 2mm wire copper wire 65mm long is used as the centre conductor. One of the ends is tapered before soldering to the BNC input connector. A ptfe bush which is 8mm long, a tight fit in the tube and possessing a 2mm hole, is fitted where the notch is filed and the centre conductor fitted. The BNC connector is then soldered to the tube. The coupling strip is

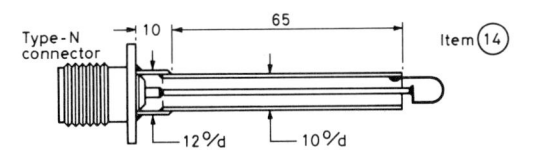

Fig 15.58. Output coupling link (part 14)

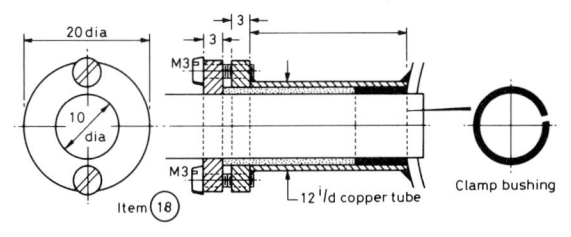

Fig 15.60. Guide tube for the input coupling link (part 18)

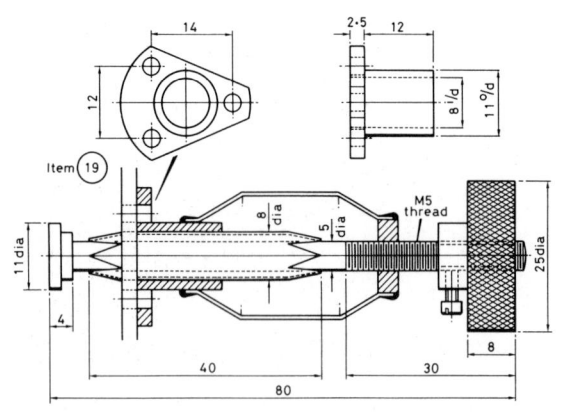

Fig 15.61. Cathode tuning capacitor assembly (part 19)

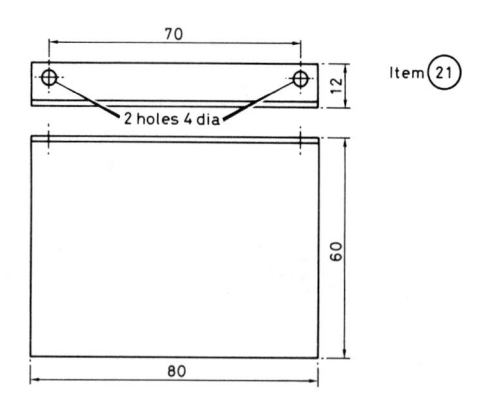

Fig 15.63. Amplifier base plate (part 21). Material: 1.6mm aluminium, two required

made from a 36mm length of copper strip 3mm wide and 0.6mm thick. A 2mm hole is drilled near one end by which means it is attached to the centre conductor. The free end of the loop is soldered into the 3mm notch previously filed. The bushing for holding the link tube is made as shown in (18), Fig 15.60. A clamp bushing is made to fit between the tubes as shown.

The cathode tuning capacitor (19), Fig 15.61, is made from a series of parts. The tuning shaft is an 80mm length of 5mm diameter threaded at one end M5 for a length of 30mm. An 11mm diameter brass disc is then soldered to the opposite end.

A 40mm length of 8mm brass tube is split into eight segments for a distance of 6mm at each end then bent to obtain a good fit on the shaft. The flange bushing is made on a lathe. It should be a good fit onto the brass tube. The 12mm diameter flange is to fit the 12mm hole in the cathode cover plate (7). The screw holes are marked out, drilled 2.5mm then used as a template to mark the holes in (7). The holes are then opened to 3mm and the holes in (7) tapped M3. The flange should be shaped so as not to interfere with the cathode capacitor plate. These pieces should then be assembled as shown. An M5 nut is then screwed on the shaft at a suitable position to enable soldering to the two support strips which are soldered in place. Finally the anode tuning piston (20), Fig 15.62, is constructed from a 75mm length of brass rod threaded M5

for 30mm at one end. An angle piece is made from 2.5mm brass strip 45mm long and 22mm wide bent 16mm from one end. The hole for the spindle is drilled 7.4mm from the angle. The end play of the spindle is taken up by two nuts which are soldered to the angle piece. The assembly is clamped to (1) in the correct position then 2.5mm holes are drilled through the angle into the partition plate, the holes in the latter being tapped M3. The angle holes are then increased to 3mm. The assembly is attached to the partition plate by means of two short M6 screws.

A base plate (21), Fig 15.63, is made to hold the amplifier vertical. This takes the form of a pair of angle brackets as shown in (17).

A good blower should be provided for the amplifier along the necessary ducting to direct the air in the right direction. Air from the blower must be directed into both the anode and cathode chambers and through the cooling fins of the valve. This latter function may be done using a perspex scoop to force air across the anodes. The cavities are cooled by positioning 10mm copper tubes in the air stream and using these to direct some of the flow into the appropriate cavity.

Photographs of the completed amplifier are shown in Figs 15.64 and 15.65.

Alignment

Initially two 250Ω potentiometers are connected, one between each cathode stem and ground, to establish the anode standing current. The heater voltage of 5.8V at around 1A for each valve is now applied and the heaters allowed to preheat for one minute, minimum. The input link is positioned to half way into the cavity.

A few watts of drive should be applied and an rf probe inserted through one of the cathode chambers' air holes. The meter should show some response which will increase as the cavity is tuned to resonance. Depending on the care taken in construction, resonance should occur with the capacitor plate three turns from its fully in position. The position of the input link can then be adjusted along with

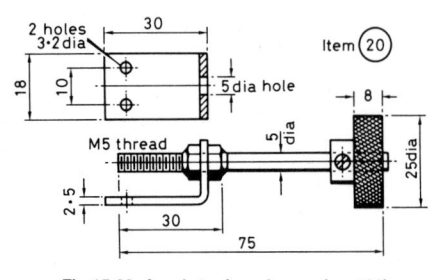

Fig 15.62. Anode tuning plunger (part 20)

Fig 15.65. Completed amplifier viewed from the input connector

Fig 15.64. Photographs of (a) Completed amplifier viewed from above, (b) Completed amplifier viewed from below

the tuning capacitor to maximise the meter reading. The probe is then removed.

A dummy load is now connected to the amplifier. This may consist of a long length of coaxial cable connected through a power meter. Alternatively a loop should be soldered to the end of the cable where the rf probe may be coupled. A 50Ω *non-inductive* 50W resistor is connected in series with the eht supply, close to the amplifier. This may consist of ten 5W carbon resistors in parallel.

Now, with initially a low eht applied (around 1,000V), the combined valve standing current should be around 60mA, shared equally by the two valves. By moving the anode plunger the anode cavity should be brought to resonance as indicated by the peak in the output indication. All adjustments are then optimised for maximum output. Different makes of tube may give more output if they are pulled slightly out of their sockets and/or rotated, *with the eht removed!* Cooling air should be applied at all times that eht is applied.

After this initial pre-tuning of the amplifier the final operating conditions can be set. With 1,600V on the anode and still with a 60mA combined standing current, the amplifier should be retuned for maximum output. In the prototype with 18W of drive, the amplifier produced around 110W of output.

15.5 ANTENNAS

The most commonly used antenna at 2.3GHz is a suitably fed parabolic dish. This popularity is mainly due to the high gain achievable with a relatively small sized antenna. A graph of dish gain against dish diameter is shown in Fig 15.66. The parabolic surface of the dish may be solid, but in some circumstances this may lead to unacceptable wind loading on the mast. As an alternative, the reflecting surface may be made from wire mesh as long as the holes are not greater than $\lambda/10$, i.e. 13mm) in the plane of polarisation. Two suitable types of dish feed are to be described

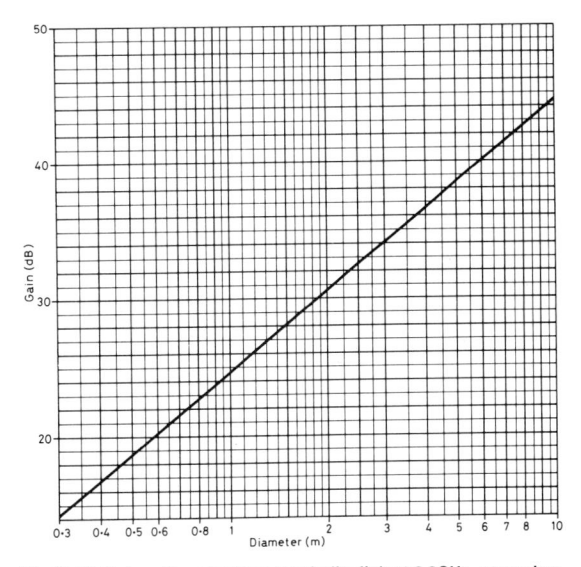

Fig 15.66. Gain v diameter for a parabolic dish at 2.3GHz, assuming 50% feed efficiency

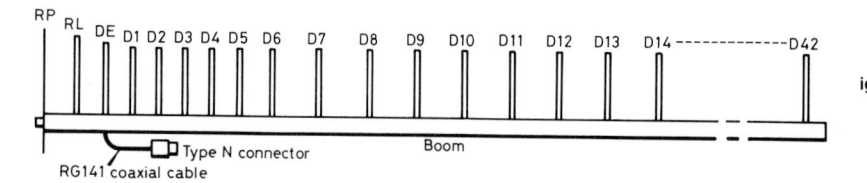

Fig 15.67. A 2.3GHz quad-loop Yagi. For dimensions see Table 15.1

here, the dipole/splashplate feed and the feed horn. This latter feed can be used on its own as a low gain, broad beamwidth radiator which can be useful for local testing. If an even wider beamwidth is wanted the omnidirectional Alford slot antenna may be used.

Where circumstances do not permit the use of a dish, a loop quad Yagi can still offer a useful gain when constructed carefully.

15.5.1 Loop quad antenna

This antenna was designed by M Walters, G3JVL. The construction of this antenna is shown in Fig 15.67, with the critical dimensions being shown in Table 15.2. The design is a scaled version of the 1.3GHz design to the 13cm narrowband segment at 2,320MHz. The number of elements is increased to 44 which means its boom length is approximately 2 metres, the same as the 1,296MHz design. It has a gain of around 22dBi.

The construction of the antenna is quite straightforward, but the dimensions should be closely adhered to. In drilling the boom for instance, measurements of the positions of the elements should be made from a single point adding the appropriate lengths. If the individual gaps are marked out, then errors may accumulate to an excessive degree. For this reason the distances shown in Table 15.2 are cumulative from the end of the boom. All elements are made from 1.6mm diameter brass welding rod cut to the lengths shown in Fig 15.68a. The driven element, as shown in Fig 15.68b, is attached to the head of an M4 × 25mm counter sunk screw which is drilled 3.2mm to accept the RG141 semi-rigid cable. All elements, screws and soldered joints should be protected with polyurethane varnish after assembly, followed by a coat of paint on all surfaces. If inadequate attention is paid to this protection, then the performance of the antenna will degrade as a result of corrosion. Provided the antenna is carefully constructed its feed impedance will be close to 50Ω. If a suitably rated power meter or impedance bridge is available the match may be optimised by carefully bending the reflector loop toward or away from the driven element. The antenna can be mounted on a vertical support, as horizontal metal work in the vicinity of

Table 15.2. Dimensions of 2.3GHz quad loop yagi

Boom diameter	12.7mm od
Boom length	2.01m
Boom material	Aluminium alloy
Driven element	1.6mm diameter welding rod
All other elements	1.6mm diameter welding rod

Element sizes:

Reflector plate	78.0mm x 63.9mm
Length L (mm)	See Fig 15.64(a)
Reflector loop	142.9
Driven element	133.6
Directors 1-12	120.8
Directors 13-20	117.0
Directors 21-30	112.7
Directors 31-40	109.7
Directors 41-50	108.3

Cumulative element spacings (mm):

RP	0.0	RL	44.0
DE	57.5	D1	73.4
D2	85.1	D3	110.5
D4	135.6	D5	153.4
D6	186.2	D7	236.7
D8	287.2	D9	337.7
D10	388.2	D11	438.7
D12	489.2	D13	539.7
D14	590.3	D15	640.8
D16	691.3	D17	741.8
D18	792.3	D19	842.8
D20	893.3	D21	943.9
D22	994.4	D23	1044.9
D24	1095.4	D25	1145.9
D26	1196.4	D27	1246.9
D28	1297.4	D29	1348.0
D30	1398.5	D31	1449.0
D32	1499.5	D33	1550.0
D34	1600.5	D35	1651.0
D36	1701.5	D37	1752.1
D38	1802.6	D39	1853.1
D40	1903.6	D41	1954.1
D42	2004.6		

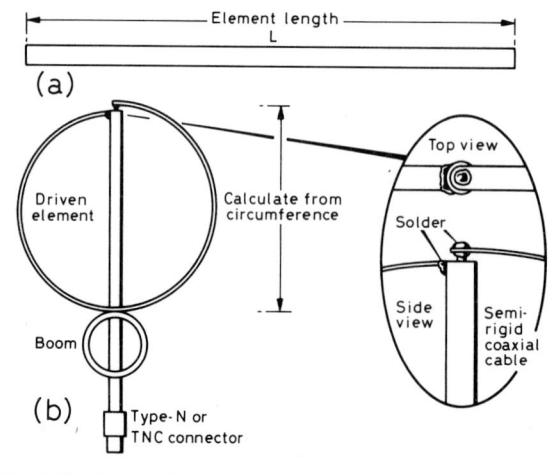

Fig 15.68. The 2.3GHz quad-loop Yagi: (a) Detail of the element construction. (b) Assembly of the driven element

the antenna can cause severe degradation in its performance. It is usually best to mount the antenna with the loops pointing downwards to reduce the likelihood of the loops being damaged by perching birds.

Two or more of these antennas may be stacked and bayed to increase the gain. A vertical stacking distance of 0.43 metres seems optimum. A suitable combining unit is described later in this chapter.

15.5.2 Alford slot antenna

Developed by M Walters, G3JVL, this antenna fulfils the need for an omnidirectional horizontally polarised antenna. This makes it particularly useful for beacons, fixed station monitoring purposes and mobile operation. Mechanical details of this antenna are shown in Fig 15.69. The prototype was made from 22mm outside diameter copper water pipe. Material is removed from one part of the tubing to produce a slotted tube with an outside diameter of 18.5mm and a slot width of 2.6mm. To ensure circularity the tube is best formed around a suitable diameter mandrel. Small tabs are soldered at the top and bottom of the tube to define the slot length of 229mm. A plate is soldered across the bottom of the tube to strengthen the structure.

The rf is fed via a length of 0.141" (3.6mm) RG141 semi-rigid coaxial cable up the centre of the tube to the centre of the slot via a 4:1 balun constructed at the end of the cable. The detailed construction of the balun is shown in Fig 15.69b. The two diametrically opposite slots are cut carefully using a small hacksaw with a new blade. The inner and outer of the cable are shorted using the shortest possible connection and the balun is attached to the slot using two thin copper foil tabs.

If suitable test gear is available, the match of the antenna can be optimised by carefully adjusting the width of the slot by squeezing the tube in a vice, or by prising the slot apart with a small screwdriver.

Typical antenna characteristics are shown in Fig 15.70. The gain of the antenna has been measured as 6.4dBi.

15.5.3 Feed horn for 2.3GHz

The feed horn is ideally suited to feed a dish having an f/D of between 0.2 and 0.4. It consists of a short length of circular waveguide with a coaxial to waveguide transition made from an N-type socket and a short probe. Constructional details are shown in Fig 15.71. The circular waveguide section is made from a cylinder closed with a plate at one end (e.g. a large coffee tin) cut to the specified length. All dimensions are fairly critical (especially the length of the probe) and hence care should be taken during construction if a good match is to be obtained without adjustment. If a different can is used, the length and position of the probe, and the length of the tin may need altering for minimum vswr. The N-type socket should be fixed to the tin by using four fixings, using washers as spacers to accommodate the curved surface of the tin. A completed feed, made from a coffee tin, is shown in Fig 15.72. See also chapter 4, "Antennas", for further information on the design of circular feed horns.

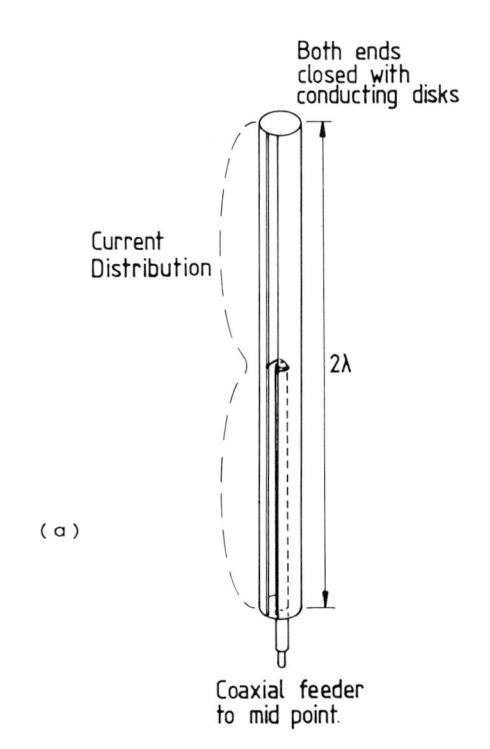

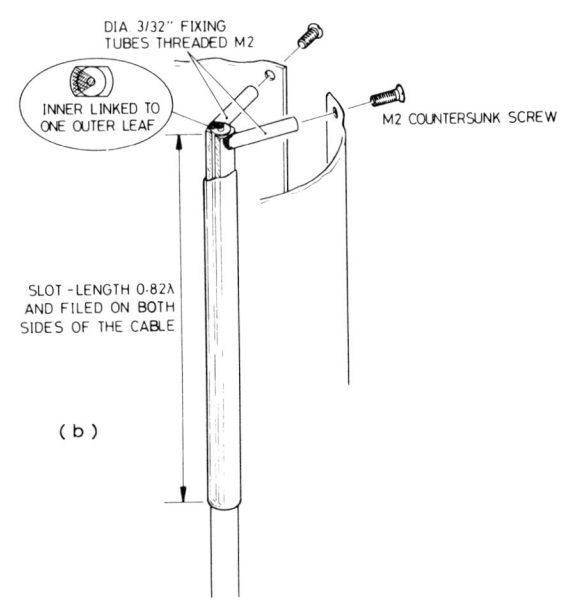

Fig 15.69. Construction of the 2.3GHz Alford slot antenna using a dual slotted cylinder. The impedance at the feedpoint is 200Ω. (a) Dimensions for 2,320MHz are: slot length 280mm, slot width 3mm, tube diameter 19mm by 18swg (0.75in). (b) Construction of a suitable balun. The balun slots are 1mm wide and 26mm long

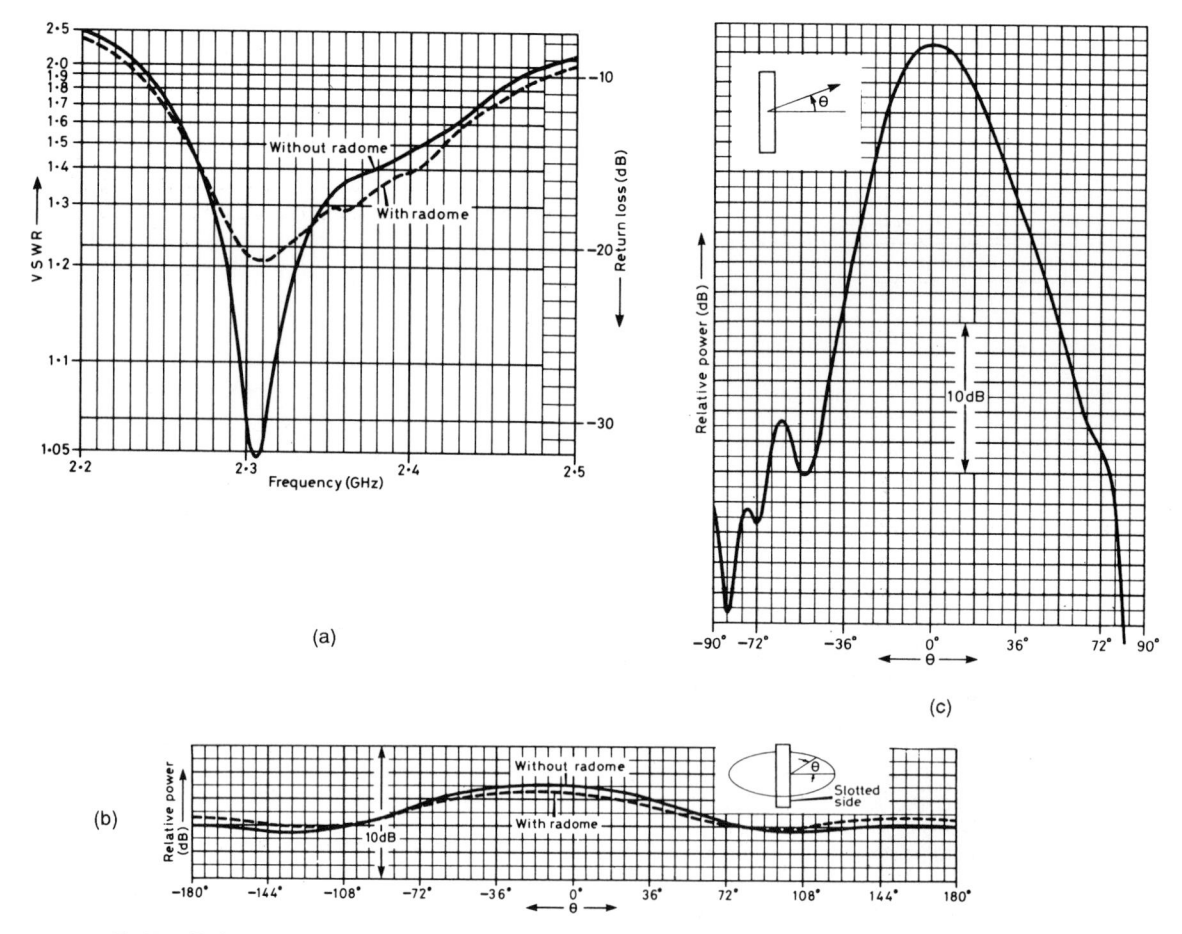

Fig 15.70. Performance of a 2.3GHz Alford slot antenna: (a) Impedance (b) Horizontal polar diagram (c) Vertical polar diagram

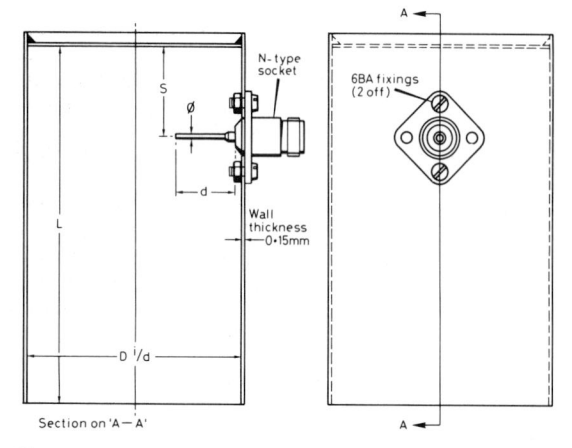

Fig 15.71. Construction of a 2.3GHz feed horn. Dimensions are: D=80.5mm, L=310mm, s=76mm, d=27mm and ϕ=3.5mm

Fig 15.72. Photograph of G4DKX's feed horn constructed from a coffee tin

THE 2.3GHZ ("13CM") BAND 15.27

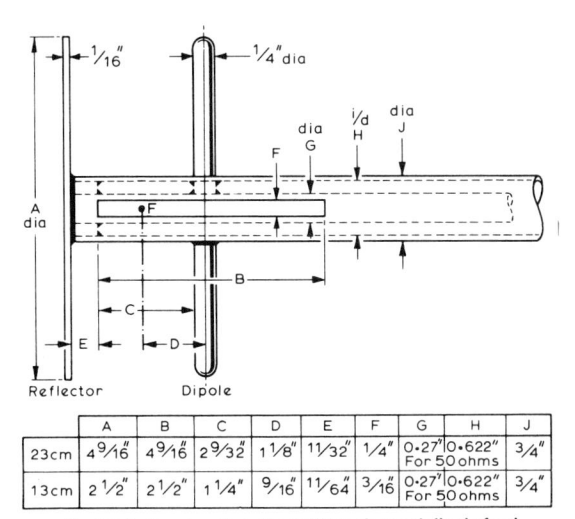

Fig 15.73. Construction of a 2.3GHz disc and dipole feed

	A	B	C	D	E	F	G	H	J
23cm	4 9/16"	4 9/16"	2 9/32"	1 1/8"	11/32"	1/4"	0·27"	0·622" For 50 ohms	3/4"
13cm	2 1/2"	2 1/2"	1 1/4"	9/16"	11/64"	3/16"	0·27"	0·622" For 50 ohms	3/4"

Fig 15.74. Photograph of the 2.3GHz feed in the author's 1.2m mesh dish

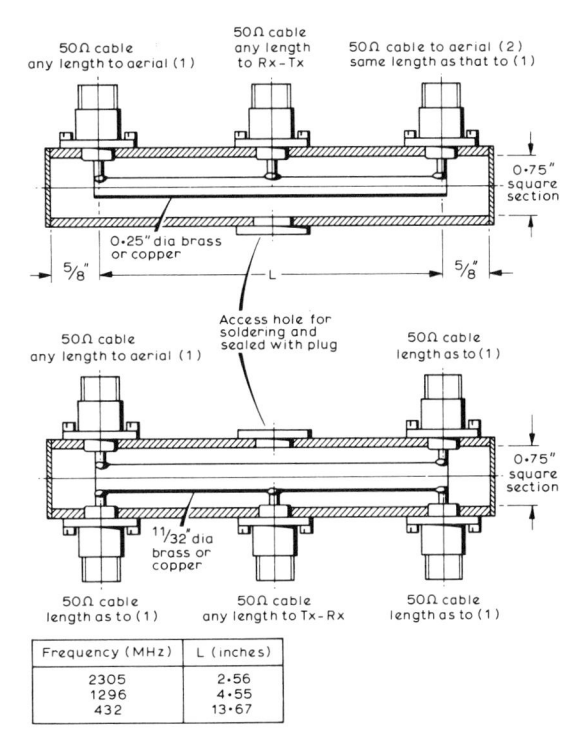

Fig 15.75. Two-way and four-way antenna combiners for 2.3GHz

Frequency (MHz)	L (inches)
2305	2·56
1296	4·55
432	13·67

15.5.4 Dipole/splashplate feed

The version of this feed to be described here is suitable for dishes having an f/d of between 0.25 and 0.45. The feed is built around a length of fabricated coaxial line. The ratio of inner conductor diameter to internal outer diameter is chosen to produce the desired impedance, in this case 50Ω. Details of the feed are shown in Fig 15.73. It is constructed from a piece of 15mm outer diameter copper water pipe, the length being chosen to allow point "x" to be positioned at the focus of the dish. The 43mm diameter disc is turned or filed from a 1.6mm thick sheet of brass or copper. The two slots may either milled or hacksawed and filed carefully. The ptfe spacers marked SP are 13.5mm o.d. by 5mm thick spacers with a 6.14mm hole drilled in the centre; they are used to ensure line concentricity.

The feed can be held in position in the dish by using a modified 15mm cold water tank compression fitting. The internal step in the fitting is filed away to leave a constant 15.2mm bore. The fitting mounted in the dish in a 35mm hole using the supplied nut. The position of point "x" can be altered by sliding the feed in the fitting. When the position of the feed has been optimised by monitoring a remote signal, the feed can be locked in place by tightening the nut over the olive. This type of fitting must be fitted before attaching the rf connector.

The N-type female line connector is fixed to the line by counterboring its collar 15mm diameter by 4mm deep and soldering the collar to the line. The centre pin is soldered to the line inner and the connector assembled with its supplied ptfe insulators. The completed feed in the author's 1.2m mesh dish is shown in Fig 15.74.

15.5.5 Multiway combiners for 2.3GHz antennas

The power splitters shown in Fig 15.75 enable either two or four 50Ω antennas to be fed from a single 50Ω cable. The units consist of a length of fabricated coaxial line which performs the appropriate impedance transformations.

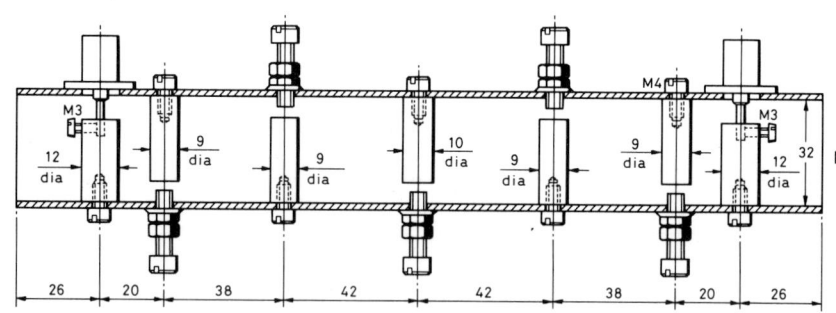

Fig 15.76. Dimensions of a 2.3GHz interdigital filter

In this case the inner is made λ/2 long between the centres of the outer connectors, and the outer is made approximately 32mm longer. In the original design, the outer was made from square section aluminium tubing, the ends of which (and the access hole for soldering the centre conductor) were sealed with aluminium plates bonded with adhesive. Alternatively, copper or brass tubing may be used and the plates soldered. Any other size of inner or outer within reason may be used provided that the ratio of the inside dimension of the outer to the diameter of the inner conductor is suitably chosen.

The cables connecting each antenna to the splitter must be 50Ω and can be of any length provided they are the same in all cases. Preferably, all the cable is taken from the same batch. Note that the two-way combiner may be used to combine a broad beamwidth antenna of moderate gain with a high gain narrow beamwidth antenna providing both antennas are 50Ω impedance.

15.6 ACCESSORIES

15.6.1 Interdigital bandpass filter for 2.3GHz

The narrowband filter to be described here was originally published in [12]. It has an insertion loss of around 1dB and a 3dB bandwidth of 48MHz. The rejection of signals 144MHz away from the centre frequency is greater than 90dB. The filter is constructed from brass material, although copper is preferable.

The sidewalls are each made from pieces 300mm × 24mm × 4mm bar, drilled as shown in Fig 15.76. The baseplate and coverplate are identical and are made from 1mm thick sheet 300mm × 32mm. The covers are attached to the sidewalls using a total of 28 M3 × 6 brass screws to ensure a good rf seal. The coupling elements are made as in Fig 15.77. The coupling elements are 15.3mm lengths of the appropriate diameter rod as shown in Fig 15.78. They each have a 8mm deep M4 tapped hole in one end. They are attached to the sidewalls with M4 × 8 brass screws, and are tuned with M4 × 12 screws each fitted with an M4 locknut. The input and output connectors are N-type with 4 fixing holes. The size of the connector mounting holes in Fig 15.76, may need altering to accommodate the various sized centre bosses encountered in different connectors.

The filter can be tuned by connecting it between a transmitter and receiver including a suitable attenuator to avoid destroying the receiver, and tuning the filter for maximum signal. For final tuning the filter should be connected in its final configuration and tuned for maximum signal.

15.6.2 Slug tuner for 2.3GHz

Frequently 2C39A type amplifiers for 2.3GHz produce rather less output than might be expected. Apart from being detrimental to overall performance, the low efficiency of such a pa stage can lead to problems with drift of the amplifier and also considerably shorten the valve life.

Efficiency lower than normally expected can arise from a number of causes, such as poor valves, a badly designed or constructed cavity, and incorrect operating/tuning conditions.

One common cause of low efficiency is incorrect loading. Most amplifiers do have some form of adjustable output coupling, but this may only by sufficient to cope with the original operating conditions (i.e. anode voltage, anode current, drive level, valve type, etc.) and not be capable of

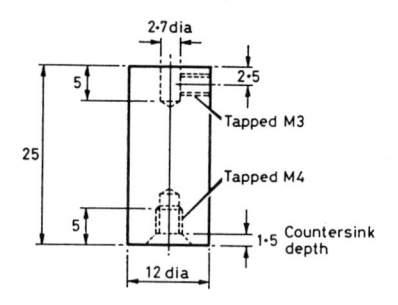

Fig 15.77. Construction of the interdigital filter input and output coupling rods

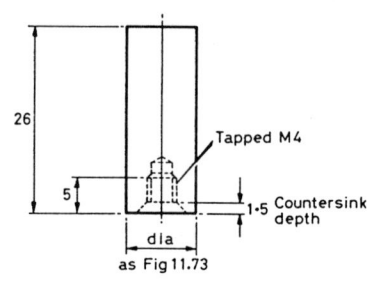

Fig 15.78. Construction of the remaining interdigital filter coupling rods

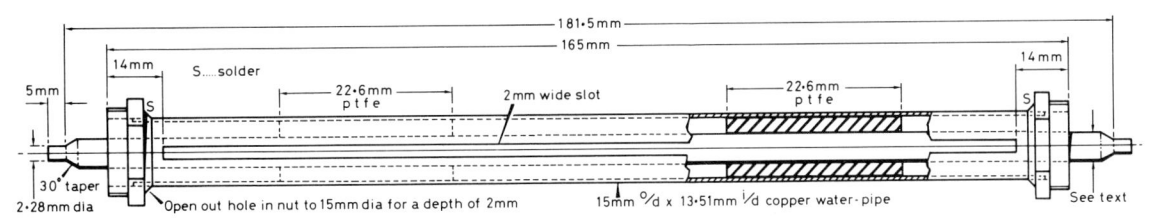

Fig 15.79. Construction of a 2.3GHz double-slug tuner

loading a dissimilar amplifier. Some amplifiers have no variable output loading which makes optimisation even more difficult. One solution to the problem of incorrect loading is to use an external impedance matching device (tuner) between the amplifier and the antenna. This can be adjusted empirically during tuning up, in conjunction with the existing load control if necessary, to produce optimum loading and thus maximum rf output.

Various types of tuner can be used for this but the preferred type, on grounds of ease of construction and lowest loss, is the double slug tuner. In this type, two dielectric slugs (usually made from ptfe), each $\lambda/4$ (taking into account the velocity factor), are moved independently inside a coaxial line to perform the impedance matching. The range of vswr which can be matched is up to 4:1, using ptfe slugs. A design for 2.3GHz is shown in Fig 15.79 and is an adaptation of the original W2IMU 1.3GHz version. The ratio of the inside dimension of the outer to the diameter of the inner conductor remains unchanged, but the overall length and the ptfe slug lengths may be reduced to the dimensions given with the diagram. For 50Ω, using 15mm o.d. water pipe as the outer conductor, 6mm diameter inner rod is near optimum. Using an 0.25" inner will result in a 45Ω line, which will limit the performance of the line.

15.7 ACKNOWLEDGEMENTS

In addition to all those giving permission to publish their equipment designs in this chapter, acknowledgement is also due to R Bates, G8TIR, and S Page for their practical assistance in realising some of the mechanical designs.

15.8 REFERENCES

[1] Balanced mixer for 13cm, *RSGB VHF/UHF Manual*, Third edition, p4.41.
[2] "A stripline converter for the 13cm band", K Hupfer, DJ1EE, *VHF Communications* 4/74 pp238–245.
[3] "Interdigital converters for 1,296 and 2,304MHz", R E Fisher, W2CQH, *QST*, Jan 74 pp11–15.
[4] "Interdigital converters for the GHz amateur bands", J Dahms, DC0DA, *VHF Communications* 3/78 pp154–168.
[5] "Broadband amplifier for 13 and 9cm", C Neie, DL7QY, *Dubus Informationen* 4/83 pp269–270.
[6] "Two stage low noise preamplifiers for the amateur bands from 24cm to 12cm", J Grimm, DJ6PI, *VHF Communications* 1/80 pp2–13.
[7] OE9PMJ correspondence subsequently published in *Der SHF Amateur*, produced by J Dahms, DC0DA.
[8] "Rauscharme Vorstufe für das 13cm Band in Kammerbauweise und in Platinentechnik", J Grimm, DJ6PI, *UKW Unterlage* Teil III, pp455–458.
[9] "A 1,152/2,340MHz varactor doubler", S Freeman, G3LQR, *Radio Communication* June 1975 p457.
[10] "A frequency doubler for the 13cm band with 6W output power", O Frosinn, DF7QF, *VHF Communications* 3/79 pp141–143.
[11] Personal correspondence subsequently published as: "Een Varactor mixer voor 13cm", H Schildt, PA0HEJ, *CQ–PA* 1982 pp423–429.
[12] "Narrowband filters for the 23cm, 13cm and 9cm band", D Vollhardt, DL3NQ, *VHF Communications* 1/78 pp 2–11.

CHAPTER 16

The 3.4GHz ("9cm") band

16.1 INTRODUCTION

The 3.4GHz band may be considered as a transitional band between the lumped circuit techniques of the lower microwave bands and the dominant waveguide techniques of the higher microwave bands. The relatively short wavelength has a big influence on the antenna designs used on the band in that relatively small dish antennas will have a very usable gain. The band also represents the highest frequency at which the Yagi may be considered a reasonable antenna prospect.

Experimentation on the 3.4GHz band in the UK began in earnest in the early 1970's. This was mainly using "polarplexer" type transmitters and receivers utilising wideband frequency modulated signals. This type of equipment remained in use for a considerable period of time, but recent interest in the band has developed the design and use of relatively advanced narrowband equipment with which many more difficult paths could be successfully tried.

16.1.1 Allocation to the amateur service

A majority of countries, worldwide, have an allocation in the 3,300 to 3,500MHz range. In the UK, the amateur service is allocated the range 3,400 to 3,475MHz on a secondary basis to the fixed, fixed-satellite and radiolocation services. Most operation at the present time centres on 3,456MHz due to the availability of sources of 1,152MHz which can be tripled to generate a signal at this frequency. The amateur satellite service has no allocation in this frequency range. Fig 16.1 shows the frequency allocation diagrammatically.

16.1.2 Beacons

A map showing the approximate locations of the 3.4GHz beacons located in Western Europe is given in Fig 16.2. For exact locations and frequencies, see chapter 9, "Beacons and repeaters", of this handbook. Unfortunately most of the European beacons are too low powered and too far away to be of real use as signal sources and propagation indicators in the UK.

16.1.3 Propagation and equipment performance

Most of the propagation modes encountered on the vhf/uhf bands are also present at 3.4GHz, with the exception of those which require ionised media such as aurora, meteor

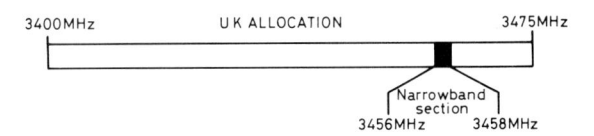

Fig 16.1. The 3.4GHz allocation in the UK

scatter and Sporadic E. The most commonly encountered propagation modes at 3.4GHz are line of sight, diffraction, tropospheric scatter, ducting and aircraft scatter. Moonbounce is also a possibility at this frequency. Further information on some of these modes can be found in chapters 2 and 3.

Using the techniques of chapters 2 and 3 it is possible to predict whether communication is possible over a given path, when the performance parameters of the equipment

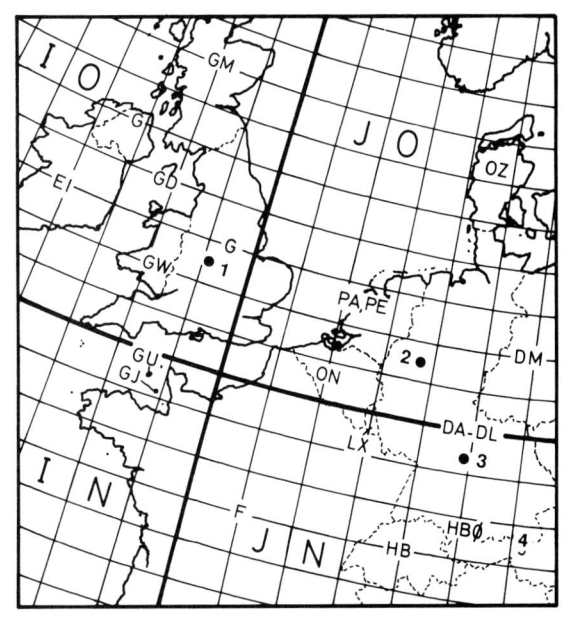

Fig 16.2. Western European 3.4GHz beacons. Key: 1, GB3OHM; 2, DC0DA; 3, DL7QY/A; 4, DB0AS. For frequencies and locators, see chapter 9, "Microwave beacons and repeaters"

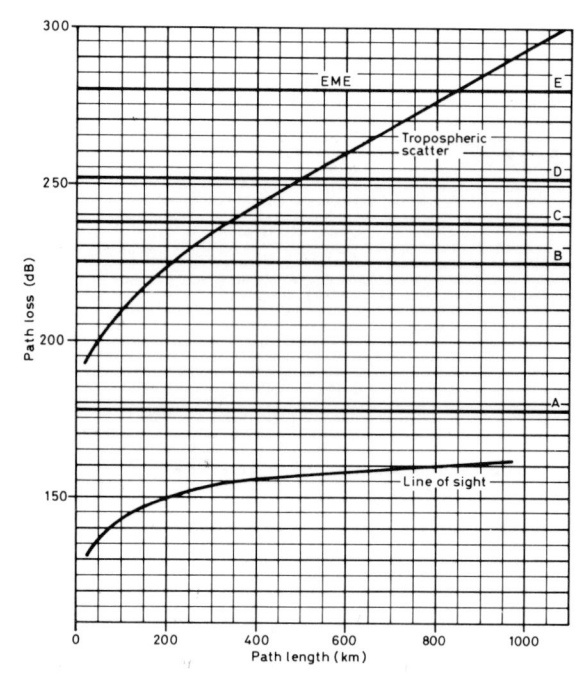

Fig 16.3. Graph of path loss versus distance for line of sight, tropospheric scatter and eme propagation, related to the performance of five sets of equipment. A: 20dBi antenna, 1W transmitter, 15dB nf, 25kHz bandwidth. B: 1.2m dish, 10W transmitter, 15dB nf, 2.5kHz bandwidth. C: 1.2m dish, 10W transmitter, 4dB nf, 2.5kHz bandwidth. D: 1.2m dish, 10W transmitter, 4dB nf, 100Hz bandwidth. E: 4.0m dish, 15W transmitter, 1dB nf, 100Hz bandwidth

are known. Fig 16.3 shows the path loss capability of five different sets of equipment ranging from a basic system of varactor tripler, interdigital converter and single Yagi antenna to a system with moonbounce capability. Also shown on the graph are the path losses associated with three different modes of propagation; line of sight (the least lossy mode of propagation), tropospheric scatter (a reliable, commercially used mode of propagation) and moonbounce.

It can be seen that relatively simple equipment such as a varactor multiplier transmitter generating 1W of nbfm, an interdigital converter and a single loop quad Yagi (A) is capable of working very long line of sight paths easily. However, very few paths are purely line of sight and propagation losses are usually noticeably higher. The tropospheric scatter loss is a reasonable upper limit for a terrestrial path and can be used to estimate the maximum range of equipment under normal propagation conditions. By adding a single stage valve amplifier producing 10W, changing the antenna to a 1.2m dish and changing to ssb or cw, (B), the equipment is capable of covering a troposcatter path of up to 210km. By adding a 3.5dB noise figure bipolar receive preamplifier, (C), the coverage distance increases to 330km. Line (D) shows the effect of the equipment of (C) but using a narrow bandwidth i.f filter, as one could when receiving cw, the coverage distance increasing to 510km. At distances over about 840km it can be seen that the terrestrial path loss is greater than the route via the moon. Line (E) shows the parameters of equipment capable of moonbounce on 3.4GHz.

Under anomalous propagation conditions path losses are considerably reduced, often approaching the line of sight loss and under these conditions even simple equipment is then capable of working very long paths.

16.2 RECEIVERS

The most used commonly used type of receive converter at 3.4GHz is of the interdigital type as described in references [1], [2] and [3]. The design has the advantage of inherent front end selectivity which makes the design less prone to out of band signals and also rejects noise power at the receiver image frequency which can degrade overall receiver sensitivity. Only this type of receive mixer will be described here.

In order to improve receiver sensitivity a preamplifier may be added to the system at a later date. Most published designs at the present time use bipolar devices due, in part, to the availability of suitable surplus devices. Such a transistor design will be described here. Preamplifiers using GaAs fets are starting to appear, such as the example in [6].

16.2.1 An interdigital converter for 3,456MHz

The converter described here is a scaled version of the 1.3GHz converter described in [3]. It can be constructed and aligned without the use of any special facilities and has a noise figure of approximately 15dB.

The heart of the unit is an interdigital network. Referring to the circuit of Fig 16.4, this consists of five rod elements L1 to L5. L1, L3 and L5 are low Q coupling elements. L2 is resonant at the signal frequency of 3,456MHz. L4 is resonant at the local oscillator frequency and selects the harmonic at 3,312MHz generated by multiplier D2. This diode produces a few milliwatts at 3,312MHz for around 200mW drive at 368MHz. A suitable uhf source designed to generate this signal was described in chapter 8, "Common equipment". It should be fitted with a 92.000MHz crystal for this application. The network consisting of L6, C4 and C5 matches the impedance of the diode to 50Ω. D1 is the mixer diode, and its 144MHz output signal is fed to a low noise preamplifier stage using a BFR34a. The preamplifier should be well screened by a small box made from copper foil or copper clad pcb material. Using an HP5082-2817 mixer diode, the converter has a noise figure of 16dB. Using an MBD102 mixer diode. this is reduced to around 13dB. The multiplier diode is a Mullard BXY28 diode which appears to be most commonly available in the "e" package. The major mechanical details of the converter are shown in Fig 16.5. The baseplate is made from 1.6mm double sided pcb.

In the prototype converter, the multiplier was soldered into position as shown in Fig 16.5. However, to enable the

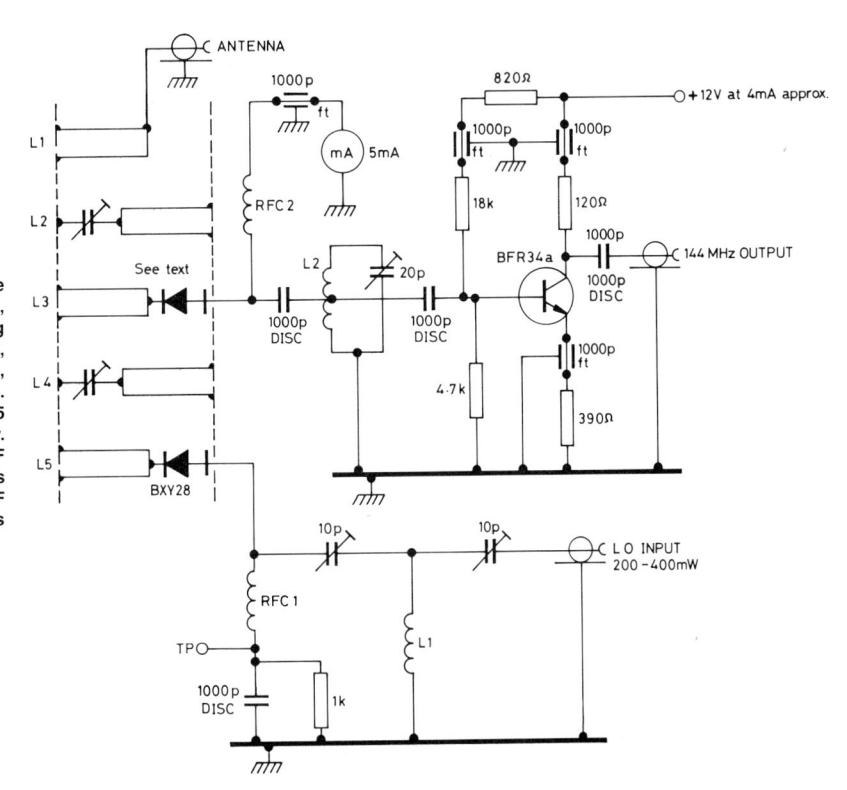

Fig 16.4. Schematic diagram of the interdigital mixer unit. L1–L5, interdigital rods. L6, 45mm 20swg t.c.w. bent to form one-turn loop. L7, 5t 20swg t.c.w. 6mm i.d. 20mm long, tapped 1.5 turns from earthy end. RFC1, 3.5t 30swg e.c.w. on FX1115 ferrite bead. RFC2, 20t 30swg e.c.w. 3mm i.d. closewound. FT, 1,000pF feedthrough capacitor. All trimmers are plastic foil types and all 1,000pF capacitors other than feedthroughs are miniature ceramic plates

diode to be more readily changed, the mounting scheme shown in Fig 16.6a was devised. The decoupling disc is tapped M3 and a brass half-nut is jigged into place and soldered into place on one side of the disc. Jigging is most easily carried out using either a chrome-plated or a stainless steel bolt which will not "take" solder and can consequently be easily removed. The requisite length of brass studding is faced off and centre-drilled 1.6mm diameter to a depth of around 3.5mm. The solid resonator is similarly centre-drilled. A TO3 insulator-bush is carefully filed out to just clear the brass studding and the resonator mounting bar drilled to just clear the bush. The assembly of the components should be clear from the figure. The diode does not need to be clamped too tightly or else there is a risk of fracturing the package. Finger tightness is quite adequate.

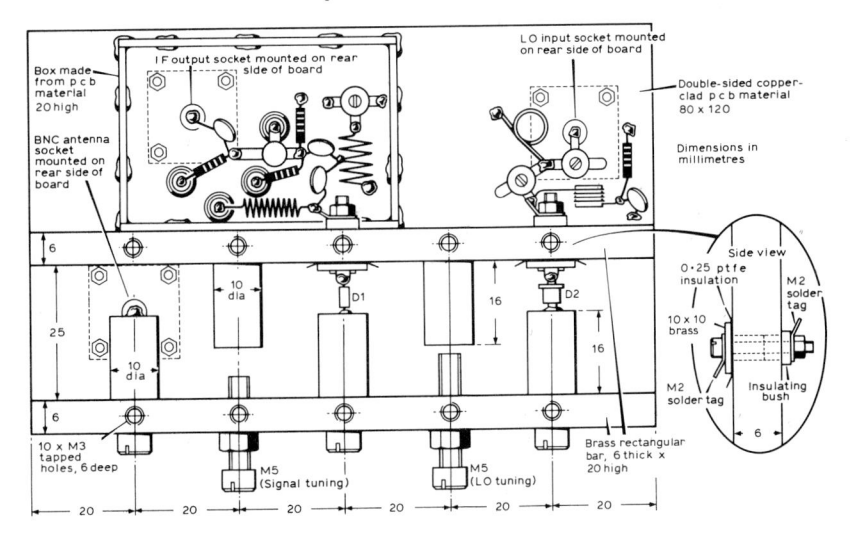

Fig 16.5. Layout of the interdigital converter

16.4 MICROWAVE HANDBOOK

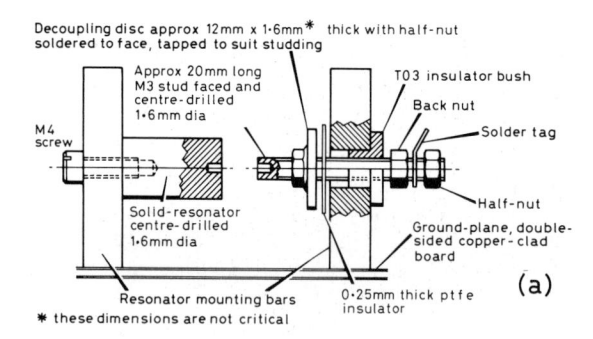

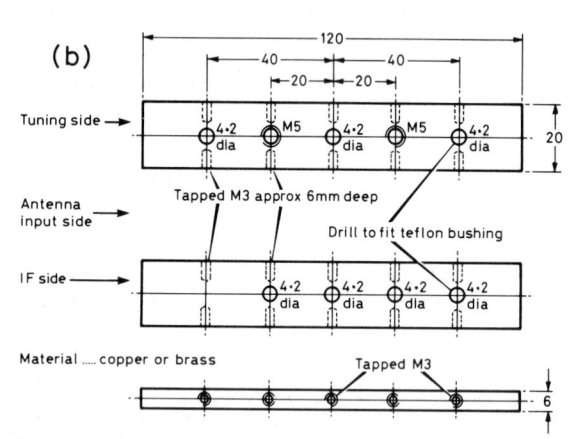

Fig 16.6. (a) Alternative method of mounting the "e" package multiplier diode. (b) Drilling details of interdigital unit sidewalls. Material: brass bar

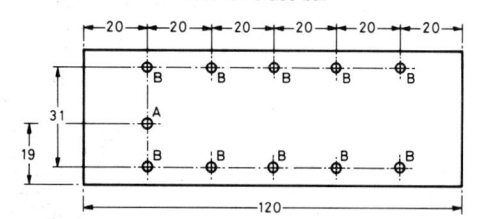

Fig 16.7. Drilling details of the interdigital unit coverplate. Hole A: 4mm diameter and holes B: 3mm diameter. Note that the mounting holes in the baseplate will have similar disposition

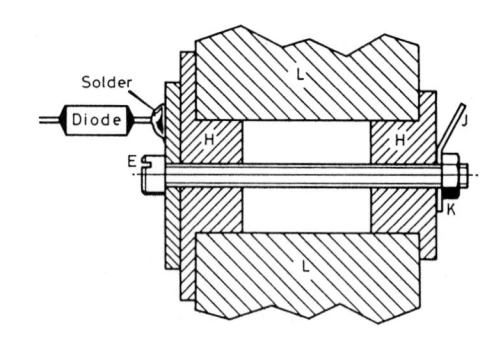

Fig 16.8. Construction of capacitors C1 and C2. E, M2 × 12 brass screw. F, 10mm × 10mm × 1mm copper plate. G, 12mm × 12mm × 0.25mm ptfe sheet. H, Two 4.2mm o.d. × 2mm i.d. × 2mm thick ptfe spacers. J, M2 solder tag. K, M2 nut. L, Interdigital unit sidewall with 4.2mm hole

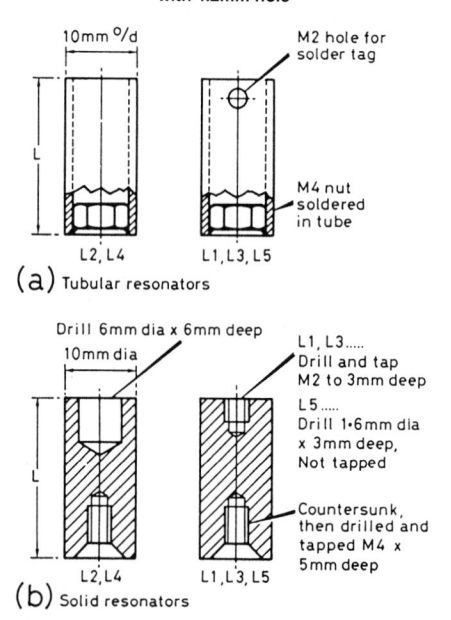

Fig 16.9. Construction of the interdigital rods. Material: L = 16mm × 10mm o.d. × 8mm i.d. brass tube as in (a), or 10mm brass rod, suitably drilled and tapped as in (b). On this band, the solid elements L2 and L4 do not require drilling with the 6mm hole at the tuning screw end of the rods

Drilling details of the interdigital unit sidewalls are shown in Fig 16.6b. The top cover drilling details are shown in Fig 16.7, the material being 1mm copper sheet or double clad pcb material. Fig 16.8 shows the construction of capacitors C1 and C2 in more detail and Fig 16.9 shows the method of mounting L1–5. A photograph of the completed interdigital converter unit is shown in Fig 16.10.

Alignment

Connect the 368MHz source to the lo input of the converter. Connect a voltmeter between the test point and ground. Adjust the two 10pF trimmers for maximum voltage at the test point (around 1.25V) The tuning is fairly critical and some instability may be encountered if the settings are not quite correct. Remove the voltmeter. Next fully unscrew the tuning screws in the interdigital assembly and adjust the lo tuning screw for maximum current observed on the 5mA meter – between 1 and 4mA should be seen. It is possible to select the wrong harmonic and the method described in chapter 8, "Common equipment", using Lecher lines, may be used to check the frequency if no other test equipment is available.

A 144MHz receiver is connected to the i.f output socket and the 20pF trimmer adjusted for maximum noise. Correct operation of the mixer can be checked by disconnecting the 368MHz drive to the mixer, when the receiver noise level should fall slightly. A large change in receiver noise level is indicative of unstable multiplier operation and the 10pF trimmers in the matching network should be readjusted to stop this. Finally the signal tuning screw should be adjusted for maximum signal when injecting a weak signal.

16.2.2 A transistor preamplifier for 3.4GHz

At 3.4GHz, bipolar transistors such as the NE64535 can give a significant improvement in receiver noise figure at relatively little expense. The preamplifier to be described here is a compromise between lowest noise figure and highest gain, as optimising one parameter usually degrades the other.

The circuit of a suitable preamplifier, developed by C Suckling, G3WDG, is shown in Fig 16.11. The function of the series capacitors on input and output is twofold. They act both as dc blocking elements and as part of the matching circuitry. In the latter function their values are fairly critical and, since suitable capacitors are not commercially available, they have to be home-made.

The capacitors in the prototype were made from pcb material (RT/Duroid type 6010, 0.63mm (0.025in) thick). The material was cut, then filed to shape. The dimensions are critical to 0.05mm and should be checked with a micrometer during the filing operation. The required dimensions are 1.60 by 1.60mm for the input capacitor (0.36pF) and 1.84 by 1.84mm for the output capacitor (0.47pF). Other constructional details should be apparent from Fig 16.12. A photograph of a completed preamplifier is shown in Fig 16.13.

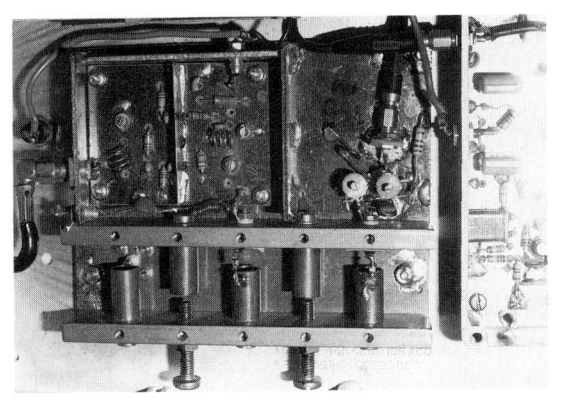

Fig 16.10. Photograph of the completed interdigital unit

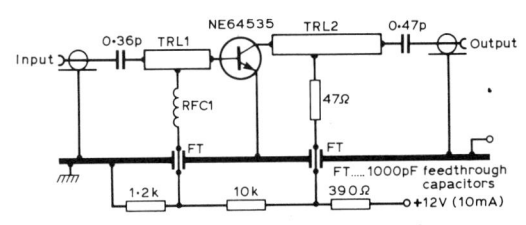

Fig 16.11. Circuit diagram of the G3WDG 3.4GHz bipolar preamplifier. TRL1: 0.128λ length of 50Ω transmission line, tapped midway for RFC1. TRL2: 0.336λ length of transmission line, tapped 0.14λ from transistor end. RFC1: 5t of 0.4mm diameter enamelled copper wire, 1mm i.d.

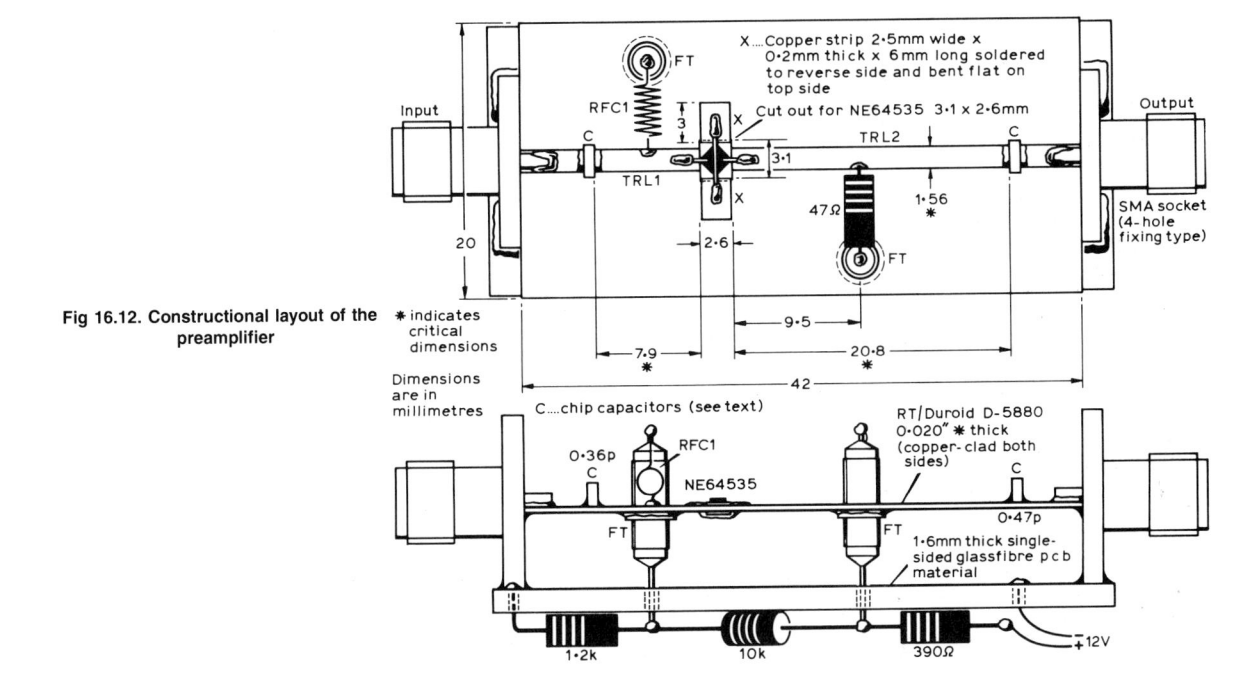

Fig 16.12. Constructional layout of the preamplifier

* indicates critical dimensions

Dimensions are in millimetres

Fig 16.13. Photograph of the completed preamplifier

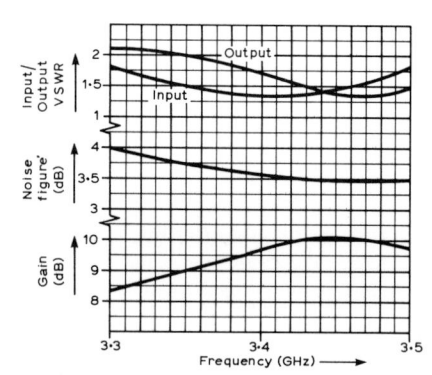

Fig 16.15. Measured performance of the prototype preamplifier

Dimensions are given for the transmission lines on 0.5mm (0.020in) thick RT/Duroid type D-5880 pcb material. This material was chosen as it is a good medium for microstrip circuits at this frequency (it is usable to beyond 10GHz). Conventional 1.6mm Teflon or epoxy glass fibre boards are *not* suitable for this application. The connectors shown (SMA) were chosen because of their excellent performance and size compatibility with the microstrip. The socket bodies are soldered to the reverse of the pcb and also to the "chassis".

It is possible to use N-type input and output connectors. The inner of the N-socket is filed down to form a small tab, as shown in Fig 16.14a. This tab is soldered to the 50Ω microstrip lines on the pcb. In order to minimise the discontinuity at this junction with the microstrip, a cut-out is filed in the pcb to accommodate the bush on the connector. In order to allow for this, the length of the pcb must be increased by 6mm. The rear side metallisation on the pcb is soldered to the body of the N-socket, as shown in Fig 16.14b. BNC connectors are *not* suitable for this application. The cut-out for the transistor is made with a sharp blade such as a scalpel.

After fitting the emitter grounding strips, made of 0.25mm copper foil, and soldering them to the reverse side of the board, they are flattened to lie flush to the board and filed slightly. This ensures that the transistor is located with a minimal gap between the base collector leads and the microstrip. When soldering in the transistor (which should

be the final operation) cut all the leads to about 3mm in length. Ensure they are soldered to the microstrip directly at its ends, after pre-tinning the lines with a hot, clean soldering iron. Note the base connection has the chamfered end.

Provided that the preamplifier has been constructed carefully, it should work immediately upon application of power. The only "adjustment" which may be necessary is to set the collector current to 10mA by making small changes to the 1.2kΩ or 10kΩ resistors, by adding a suitable resistor in parallel with one or the other. Do *not* alter the value of the 390Ω resistor. The performance of the prototype is shown in Fig 16.15. It can be seen that the preamplifier has low input and output vswrs. The noise figure of the whole preamplifier is only 0.75dB higher than the minimum possible for the device when designed for lowest possible noise figure.

When a single preamplifier stage is used with the previously described interdigital converter using an MBD102 diode, an overall noise figure of 6.2dB is obtained. When two of these preamplifiers are cascaded the overall noise figure is reduced to 4.2dB.

16.3 TRANSMITTERS

The easiest way to become operational on the 3.4GHz band is to use a varactor tripler to generate a few watts of fm or cw. This is driven by an 1,152MHz source. The whole system may take the form of Fig 16.16.

Once experience of the band has been gained with this arrangement, the constructor can progress to a linear transmit converter to allow all mode operation on the band. There is, at the present time, a choice between two main types of mixer. The first is a varactor mixer which has the advantage for portable operation of being passive and therefore needing no power supply [7]. The second is a valve mixer usually using a 2C39a type valve such as described in [8]. This has the disadvantage of requiring a high voltage and blower whilst producing a similar power output. The local oscillator for either of these can be obtained by retuning the original varactor multiplier transmitter chain to 3,312MHz.

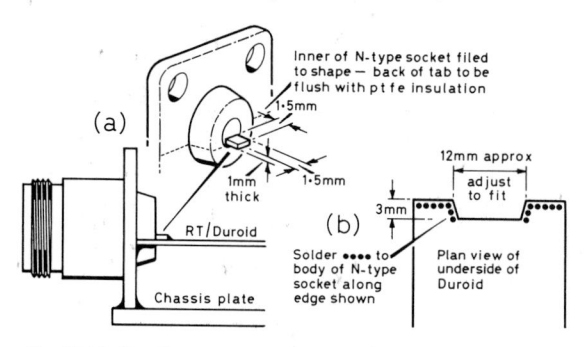

Fig 16.14. Details of the N-socket to RT-Duroid transition. (a) modification of N-connector. (b) modification of board. Note overall length of board must be increased by 6mm

THE 3.4GHZ ("9CM") BAND

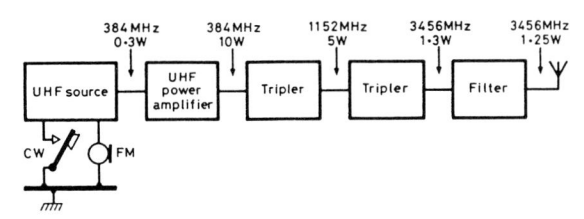

Fig 16.16. Schematic diagram of 3.4GHz crystal controlled varactor transmitter

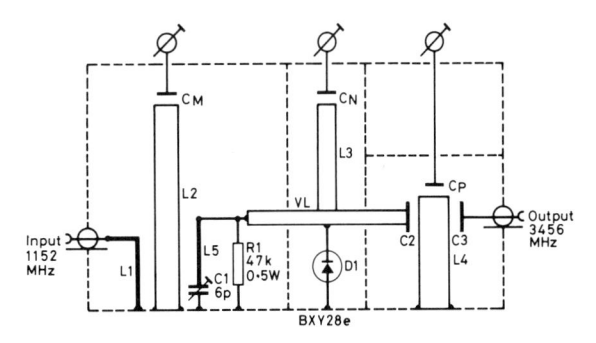

Fig 16.17. Circuit diagram of the 3.4GHz tripler

The final stage of equipment development is to increase the output power. Due to the lack of suitable transistors at this frequency, the choice is limited to valves, mainly of the 2C39a family. A suitable design claiming 6W output was described in [9].

16.3.1 A varactor multiplier for 3.4GHz

The varactor tripler from 1,152MHz to 3,456MHz to be described here was developed by M Johnson, G0BPU, from a design originally published in [10]. Provided the given dimensions are closely followed, the tripler is capable of producing 1.25W output at an efficiency of 30% when using a BXY28e varactor diode.

Operation

A diagrammatic representation of the tripler is shown in Fig 16.17. Power input at the BNC connector is coupled to the input line L2 by the coupling loop L1. L2 is tuned to 1,152MHz by CM. RF is coupled to the varactor by L5 and matched by C1. The doubled component generated by the diode is suppressed by an idler circuit, L3 and CN, tuned to 2,304MHz. The tripled component is coupled to the output line L4 by capacitor C2. The line is tuned to 3,456MHz by CP. The output signal is coupled to the connector by C3.

Construction

The tripler is constructed from brass or copper sheet and bar. Figure 16.18 shows an overview of the completed tripler. Drilling diagrams for the individual parts are given in Figs 16.19 to 16.22.

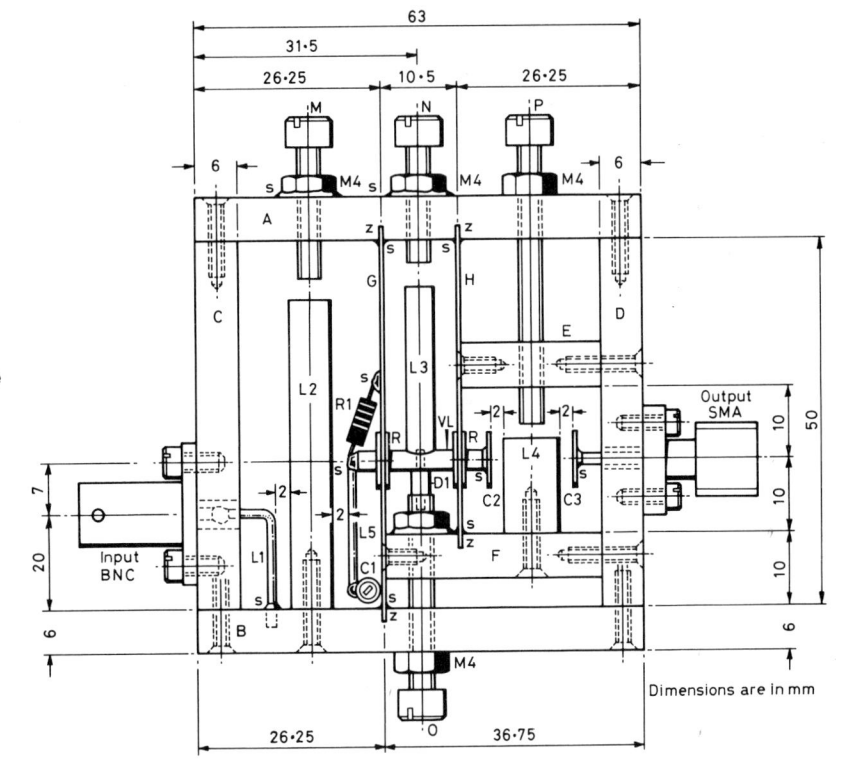

Fig 16.18. Overall dimensions of the varactor tripler

Table 16.1. List of parts for 3.4GHz tripler

A,B	63 x 25 x 6mm bar
C,D	50 x 25 x 6mm bar
F	30.5 x 25 x 6mm bar
G	54 x 25 x 0.5mm sheet
H	44 x 25 x 0.5mm sheet
I,J	63 x 62 x 0.5mm sheet
M	M4 x 20mm brass screw
N	M4 x 20mm brass screw
O	M4 x 30mm brass screw with 1.6mm dia by 3mm deep hole in one end
P	M4 x 40mm brass screw
Q,R	8mm o.d. x 3mm i.d. x 2mm thick ptfe spacer
L1,L5	16 swg silver plated copper wire
L2	42 x 6mm diameter rod
L3	23 x 4mm diameter rod
L4	13 x 8mm diameter rod
VL	18 x 3mm diameter rod
C1	6pF ceramic trimmer 3mm maximum diameter
C2,C3	8mm diameter x 0.5mm thick disc
R1	47kΩ 0.5W carbon film resistor
D1	BXY28e varactor diode

Not shown in full on the figures are the drilling details of the cover and baseplate (parts I and J). The 3mm holes in these should be drilled after assembling the tripler using holes F as guides. All slots shown are made 2mm deep using a small hacksaw blade. The case is assembled with M2.5 × 10 brass screws. A "U" shaped slot should be filed in the end of L3 to allow a good fit to varactor line VL. L1 and L5 should be positioned 2mm from L2. Likewise the discs forming C2 and C3 should be spaced 2mm from L4.

Assembly

Firstly, screw together parts B and C. Solder a 28mm length of 16swg copper wire to the centre pin of the input connector, then screw the input connector to part 3. L1 is formed by bending this wire so that it passes through hole E. Screw together parts A, B, C and D to form a square frame then mount the output connector on part D. It will probably be necessary to extend the centre pin of the output connector with a piece of 1mm diameter tinned copper wire.

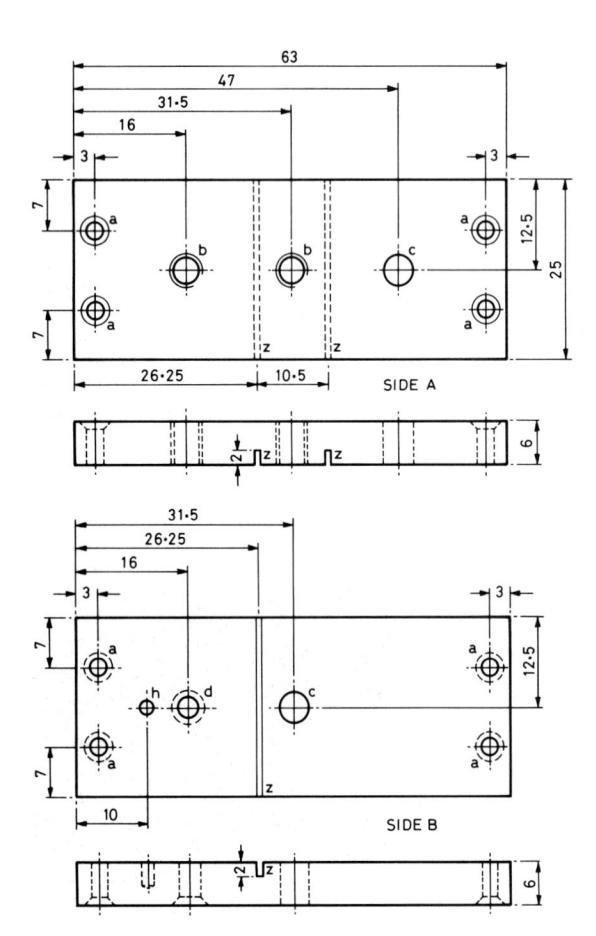

Fig 16.19. Drilling pattern for parts A and B. Material: brass bar

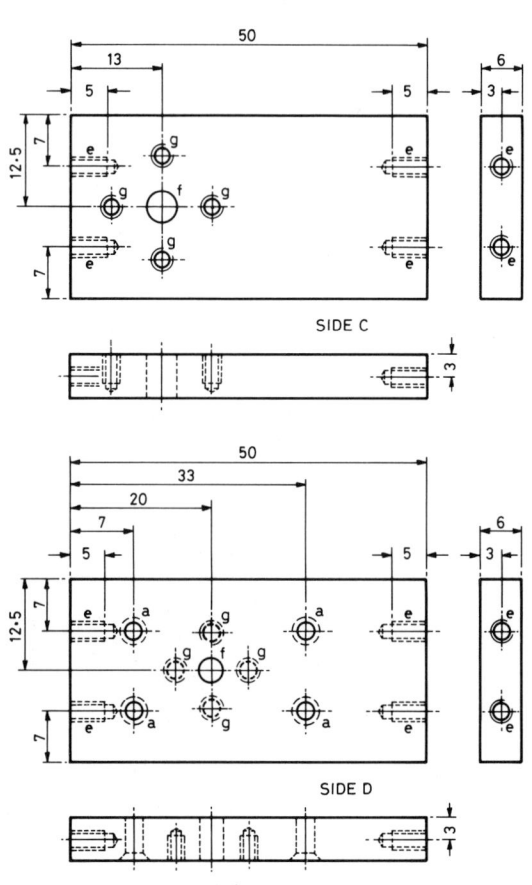

Fig 16.20. Drilling pattern for parts C and D. Material: brass bar

Table 16.2. Key to labels on Figs 16.18 to 16.22

A	drill 2.5mm & countersink
B	drill 3.3mm & tap M4
C	drill 4mm
D	drill 3mm & countersink
E	drill 2mm & tap M2.5
F	drill to match socket centre hole: for BNC 4mm diameter; for SMA 3mm diameter
G	drill to match socket fixing holes: for BNC four 3mm holes on 7mm radius; for SMA four 2mm holes on 5mm radius
H	drill 1.6mm
I	drill 8mm (to take parts Q and R as press fit)
S	solder
Z	slot 2mm deep made with small hacksaw

The varactor support nut (part Z) is soldered to part F using a rusty bolt or similar to hold it in position. Screw part G to part F and part H to part E, then place the varactor end of part H into the slot of F. This should be a tight fit; if not, the joint will need to be soldered. Mount varactor line VL in its ptfe bushes Q and R and jig into position with a dummy diode to maintain alignment of the diode hole. The dummy diode consists of a 1.6mm pin turned on the end of an M4 bolt. L3 is then soldered into position in the centre of the varactor line. Resistor R1 is next soldered, with the shortest possible leads between the input end of VL and inner wall G. Solder the copper disc of C2 into position on the end of VL, then screw L4 into position on part F. This centre unit can now be placed into the mainframe by sliding the ends of the partitions G and H into their respective slots, then screwing parts E and F to the frame. Solder C3 copper disc to the centre pin of the output connector, ensuring a 2mm gap between disc and line. L2 is then screwed to sidewall B.

C1 is next mounted on the baseplate. Its position will depend on the type of trimmer used. The base plate is screwed to the frame, then L5 is soldered between C1 and VL. L1 is soldered into position with its 2mm spacing from L2.

To ensure correct operation of the output line it has been found necessary to use a piece of 0.25mm thin copper foil

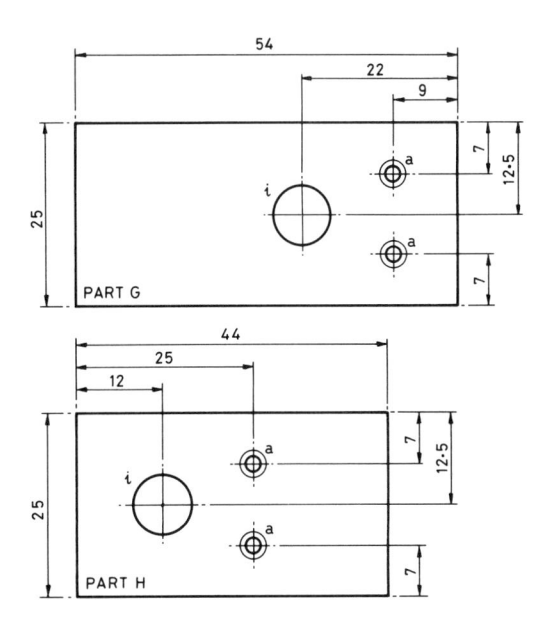

Fig 16.22. Drilling pattern for parts G and H. Material: brass sheet

between the tops of the sidewalls of the output chambers and the two cover pieces. A photograph of a completed tripler with the top cover removed is shown in Fig 16.23.

Alignment

For testing purposes the varactor input is connected via a reflectometer and to an adjustable 1,152MHz power source. This power source may consist of modules from chapter 8, "Common equipment". The reflectometer can be the printed circuit design of chapter 10, "Test equipment", configured to monitor reflected power. The output of the tripler is connected via a 3,456MHz filter and power detector to a suitable 50Ω load as shown in Fig 16.24. This load

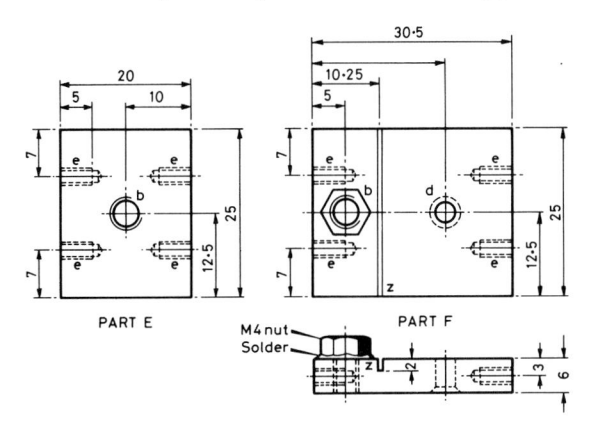

Fig 16.21. Drilling pattern for parts E and F. Material: brass bar

Fig 16.23. Photograph of the completed varactor tripler

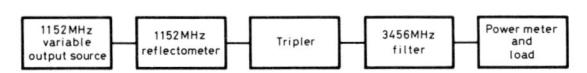

Fig 16.24. Equipment for aligning the varactor tripler

may be a long length of 50Ω coaxial cable (e.g. 10m of RG174) terminated in a 50Ω resistor.

Initially apply 500mW of drive to the varactor and adjust the tuning screws and capacitor C1 for minimum reflected power at the varactor input, combined with maximum output power. The drive can then be increased to its final value and the components slightly adjusted as necessary. The idler circuit is aligned by tuning L3 for minimum detected 2,304MHz output. The detector for this may be a wavemeter or Lecher line (see chapter 8, "Common equipment"), coupled to the output load. Finally the varactor is assembled with its final configuration of antenna and coaxial cables and any final adjustments made.

Typically, for a drive power of 5W a minimum output power of 1.3W has been measured using a BXY28e diode. The maximum drive power for this diode is around 12W.

16.3.2 A varactor mixer for 3.4GHz

The interdigital network which has been described as a receive mixer may also, in a slightly modified form, be used as a transmit mixer. The i.f preamplifier and associated matching network are removed, along with the 3,312MHz multiplier components. The new configuration is shown in Fig 16.25. The design was described in [7]. The mixer diode is replaced with a pair of higher power dissipation diodes. The maximum output power has been realised in this design using a BXY39e and BXY28e combination. This generated 0.9W of output. A pair of BXY27e or BXY28e or BXY39e diodes produced 0.5W. By comparison, a single diode of these types produced only 100mW. All these figures are for an input of 1.5W at 3,312MHz and 10W of 144MHz.

The line lengths and cavity sizes are exactly the same as in the receive mixer, except for the middle element. This should be made from 10mm solid brass rod 16mm long. An additional 4mm length of this rod, with 1.6mm holes in each end, is used to connect the two diodes. The details of the diode mounting assembly are shown in Fig 16.26. The attenuator on the 144MHz input is essential to isolate the 144MHz transmitter from the mixer and ensure stability.

Alignment

A source of 3,312MHz producing about 1.5W is should be connected to the lo port of the mixer. This may consist of the varactor transmitter described above, retuned to 3,312MHz. A 144MHz transmitter capable of producing variable output power of up to 10W should be connected to the i.f port. A power detector should be connected via a 3,456MHz filter to the output port. The 50kΩ preset should be set to maximum resistance. Apply 5W of 144MHz drive to the mixer and adjust the variable capacitors for maximum voltage measured across the potentiometer. Remove the 144MHz drive and turn on the local oscillator. Adjust the lo tuning screw for maximum voltage, again measured across the 50kΩ potentiometer. Remove the voltmeter from across the potentiometer. Reconnect the 144MHz drive and adjust the output tuning screw for maximum output power. All adjustments should next be re-optimised for maximum output. The 144MHz drive should be increased to its final 10W level and the adjustments, including the potentiometer, re-optimised.

The local oscillator suppression has been measured at 23dB down on the 3,456MHz signal, so a filter should be connected to its output when feeding an antenna directly.

16.4 ANTENNAS

The most commonly used antenna at 3.4GHz is a suitably fed parabolic dish. This popularity is mainly due to the high gain achievable with a relatively small sized dish. A graph of dish gain against dish diameter is shown in Fig 16.27. The parabolic surface of the dish may be solid, but in some cases this may lead to excessive wind loading on the mast. As an alternative, the reflecting surface may be fabricated from wire mesh, as long as the holes are not greater than $\lambda/10$ (9mm) in the plane of polarisation.

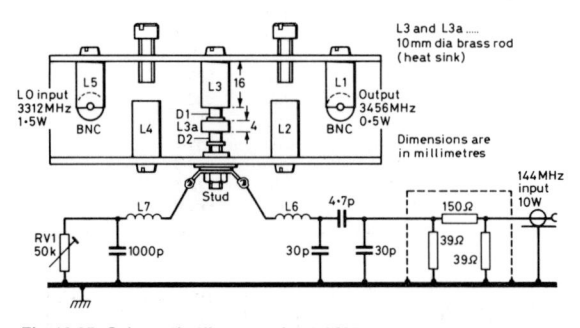

Fig 16.25. Schematic diagram of a 3.4GHz varactor mixer. D1, D2, see text. RV1, 50kΩ carbon preset. L1, L2, L4, L5, 16mm long × 10mm o.d. × 9mm i.d. copper tube. L6, 5t 18 swg t.c.w. 8mm i.d. 10mm long. RFC1, 3t 26 swg e.c.w. on FX1115 bead. L3A, 4mm long × 10mm dia copper bar. L3B, 16mm long × 10mm dia copper bar

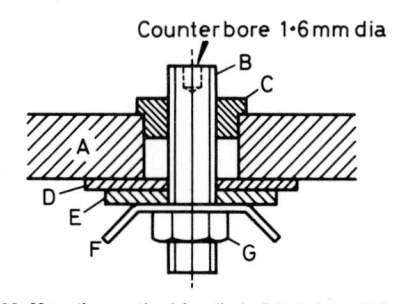

Fig 16.26. Mounting method for diode D2. A: interdigital side wall. B: M4×12 brass screw. C: M4 nylon nut filed to form a lip. D: double thickness of Sellotape. E: 15mm diameter × 1mm thick copper disc. F: M4 double solder tag. G: M4 brass full nut

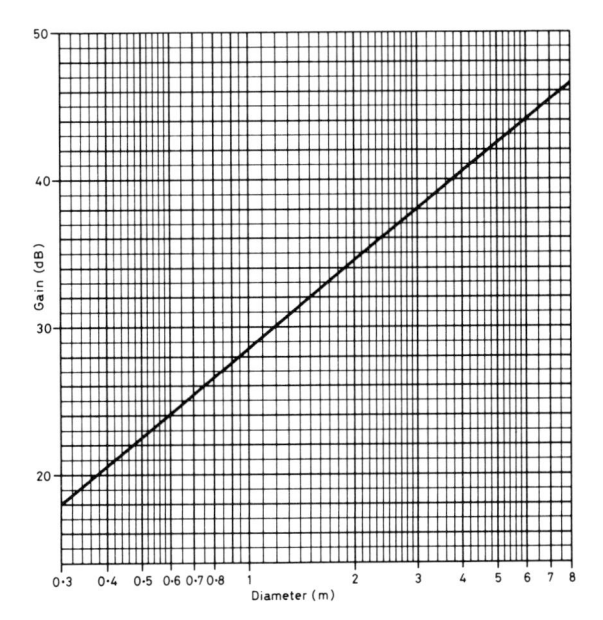

THE 3.4GHZ ("9CM") BAND 16.11

Fig 16.27. Gain vs diameter for a parabolic dish at 3.4GHz, assuming a 50% feed efficiency

Table 16.3. Dimensions of 3.4GHz loop quad

Boom:
Boom diameter	12.5mm o.d.
Boom length	2.0m
Boom material	Aluminium alloy

Elements:
Driven element	1.6mm diameter welding rod
All other elements	1.6mm diameter welding rod

Reflector size:
Reflector plate	52.4 x 42.9mm

Element lengths:
Length l(mm) (see Fig 16.29a):
Reflector loop	99.2
Driven element	92.7
Directors 1–12	83.9
Directors 13–20	81.2
Directors 21–30	78.2
Directors 31–40	76.1
Directors 41–50	75.1
Directors 51–60	74.6

Cumulative element spacings (mm):

RP	0.0	RL	29.5
DE	38.6	D1	49.2
D2	57.2	D3	74.1
D4	91.1	D5	103.0
D6	125.0	D7	158.9
D8	192.8	D9	226.7
D10	260.6	D11	294.5
D12	328.4	D13	362.3
D14	396.2	D15	430.1
D16	464.1	D17	498.0
D18	531.9	D19	565.8
D20	599.7	D21	633.6
D22	667.5	D23	701.4
D24	735.3	D25	769.2
D26	803.1	D27	837.0
D28	871.0	D29	904.9
D30	938.8	D31	972.7
D32	1006.6	D33	1040.5
D34	1074.4	D35	1108.3
D36	1142.2	D37	1176.1
D38	1210.0	D39	1244.0
D40	1277.9	D41	1311.8
D42	1345.7	D43	1379.6
D44	1413.5	D45	1447.4
D46	1481.3	D47	1515.2
D48	1549.1	D49	1583.1
D50	1617.0	D51	1650.9
D52	1684.8	D53	1718.7
D54	1752.6	D55	1786.5
D56	1820.4	D57	1854.3
D58	1888.2	D59	1922.1
D60	1956.1	D61	1990.0

Two suitable types of dish feed are described here, the dipole/splashplate feed and the feed horn. This latter design may be used independently of the dish as a low gain, broad beamwidth radiator which can be useful for local testing.

Where circumstances do not permit the use of a dish, a loopquad Yagi can still offer a useful gain when constructed carefully. A loopquad design scaling program was given in chapter 4, "Microwave antennas", which will enable suitable dimensions to be calculated. Alternatively the dimensions given below may be used.

16.4.1 The loop quad antenna

This antenna was originally designed by M Walters, G3JVL. The design of the antenna is shown in Fig 16.28 with the critical dimensions being shown in Table 16.3. The design is a scaled version of the 1,296MHz design adapted to the narrowband segment at 3,456MHz. The number of directors is 61 which means its boom length is 2m. The construction of the antenna is quite straightforward providing care is taken in the marking-out process. Measurements should be made from a single point or datum. In marking the boom for instance, measurements of the position of the elements should be made from a single

Fig 16.28. The 3.4GHz quad loop Yagi for which dimensions are shown in Table 16.3

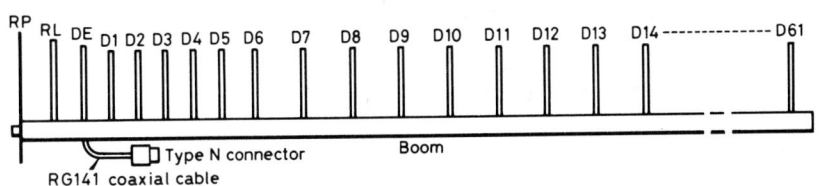

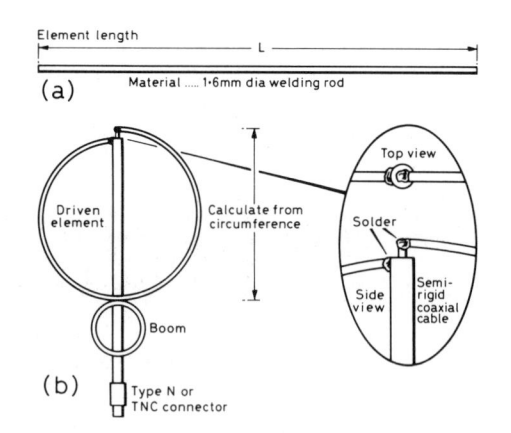

Fig 16.29. (a) Detail of the element construction. (b) Assembly of the driven element

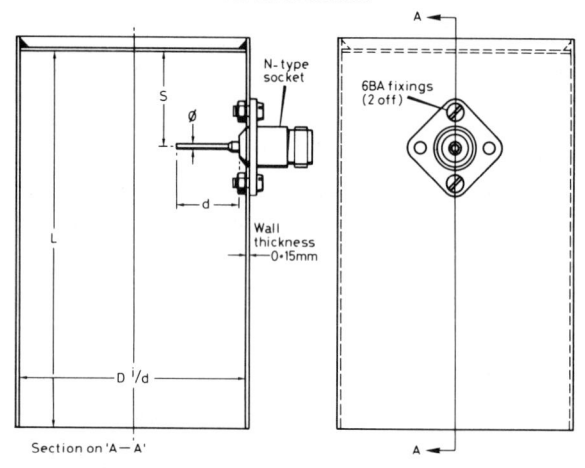

Fig 16.30. Constructional details of a 3.4GHz feed horn. Dimensions are: D=56mm, L=208mm, s=51mm, d=18mm and φ=2.1mm.

Fig 16.31. Photograph of completed 3.4GHz feed horn. Note that this version is constructed from brass tube and plate and uses an SMA socket

point (as tabulated in Table 16.3) rather than marking out individual spacings.

All elements are made from 1.6mm diameter welding rod cut to the lengths shown in the table then formed into a loop as shown in Fig 16.29a. The driven element is brazed to an M6 × 25 countersunk screw drilled 3.6mm to accept the semi-rigid coaxial cable. All other elements are brazed onto the heads of M4 × 25 countersunk screws.

All elements, screws and joints should be protected with a coat of polyurethane varnish after assembly. If inadequate attention is paid to weatherproofing the antenna, then the performance of the antenna will gradually deteriorate as a result of corrosion. Provided the antenna is carefully constructed, its feed impedance will be close to 50Ω. If a suitably rated power meter or impedance bridge is available, the match may be optimised by carefully bending the reflector loop toward or away from the driven element. The antenna can be mounted using a suitable antenna clamp. It is essential that the antenna be mounted on a vertical support, as horizontal metalwork in the vicinity of the antenna can cause severe degradation in its performance (see chapter 4, "Microwave antennas").

16.4.2 Feed horn for 3.4GHz

The feed horn described here is ideally suited to feed a dish having an f/D ratio of between 0.2 and 0.4.

The feed horn consists of a short length of circular waveguide with a coaxial to waveguide transition made from an N-type socket and a short probe. Constructional details are shown in Fig 16.30. The circular waveguide section can be made from an empty food tin closed with a plate at one end. All the dimensions are fairly critical (especially the length of the probe) and, hence, care should be taken during construction if a low vswr is to be obtained without adjustment. If a different diameter can is used the length and position of the probe and the length may need altering for best match. See also chapter 4, "Antennas", for further information on the design of circular feed horns.

The N-type socket should be fixed to the horn using four fixings, using washers as spacers to accommodate the curved surface of the tin. Alternatively an SMA socket may be used. Fig 16.31 is a photograph of the finished feed horn. The vswr response of the feed horn is shown in Fig 16.32, from which it can be seen that the feed is an excellent match at 9cm.

16.4.3 Dipole/splashplate feed

The version of the feed to be described here was developed by G Coleman, G3ZEZ. It is suitable for use in a dish having an f/D ratio between 0.25 and 0.45.

The feed is constructed around a length of fabricated coaxial line, the ratio of the inner diameter of the outer conductor to the outer diameter of the inner conductor being chosen to produce the desired impedance in this case 50Ω.

Details of the feed are shown in Fig 16.33. It is constructed from a length of 15mm diameter copper water pipe, the length of which is chosen to enable point X to be positioned at the focus of the dish. The 43mm diameter disc is turned or filed

THE 3.4GHZ ("9CM") BAND 16.13

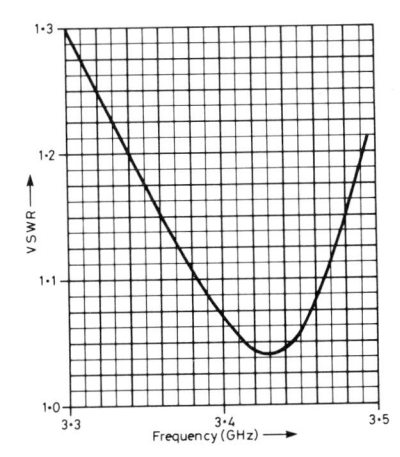

Fig 16.32. VSWR response of the 3.4GHz feed horn

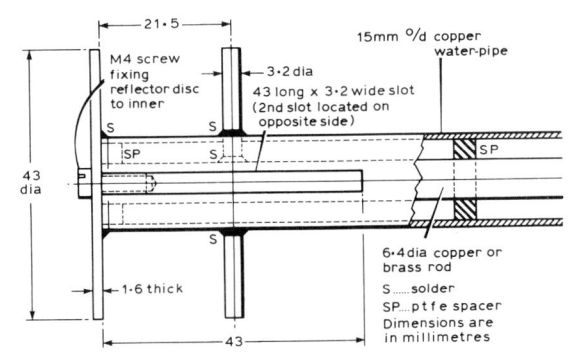

Fig 16.33. Construction of a 3.4GHz disc and dipole feed

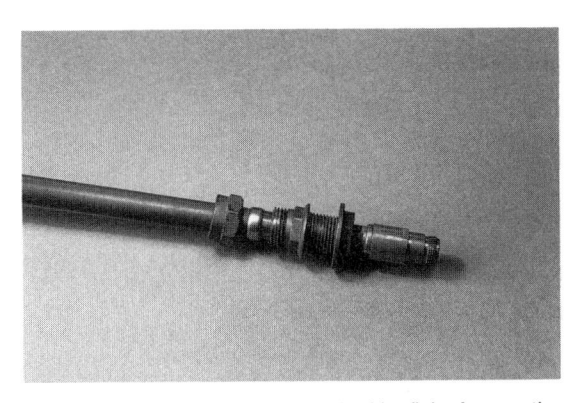

Fig 16.34. Photograph of the modified plumbing fitting for mounting disc and dipole feed

from a 1.6mm thick sheet of brass or copper. The two slots may be either milled to shape or carefully hacksawed and filed. The ptfe spacers marked SP are 13.5mm o.d. by 5mm thick spacers with a 6.35mm hole drilled centrally. They are used to maintain line concentricity.

The feed is held in position with a modified 15mm cold water tank compression fitting (Fig 16.34), preferably one with a mounting nut on either side of a central flange. The internal step in the fitting is filed away to leave a constant 15.2mm bore. The fitting is mounted in the dish in a 35mm diameter hole using one of the supplied nuts. The position of point X can be altered by sliding the feed in the fitting. When the position of the feed has been optimised by monitoring a remote signal, the feed can be locked in place by tightening the nut over the olive. This type of fitting must to fitted on the line *before* attaching the N-type connector.

The N-type female connector is fixed to the line by counterboring the collar 15mm diameter by 2mm deep and soldering the collar to the line. The centre pin is soldered to the line inner and the connector assembled with its supplied ptfe insulators.

16.5 ACCESSORIES

16.5.1 A 3.4GHz interdigital bandpass filter

The narrowband filter to be described here was originally published in [11]. It has an insertion loss of around 1dB and a 3dB bandwidth of 48MHz. The rejection of signals 144MHz away from the centre frequency is greater than 90dB.

The filter is constructed from brass, although it could, with advantage, be made from copper. The sidewalls are each made from pieces of 300 × 24 × 4mm bar drilled as shown in Fig 16.35. The base and coverplate are identical and made from 300 × 32mm pieces of 1mm sheet. The

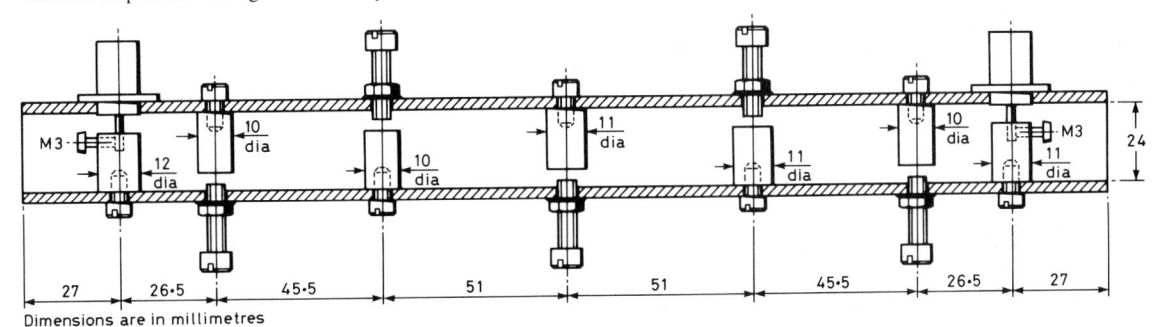

Fig 16.35. Dimensions of a 3.4GHz interdigital filter

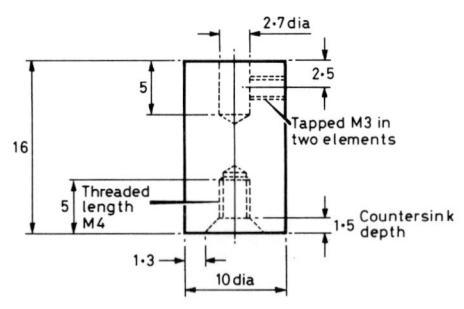

Fig 16.36. The construction of the interdigital filter input and output coupling rods. Diameter D is 17mm

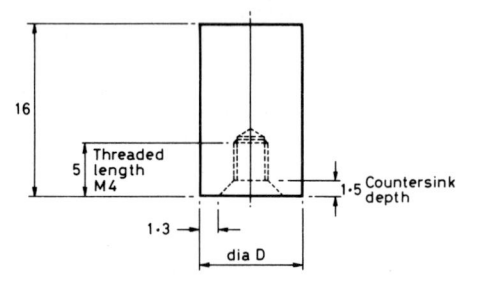

Fig 16.37. The construction of the remaining interdigital coupling rods. Dimension D is 15.3mm

covers are attached to the sidewalls using a total of 28 M3 × 6 brass screws to ensure a good rf seal. The input and output coupling elements are made as shown in Fig 16.36. The remaining coupling elements are shown in Fig 16.37. Each have an 8mm deep tapped M4 hole in one end. They are attached to the sidewalls using M4 × 8 screws and tuned at the opposite end with M4 × 12 screws, each fitted with an M4 locknut. The input and output connectors are four hole fixing N-types. The size of the central hole drilled for mounting these connectors may need altering, depending on the size of the boss of the connector.

The filter can be aligned by connecting it between a 3,456MHz transmitter and receiver, including suitable attenuators on the filter input and output to provide both a 50Ω termination for the filter, but also to avoid damaging the receiver. The filter should then be tuned for maximum received signal. The filter may need slight retuning in its final configuration to accommodate any slight source and termination impedances.

16.6 REFERENCES

[1] "3,456/28MHz converter", C Neie, DL7QY, *Dubus Technik,* pp158–162.
[2] "Interdigital converters for 1,296 and 2,304MHz", R E Fisher, W2CQH, *QST,* January 1984, pp11–15.
[3] "Interdigital converters for the GHz amateur bands", J Dahms, DC0DA, *VHF Communications,* March 1978, pp154–168.
[4] "A 9cm preamplifier with two stages", DB6NT and PA0JGF, *Dubus Informationen,* January 1980, pp16–17.
[5] "Broadband amplifier for 13 and 9cm", C Neie, DL7QY, *Dubus Informationen,* April 1983, pp269–270.
[6] "Einstufiger vorverstarker fur das 3.5GHz Band", DJ6PI, *UHF Unterlage,* Teil IV, pp641–643.
[7] 9cm high level mixer, *RSGB Microwave Newsletter,* April 1981, p5.
[8] "A local oscillator, transmit mixer and linear amplifier for the 9cm band", H J Senckel, DF5QC, *VHF Communications,* April 1980, pp236–245.
[9] "6W PA for 9cm", K D Broker, DK1UV, *Dubus Informationen,* April 1981, pp257–261.
[10] "3,456MHz tripler", C Neie, DL7QY, *Dubus Technik,* p215.
[11] "Narrowband filters for the 23cm, 13cm and 9cm band", D Vollhardt, DL3NQ, *VHF Communications,* January 1978, pp2–11.

CHAPTER 17

The 5.7GHz ("6cm") band

17.1 INTRODUCTION

The 5.7GHz band is the lowest frequency amateur microwave band to be dominated by waveguide techniques. It is also the first band on which dish antennas are extensively used in preference to Yagi antennas.

Over the past few years the 5.7GHz band has been rather ignored, with activity concentrating on the 10GHz band. This is mainly due to the equipment being physically much larger than corresponding 10GHz equipment which is based on a smaller waveguide size.

17.1.1 Frequency allocation to amateur services

The majority of countries, worldwide, have an allocation in the 5.7GHz area. In the UK the 5.7GHz allocation is very fragmented, as shown in Fig 17.1. It will again be noted that the narrowband segment, at the moment, is a multiple of 1,152MHz. However, from 1 January 1991, IARU Region 1 recommended a change to 5,668 to 5,670MHz for narrowband activities, this being the only area of the allocation common to the majority of countries within the region, although the harmonic relationship to 1,152MHz is lost. The disposition of activities within the new sub-band will not change. For example, the "centre of activity" will become 5,668.20MHz instead of 5,670.20MHz and so on. This change has, however, not yet taken place and is expected to take some years to fully implement. The Amateur Satellite Service has secondary allocations at 5,650 to 5,680MHz (earth to space) and 5,830 to 5,850MHz (space to earth).

17.1.2 Beacons

In the United Kingdom it is now possible to obtain a licence for a 5.7GHz beacon as a result of the licence changes made at the beginning of 1989 and mid-1990: see appendix 3, chapter 9, "Microwave beacons and repeaters". It is expected that beacons in this band will appear in the UK in due course. There are, however, a few beacons in the rest of Europe (see chapter 9), most of which are too far away to be of any real value other than under very exceptional conditions.

17.1.3 Propagation

Most of the propagation modes encountered on the vhf/uhf bands are also present at 5.7GHz, with the exception of those which require ionised media such as aurora, meteor scatter and sporadic E. The most commonly encountered propagation modes at 5.7GHz are line of sight, diffraction, tropospheric scatter, ducting and aircraft scatter. Moonbounce is also a possibility at this frequency. Further information on some of these modes can be found in chapter 2, "Operating techniques" and chapter 3, "System analysis and propagation".

Using the techniques of chapter 3, it is possible to predict whether communication over a given path is feasible, when the performance parameters of the equipment are known. Fig 17.2 shows the path loss capability of three different sets of equipment ranging from a basic system of low power transverter to a system with moonbounce capability. Also shown on the graph are the path losses associated with three different modes of propagation, line of sight (the least lossy mode of propagation), tropospheric scatter (a reliable, commercially used mode of propagation) and moonbounce. It can be seen that relatively simple narrowband equipment is capable of working very long line of sight paths easily. However, very few paths are purely line of sight and propagation losses are usually noticeably higher.

The tropospheric scatter loss is a reasonable upper limit for a terrestrial path and can be used to estimate the maximum range of equipment under normal propagation conditions. From the graph it can be seen that the simplest equipment (A) might be expected to work paths of up to 30km long. By employing a larger dish and separate transmitter based on a varactor multiplier, (B), the coverage distance increases to 120km. By adding a receive preamplifier, (C), the coverage distance increases to 200km. Line (D) shows the effect of the equipment of (C), but using a narrow bandwidth i.f filter, as one could when receiving cw and adding a small travelling wave tube (twt) amplifier, the coverage distance increases to 425km. At distances over

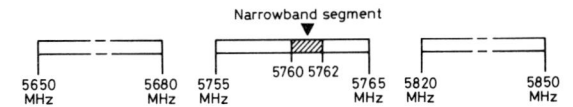

Fig 17.1. 5.7GHz allocation in the UK. Note: since 1 January 1991, the narrowband segment recommended by the IARU in region 1 is 5,668 - 5,670MHz, but this change has not yet taken place (Nov 91).

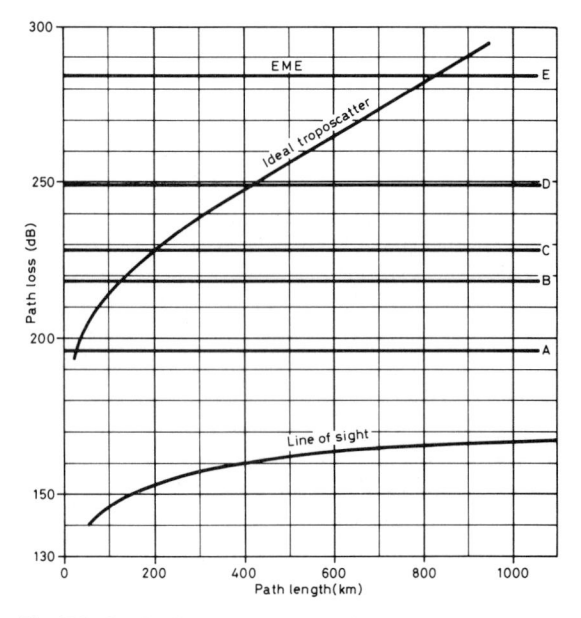

Fig 17.2. Graph of path loss versus distance for line of sight, tropospheric scatter and eme propagation, related to the performance of five sets of equipment:
  A: 0.7m dish, 4mW transmitter, 10dB nf, 2.5kHz bandwidth.
  B: 1.2m dish, 100mW transmitter, 10dB nf, 2.5kHz bandwidth.
  C: 1.2m dish, 100mW transmitter, 3dB nf, 2.5kHz bandwidth.
  D: 1.2m dish, 1W transmitter, 3dB nf, 250Hz bandwidth.
  E: 4.0m dish, 25W transmitter, 3dB nf, 250Hz bandwidth

about 830km it can be seen that the terrestrial path loss is greater than the route via the moon. Line (E) shows the parameters of equipment capable of moonbounce at 5.7GHz. Under anomalous propagation conditions path losses are considerably reduced, often approaching the line of sight loss, and even simple equipment is then capable of working very long paths.

17.2 RECEIVERS

There have been designs for 6cm receive mixers published, references [1], [2], [3] and [4]. These designs were based on waveguide techniques, with the exception of [1] which is an interdigital design having inconsistent performance at this frequency, mainly due to the accurate machining needed during construction. The mixers described in [1] and [2] were designed for an intermediate frequency of 28MHz which does not help their performance, as the filtering requirements for such a low i.f are quite stringent and not easily met. Hence, a 144MHz i.f is popular in amateur equipment for the band, making the filter design have less of an effect on receiver performance.

Most receive mixers employ the system shown in Fig 17.3. A diode multiplier is driven at uhf to generate a comb of signals. A filter is used to select the desired local oscillator signal at 5,616MHz. The received signal is passed from the antenna through a filter at the signal frequency of 5,760MHz. The two signals meet in a mixer diode which produces the difference signal at 144MHz. This signal is then passed through an i.f amplifier to the 2 metre receiver.

To improve the receiver sensitivity, a preamplifier may be added to the system at a later date. Few transistor designs have been published previously due to the high price of suitable devices. However, the emergence of GaAs devices has brought low noise figure and cost of acceptable performance within the reach of many. Examples of suitable designs may be found in [5] and [6]. These latter devices have a relatively lower noise figure and therefore greater sensitivity than bipolar devices. Unfortunately the devices need careful handling to avoid damage by stati c.

17.2.1 Receive mixer for 5.7GHz

The receive mixer presented here is based on a design by T Morzinck, DD0QT, [4] and follows the general design outlined in Fig 17.3. It is made from a 210mm length of brass or copper WG14. The major metalwork dimensions are shown in Fig 17.4.

The multiplier diode D1 produces a few mW at 5,616MHz for around 100mW drive at 1,123.5MHz. This signal may be generated using modules described in chapter 8, "Common equipment". The source should be fitted with a 93.6MHz crystal for this application. The matching network that transforms the multiplier diode impedance to 50Ω is shown in Fig 17.5. The 144MHz output from the mixer diode, D2, is fed to a low noise preamplifier stage using a BFR34a as shown in Fig 17.6.

Construction

The waveguide should be marked out and drilled in accordance with Fig 17.4. The eight filter posts, made from 2mm diameter silver plated copper wire, are then soldered into place. The holes for the tuning screws are tapped M5, then the brass nuts are jigged in place using non-solderable chrome-plated or stainless steel M5 screws and finally soldered.

The preamplifier and the multiplier matching network are mounted on pieces of brass plate soldered to the waveguide wall. Both items should be contained in screened boxes made from thin sheet metal or pcb material.

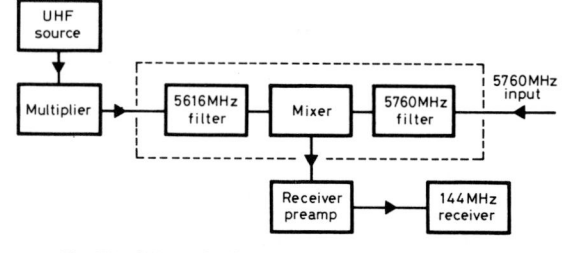

Fig 17.3. Schematic diagram of a typical 5.7GHz receiver

THE 5.7GHZ ("6CM") BAND

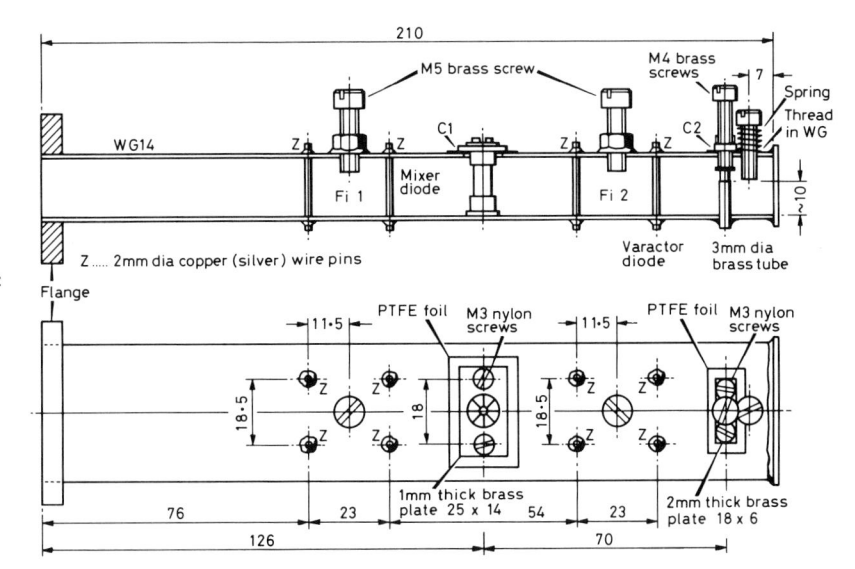

Fig 17.4. Dimensions of the 5.7GHz receive mixer

Alignment

As an aid to alignment, a milliammeter is first connected between TP1 and ground. The local oscillator is connected to the local oscillator input and all the variable components in the matching network adjusted for maximum current indication at TP1. While monitoring the mixer current meter, the local oscillator filter tuning screw is adjusted for maximum indicated current. This should occur with very little penetration of the tuning screw into the waveguide.

A weak signal, for instance derived from a harmonic generated by a lower frequency source, should be injected at the input of the mixer and the other filter tuning screw, along with the variable capacitor in the i.f preamplifier, adjusted for best possible signal to noise ratio.

17.3 TRANSMIT/RECEIVE CONVERTER

The schematic of the G3JVL 5.7GHz transmit and receive converter is shown in Fig 17.7. A 93.6MHz crystal is used in the uhf source to generate 200mW at 374.4MHz. This is fed to a power amplifier generating 2.5W at 374.4MHz. This may be a discrete transistor design but, perhaps more conveniently, may employ a Mullard BGY22 power module, which is capable of giving 2.5W output for 50mW of drive. This signal is then fed to a step recovery diode multiplier which generates a comb of frequencies at 374.4MHz spacing. A filter selects the appropriate harmonic at 5,616MHz which acts as local oscillator on both receive and transmit. On receive, the incoming signal is passed through the rf filter and mixed in the mixer diode, D2. The difference signal passes through relay RL1 to the preamplifier, thence to the 144MHz receiver. On transmit,

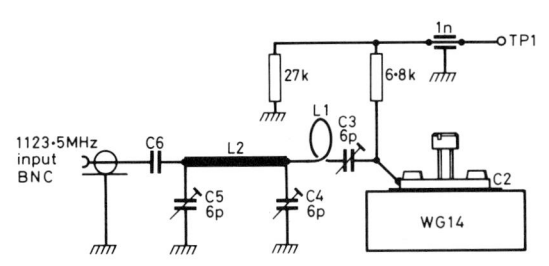

Fig 17.5. Details of the local oscillator matching circuit. C2: homemade bypass capacitor, plate size 18 × 6mm dielectric 0.13mm thick ptfe foil. C3: 6pF plastic foil trimmer, 7.5mm diameter. C4, C5: 6pF tubular ceramic trimmer. C6: 100–1,000pF leadless disc capacitor. L1: 1t, 1mm diameter silver plated copper wire 4mm id air spaced. L2: 14 × 5 × 0.5mm brass strip mounted between C4 & C5, 5mm above ground surface. D1: BXY28 or BXY38

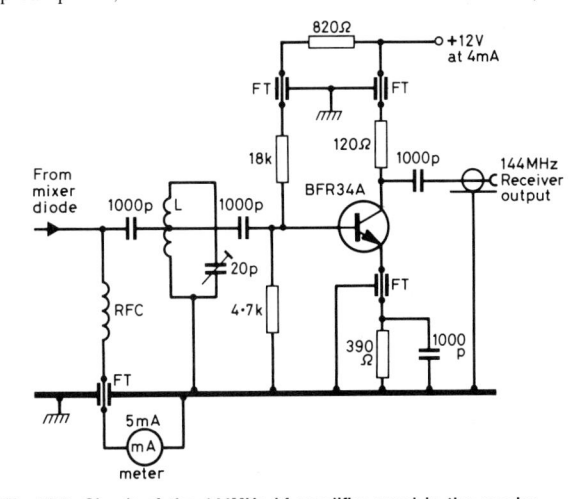

Fig 17.6. Circuit of the 144MHz i.f amplifier used in the receive converter. All 1,000pF capacitors are miniature ceramic plate types. L: 5t 18swg t.c.w., 13mm long, 6mm id tapped at 2.5t. RFC: 20t 26swg e.c.w., closewound 3.5mm id. RLA: RS type 349–591 or equivalent. FT: 1,000pF feedthrough capacitors

17.4 MICROWAVE HANDBOOK

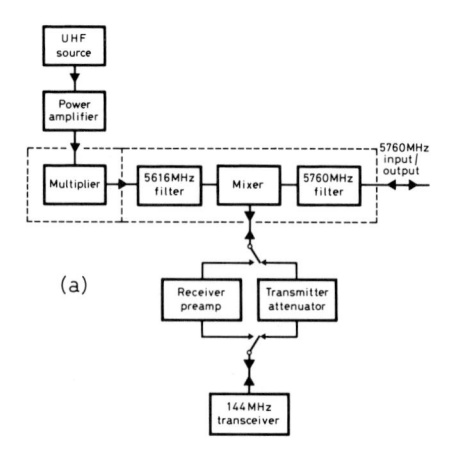

Fig 17.7. (a) Schematic of the 5.7GHz transverter (b) Physical configuration of the 5.7GHz transverter

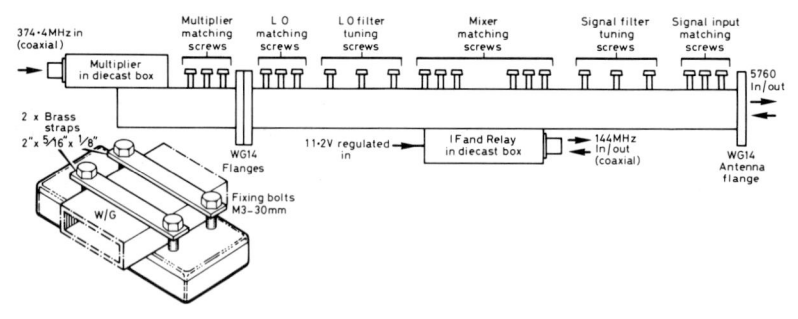

the 144MHz signal is passed through an attenuator and RL1 to present around 10mW to the mixer diode.

Construction

The waveguide parts of the transverter are split into two units for ease of alignment. The two filters and mixer are built into one unit and the multiplier is built separately. The dimensions of the mixer assembly are shown in Fig 17.8.

First, saw off a length of waveguide roughly to length and square off the ends by filing. Mark out the positions of the iris slots carefully (aim for ±0.2mm accuracy). The slots may then be cut out using a small hacksaw fitted with a new blade, starting at each corner and working inwards to maintain the accuracy of each cut. The tuning and matching-screw holes are then marked out and drilled appropriately, 2.5mm for the M3 holes and 1.6mm for the M2 holes, then they are tapped.

The whole assembly is carefully deburred by running a hacksaw blade carefully (like a file) through the waveguide. This tends to push metal back into the slots, which are then cleared again using the appropriate drill and taps.

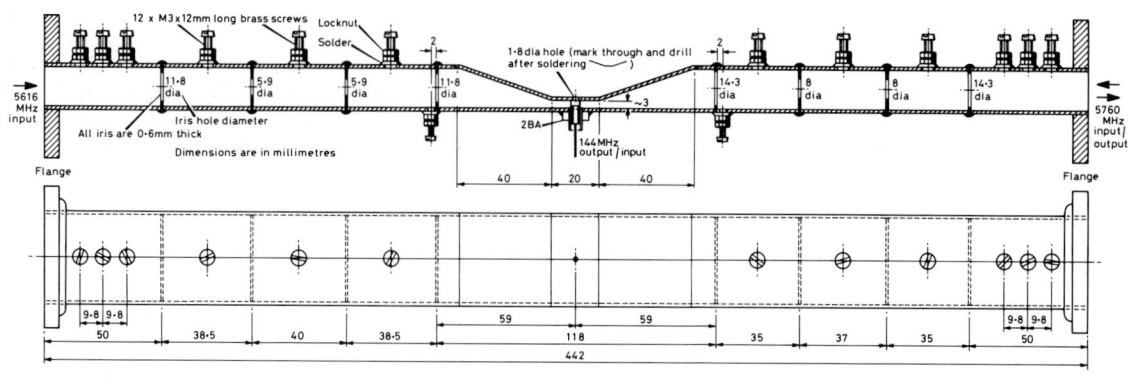

Fig 17.8. Dimensions of the 5.7GHz transverter mixer assembly

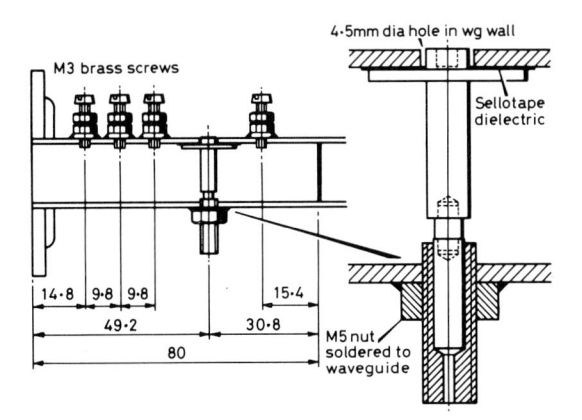

Fig 17.9. Dimensions of the 5.7GHz transverter multiplier assembly

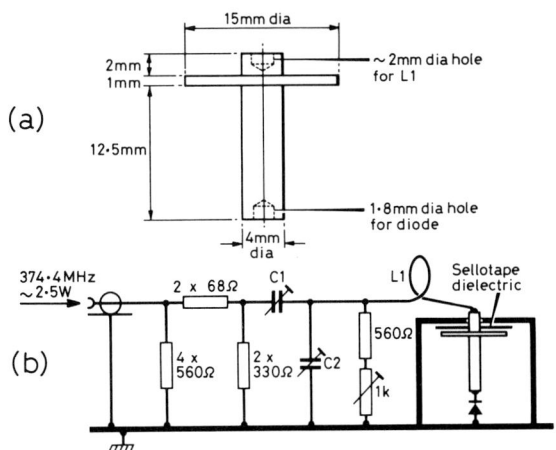

Fig 17.10. (a) Dimensions of the multiplier diode mounting post (b) input matching circuit

Further "filing" and clearing out is repeated until all traces of stray metal are removed.

The iris plates are made from 0.6mm thick copper sheet sawn and filed to be 34.8mm by 18.3mm. The iris holes are drilled centrally in the plates. This is greatly assisted by sandwiching the plates between two 16 swg brass or aluminium sheets – this keeps the irises flat and reduces burring. After deburring and flattening the plates between two smooth pieces of metal in a vice, they may be assembled into the waveguide. The guide nuts for the tuning screws and the mixer mount are jigged in place using non-solderable screws. The position of the tapered section is then marked out and only one of the broad faces removed between its limits. A piece of 1.6mm brass sheet is roughly cut to size and bent to the shape shown. It is then filed to be a tight fit between the waveguide narrow faces.

The assembly is then soldered using a hotplate to provide the background heating of the waveguide, ie. not hot enough to melt the solder until a soldering iron is applied locally. A suitable hotplate may consist of a 3mm aluminium plate on a domestic cooker ring. Perhaps the most critical processes in the construction of the transverter are the soldering operations. Too much solder in the cavities will make them excessively lossy and difficult to tune. Thus, the cleaning of all parts to a bright finish prior to soldering is vital. If all the parts are clean, it should require only the application of a small quantity of solder to each side of the irises to solder them. However, be careful that no gaps remain after soldering!

The position of the mixer mount is marked out, then a 2mm pilot hole drilled vertically through *both* broad faces of the waveguide in one operation. The hole in the upper (flat) face is then opened up to 3.3mm and tapped M4. An M4 nut is then jigged in place, as before, and soldered into position. In addition, two solder tags are soldered to this face to facilitate attachment of the i.f amplifier assembly.

The multiplier is next to be constructed. A 100mm length of waveguide is cut and the ends squared off. The waveguide is then marked out and drilled according to Fig 17.9.

The closed end of the waveguide may be constructed by cutting two slots in the broad faces of the waveguide using a small hacksaw and inserting a piece of copper plate 100 × 50mm exactly as per the iris-plate construction, except this time omitting to make the hole. The waveguide is thoroughly cleaned and the necessary nuts jigged in place with the usual "rusty" screws. The whole assembly, with a waveguide flange, is then soldered as before. Whilst cooling, the components for the attenuator and matching circuit should be soldered to the waveguide broad face with a large soldering iron.

The multiplier diode mounting post should be made preferably from copper to the dimensions of Fig 17.10a. Sellotape or ptfe tape is positioned as shown to act as the post decoupling capacitor. The post and multiplier diode (type MD4901 or similar) are then placed in position and L1 soldered *carefully* to the top of the post. The remainder of the matching components shown in Fig 17.10b are then soldered in position on the free waveguide, behind the multiplier assembly.

The mixer diode mounting post is made as illustrated in Fig 17.11. Again Sellotape or thin ptfe sheet is used as the dielectric in the capacitor. The wire connecting to the centre conductor is soldered in position first and a single layer of tape is applied. The part is then slid *carefully* into the outer conductor avoiding damaging the dielectric. The post and the mixer diode are positioned in the centre of the filter assembly. All the tuning and matching screws with locknuts are then positioned as shown.

The last remaining constructional item is the i.f preamplifier and changeover system. The circuit of Fig 17.12 is built to the layout of Fig 17.13 and attached to the waveguide using the previously installed solder tags.

It is possible to reduce the overall length of the mixer by folding the design at the mixer diode. A suitable method was described in [7]; however this is not an easy exercise and should not be undertaken lightly.

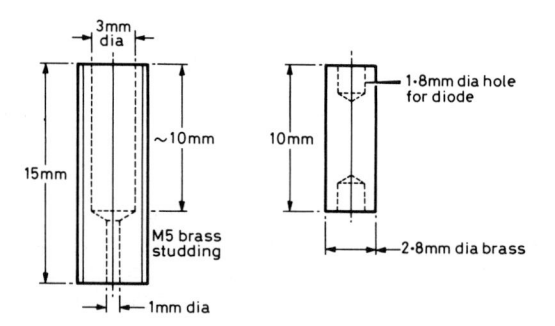

Fig 17.11. Arrangement of the mixer diode mount

Alignment

For initial alignment, connect a milliammeter with short leads between the 560Ω and the 1kΩ potentiometer (multiplier diode dc return) of Fig 17.10, with the potentiometer set at minimum. The output of the multiplier is connected to a detector, e.g. a detector diode mounted on the N-type socket of the waveguide transition described later.

A suitable uhf source is assembled, capable of 2W at 368.8MHz (for a 144MHz i.f). This may consist of the source described in chapter 8, fitted with a 94MHz crystal and followed by a suitable amplifier. Upon application of the uhf source output to the multiplier, some multiplier current should be seen on the meter. This is maximised using the two variable capacitors on the multiplier. Some detector current should be seen at this point. Maximise this by repeated adjustment of the tuning and matching screws, the two trimmer capacitors and the variable resistor on the multiplier. When several mA of detector current has been obtained the multiplier may be transferred to the local

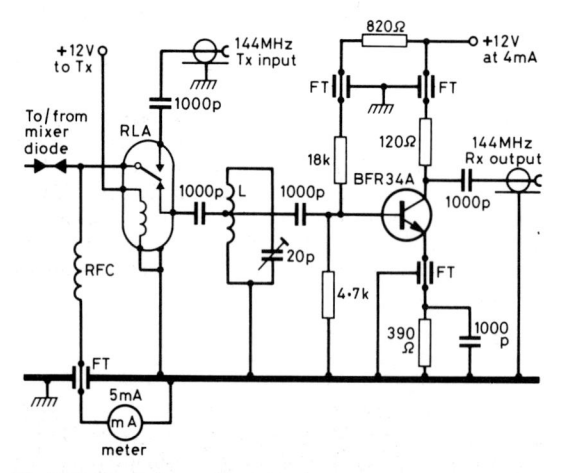

Fig 17.12. Circuit diagram of a suitable i.f preamplifier and changeover system. All 1,000pF capacitors are miniature ceramic plate types. L: 5t 18swg tinned copper wire, 13mm long, 6mm id tapped at 2.5t. RFC: 20t 26swg enamelled copper wire, closewound 3.5mm id. RLA: RS type 349-591 or equivalent. FT: 1,000pF feedthrough capacitors

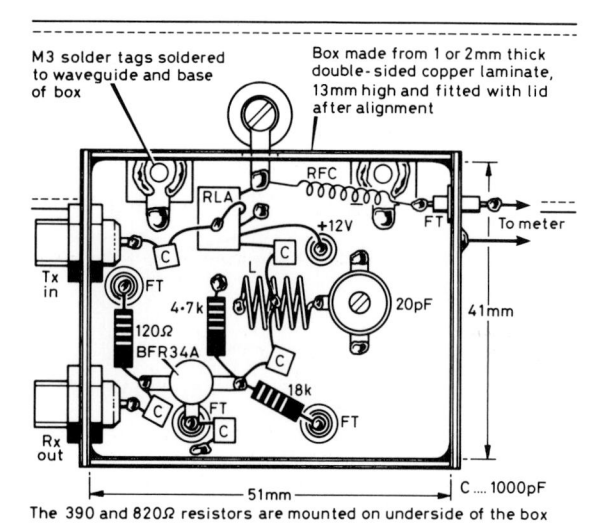

Fig 17.13. Layout of the i.f preamplifier
The 390 and 820Ω resistors are mounted on underside of the box

oscillator input of the mixer unit. With the matching screws fully out of the waveguide, screwing in the first filter tuning screw should produce a marked *dip* in the multiplier diode current as the cavity resonates. Tuning the second filter screw should show a *peak* in the multiplier current. Tuning the third screw should again show a dip but should also produce some mixer diode current: this is normal "tune-up" behaviour. At this point the multiplier diode current meter is removed, the potentiometer re-connected to the 560Ω resistor (Fig 17.10) and the tuning and matching screws readjusted to secure maximum mixer current. It should be possible to obtain several mA of current.

A source of 144MHz, at around 5mW, is applied to RLA with 12V applied to the relay coil. This should approximately double the mixer current. As before, tuning the filter screw nearest the mixer diode should produce a drop in the current, the next tuning screw should produce an increase and the final tuning screw should produce an indication on the detector. All the tuning and matching screws should then be adjusted for maximum output. If a weak signal source for 5,760MHz is available, the tuning of the filters may be optimised using this although, as the settings for maximum transmit output power and best receive noise figure do not coincide, it will be necessary to compromise if real transceive mode operation is intended. However, if it is expected to use the unit mainly as a receiver (perhaps using a separate multiplier-type transmitter), then optimise for best sensitivity only.

If full transceive operation is intended it will be necessary to add a coaxial relay switched by the 144MHz transceiver ptt line to switch the transceiver from the receiver output socket to the transmitter input socket, as shown in Fig 17.14. The value of the attenuator required in the transmit line should be such as not to overdrive the mixer diode. Suitable starting values are 20dB for a 3W output 144MHz

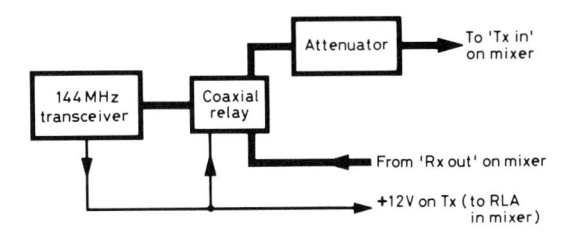

Fig 17.14. Block diagram of the 144MHz transceiver interface circuit

exciter and 27dB for a 15W exciter. Extra attenuation should be inserted until the 5.7GHz output just starts to decrease; the attenuator can then be rebuilt to that value. It is important to ensure that the attenuator is well screened to prevent rf leakage to the mixer diode but, perhaps more importantly, that the resistors comprising the attenuator are adequately rated to carry the intended power, since any failures could result in a large drop in attenuation with the consequent risk of damage to the mixer diode. A photograph of the author's mixer unit is shown in Fig 17.15.

17.4 TRANSMITTERS

Whilst the low power transverter will enable many line of sight paths to be covered under normal conditions, eventually the need to increase the output power will arise.

The choices for increasing the output power at this frequency are severely restricted and fall into two main categories. The first option is to amplify the signal already available. Bipolar transistor amplifiers are "non-starters" on this band, lacking in gain and output power. GaAs fet amplifiers lack gain but can produce the watts; however, their cost at the present time makes them unattractive to the average amateur. The ubiquitous 2C39 valve is well above its maximum frequency ratings and therefore unusable. Although there are a few valves which can give gain at this frequency, such as the YD1060, they are generally scarce but, for those possessing them, a suitable amplifier design was published in [8]. The most useful thermionic device at 5.7GHz is the travelling wave tube (twt). The operation of this device is fully described in chapter 6, "Semiconductors and valves", with further information in chapter 18, "The 10GHz band". Suffice to say here that typical performance figures are: 1mW input produces 10W output. They are very useful indeed, if they can be obtained. However, if none of these devices are obtainable a second method exists, that of employing a dedicated transmit multiplier using a microwave multiplier diode. These are cheaper and generally easier to obtain than any of the above devices and, consequently, a suitable design is described here.

17.4.1 A varactor multiplier for 5.7GHz

The dimensions for a ×5 multiplier (quintupler) capable of over 100mW output power are given in Fig 17.16. It is made in a 120mm long piece of brass or copper WG14.

Construction

First, cut the waveguide to length and square off the ends. Make a mark 7mm from one end and scribe a line, through this mark, across the top face of the waveguide and then continue the line around the other three faces (see Fig 17.16). Cut a slot in the top and bottom faces using a small hacksaw. This is best done by starting to cut at one side and then continuing the cut across to the other side, rather than trying to cut the slot all at once. Next, mark out the positions for all the holes (except for the holes 'A' and hole 'F') relative to the centre of the slot. The diode bypass plate should then be fabricated, as shown in Fig 17.17. Initially, the holes in the plate should be drilled 2.5mm. Using the plate as a template, drill the holes designated 'A' and 'E' in the waveguide 2.5mm. Open out the holes to the sizes/threads shown. All the other holes in the waveguide can then be drilled and tapped as appropriate.

The end plate, which fits into the slots previously cut, can then be made: 34.8 × 18 × 0.6mm brass or copper. Four lengths of 2mm copper wire (preferably silver plated) are next fitted through the waveguide using the 2mm holes. Cut off, leaving about 1mm of wire protruding either side. The diode posts are made next (Fig 17.18 and 17.19). If a lathe is not available, take great care to drill the 1.6mm holes centrally, to avoid the possibility of damaging the diode during assembly.

Deburr the inside faces and fit the end plate into position. Using stainless steel or rusty screws, jig nuts into position at holes F, C, and D in the top wall, and hole A in the sidewall. The waveguide assembly can then be soldered (including the end plate), using a hotplate or a small gas torch. Before the assembly cools, solder the 1mm plate (which can be 'L' shaped for strength) to the sidewall, as shown in Fig 17.20.

Fig 17.15. Transverter mixer unit (photo: G4FRE)

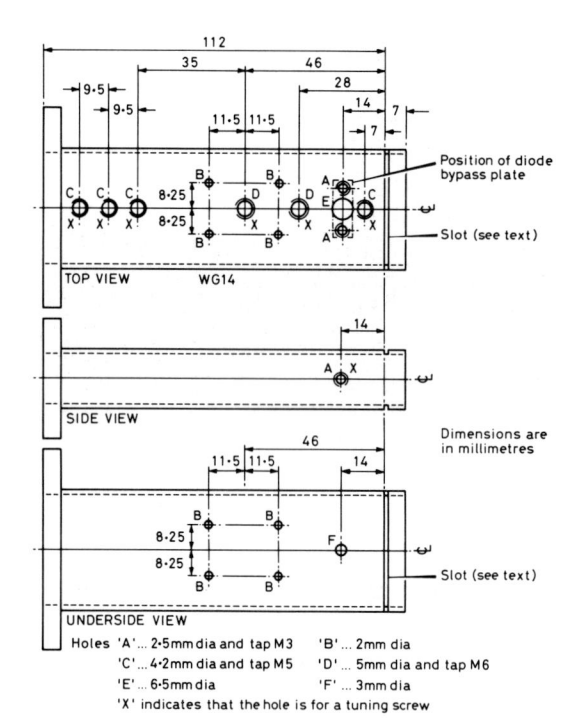

Fig 17.16. Mechanical details of the 5.7GHz multiplier waveguide assembly

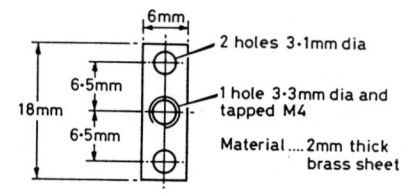

Holes 'A'... 2·5mm dia and tap M3   'B'... 2mm dia
      'C'... 4·2mm dia and tap M5   'D'... 5mm dia and tap M6
      'E'... 6·5mm dia   'F'... 3mm dia
      'X' indicates that the hole is for a tuning screw

Fig 17.17. Diode bypass plate details

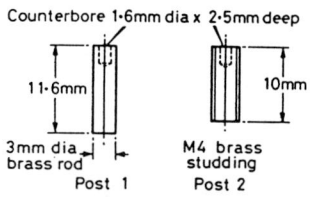

Fig 17.18. Diode post details

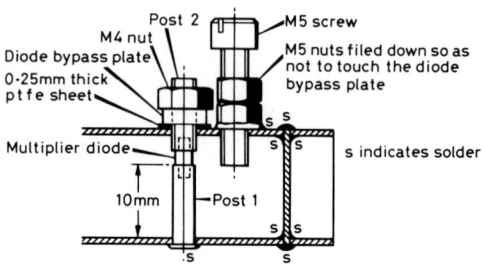

Fig 17.19. Diode mount details

The final stage of assembly is to build the input matching network, details of which are given in Fig 17.20. First, fit the diode bypass plate and the ptfe insulation. Use short nylon screws to fix these to the waveguide and include a solder tag under the screw nearest to the plate on the sidewall. After building the rest of the matching network, fit the diode into its mount (see Fig 17.19) and fit the tuning screws (each with a locknut). A photograph of the author's completed multiplier is shown in Fig 17.21.

Alignment

Initial alignment should be carried out with a drive power of approximately 500mW applied to the input of the multiplier. The drive source should be connected to the multiplier via a 3dB attenuator (e.g. 6m of UR43 coaxial cable) to ensure stable operation. Set R2 to maximum resistance and adjust C2, C3 and C4 for maximum dc voltage between TP and ground. Using a suitable detector coupled to the output of the multiplier, the tuning screws can then be adjusted for maximum output power. The drive power can then be increased and all adjustments (including R2) re-optimised for maximum output. With 3W drive (at the input of the 3dB attenuator) an rf output of at least 100mW should be obtained with a BXY28E diode. With 5W drive, it should be possible to achieve 150mW with a BXY28E, 200mW with a VSC64J and 275mW with a BXY39E diode.

17.5 ANTENNAS AND ACCESSORIES

17.5.1 A simple waveguide feed for short focal length dishes

This type of feed, often referred to in the United Kingdom as the "penny-feed", is suitable for dishes having an f/D ratio of 0.25 to 0.3, although professional practice seems to be to use an f/D ratio of about 0.33 to 0.35. Whilst it is not the ultimate in dish feed designs, it does have the advantage of ease of construction. It is shown diagrammatically in Fig 17.22.

Construction begins by squaring off the end of a suitable length of waveguide to reach the focal point of the dish. Two grooves are then filed in this end, each 26 by 2.6mm. A 52mm diameter disc is made, either on a lathe or by carefully cutting and filing a piece of brass sheet. The ends of the waveguide are cleaned, the disc is soldered to the slotted end and a waveguide flange soldered to the other.

The slot should be positioned at the focus of the dish and the optimum position may be found by adjustment whilst receiving a *distant* signal; the use of a near source can give misleading results. Signals having the standard horizontal polarisation are produced when the broad faces of the waveguide are mounted vertically.

THE 5.7GHZ ("6CM") BAND

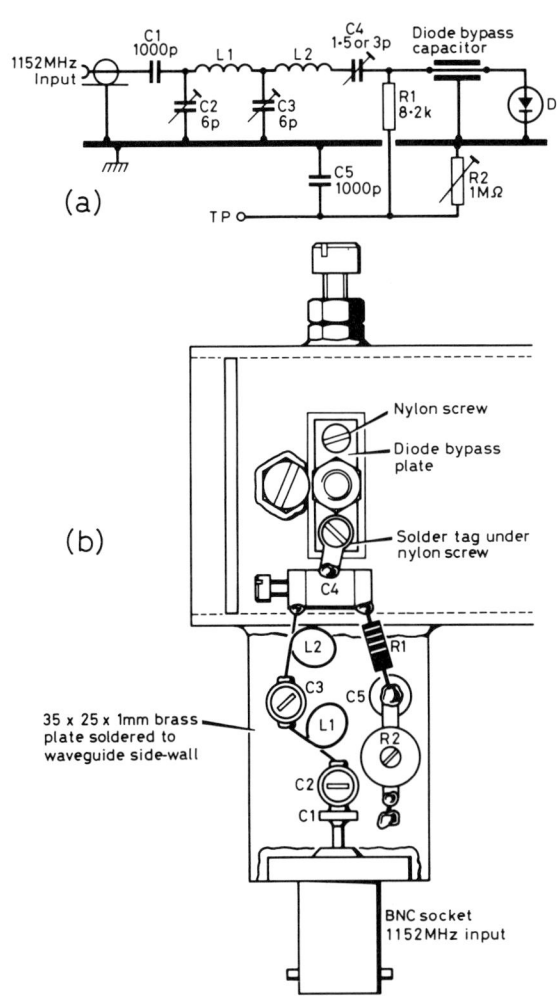

Fig 17.20. Input matching network. (a) Circuit diagram. (b) Layout. C1, C5: 1,000pF leadless disc; C2, C3: 6pF tubular trimmer; C4: 1.5 or 3pF ptfe tubular trimmer; L1: 1t 1.6mm t.c.w. 6mm dia; L2: 1t 1mm t.c.w. 6mm diameter; R1: 8.2kΩ 1/8W; R2: 1MΩ preset pot; D: BXY28E, BXY39E or VSC64J varactor diode

Fig 17.21. Prototype multiplier unit (photo: G4FRE)

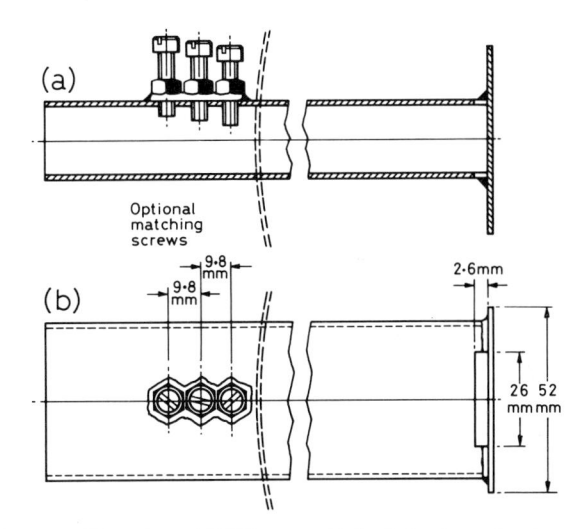

Fig 17.22. Dimensions of 5.7GHz waveguide feed, showing optional matching screws. (a) Side view (b) Top view

The feed can be used without any attempt to improve the match – the vswr typically is around 1.5:1. The match can be improved by conventional matching screws mounted 9.7mm apart, preferably mounted behind the dish so as to avoid unwanted resonances.

17.5.2 Feed horn for 5.7GHz

This feed horn is ideally suited to feed a dish having an f/D ratio of between 0.2 and 0.4. The feed horn consists of a short length of circular waveguide with a coaxial to waveguide transition made from an N-type socket and a short probe. Constructional details are shown in Fig 17.23. The circular waveguide section is made from a cylinder closed with a plate at one end (e.g. a small food tin cut to the specified length) or turned on a lathe. All the dimensions are fairly critical (especially the length of the probe) and hence care should be taken during construction if a good match is to be obtained without adjustment.

If a different diameter can is used (which should not differ by more than ±3mm from the specified dimension), the length and position of the probe, and the length of the tin may need altering for minimum vswr. The N-type socket should be fixed to the tin by all fixings, using washers as spacers to accommodate the curved surface of the tin.

17.5.3 Coaxial N-type to WG14 transition

Although most 5.7GHz equipment makes use of waveguide techniques, occasions may arise when it is necessary to interface waveguide to coaxial cable; for example when

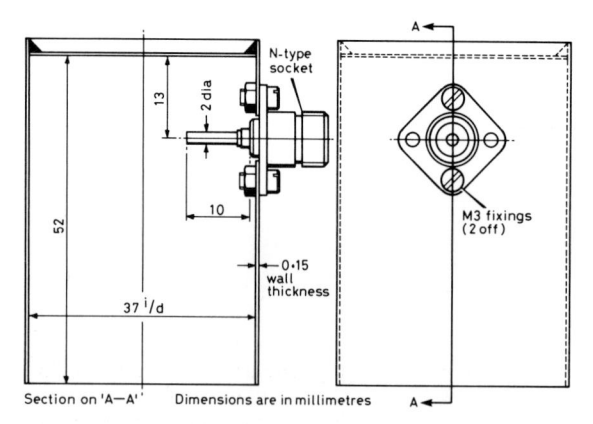

Fig 17.23. 5.7GHz feed horn. Although an N-type socket is shown, an SMA socket could equally well be used

laboratory equipment is being used to test equipment performance or if coaxial cable is to be used as the feeder.

Constructional details of a suitable transition are shown in Fig 17.24. The N-type connector is a standard four-hole fixing type with a small extension piece of copper wire soldered into its centre connection. The short-circuit plate is made by sawing opposing slots across the broad faces of the waveguide using a small hacksaw, inserting a 34.8 by 19 by 0.6mm piece of brass sheet across the waveguide and soldering this into position.

The transition has a vswr of better than 1.2 across the 5,650 to 5,850MHz band, with a total loss of less than 0.1dB.

17.6 REFERENCES

[1] "5,760/28MHz converter (6cm)", C Neie DL7QY, *VHF UHF DUBUS Technik*, pp163–168.

[2] "A 6cm waveguide converter". M Kuhne DB6NT, *Dubus Informationen*, 2/79, p74–75.

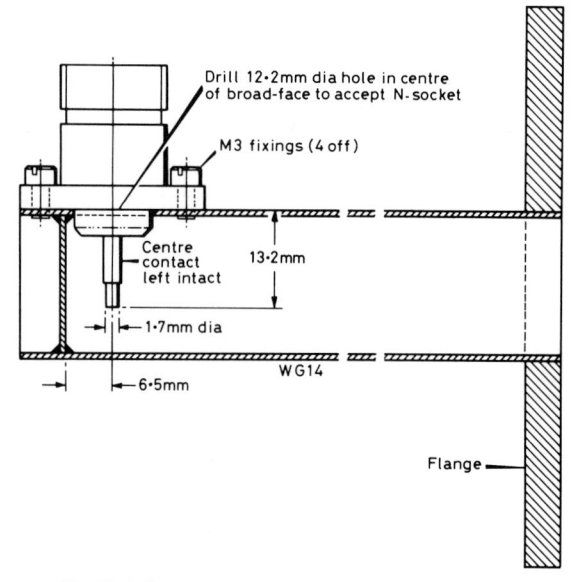

Fig 17.24. Waveguide 14 to N-type transition for 5.7GHz

[3] "Receive mixer for the 6cm band", R Heidermann, DC3QS, *VHF Communications*, 1/80, pp46–50.

[4] "A receive converter for the 6cm band", T Morzinck, DD0QT, *VHF Communications*, 2/82, pp89–93.

[5] "6cm GaAs Fet Vorverstärker", D van Delft, PA2DOL, *Der SHF Amateur*, Heft 1, 1983.

[6] "A 6cm preamplifier equipped with the MGF1400 and a push-pull mixer for transmit and receive", Hans Wessels, PA2HWG, *VHF Communications*, 4/83, pp210–217.

[7] *RSGB Microwave Newsletter*, 03/82, pp4, 5.

[8] "5,760MHz Linearverstärker mit der Schiebentriode YD1060", H Senckel, DF5QZ, *Der SHF Amateur*, Heft 5, 1984, pp170–179.

CHAPTER 18

The 10GHz ("3cm") band

18.1 INTRODUCTION

18.1.1 History

The 10GHz band has been the subject of probably one of the longest development histories of any of the amateur microwave bands, extending to some 40 years or more in the UK and, therefore, must occupy a special place in the annals of UK amateur radio. The learning and experimental phase continues to this day, for the 10GHz band is still, in many respects, an ideal beginner's band.

References to the theory and use of waveguides and magnetron oscillators to generate centimetre waves, then referred to as "micro-rays", started to appear in the early 1930s and klystron oscillators were described in 1939, just before the second world war. Most of the microwave tubes developed in the '30s and '40s are still, with some refinements, in use today.

The *RSGB Bulletin* carried a series of articles during 1943 entitled "Communication on centimetre waves", although the earliest recorded amateur 10GHz experiments, over a distance of two miles (a little over three km), took place in 1946 between W2RJM and W2JN. In October 1947, the RSGB published a 54 page booklet entitled *Microwave Technique* which sold for the princely sum of two shillings and threepence (including postage) or, in todays' terms, eleven pence!

UK 10GHz experimentation appears to have begun in earnest some two years later, in 1949, with experiments between G3BAK and G3LZ who used the 430MHz band to provide talkback, no mean feat in its own right at that time. The culmination of these experiments was the first recorded UK amateur 10GHz contact in January, 1950.

The band available in the UK on a shared, secondary basis is 500MHz wide, extending from 10.0GHz to 10.5GHz. The first 450MHz is allocated to the amateur service and the top 50MHz to the amateur satellite/space service. Common band usage is illustrated in Fig 18.1 and this "bandplan" has come about, over the years, by evolutionary growth rather than active band planning as such.

Amateur exploitation started with the availability of low powered ex-radar klystrons such as the 723A/B and its derivatives. Such devices were originally designed to work in the region of 8.0 to 9.5GHz (i.e. the lower part of "X-band") and with ingenuity and mechanical modification some klystrons could be persuaded to yield a few milliwatts up to about 10.10GHz, but more commonly no higher in the band than 10.05GHz. Valuable archival material can be found in [1], a then-definitive three-part article by G3BAK (now VK5ZO and still active, in 1991, on 10GHz). It was natural that the earliest established practice was to operate in this part of the band using modified klystrons as both receiver local oscillator and transmitter. The stability left much to be desired and the design of power supplies, particularly for portable use, to provide the several operating voltages needed had quite a bearing on the use of klystrons; problems arose from both thermal drift of the inter-electrode spacings of the device and the frequency pushing effects of short-term voltage drift on those electrodes. Supplies of between 450 and 1,000V were needed to operate them and this was potentially dangerous in portable operation. Nevertheless, working systems were established and used by a handful of pioneering amateurs. Those who wish to continue to use klystrons, for

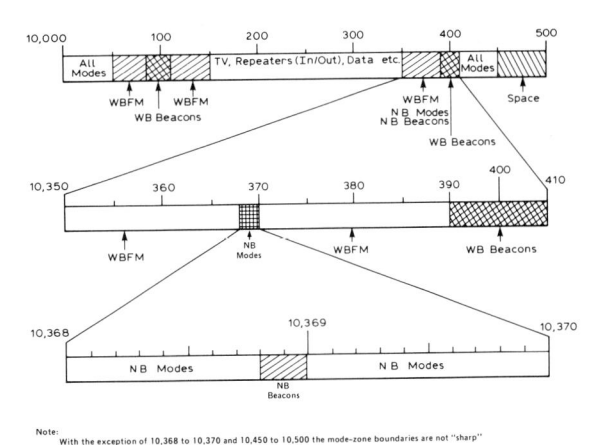

Fig 18.1. Band plan and "common" usage. For the sake of clarity, some sub-band allocation details have been omitted, e.g. packet (digital), repeater links and control (10,006 to 10,026MHz and 10,150 to 10,170MHz) and atv repeater channels (RMT101, 10,200/10,040MHz; RMT102, 10,225/10,065MHz; RMT103, 10,250/10,150MHz). In-band speech repeaters are recommended to use inputs around 10,100 or 10,440MHz and outputs in the slot 10,270 to 10,300MHz

18.1

historical or sentimental reasons, should read [1], [2], [3] and [4]. Wideband beacons were established at around 10.1GHz once activity demanded them, although this was much later than the early work mentioned.

In the mid to late 1970s, Gunn-effect devices started to appear on the "surplus" market at attractive prices in the form of Gunn diodes, complete oscillators capable of acting as self-oscillating mixers, dual-cavity oscillator/mixers and finally in-line oscillator/mixers, all designed for doppler intruder alarms or speed measuring modules. Most of these modules operated at around 10.67GHz and it was soon found that optimum efficiency could be obtained by retuning the modules into the area around 10.35GHz to 10.40GHz. This part of the band was thus adopted as preferred for wideband operation, being adjacent to both the narrowband segment and to beacons at 10.40GHz.

Gunn oscillators offered much improved stability compared with klystrons plus the added benefit of very simple, low voltage power supplies. The ready and inexpensive availability of such modules has made the 10GHz band one of the easiest and most accessible of the amateur microwave bands for the beginner in the field of low power (QRP) microwave communication.

Concurrent with the availability of Gunn devices, a few dedicated microwave operators started to develop narrowband equipment and most of the work in the UK centred around the design work of Mike Walters, G3JVL, who pioneered construction and use of waveguide 16 iris-coupled multicavity filters for transmitting, receiving and transverting. Finally, in the early 1980s and up to the present, amateur radio has seen the arrival of relatively inexpensive GaAs fets usable at 10GHz (and beyond) and GaAs fet dielectric resonator stabilised oscillators (dros). These devices, designed for use in satellite tvro (tv receive only) systems at 11 to 12GHz, can be modified to operate at the high end of the band. They are proving to be intermediate between the accepted current amateur standards of wideband and narrowband operation. That is to say, power levels comparable with low powered Gunn oscillators can be achieved with similarly simple power supplies, but with the advantage that the GaAs fet dro is more efficient (more output for less power dissipation), considerably more stable (allowing the use of narrower receiver bandwidths) and has a cleaner spectrum than the Gunn which, in its turn, was better than the klystron.

Further development of 10GHz narrowband techniques will inevitably centre around the use of pcb technology and active mixers, "printed", dielectric resonator or miniature on-board cavity filters, and amplifiers as device prices continue to fall and their application is mastered by amateurs. These features, coupled with the techniques becoming accessible to amateurs, point towards the 10GHz band growing in importance as an amateur allocation, particularly from the point of view of the beginner in microwaves. It should be pointed out, if it is not immediately obvious to the reader, that the acquisition of skills at 10GHz will also give a firm foundation to those amateurs who may subsequently become interested in satellite tv reception; the rf techniques are very similar and much amateur-built equipment for the 10GHz band can be "stretched" to cover the 11 to 12GHz band.

18.1.2 Nature of the band

All the tropospheric modes of propagation described in chapters 2, 3 and the individual band chapters can be expected to manifest themselves at 10GHz. As with the other microwave bands, there is no possibility of ionospheric modes operating.

Atmospheric phenomena, such as heavy rain, can be expected to play a significant part in propagation effects, although less so than in the higher bands where water, water vapour or oxygen absorption are of high significance. Contacts via reflection and scattering from heavy rainstorms are well known to amateurs, as are the effects of frontal systems, inversions and ducting. Under normal weather conditions, forward troposcatter is always present, as described in [5].

The wavelength is such that significant diffraction effects can be observed over and around sharply defined objects and usable reflections can occur from buildings, gasholders, cooling towers, islands and other similar objects. Significant refraction effects can be observed under ducting conditions and particularly over water by the formation of a "super-refractive" duct of limited depth which readily propagates these frequencies much in the manner of a waveguide. Much amateur communication takes place over "line of sight" paths but even here atmospheric effects may be significant.

Path losses for various propagation modes are given in Fig 18.2, together with an indication of the capability of different types of equipment in terms of dB of path loss capability (plc – see chapter 2). If sufficient power can be generated by amateur equipment then, when this level of power is coupled with very high antenna gains, even the eme path is surmountable within the limits allowed by the amateur licence, as was shown during 1989 by WA5VJB and WA7CJO who made an 888.5 mile (1,421km) contact from Texas to Arizona, via the moon.

18.1.3 Modes

All the modes used on the lower bands are now in use at 10GHz. Both wideband modes (wbfm, fast-scan atv and data) and narrowband modes (nbfm, fsk, afsk, ssb and cw) are in use.

Wideband (fm) modes, using bandwidths of between about 50kHz and many MHz, are most easily and directly generated in simple equipment using the voltage-frequency characteristics of Gunn devices. Similar characteristics exist with klystrons, but these have now been totally superseded in amateur usage. The use of narrowband modes implies a much more stable source so that the signal generated will remain in the passband of the receiver tuned to it. Thus, the source is almost invariably crystal-controlled and derived from a lower frequency by multiplication, phase locking or by a combination of multiplication, mixing and filtering. The "architecture" of a narrowband system is

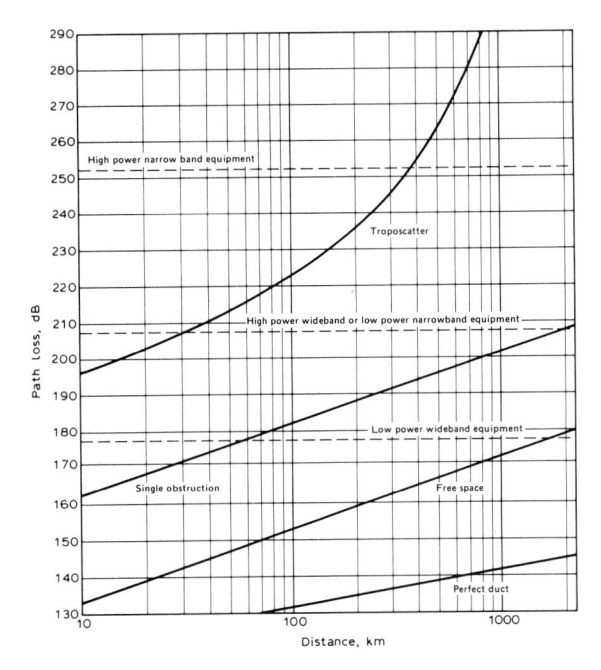

Fig 18.2. Path loss and equipment capability

Table 18.1. Receiver sensitivity for different modes

Mode	Detector Threshold(1)	Bandwidth	System Gain	Typical Sensitivity	Notes
wbfm	+10dB	250kHz	–	–110dBm	(2)(3)
wbfm	+10dB	25kHz	10dB	–120dBm	(2)(3)(4)
nbfm	+10dB	5kHz	7dB	–127dBm	
ssb	0dB	2.5kHz	16dB	–143dBm	(5)
cw	0dB	0.5kHz	7dB	–150dBm	
cw	0dB	0.1kHz	7dB	–157dBm	

Notes:
1. Detector thresholds assumed: ssb/cw 0dB; fm +10dB
2. Basic optimised in-line Gunn system with no antenna gain and negligible feeder loss, as might be measured using a professional signal generator/attenuator straight into the microwave head.
3. Assumes no image rejection for wbfm equipment.
4. To utilise the gain from a narrower receiver bandwidth, it may be necessary to employ effective afc.
5. Assumes 3dB additional gain from image recovery in the change from wbfm to nbfm. Image recovery is assumed for ssb and cw modes.

Narrowband equipment set up to receive ssb or cw will normally reject the unwanted image and employ a product detector, giving the receiving system an immediate additional gain of about 13dB. The more stable signal will also allow the use of narrower bandwidth receiver filters, thus enabling further system sensitivity to be realised. The cumulative effects of all these changes are summarised in Table 18.1. The figures given are not absolute but serve as a fair representation of the *order* of system improvement which is likely to be achieved in moving from a simple wideband system to a narrowband system. Once this level of sophistication has been reached, it is well worth the operators' efforts to squeeze the last dB out of the system. This might be achieved by optimising antenna feeds, tuning out all mis-matches, optimising the mixer and so forth.

18.1.4 Power

Most amateur work uses powers of between a few hundred microwatts (direct passive narrowband mixing techniques) and 300mW using Gunn devices. Many UK operators currently use powers of 1 to 2mW derived from the simple and popular "in-line" mixer/oscillator transceiver described later. Others, using separate transmit and receive oscillators, are using between 10 and 300mW, a worthwhile increase in power. A few fortunate amateurs with the necessary skills and access to surplus travelling wave tube amplifiers (twtas) have succeeded in generating and using powers of up to about 10W. At this power level, using high antenna gain, narrowband signals and a receiver employing the narrowest practicable bandwidth, it is possible to explore and exploit the characteristics of the ever-present forward troposcatter propagation mode [4].

System gain realised by increasing transmitter power is directly related to the increase in power expressed in dB. Going up in power from 1mW to 1W is an increase of a factor of 1,000 or 30dB and the increase from 1W to 10W is a further 10dB. In total this is an effective gain in potency of 10,000-fold.

much more complex than that of a wideband system. It is now possible to use dros to provide a signal stable enough to class as "pseudo-narrowband" whilst retaining most of the simplicity of the Gunn transceiver. With careful design and application of the dro it should be possible to generate nbfm of stability adequate to support the use of receiver bandwidths of 25kHz or less.

Equipment designed to take advantage of the simplicity of wbfm systems seldom, if ever, has any protection against image reception. To give an example, if the receiver local oscillator is set to 10.380GHz and the receiver intermediate frequency is 30MHz, then the receiver will "hear" two signals at F(osc) + F(i.f) and F(osc) − F(i.f); in this case at 10.410GHz (upper image or "sum" product) and 10.350GHz (lower image or "difference" product). If the wanted signal is, in fact, on 10.350GHz then what the receiver "hears" is the sum of the wanted signal plus the noise in the image channel. In theory, removing this image noise will result in a system sensitivity gain of 3dB.

A further penalty to be paid in using fm is that the detector threshold is 10dB above that of a product detector such as that used to detect ssb or cw. This means that the audio signal produced tends to be present at many dB above this threshold, or completely inaudible in the noise. This is in contrast to cw where the operator with a "trained ear" can hear signals quite distinctly even when almost down in the noise. Some of this disadvantage can be recovered by employing higher oscillator stability, narrower deviation and therefore narrower receiver bandwidths. A full discussion of the "fm effect" is given in section 18.6.2.

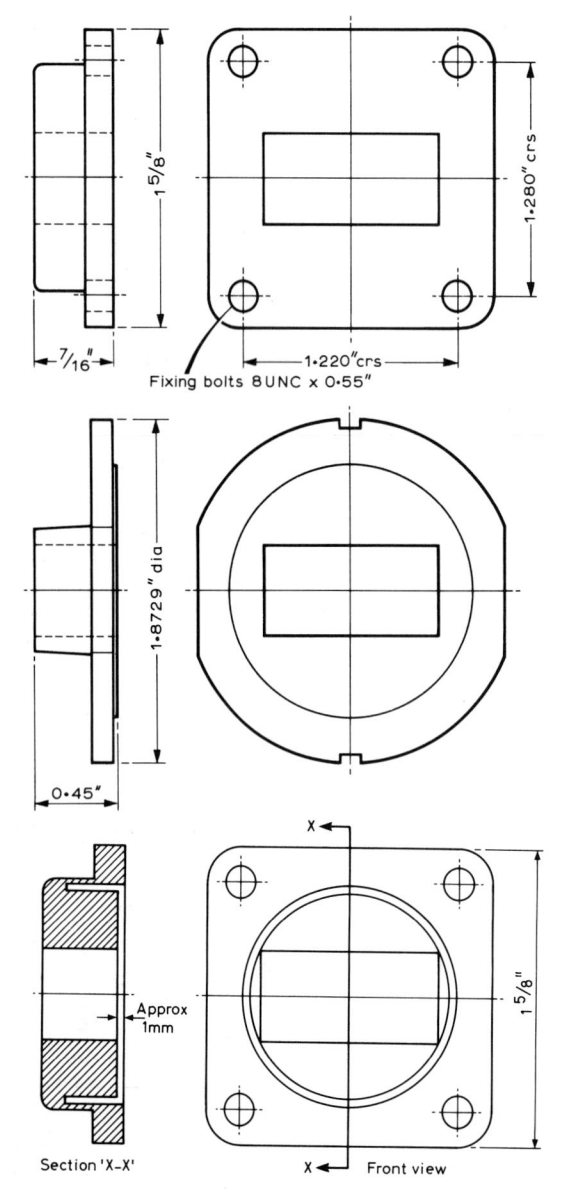

Fig 18.3. Standard WG16 flanges

18.2 WAVEGUIDE COMPONENTS

The theory of many of both the passive waveguide components and other, active, components commonly used at 10GHz has already been described in chapters 5 and 6. The construction of many components for 10GHz which are of particular relevance to test equipment were described in chapter 10. Other generally useful components for 10GHz are given in this chapter, often in some detail (where this is appropriate), and sometimes as "ideas" where the design is

either difficult to make without an elaborate workshop, is not well proven, or where it is expected that use will be made of surplus professional components when these are available. Waveguide loads, calibrated variable attenuators, directional couplers, detectors and wavemeters when assembled together as a "test bench" and used properly, enable many measurements to be made at home with sufficient accuracy to be meaningful without continual recourse to the use of expensive professional test-gear. It will also become apparent that some components which cannot be readily made or obtained from surplus sources (such as circulators and isolators) are also valuable adjuncts to testing as well as operating.

18.2.1 Waveguides and flanges

10GHz equipment with components built in waveguide 16 (WG16) is most familiar. There is nothing "magic" about this particular size of guide and it is possible to use other similar sizes, other materials (such as copper water pipe) and quite wide tolerances, because the dimensions are often not very critical and the precision matching of professional components is not essential for amateur purposes. Good results can be obtained from components constructed by hand, almost literally, "on the kitchen table"!

The most suitable rectangular guide for use at 10GHz is WG16 which has external dimensions of 1in (25.4mm) by 0.5in (12.7mm). The internal dimensions are 0.9in (22.9mm) by 0.4in (10.2mm). It is available most commonly in copper, brass and aluminium. It can be obtained from a number of sources either new or secondhand on the surplus market. Copper guide should always be used where high Q is required, for instance in the construction of filters, or where high thermal conductivity is required to assist in the heat-sinking of higher-powered Gunn diodes than those commonly used in simple amateur equipment. For most other purposes brass, which is more rigid, easier to "work" and less expensive, is quite adequate. Aluminium or aluminium alloy guide should be avoided as it is very difficult for the amateur to solder parts together, whereas this is very easy with either copper or brass, provided that the metal is clean.

Two shapes of standard flange are available, with the dimensions of each given in Fig 18.3. Rectangular flanges are less expensive than the round type, and require only four bolts to fasten them together. The constructor can make substitute flanges of this type from thick brass plate suitably cut, drilled and filed to fit the waveguide. The "boss" on the rear of commercially-produced flanges assists accurate fitting but is not essential to their function. The round types require a pair of special screw-coupling/locating rings which make them "quick-release" and are particularly useful where the operator may need to repeatedly attach or detach different pieces of equipment – to or from a common antenna feed, for example.

Both types of flange are available in plain or choke format. Plain flanges normally have a shallow groove machined into them which takes an O-ring gasket to provide a weather seal. The plain type is designed to fit flush with the

THE 10GHZ ("3CM") BAND

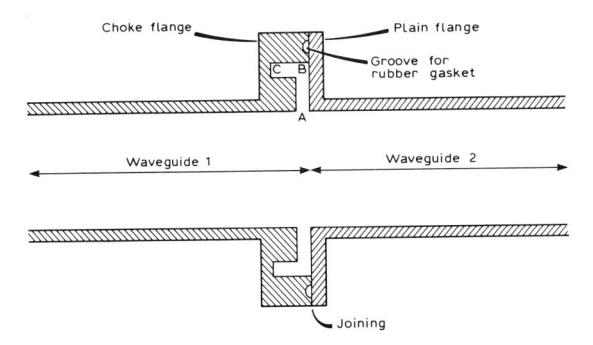

Fig 18.4. Correct use of a choke/plain flange combination – see text

end of the waveguide and, when using this type, the constructor should try to make the ends of the waveguide flat and square with the waveguide neither protruding from, nor recessed in, the flange. Simple methods for doing this were described in the constructional chapter. The consequences of improper fitting might be either some rf leakage or mis-match at the joins.

A choke flange has a recessed face, relative to the "mating" surfaces, and another groove (other than that for the O-ring seal) machined into it. This groove and the recessed face forms a quarterwave choke which reduces leakage from the joint and does not rely on metal to metal contact for its operation. It is designed to be matched with a plain flange to provide a "leak-proof" rf connection which should be better matched than possibly mis-aligned plain flange joints. Fig 18.4 shows such a pair of flanges diagramatically in cross section; the distance CBA is $\lambda/2$, the short circuit at C being reflected as a short circuit at A, effectively joining the ends of the waveguide electrically, even if there is no close physical connection between the two. A sound mechanical connection between the flanges is thus not essential. It should be obvious that a choke flange should *always* be matched to a plain flange to be effective in this way.

Brass WG16 will exhibit a loss of about 5.5dB per 100ft (approx 31 metres) whilst copper guide should be about 40% of this, i.e. about 2.2dB for a similar run.

Fig 18.5 gives the relationship between the waveguide wavelength (λ_g) and the frequency in GHz for WG16 to enable calculation of critical dimensions. With the ever increasing cost of new rectangular waveguide it seemed useful to investigate a cheaper alternative – copper water pipe – for long runs such as might be used in a beacon or fixed home station installation.

For minimal losses the wave needs "launching" from conventional rectangular guide into the pipe (or vice versa), since the propagation modes are different in the two forms. The fabrication of waveguide components such as oscillators and mixers in circular guide is much more difficult than with rectangular guide and this justifies the time spent in making transitions. Typical waveguide modes in rectangular and circular guides are given in chapter 20, entitled "Bands above 24GHz".

Table 18.2. Standard copper water pipes used as waveguide

OD (mm)	Wall Thickness (mm)	Cutoff Frequency (MHz)	Next mode Cut-off (MHz)
15	0.5	12557	16404
22	0.6	8452	11041
28	0.6	6560	8569
35	0.7	5232	6835
42	0.8	4351	5684
54	0.9	3368	4400

Water pipe is available in a range of standard sizes as shown in Table 18.2, the first few smaller sizes being most readily available at d.i.y. supermarkets, the larger sizes from builders' merchants.

It can be seen that 22mm pipe is usable at 10GHz, 35mm pipe at 5.7GHz, and possibly 54mm pipe at 3.4GHz, though the latter would be operating rather close to cutoff. Since this chapter covers 10GHz, attention was concentrated on assessing the suitability of 22mm pipe, which costs about 50p per foot compared with at least ten times this for new WG16. Before any attenuation measurements could be carried out, some transitions from normal

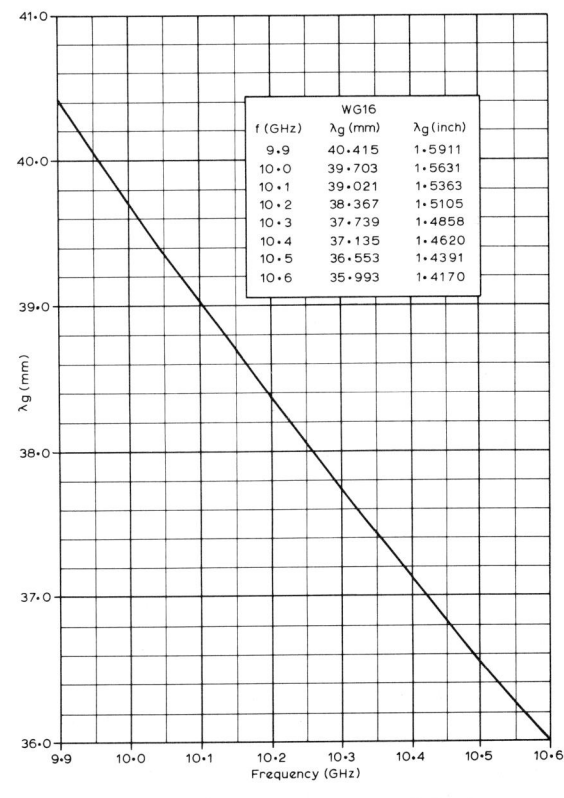

f (GHz)	λ_g (mm)	λ_g (inch)
9.9	40.415	1.5911
10.0	39.703	1.5631
10.1	39.021	1.5363
10.2	38.367	1.5105
10.3	37.739	1.4858
10.4	37.135	1.4620
10.5	36.553	1.4391
10.6	35.993	1.4170

Fig 18.5. Relationship of waveguide wavelength (λ_g) to frequency

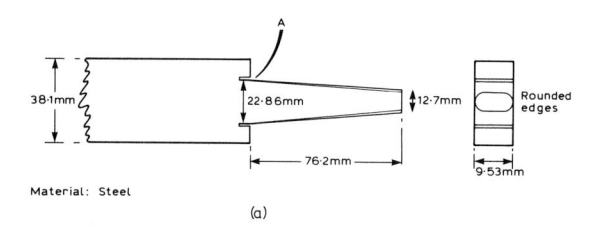

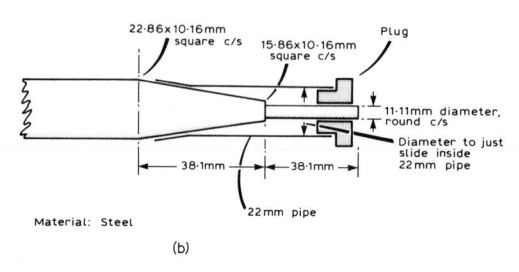

Fig 18.6. Two types of tool for making rectangular to circular waveguide transitions (G8AGN). Tool (b) has a tight fitting plug which prevents the round end of the transition distorting whilst forming the transition

rectangular waveguide had to be built. This was achieved by using short pieces of pipe, approximately 3in (75mm) long and deforming one end of each pipe into an approximately rectangular shape. Initial attempts to do this simply by squeezing the pipe in a vice were not very successful; better results were obtained by gently hammering a tapered plug, as shown in Fig 18.6a, into the pipe and then gently squeezing the rectangular/elliptical cross section of the pipe right up to the plug at point A. The resulting cross-section is then roughly the same as standard WG16. Although the plug tool shown in Fig 18.6a is adequate if used with care, a second design was conceived which has specific provision to ensure that during the deformation of one end of the pipe from a circular to a rectangular cross-section, the other end of the pipe is kept circular by means of a separate plug which is a tight fit into the pipe. The general arrangement is a shown in Fig 18.6b. This second plug tool has the minor disadvantage of being able only to be used on short lengths of pipe; it does, however, enable a better (i.e. more smoothly varying) cross-section transition to be made.

Normal square waveguide flanges were adapted to fit onto both circular and rectangular ends of the transitions. Standard self-soldering "Yorkshire" fittings can be used to join sections of permanently fixed pipe together, for instance in a home station feeder run. Such a run should, preferably, be kept as straight as possible. Bends are permissible provided that they can be made without the walls of the tube collapsing or becoming corrugated and that they take place over a number of wavelengths, although there is still a risk of mode change. It is stressed that transitions should be made in copper, since this is malleable and ductile, allowing the metal to be "stretched" without fracture, provided the operation is carefully carried out.

The measured insertion loss of a pair of such transitions connected back-to-back was 0.8dB at 10.368GHz.

Sections of flexible or flexible and twistable waveguide are occasionally available from surplus sources and could be used to correct guide misalignments in an otherwise rigid system. However, such waveguide can introduce significantly higher losses into the system together with mismatches which may need to be subsequently tuned out. Such waveguide was illustrated in chapter 6, Fig 6.41 and 6.42.

The reader may also occasionally come across elliptical corrugated semi-flexible waveguide such as that illustrated in chapter 6, Fig 6.43. Due to its cost and difficulty of supply, it is seldom used in amateur installations. Suitable flanges are available but are rarely obtainable at amateur prices from surplus sources. They can, with a little patience, be fabricated from thick brass plate or even from standard square, plain flanges. Terminating such waveguide with these flanges is difficult, but not impossible, for the amateur to do. Again, suitable rectangular to elliptical transitions are not too difficult to make using similar techniques to those described above for rectangular to circular transitions. The starting material should be annealed copper WG16. Such transitions are shown in Fig 18.7. Measured losses are similar to the rectangular to circular transitions.

Another type of waveguide, these days rarely seen, is the so-called "Old English" (OE) guide. This has *internal* dimensions which are the same as the *external* dimensions of WG16, viz. 1in by 0.5in. Thus the OE waveguide is a neat sliding fit over WG16. This can be useful in a number of ways: as a support to join a horn antenna to WG16, as a crude but effective "plug and socket" to join pieces of WG16 together or as a rudimentary form of waveguide switch. A method of joining OE and WG16 together, whilst avoiding significant mismatch, is given in Fig.18.8. Standard flanges to fit OE waveguide do exist but are likely to be even rarer than the waveguide itself.

18.2.2 Tuning and matching waveguide components

One of the attractive features of working in waveguide is the ease with which certain operations can be carried out compared with the lower (non-waveguide) frequencies. An example of this is the tuning and matching of waveguide components. Two methods are used, the (sliding)

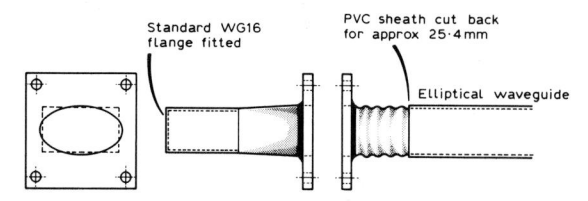

Fig 18.7. Elliptical to rectangular waveguide transitions. These can be made in a similar manner to the rectangular to round transitions, but starting with annealed copper WG16

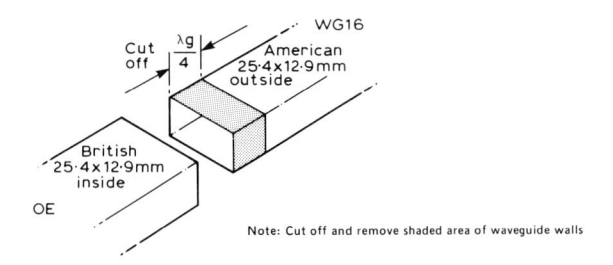

THE 10GHZ ("3CM") BAND

Fig 18.8. WG16 to "Old English" (OE) waveguide transitions

FREQUENCY	CUT OFF	CUT OFF	FREQUENCY	CUT OFF	CUT OFF
9,373MHz	0.423"	10.7mm	10,100MHz	0.374"	9.5mm
10,000MHz	0.378"	9.6mm	10,380MHz	0.364"	9.2mm

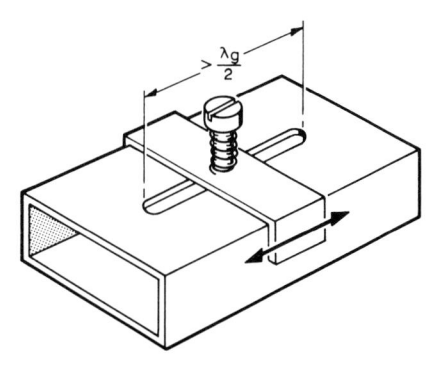

Fig 18.9. A single, critically placed screw may be used for "tuning-out" mismatches (see text). Alternatively, the sliding version shown here will give a wider range of matching

single-screw tuner or the electrically similar three-screw tuner. The three-screw tuner is simpler to construct and therefore is the one most often used in amateur-built equipment. A single screw, if fitted in the correct position along the centre line of the broad face of the waveguide, can be made to tune out almost any mismatch in the system simply by adjusting the depth of penetration of the screw. This adjustment will hold over a range of frequency which, although narrow in absolute terms (in the region of 1%), is usually more than adequate by amateur standards.

The matching screw is preferably fitted close to the component which requires matching. The further the screw is from the mismatch, the narrower the bandwidth of the matching, so that the screw should ideally be placed within a wavelength or so of the mismatched component. Its position is fairly critical within the range $\pm \lambda_g/4$ and a method of optimising this is to mount the screw on a carriage which slides along the waveguide as shown in Fig 18.9. Both the carriage and the screw should be spring loaded to eliminate mechanical play. This design is not easy to make and use, so the constructor will normally use the three-screw tuner. One possibility is the use of a short length of OE waveguide as the carriage. The $\lambda_g/2$ slot is best milled into the waveguide face, but it is possible to create a sufficiently accurate slot by drilling a line of small holes along the centre line and opening this into a slot by means of a hacksaw blade followed by a fine file and considerable patience.

The three-screw tuner is illustrated in Fig 18.10 and consists, simply, of three screws mounted centrally along the broad face of the waveguide. During experimental work the screws can be spring loaded to eliminate play, but after adjustment they should be secured in place by locknuts. As with the sliding screw tuner, the screws may be mounted anywhere along the waveguide but preferably near to the component to be matched. Additional support for the screws can be provided by soldering a piece of thick brass plate to the face of the waveguide before drilling and tapping the screw holes.

Setting up either the single screw or the three-screw tuner can be done using a vswr indicator, if one is available. Alternatively, they can be set during final testing of the equipment. For example, if fitted to a mismatched antenna, they are tuned to maximise signals received or transmitted and, if fitted to a mixer, to optimise the signal/noise ratio on a received signal.

It will usually be found that only one, or perhaps two, of the three screws will have a significant effect. The screws should be inserted the minimum amount necessary to achieve a good match. If the screws have to be inserted well into the waveguide then the component is probably a bad match; if none of the screws have any beneficial effect, then the component is probably already well matched.

Waveguide slotted line for vswr measurement

A rudimentary but satisfactory vswr indicator can be made by modifying a sliding-screw tuner design to take a diode mount, the diode being coupled into the waveguide field by means of a small probe. The probe should be made from thin, stiff wire so that it causes minimum field disturbance whilst making measurements and it will be necessary to arrange for the probe penetration to be adjustable for the same reason. Minimum penetration consistent with an

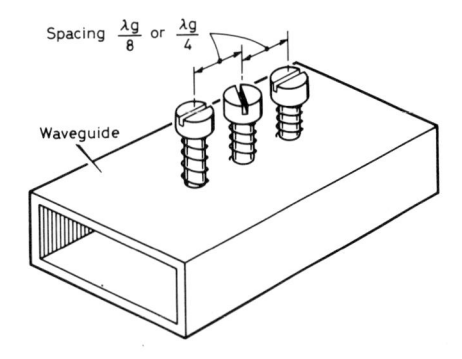

Fig 18.10. Three-screw tuner. This type of tuner can be used instead of the adjustable single-screw tuner in Fig 18.9

Fig 18.11. Photograph of a home made slotted line vswr detector (G3PFR). Photo: G6WWM

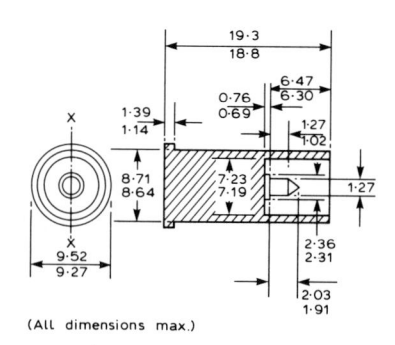

adequate indication should always be used. The coupling slot should be made at least one guide-wavelength long in order to ensure that several maxima and minima of the standing wave can be detected. An example of a simple amateur-built vswr slotted waveguide and probe is given in Fig 18.11. In this instance the sliding carriage is made from a split, milled block of brass, the two parts of which are held together by four countersunk screws. These can be tightened to make the carriage a smooth sliding fit on the waveguide. Probe penetration can be adjusted by sliding the diode mount up or down in the block and clamping in place by means of a knurled screw. To minimise the effects of variable metal-to-metal contact whilst sliding the block along the waveguide it was found necessary to glue a thin ptfe sheet to the carriage to isolate it from the slotted top surface of the waveguide. This completely eliminated the effects of contact problems, which manifested themselves as erratic readings on the indicating meter, and allowed "smooth" detection of maxima and minima.

For quick indications of vswr a directional coupler and diode detector would be more useful, as it would give continuous readout.

18.2.3 Waveguide diode mounts – mixers/detectors

Mixers and detectors can be constructed in many ways, some complex, some simple. Three main diode packages commonly used at these frequencies and the construction of the mixer or detector may be determined by the outline of the diode to be used.

The most familiar package is the DO22/SOD47, illustrated in Fig 18.12, which is used for the 1N23 type point-contact diode and for some of the recent Schottky barrier diodes such as the BAV92 series. Older versions had a fixed anode collet whilst more modern versions, such as the 1N415 series, are supplied with a removable collet which will fit either the anode or cathode spigot of the device. This package is designed to fit within the narrow dimension of WG16. The "cartridge" package, SO26, shown in Fig 18.13, is typical of the CV2154/55 and SIM2/3 series and is designed to be mounted external to the waveguide and coupled into it by means of a probe or "cross-bar" coupling.

The third package is the SOD31 outline illustrated in Fig 18.14 and is familiar as the package used for low to medium-power Gunn diodes. Like the Gunn diode, it is designed for post mounting within the waveguide and the methods of mounting Gunns described later should be used if the constructor chooses this type of diode for a mixer or detector.

A simple detector for 10GHz

The design of a simple detector is shown in Fig 18.15 and consists of a 1N23 type diode mounted centrally across the waveguide. The "live" end of the diode is decoupled by the

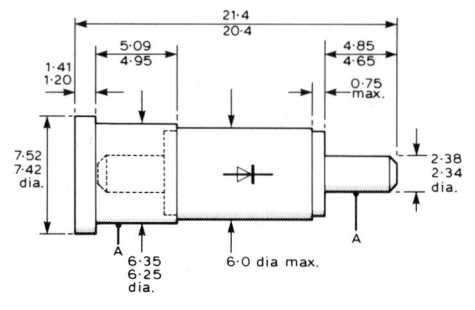

Fig 18.13. SO26 diode package outline

Fig 18.12. DO22/SOD47 diode package outline

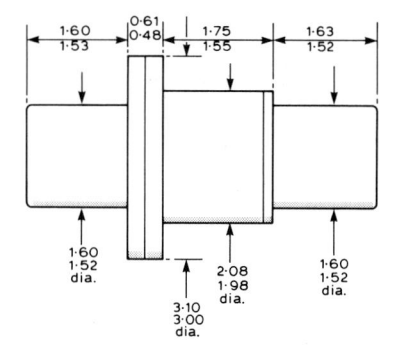

Fig 18.14. SOD31 diode package outline

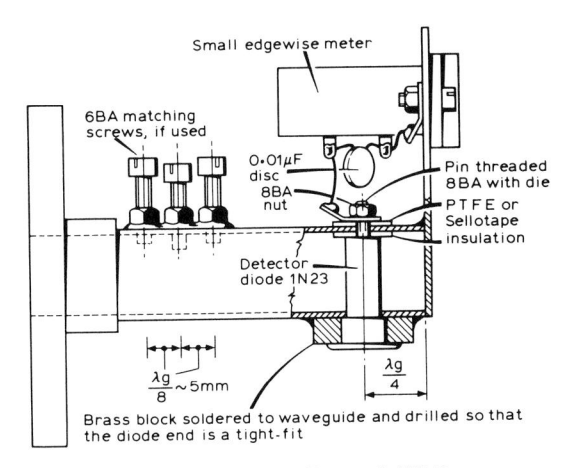

Fig 18.15. Simple mixer/detector in WG16

drill bit (7/64in or 2.8mm, whichever has been used) passed through it and the waveguide. The end-plate can be held in position by a small G-clamp, whilst the flange should be self-jigging. Solder all the joints using a small, hot gas flame and a minimum of solder. Allow the assembly to cool thoroughly before disturbing it. Trim off any excess from the end plate, open the hole in the choke block and upper waveguide wall to 3/8in (9.5mm) diameter and then remove all traces of flux and any excess solder, especially from inside surfaces.

The rf choke is obviously best turned in a lathe but can also be fabricated in one piece using hand tools or a stand drill as indicated in the constructional techniques chapter. A suitable choke can sometimes be salvaged from ex-doppler units of a type not really suited to amateur use (for example the side-by-side dual cavity oscillators and mixers), in which case the choke block should be drilled out to suit the diameter of the salvaged choke. Such chokes are usually already insulated, either with a coating of some durable epoxide or by anodising, and are typically 1/4in (6.3mm) in diameter. When installing such chokes, care should be taken that the insulation is not punctured by roughness inside the choke block or by too tight a fit into the block. If damaged, Sellotape, ptfe tape or Mylar tape can be used to re-insulate the choke before use.

small capacitor formed between the end of the diode package and the inside wall of the waveguide and between the nut/solder-tag and the outside surface of the guide wall.

The diode is matched into the waveguide by adjusting the three matching screws for maximum current output with rf input at the chosen operating frequency.

Offset mixer using choke decoupling and a 1N415 type diode

The construction of this mixer is shown in Fig 18.16. Note that the diode is offset from the centre-line of the broad face of the waveguide in order to improve matching. The recommended techniques of construction are described in more detail in the constructional techniques chapter and for the sake of clarity only the general methods are outlined here.

Drill a 7/64in (2.8mm) hole in the choke block, in the correct position, and also through the two broad faces of the waveguide in the position shown. Drill and tap 6BA (or 2.5mm) holes for the matching screws. Deburr and remove swarf. Ensure all metal parts are bright, clean and grease-free. Jig the bearing nuts in place with non-solderable screws and locate the choke block by using the shank of the

A mixer using a "cartridge" diode

A mixer using the CV2154/55 or SIM2/3 type diode is shown in Fig 18.17. It is tuned by an adjustable rf short and by raising the diode within its mount. Construction can be simplified by replacing the sliding short by a fixed short as in the preceding designs and the diode centre-line to shorting plate distance can be made any *odd* number of $\lambda_g/4$ if this is convenient to the form of construction adopted. A three-screw tuner can then replace the sliding short.

Alternative forms of construction

Construction of an "in-line" oscillator and mixer is described in a later section. This suggests alternative methods of

Fig 18.16. Improved mixer using offset diode mounting and choke decoupling. Note the three-screw tuner for matching, immediately in front of the diode mount. A simple type of 75Ω coaxial connector is used at the i.f output and is quite adequate up to uhf

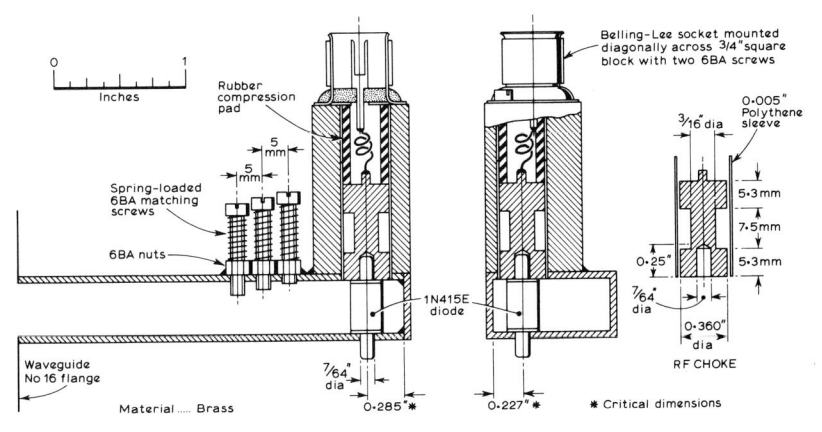

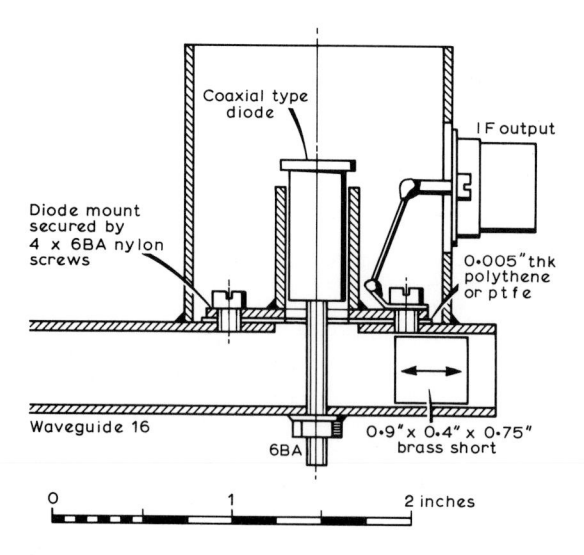

Fig 18.17. Mixer using a "cartridge" diode. The mixer may be "tuned" by sliding the diode up or down in its mount and by adjustment of the sliding short-circuit

construction for any mixer mount. One which is singularly easy to build utilises the 1/4in (6.3mm) brass bush salvaged from an old control knob. The diode can be held in place by means of a short length of 1/4in (6.3mm) brass rod with a 3/32in (2.4mm) hole drilled concentrically through its axis. This substitutes for the removable collet supplied with the SOD47 outline diode.

If the constructor has access only to fixed-collet types, the rim of the collet can be carefully filed away to allow mounting as shown in the relevant figure (Fig 18.42). Yet another simple method of capacitive decoupling is shown in Fig 18.43, construction of which should be obvious from the figure. Construction of mixers using the SOD31 outline can use the pillar mounting and decoupling arrangements shown in Fig 18.42, which shows the construction of a Gunn oscillator; the diode centre-line to back-plate distance can be as shown for the simple mixer/detector described above.

Approximate power measurement using a mixer/detector

This topic was discussed in chapter 10, "Test equipment", under the heading "Waveguide power meters" and thus will not be discussed further here, except to say that used with care and due regard to the limitations, the technique can provide a very useful indication to the constructor that power levels and the performance of other waveguide components in a system is close to that predicted.

Modifying surplus mixers

Most surplus mixers, provided they are built in WG16, are suitable for use in the amateur 10GHz band. The only modification which is normally needed is the addition of a three-screw tuner, a simple operation, followed by optimisation at the chosen frequency.

18.2.4 Matched loads and attenuators

Professional matched loads are usually made from a lossy ferrite material, the form of which is designed to absorb virtually all the incident power. This is usually achieved by step matching or by tapering the ferrite over several wavelengths in both the narrow and the broad dimensions of the waveguide until the load occupies the whole cross-sectional area of the guide. Occasionally external cooling fins may be attached to the waveguide in order to increase the power dissipation capability of the load. However, power dissipation is seldom a problem with most amateur 10GHz equipment which usually develops only tens or at most a few hundred mW. Low power matched loads can be made using everyday materials, such as wood, without resort to the more exotic materials used professionally. Professional loads are often obtainable from surplus sources. Fixed attenuators can be constructed in a similar manner, although even more useful variable attenuators are more difficult to construct. The attenuating material consists of a resistive card tapered at both ends in order to provide a good match in either direction. Further information on resistive card, loads and attenuators was given in chapter 10 and general details were shown in Fig 5.75, chapter 5.

18.2.5 Directional couplers

Directional couplers are widely used at lower frequencies and most amateurs will be familiar with them as so-called "swr meters" where they are used to measure the forward and reflected power sampled from the transmission line in an antenna system.

The theory of waveguide directional couplers has been covered in chapter 5, "Transmission lines and components" and so will not be discussed further here, except to say many designs exist, that they are relatively simple to make and, again, are very useful components. The construction of three of the commonest types of waveguide coupler – a 3dB sidewall coupler, a 9 to 30dB Moreno cross-coupler and a round hole cross-coupler – were described in chapter 10.

18.2.6 Adjustable waveguide short circuits

It is sometimes useful to terminate a waveguide component with an adjustable short circuit as distinct from a fixed $\lambda_g/4$ short circuit, e.g. when matching a detector diode. The short circuit is more versatile, if adjustable, when experimenting with waveguide component design and optimisation. One component, for example, which will benefit from the inclusion of an adjustable short circuit is the multi-element, stacked slot, omnidirectional antenna described in chapter 4, "Microwave antennas".

An adjustable short circuit can also be used as a wavemeter as described in chapter 10 where the design used a combination of a cross-coupler and adjustable short as a simple form of self-calibrating wavemeter. Although such a wavemeter does not offer *very* good frequency resolution, it represents one of the most fundamental measurements which can be made: the *direct* measurement of guide wavelength.

The simplest form of adjustable waveguide short consists of a block which is a firm press-fit in the waveguide and which can be clamped in place after adjustment. The rear face of the block can be tapped to take a screw which can be used to adjust the position of the block and subsequently removed. If the block is a good fit in the waveguide, then its length is uncritical. If not, then it should be about $\lambda_g/4$ in length so that it becomes, in effect, a choke. A slightly smaller block, $\lambda_g/4$ long, insulated with Sellotape (or self-adhesive ptfe or polyester tape) will provide more reproducible adjustment. One form of choke block uses fingering to ensure contact, as shown in Fig 18.18. The ends of the fingers should be radiussed and smoothed using fine emery paper so that adequate contact pressure can be used without scoring the walls of the waveguide during repeated adjustment. The length of each of the choke sections is made $\lambda_g/4$, i.e. approximately 10mm at 10GHz. This would *not* be suitable for tuning an oscillator as intermittent contact of the fingering would cause frequency jumps.

A highly recommended form of adjustable short is shown in Fig 18.19. The choke is "floated" on "pips" of an insulator which protrude about 0.01in (0.254mm). A method for making these inserts is to drill 1/8in (3.174mm) diameter holes to about the same depth in the appropriate positions on the choke and to press a suitable piece of insulator into each hole. The ptfe body of a standard feedthrough insulator is ideal. The ptfe is cut off flush with the surface of the choke and then removed with a needle. The bottom of the hole is then packed with material of thickness equal to half the difference between the size of the choke and the dimensions of the waveguide to be fitted and the insert replaced. A second method is to tap a hole in the choke of just sufficient length to take a screw and then to fit a nylon screw. Using a micrometer to monitor the process, the nylon screw is cut back so that the height of the pip is equal to half the difference between the width of the choke and the inside dimension of the waveguide.

The insert on the opposite face is fitted in the same way and trimmed so that the total width over the inserts is equal to the internal dimension of the waveguide. With a little care and patience it is possible to eliminate play almost completely and obtain very smooth, jump-free tuning.

A third alternative is to make a block 0.88in (22.35mm) wide by 0.38in (9.65mm) high by about 0.75in (19mm) long and to create a v-shaped groove about 0.1in (2.5mm) from one end of it. This groove, which should be about 2mm wide and the same depth at its centre, should be cut into all four faces of the block by marking and filing. The groove does not need to be particularly accurate in profile as irregularities are not too important when the groove is subsequently filled with the contacting medium. A short length of subminiature coaxial cable of 1.5mm o.d. has its outer sheath removed leaving the bare braid, dielectric and centre conductor, the overall diameter of which is about 1.3mm. This is formed and cut to fit the groove. Insertion of the block into the waveguide will compress the cable somewhat and the braid will act as a multi-contact "wiper" against all four walls of the guide, much like the fingering of the earlier design.

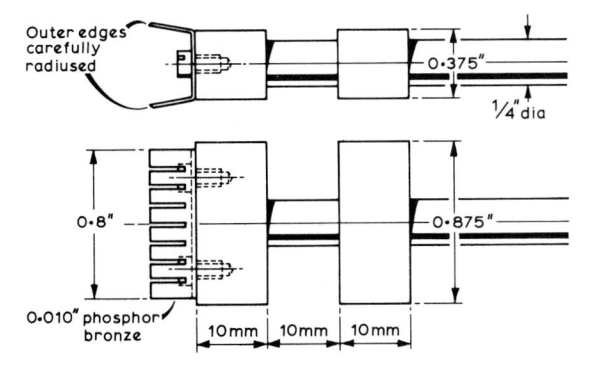

Fig 18.18. Adjustable waveguide short using contacting fingering, backed by two quarter-wave spaced, quarter-wave long choke blocks

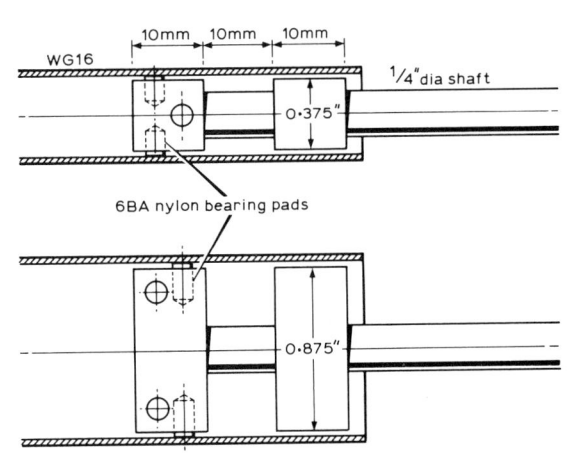

Fig 18.19. Alternative form of adjustable waveguide short using non-contacting choke blocks

18.2.7 Wavemeters

Most simple equipment in current use on 10GHz employs free-running oscillators based on Gunn devices and whilst their stability is remarkably good, it is poor by comparison with crystal controlled standards. It is very desirable that the constructor (and more particularly the operator) of such equipment has a ready means of frequency calibration available, either as part of a "test bench" or, better still, built into the system. Better frequency setting "in the field" will considerably improve the operator's chances of making successful contacts. Details of the construction of four different designs of absorption wavemeter were given in chapter 10, "Test equipment".

18.2.8 Coaxial to waveguide transitions

Coaxial "cables" are not much used at 10GHz, other than in very short lengths, because of the high losses involved compared to waveguide. "Heliax" can be useful, e.g. Andrews

Work carried out by Mike Walters, G3JVL, at 10GHz gave the optimum dimensions for a transition from waveguide to 50Ω coaxial line as being:

$$D = 0.027 \lambda_g$$
$$L = 0.160 \lambda_g$$
$$S = 0.120 \lambda_g \text{ (theoretically } \lambda_g/4)$$
or $$S = 0.620 \lambda_g \text{ (theoretically } 3\lambda_g/4)$$

These relationships can be scaled for other frequencies. For instance, the successful construction of such transitions for the 3.4GHz band in WG11 was reported by G4KNZ [6].

For 10.369GHz λ_g is 37.32mm and, from this, the following dimensions are derived:

$$D = 0.995\text{mm (1mm)}$$
$$L = 5.970\text{mm (6mm)}$$
$$S = 4.478\text{mm (5mm)}$$
or $$S = 23.14\text{mm (23mm)}$$

It is quite in order to use the "rounded" figures given in brackets; no really noticeable change in performance results from this. An exact match can be obtained by fitting a single screw tuner between the probe and the waveguide short to alter the centre frequency and, if necessary, a three-screw tuner between the probe and the flange.

A coaxial (N type) to WG16 transition

N-type connectors are specified for use only up to about 10GHz and can be used successfully although there is a slight risk of over-moding in the connectors or the cable (typically a short length of FHJ4–50) used with them. The use of conventional cables is not advised because the losses involved are high and the flexible nature of the cable can lead to unpredictable impedance changes. However, some of the older surplus X-band equipment is fitted with these connectors.

A square flange N-type socket should be selected. The dimensions of the flange are 18 by 18mm with 13mm fixing centres for 3mm screws. The solder spill is usually 3mm in diameter and hollow. The dielectric insulation carrying the centre pin is normally held in place by a spun collar or rim of metal which protrudes below the flange and goes through a clearance hole in the panel on which the socket is mounted. This collar is normally deeper than the thickness of the waveguide wall (0.05in, 1.27mm) and would protrude into the waveguide if mounted with the flange flush on the waveguide surface. The diameter of the solder spill is also larger than the optimum diameter discussed above. Thus, the socket needs modification before it can be used easily. Occasionally, surplus sockets can be found which do not have this rim; they may also have an elongated, solid metal "spill". This type of socket is very easy to modify, as all that is required is that the spill is turned down to the required diameter and then cut to length.

First the "rim" around the insulation on the back of the socket is removed, by careful filing or sawing, to enable the socket to fit flush with the waveguide face. This may

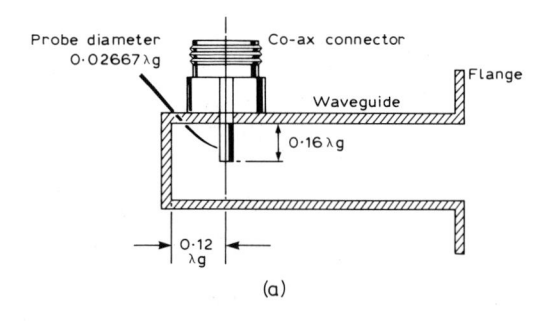

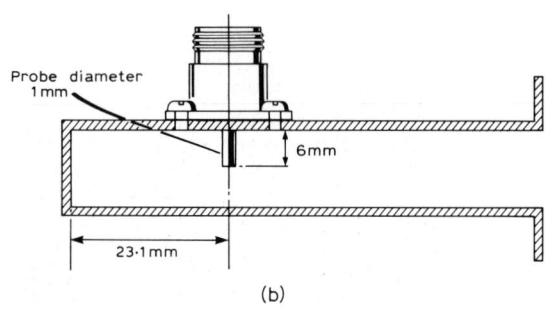

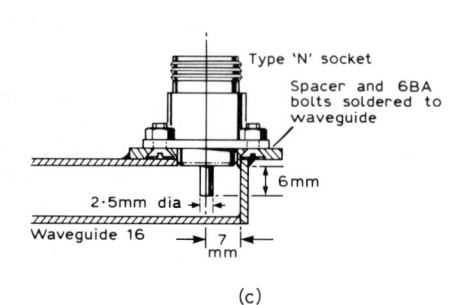

Fig 18.20. (a) General form of a WG to coaxial transition (b) WG16 to N-type transition (schematic, non optimised) (c) optimised WG16 to N-type transition

FHJ4–50 exhibiting a loss of about 4dB per 20ft and short lengths of conventional flexible cable terminated in N-type connectors may sometimes be found, particularly in older X-band equipment, but it is more likely that short 0.141in (3.58mm) semi-rigid coaxial leads terminated in SMA connectors will be found on more modern equipment. It is useful, therefore, to have coaxial to waveguide transitions available, when experimenting with pcb mounted circuits or some of the more recent professional equipment appearing on the surplus market. Two types are described.

General design dimensions

The general form of a coaxial to waveguide transition is shown in Fig 18.20a, where D is the probe diameter, L is the probe length and S is the distance between the probe centreline and the waveguide short.

cause the dielectric bearing the pin to become loose in the socket, so care is needed. In the final assembly, the dielectric will be held in place by compression against the outer surface of the waveguide to which the socket is attached. The dielectric is cut back flush with the flange surface, using a sharp knife or scalpel.

Next the spill diameter is modified by cutting it almost flush with the dielectric and soldering a 1mm diameter probe of stiff copper wire of the required length in place of the original spill.

In order to accommodate the socket mounting flange on the waveguide without complicating construction, the probe spacing from the waveguide short has been increased by $\lambda_g/2$, from $0.12\lambda_g$ to $0.62\lambda_g$, i.e. to 23.1mm. This allows the use of a simple, soldered end plate as the waveguide short. The probe enters the waveguide through a hole a little smaller in diameter than the measured diameter of the dielectric, so that the insulation carrying the centre pin of the socket is firmly held in place between the socket body and the waveguide wall. The modified socket can be fixed to the waveguide by soldering or, better, by means of 3mm screws inserted into tapped holes in the waveguide wall. If this method of fixing is used, then the fixing screws should be trimmed in length so that they end up flush with the inner face of the guide wall. This has the advantage that the socket is easily removable for trimming the length of the probe during testing.

In general the use of a larger than optimum diameter probe will increase the effective bandwidth of the transition, which is of little consequence in amateur installations, but may lead to the need to experiment with the length of the probe for best match and signal transfer.

The two alternative forms of construction of an N-type transition are given in Fig 18.20b, optimised as described above, and 18.20c, not optimised, but of acceptable performance.

A coaxial (SMA) to WG16 transition

SMA connectors are specified for use up to 18GHz and are used with 0.141in (3.58mm) semi-rigid coaxial "cables". This series of connectors should be the first choice for this band and special versions are available, with a carefully dimensioned "spill", to act as a microstrip "launcher". However, for the purposes of a WG16 transition, the ordinary SMA socket should be used. This normally has a solder spill which is 1.3mm diameter and 6.5mm long.

There is no need to make the diameter of the spill smaller as this diameter will still provide a reasonable match. The dielectric does not protrude behind the mounting flange and needs no modification. The hole necessary to match the dielectric of the socket is 4mm diameter.

It is possible to mount the socket on the waveguide with a 4.25mm spacing from the end short, provided that it is soldered in place. If four fixing screws are used then this distance must be increased to about 5.5mm to accommodate the rear two screws. Although this is slightly greater than the optimum, it has been found to make little

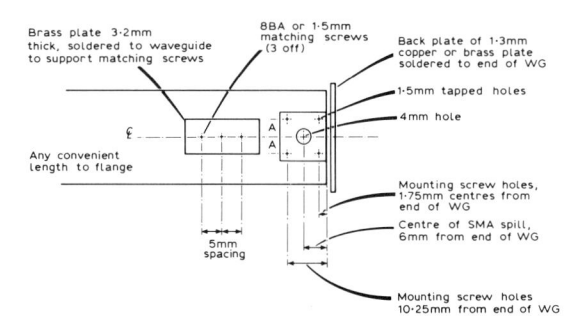

Fig 18.21. (a) Diagram of WG16 to SMA transition (b) WG16 to SMA transition (G3PFR). Photo: G6WWM

difference in practice. If screws are used they should be 1.5mm, cut so that they do not protrude into the waveguide. Dimensions and a photograph for 10.380GHz are given in Fig 18.21. None are unduly critical and slight mismatching can be corrected by adjustment of the three-screw tuner. A good match with wider bandwidth was still obtained with the probe diameter increased to 3mm by means of a soldered sleeve.

18.2.9 Waveguide switches

Waveguide switches fall into two categories: mechanical and electronic. The former consist of some mechanism which effectively moves one section of waveguide relative to a fixed section. This could be as simple as a "plug and socket" type of arrangement, a "flapper" across a waveguide junction or a finely engineered rotary switch, such as those produced commercially and occasionally found on the surplus market. The change-over might be effected manually or by means of a remotely switched solenoid.

The electronic switch usually consists of strategically placed, biased pin diodes. Although simple in principle, this type of switch has been little used by amateurs, largely because it offers performance inferior to mechanical switches, both in terms of isolation and insertion losses.

The most common application is to switch an antenna between a transmitter and receiver. Ideally, operation of the switch should not require the movement of any of the guides attached to it and for this reason the flapper or professional mechanical types are preferred.

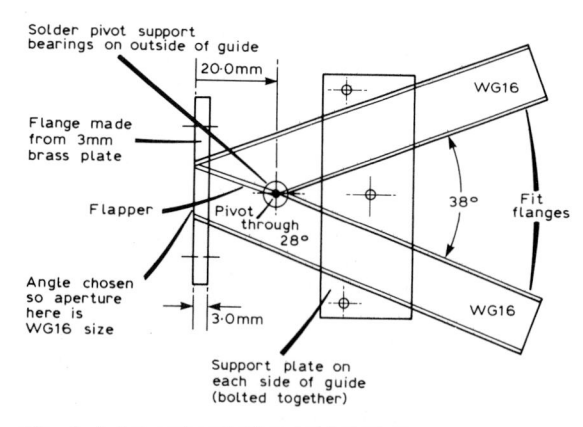

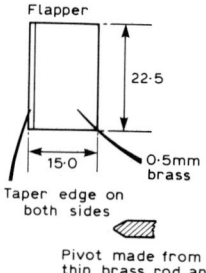

Fig 18.22. WG16 "flapper" type switch. A detailed design for such a switch, by DC8UG and DB1PM, appeared in *VHF Communications*, volume 13, issue 1/1981

Simple "plug and socket" switching

Manual switching can be accomplished, albeit a slow and cumbersome process, by the use of round flanges and locking rings of the type described in section 18.2.1. Change over can be made easier if the "male" locating ring is fixed in position on its flange by the use of "Superglue", for it is often difficult, in the field with cold fingers, to manipulate both the rings and the equipment simultaneously! An alternative is to use a "plug and socket" arrangement made from a section of OE waveguide cut as shown in section 18.2.1. It is left to the ingenuity of the individual to devise his/her own methods of changeover using such a switch.

A waveguide three-port "flapper" switch

This type of switch was described in outline in [7] and in more detail [8]. It consists of two sections of WG16 suitably joined to form a "Y" junction. At the apex of this junction a hinged "flapper" is introduced which can lie across either switched waveguide port and actuated by a lever, as shown in Fig 18.22. A coupling hole is cut in the flapper plate so that oscillator injection to the mixer is optimised with the switch in the receive position. This will need to be adjusted on test, to give the optimum mixer current for the particular diode in use. Thus the size of the hole should start small and be progressively increased on test until the optimum is reached. If separate receive and transmit oscillators are to be used, then this coupling hole can be omitted.

An interesting variant of the "flapper" three-port switch was briefly described in [9], in turn derived from [10]. It consisted of a Y-shaped assembly of OE waveguide as shown in Fig 18.23. The original design was for a frequency of 9.375GHz but the dimensions of the loop should scale linearly with the free space wavelength (λ_0), i.e. 1.15 × 9,375/10,400 at 10.400GHz. Unfortunately, this is just too large to fit into WG16, but an oval ring should work just as well as a circular one. No work has yet been done on this design but the original switch was claimed to have an input vswr better than 1.06, a bandwidth of ±3% and an isolation between switched arms of at least 25dB.

Two alternative simple switches

The first, shown diagrammatically in Fig 18.24, consists of two sectors of a disc hinged together, one being fixed and

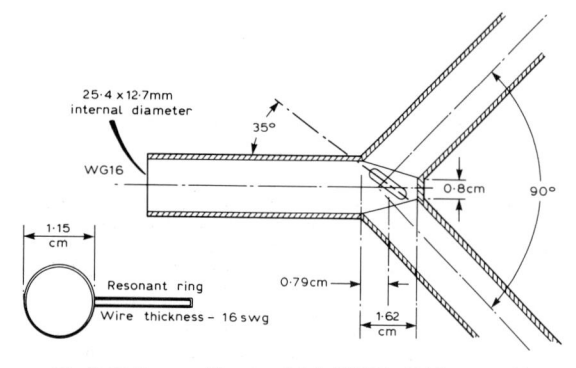

Fig 18.23. Resonant loop switch in "Old English" waveguide

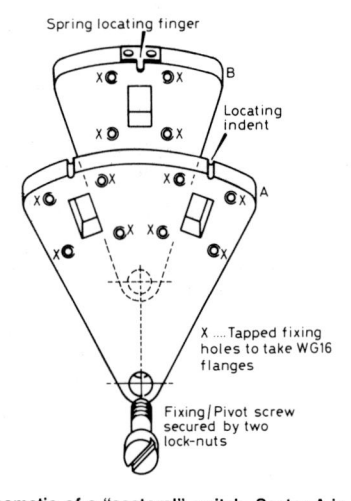

Fig 18.24. Schematic of a "sectoral" switch. Sector A is fixed and sector B moves relative to it. The antenna is connected to sector B by a short length of flexible waveguide and the transmitter and receiver to sector A. The only critical dimensions are the "windows" and flange mounting bolt-holes which must match WG16

flange and into tapped holes in the sector. The fixed sector has two holes cut in it as shown in the diagram. Some kind of locating mechanism will need to be incorporated in the switch and could consist of two v-shaped grooves cut into the periphery of the fixed sector. A suitably profiled spring could then be attached to the moving sector by means of tapped screws and positioned so that it "clicks" the movable sector into position when a changeover is made. Coupling of the fixed sector to the equipment can be made with conveniently shaped lengths of WG16.

Another mechanical arrangement is shown in Fig 18.25, a design used by ZL2AQE on 5.7GHz. There is no reason why this concept should not work at 10 or 24GHz. Provided there is reasonable alignment of the flanges there should be little insertion loss. Leakage from the flanges and losses by slight misalignment can be minimised by using plain/choke flange combinations as described earlier.

Professional mechanical waveguide switches

These usually consist of a cylindrical "barrel" into which are formed two WG16 sized 90° radiussed bends, as shown in Fig 18.26a. The barrel is mounted, together with a locating mechanism, within an outer cylinder machined to fine tolerances. The barrel can rotate within the cylinder and therefore acts as a four-port changeover switch. Isolation between ports is often improved by slabs of ferrite incorporated in chokes in the barrel, between the adjacent cavities. Insertion losses in a well designed professional switch will be fractions of a dB and the isolation between ports may well be in excess of 60dB.

One advantage to the four port switch is that the extra port can be used to take a waveguide load. On receive, a

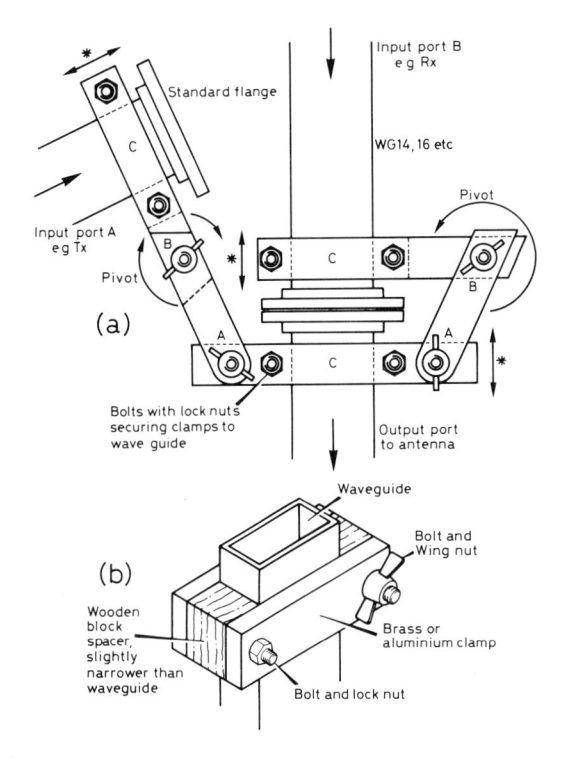

Fig 18.25. (a) Simple mechanical arrangement allowing changeover by moving sections of waveguide. Ports A and B are pivoted around the wing-nuts on the swinging arms. The swinging arms are attached to clamps (C) secured to the waveguides as shown. Losses, mismatch and leakage are minimised by use of choke/plain flange pairs. The clamps and wing-nuts are adjusted for correct alignment and close fit of the flanges. Then tighten wing-nut A and use B as a pivot. Wooden spacers are not needed in the A–B section of the clamps. (b) Detail of the clamp blocks. (c) Photograph of the switch shown in Fig 18.25 (a). (ZL2AQE)

the other movable. Into the movable sector is cut a rectangular hole whose dimensions correspond to the inside dimensions of WG16. The antenna port waveguide is attached to this hole and is a short length of flexible, twistable WG16 held in place by screws fitted through its

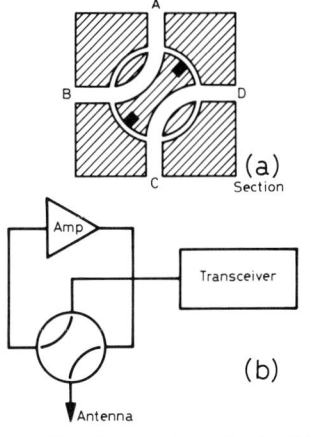

Fig 18.26. (a) Simplified diagram of a professional four-port rotary changeover switch. Waveguide flanges bolt on to the body of the switch at A, B, C and D. The central barrel rotates in 90° steps, alternately connecting A–B and C–D (as shown) or A–D and B–C. The black bars are ferrite chokes between cavities to improve isolation. A photograph of such a switch was given in Fig 5.96, chapter 5. (b) Using a four port switch to enable use of an amplifier on both receive and transmit. As shown, the switch is in the "transmit" position

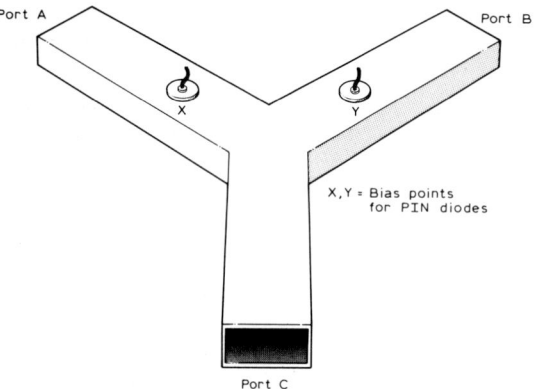

Fig 18.27. (a) A pin switch/modulator (Sivers Lab. data sheet) Insertion loss typically 2dB; isolation minimum 10dB; switching time 10μs; maximum rf power 1W average; bias –1V at 20mA (insertion) and 0V (isolation). (b) A three-port changeover pin switch

waveguide, as shown in figure 18.27a. When reverse biased (or to zero voltage), the diode presents a very high impedance and when forward biased by a dc current, a very low impedance. A single diode mounted in this way can be used as a modulator, if a modulating signal is superimposed upon the dc bias. Provided that the Gunn source is suitably isolated from the pin diode, the modulation so-produced is a.m rather than the customary fm and this method of modulation is often used professionally in test benches to produce square wave modulated signals. Typical performance of a single pin switch/modulator might be:

Insertion loss	2dB
Isolation (min)	10dB
Switching time	10 microseconds
Power handling	1W average
Switch off voltage	0V
Switch on voltage	–1V

In order to act as a changeover switch, it is necessary to employ pin diodes mounted in two of the "legs" of a Y-junction (Fig 18.27b), much in the manner of the mechanical switch discussed above. The third leg, without the diode, is the antenna port. In use forward bias is applied to, say, diode A and a zero voltage or reverse bias to diode B. Under these conditions, diode A will be low impedance and because it is a quarter wave from the junction, the A leg will present a low impedance and will be "off". The reverse will be true for leg B which will be "on" as the diode is a high impedance across the guide. It is possible that increased switching isolation could be obtained, at the expense of increased insertion loss, by placing additional pin diodes, spaced at $n \times \lambda_g/4$ (where n is an odd number) apart in each switched leg, although this has not been tried. Decoupled mountings suitable for use with pin diodes are described in the earlier sections on waveguide diode mounts.

18.2.10 Waveguide filters

10GHz filters usually take one of two forms. These are the "post" type and the iris-coupled type. Both are comparatively easy to make, giving acceptable performance, with a minimum of tools and precision metal working. Design mathematics for filters and a computer program for bandpass waveguide filters were given in chapter 12, "Filters".

Waveguide "post" filters

The design of two filters (due to G3WJG) which, between them, cover the whole of the 10GHz allocation is shown in Fig 18.28. Eight posts are fitted across the guide in fairly precise positions and two tuning screws are used to set the frequency. The filters have an 80MHz equal ripple bandwidth, a 3dB bandwidth of 120MHz and an insertion loss of 0.3dB. Frequencies 1GHz away are attenuated by about 40dB. Compared with some other filter designs, this type is fairly tolerant in terms of construction errors. Perfectly satisfactory filters have been made by marking out using a magnifying glass and an accurate rule and drilling instead

separate transmit oscillator is best switched off since there may be enough leakage of the transmit source to partially de-sensitise the receiver. However, only trial and error will reveal whether this is necessary. Using a commercial, ferrite isolated switch it was found possible to allow a 20mW transmit Gunn to run continuously into the load without undue problems whilst receiving. The receiver oscillator should, ideally, be left running continuously (for stability) and when in transmit mode its escaping power will be absorbed into the load. Another advantage is that it can allow an amplifier to be used in front of a low power transceiver – on receive as a preamplifier and on transmit as a power amplifier, by reversing its direction as in Fig 18.26b.

PIN switches and modulators

A pin switch consists of a pin diode, usually in a 1N23-type package mounted centrally across a short length of

THE 10GHZ ("3CM") BAND 18.17

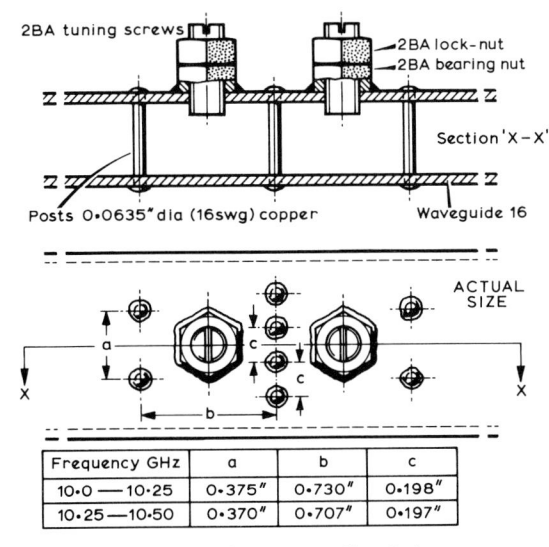

Fig 18.28. WG16 post-type filter design

Frequency GHz	a	b	c
10·0 — 10·25	0·375"	0·730"	0·198"
10·25 — 10·50	0·370"	0·707"	0·197"

of the more precise use of a milling machine. A recommended method of assembly is as follows:

1. Straighten the copper wire by stretching it slightly. Cut oversized lengths and squeeze one end sufficiently to stop the wires passing through the holes in the waveguide.

2. Fit the wires and flanges and jig the 2BA bearing nuts in place with non-solderable screws.

3. Using solder with a water soluble flux (discussed in chapter 7, "Constructional techniques"), solder the flanges in position using a gas blow-lamp. Solder the post joints on the *underside* of the waveguide only, which will prevent solder running into the inside of the guide.

4. When the solder has solidified, invert the filter and solder the two bearing nuts and the remaining post joints, again from the underside, allow to cool and then wash well in running water.

The filter can be aligned using a Gunn source, a directional coupler, variable attenuator and a mixer used as a detector, as described later.

Waveguide iris-coupled cavity filters

The response of some higher performance bandpass filters (due to G3JVL) having a slightly different method of construction is shown in Fig 18.29. The filters have 3dB bandwidths of either 20 or 60MHz depending on their dimensions as indicated in the caption to Fig 18.30a. They are particularly suitable for use in the local oscillator chain of a receiver to attenuate the noise, generated at signal frequency, which would otherwise degrade the receiver performance. Even when a relatively low i.f is used, significant rejection of noise can be obtained; for example, at 30MHz the rejection is 27dB. The 60MHz filter can be used to effectively eliminate the image (second channel) response of the receiver. Either type of filter could also be used to remove the unwanted harmonics from the output of an oscillator/multiplier chain to be used as a beacon.

As shown in Fig 18.30a, the filter consists of a length of WG16 in which iris plates are used to define three resonant cavities which are coupled by centrally placed holes of specified dimensions. The basic design frequency is 10.5GHz but the tuning screws fitted may be used to tune the filter down to 9.5GHz at least. Matching screws are fitted at each end of the filter as necessary to tune out any mismatch with the external circuitry.

In order to maintain the Q of the filter, it should be made entirely from copper. Brass should only be used if it can be copper or silver-plated after assembly and cleaning. The tuning and matching screws can be of brass only if little of each projects into the cavities, otherwise threaded copper rods should be used. The filter is constructed, in outline, as follows (a detailed work schedule was given in the construction chapter, which suggested alternative methods and a systematic approach):

1. Scribe grooves deeply into the top and bottom broad faces of the guide, corresponding to the iris positions.

2. Using a fine saw such as a junior hack-saw blade, first cut slots at each corner and then use these as a guide to extend the cuts across the full width of the top and bottom walls of the waveguide.

3. Fit the iris plates in their correct positions and fix them in place by bending the corners of the iris plates. Jig the tuning and matching screw support nuts into place using non-solderable screws. Fit any flanges required and solder

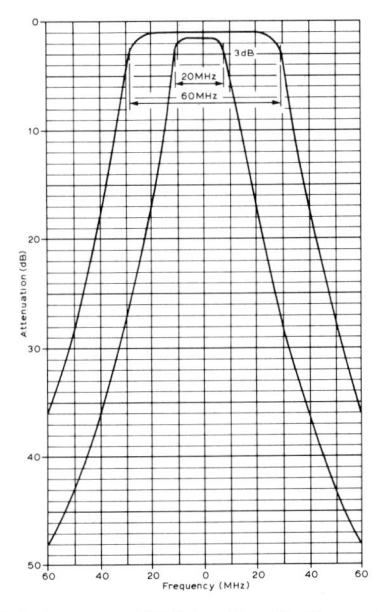

Fig 18.29. Performance of G3JVL multi-cavity iris-coupled filters

18.18 MICROWAVE HANDBOOK

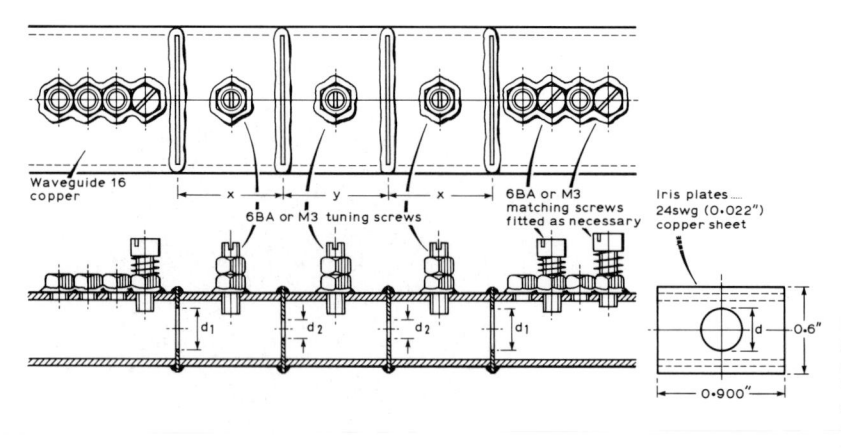

Fig 18.30. (a) Construction of G3JVL multi-cavity filters. The critical dimensions are: 60MHz filter, X= 17.8mm, Y= 18.2mm, d1= 9.0mm, d2= 4.3mm; 20MHz filter, X= 18.2mm, Y= 18.7mm, d1= 6.2mm, d2= 2.95mm (b) construction of G3JVL single cavity filters

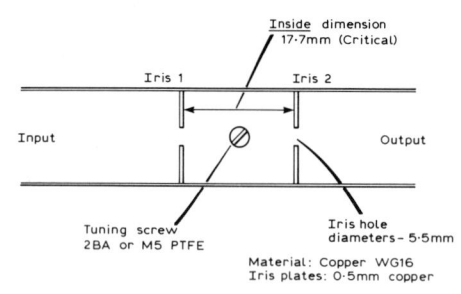

all joints at one operation using a gas flame. Alternatively, use the "bottom only" method described for the assembly of post filters.

Note that it is not necessary for the iris plates to make good contact with the sidewalls of the waveguide. Alignment of such filters is dealt with in a later section.

If it is necessary to *raise* the filter frequency, then large screws (0BA or 6mm) can be fitted centrally through the sidewalls of each cavity, although it would be better to make another filter specifically for the higher frequency.

The design of "wholeband" single cavity iris-coupled filters is given in Fig 18.30b. The main features are:

1. No matching screws are needed or intended.
2. Design centre frequency is 10.5GHz, bandwidth 10MHz, ripple 1dB.
3. Calculated insertion loss (copper) 1.9dB.
4. Tuned to 10.368GHz, adjacent products 1,152MHz away are attenuated by 40dB.
5. The filters were not designed for highest possible performance, but as a reliable, easily made piece of test equipment suitable for aligning multipliers.

18.2.11 Miscellaneous components

A number of components which are difficult or impossible for the amateur to construct are extremely useful and may, with experience, be almost indispensible. They can occasionally be found on the surplus market and should be obtained if available!

Non-reciprocal and other devices

Several forms of non-reciprocal or uni-directional device exist. Non-reciprocal isolators and circulators rely upon the special properties of ferrites for their operation. The hybrid-T, which *is* a reciprocal but unidirectional device, relies solely on waveguide symmetry.

Non-reciprocal devices: isolators

Isolators are two port devices exhibiting low insertion loss in the forward direction (typically 0.5dB) and very high attenuation in the reverse direction (typically 30dB or more).

They are particularly useful when interposed between a Gunn oscillator and its load, particularly if the load can exhibit a varying impedance which would "pull" the oscillator frequency. Isolators will commonly provide between 20 and 30dB isolation, as measured on surplus isolators of unknown design frequency but constructed in WG16. They are usually quite broad-band devices covering about ±500MHz around their design centre-frequency in X band. Thus isolators for 10.6GHz still offer acceptable performance at 10.3GHz. It may prove possible to improve performance by fitting matching screws at the input side of the isolator and tuning these for minimum insertion loss, although this has not be found necessary on several isolators of different designs which were used in both experimental transceivers and prototype wb (Gunn based) beacons.

Since the performance of the isolator depends on the degree of magnetisation of the ferrite within the guide, care should be taken not to damage the external magnet by either mechanical shock (dropping the device), overheating, exposure to strong external magnetic fields (loudspeakers and the like) or by proximity to ferrous objects which could cause either temporary magnetic distortion or, at worst, more permanent damage.

The field displacement isolator consists of a short length of waveguide with an externally mounted permanent magnet, looking something like Fig 18.31. Inside the waveguide there will be a slab of ferrite, coated with what looks like plastic or ceramic. This is an absorbing layer and it is this layer which limits the power which can be handled by the isolator. Fig 18.32 is a photograph of such an isolator.

The magnetic field is arranged so that the ferrite is magnetised transversely well below resonance and, because of its high permittivity, the rf field within the waveguide is distorted. The permeability further distorts the field in a non-reciprocal manner and the absorbing layer is placed at a point where the electric (E) field is zero for the forward wave and maximum for the reverse wave. Thus sited, the device offers minimum insertion loss and maximum reverse attenuation. The insertion sense will usually be indicated by an arrow somewhere on the external surface of the component, but is easy to determine by trial and error.

Non-reciprocal devices: circulators

The circulator is similar in concept and function to the isolator except that it is a three-port junction, usually arranged in the form of an equi-angular (120°) "Y", although sometimes in a T-arrangement. At the junction of the three "legs" there is a magnetised post or disc of ferrite and, usually, three screw tuners are present in each leg. The magnetised ferrite functions in such a way that rf energy is transferred unidirectionally to one adjacent port, much in the manner of a traffic roundabout on a road.

Fig 18.33 shows this in diagrammatic form; power applied to port A will be coupled mainly into port B. A small amount of "leakage" into port C might occur, to some degree dependent on the match of whatever load is attached to port B. Isolation between adjacent ports in the reverse direction to that indicated as "forward" (for example from A to C in the direction opposite to the arrow indicating the forward direction) may lie between about 15 and 30dB, again measured on surplus circulators of unknown frequency specification.

Circulators are, again, quite wideband devices. One circulator, known to have been designed for 10.68GHz, measured 30dB at its design frequency and was still about 20dB at 10.3GHz. It is thus often unnecessary to try to alter the degree of magnetisation as was described in [11]. Isolation can be improved, at the expense of bandwidth (seldom

Fig 18.32. Photograph of WG16 field-displacement isolator (G6WWM)

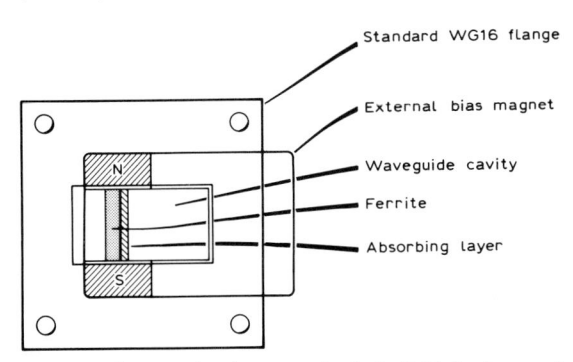

Fig 18.31. Construction (cross section) of a field-displacement isolator. Not to scale

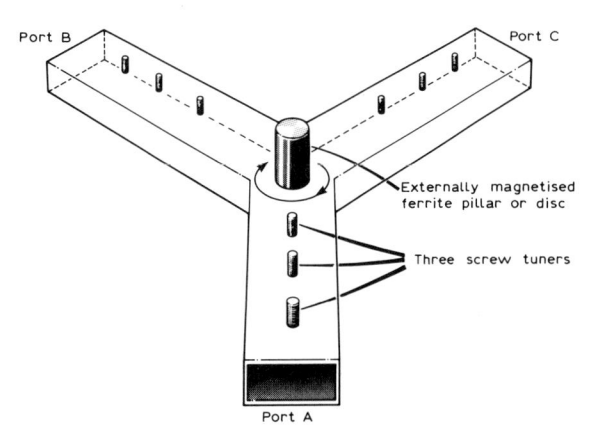

Fig 18.33. Diagram of a waveguide circulator

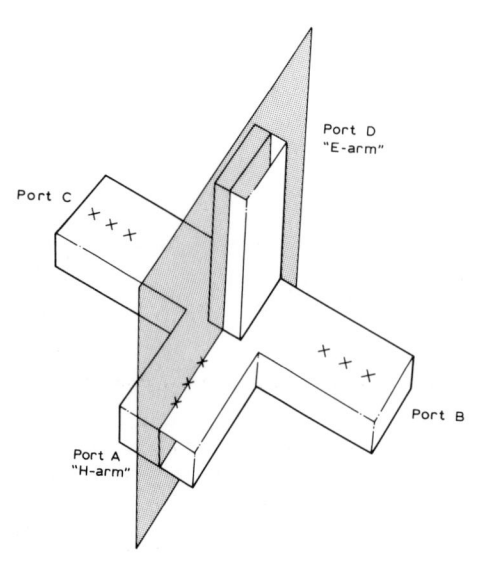

Fig 18.34. Diagram of a waveguide Hybrid-T. The shaded area shows the plane of symmetry on which the device depends. Three screw tuners may be fitted at the points marked xxx

important in the amateur context) by adjusting the three three-screw tuners normally present in the device. By this means the reverse isolation in the circulator mentioned above was improved to nearly 28dB with an insertion loss between ports of about 0.3dB. Again similar care should be taken against damaging the magnetic field when using such a device. Ways in which such a circulator can be used are discussed in section 18.5.4. and elsewhere in this chapter.

The hybrid-T

This is a four port device which is suited to making balanced mixers. Unlike the previous two devices, it *is* reciprocal and does not rely upon ferrites for its function, being constructed entirely from waveguide. It is shown diagrammatically in Fig 18.34.

In theory, it is a combination of a shunt-T network and a series-T network and is analogous to the balanced bridge

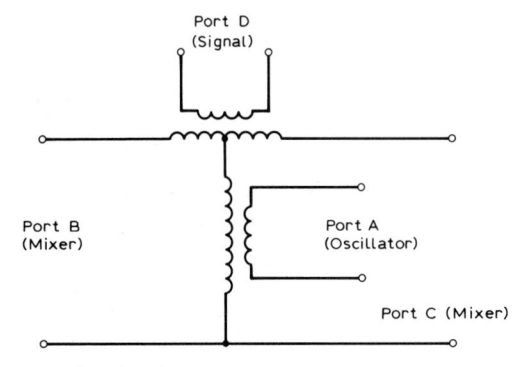

Fig 18.35. Equivalent electrical circuit of a Hybrid-T

circuit, used commonly in line telephony, which produces the sum and difference of two signals fed into it. The equivalent circuit is shown in Fig 18.35. When a bridge circuit of this configuration is properly terminated by matching impedances, the signal introduced into port A divides equally between ports B and C and not at all into port D. Similarly if the signal is applied into port D, it will be equally divided between ports B and C, but not into port A.

The same happens in the waveguide hybrid. A signal introduced into port A will produce equal amplitude, in-phase signals in ports B and C, but no (or very little, if properly matched loads are attached to each port) signal at port D. Any output appearing in port D is the result of asymmetry in the B/C arms or reflections from the loads attached to them.

If the signal is introduced into port D, there is little or no coupling into port A, but the signals produced in ports B and C will be equal in amplitude and 180° out of phase. These effects are due solely to the device symmetry and are independent of frequency within the passband of the dominant (TE10) mode in the waveguide.

If signals are applied to both the B and C ports, the vectorial sum of the signals will appear at port A and the vectorial difference at port D. When used in this way the output signals are 0° (in phase) or 180° (out of phase) relative to the input. This is then known as a "180° hybrid".

One disadvantage of the hybrid-T is that it presents a considerable vswr even if all ports are correctly terminated. To avoid this, three-screw tuners may be placed in the H-plane and E-plane arms of the hybrid and used to match out the vswr without destroying the inherent symmetry. This will limit the bandwidth of the device although this will be of little consequence to amateurs. A hybrid-T matched in this way is known as a "Magic-T". Its use as a balanced mixer is described later.

18.3 OSCILLATORS

Oscillators form the "heart" of any communications system. They are central to the function of both transmitters and receivers and thus are discussed in some detail in this section.

In the 10GHz band, both free running fundamental frequency and crystal controlled multiplied sources are commonly used, the former for simple wideband systems and the latter for more sophisticated narrowband systems. The simple, fundamental oscillator is often associated with an in-line mixer and thus it seemed natural to describe these in a section which is otherwise devoted to frequency generators. Occasionally a combination of a fundamental frequency oscillator locked to a lower power crystal controlled source has been used. There have been successful amateur developments in the use of phase locking techniques at both 10 and 24GHz using discrete ics and transistors. A design for 24GHz (by G3BNL) is given in chapter 19, "The 24GHz band". Similar methods can be used at 10GHz, provided that suitable dimensions are calculated for the waveguide components and the division ratios for the vco and pll are recalculated. Future developments are likely to centre around the use of frequency synthesis techniques using ics capable of operating directly in the 1 to 2GHz region. Such techniques are discussed in section 18.9.

18.3.1 Gunn oscillators and associated mixers

The theory of Gunn devices was given in chapter 6, "Microwave semiconductors and valves". From this it will apparent that a Gunn device, when mounted in a well designed cavity, can provide a simple, inexpensive and reliable source of rf energy which is sufficiently stable and clean to allow effective wideband communication, at least over short obstructed paths and longer unobstructed paths. Moreover, Gunn oscillators only require simple, low voltage supplies which are easily modulated to produce good quality fm signals. The simplicity and low cost of such equipment has formed the basis for the popularity of the band, especially with novices to microwave techniques.

It is worth reminding the reader of some of the fundamental characteristics of Gunn oscillators at this point. First, the Gunn device junction dimension determines its natural frequency, in that this defines the transit time of the electron domains which produce the current pulses through the device. These current pulses excite the cavity which, by providing the "flywheel" effect, translates the current pulses into sinusoidal oscillations. Control of frequency is determined mainly by the cavity dimensions. The voltage/power characteristics of the Gunn device is shown in Fig 18.36. The actual power output is determined by the type of diode employed, the Q of the cavity and the degree of coupling between the oscillator cavity and the "outside world". Tuning of the cavity can be mechanical or electronic as discussed next, or tuning over a limited range can be achieved by utilising the voltage/frequency characteristics of the Gunn which is discussed later in the section on power supplies and modulators.

Tuning Gunn oscillators

Three commonly used methods of tuning Gunn oscillators were briefly reviewed in chapter 6. These are:

1. Tuning by metallic rods
2. Tuning by dielectric rods
3. Tuning by varactors

Using an iris coupled oscillator, of the general design described later, investigation of the tuning resolution and range yielded the following results:

Metallic tuning

Tuning with a micrometer whose spindle was machined down to approximately 1mm in diameter, was found to give a tuning rate of between 10 and 60MHz/turn, depending on the depth of penetration of the screw. With care it was possible to set the oscillator frequency to within 1MHz but, due to uncertainty in the precise point of contact of the screw thread, the tuning was rather erratic and

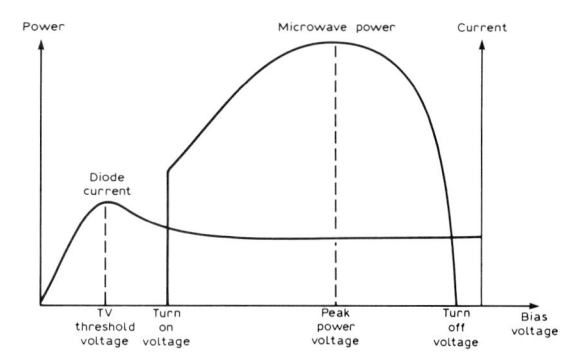

Fig 18.36. Gunn device characteristics – bias voltage vs power output

the mechanical stability poor. When the screw was inserted to substantial depths there were variations in output amplitude of the order of 25dB. A quarterwave choke on the micrometer spindle was tried, but this resulted only in a marginal improvement. The overall tuning range with this arrangement was 1.2GHz. With these findings, metallic tuning elements are definitely not to be recommended except, perhaps, for coarse "bandsetting" where they can be locked in place after adjustment.

Dielectric tuning

The tuning rate of the dielectric screw was found to be linear, except at the extremities of the cavity ie. the screw first entering the cavity or approaching the opposite wall. Initially a 6BA nylon screw was introduced instead of the metal screw and this gave greatly improved tuning characteristics, although the tuning range was restricted to 170MHz. Torsion effects leading to tuning backlash were noted with the nylon screw, so this was next replaced with a 4BA screw made from Perspex and this gave excellent backlash-free results.

With this screw the oscillator could be easily be tuned to within 300kHz and with care could be tuned to within 30kHz, the mechanical limit of resolution. The tuning characteristic was again linear (except at the extremities of travel) and very smooth, providing a tuning range of 230MHz with a maximum amplitude variation of 3dB. The sensitivity to lateral force on the tuning knob was ±3MHz and leakage from the dielectric tuner was −45dBm. The lf jitter on the oscillator frequency was less than 5kHz and drift from switch-on was less than 3MHz after one hour.

A second oscillator was built in which mechanical tuning was achieved by a Pyrex glass spring-loaded plunger driven by a standard micrometer. A Pyrex plunger of 4.2mm diameter gave the following results:

Output amplitude variation:	4.6dB (maximum)
Tuning rate:	50MHz/turn (almost linear)
Sensitivity to lateral force:	±1.8MHz for 500g force
Setting resolution:	easily within 500kHz
	with care, within 50kHz

18.22 MICROWAVE HANDBOOK

Table 18.3. Dielectric constants for various materials

Material	ε_r	tan.d
Nylon	3.6	0.02
PTFE	2.1	0.0001
Perspex	2.65	0.015
Glass (soda-borosilicate)	4.38	0.0045
Glass (Pyrex)	5.1	0.0085
Quartz (Vitrosil)	3.78	0.0001
Quartz (Spectrosil)	20	
Aluminosilicate	5.8	0.008

Table 18.3 giving dielectric properties of materials may be useful; ε_r is relative permeability and tan.d is the loss tangent.

Recent experience with iris coupled Gunn oscillators tuned with 2BA ptfe screws has given similar results to those described above, but with a tuning range of between 500 and 600MHz which is enough to cover the entire amateur band. In this case the length of the cavity was chosen so that the frequency was around 10.5GHz with no screw insertion, which resulted in optimum output in the 10.3 to 10.4GHz range. A full description of such oscillators together with an in-line mixer is given later in this section.

Electronic tuning

In the third method, the cavity may also include electronic tuning by means of a varactor tuning diode placed alongside the Gunn device. The tuning now resembles mechanical tuning obtained by dielectric loading of the cavity. Varactor diodes have a capacitance characteristic which is continuously variable with bias voltage, from a small value at large reverse (negative) bias to a large value at forward bias. The Q of the diode is maximum at high reverse bias, decreasing steadily towards zero as the bias is reduced and becomes forward bias.

The varactor feed post, which can be similar to that used to mount the Gunn, is an inductive element in series with the diode; the rf fields are in part looped around the post and coupling is by mutual inductance. The Gunn device is itself coupled to the cavity in a similar way.

The change in frequency is proportional to $1/C_v$ which is, in turn, approximately inversely proportional to the square root of the bias voltage applied, where C_v is the varactor tuning voltage. However, a proportion of the rf power generated by the Gunn is dissipated in the varactor tuning diode and may be as high as 60%.

The varactor gave a maximum tuning range of 35MHz with less than 1dB variation in output. The oscillator frequency could be tuned easily within 100kHz and with care to within 10kHz. The ability to tune the oscillator electronically enabled afc to be applied to the diode bias circuit and this worked very satisfactorily, although equivalent results can usually be obtained by varying the Gunn bias. The varactor used was a Mullard Series 821 CXY/D type but there is no reason to believe that surplus diodes should not perform as well. When a varactor was included in an oscillator designed for use as a local oscillator to produce an i.f of 432MHz, the following results were obtained:

Tuning range:	35MHz max at 9.8GHz
	24MHz min at 9.35GHz
Max output variation:	1dB pk–pk
Tuning rate:	approx 40MHz/turn (270° pot)
AFC capture range:	±6MHz
LF jitter:	less than 18kHz

Practical amateur Gunn designs

In the mid 1970s several amateur Gunn designs appeared, varying between very simply constructed units and more complex units requiring quite elaborate machining. A few of the early designs are given, with dimensioned diagrams, to illustrate the evolution of amateur designs, culminating in the almost universally used dielectric-tuned, iris-coupled design which is the most reliable and reproducible design to have appeared in amateur literature.

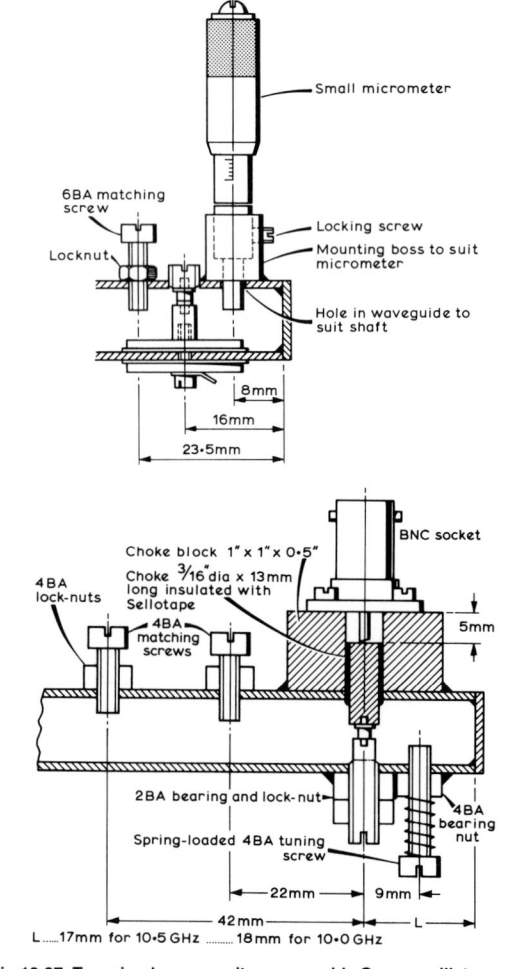

Fig 18.37. Two simple rear-cavity, open-guide Gunn oscillators

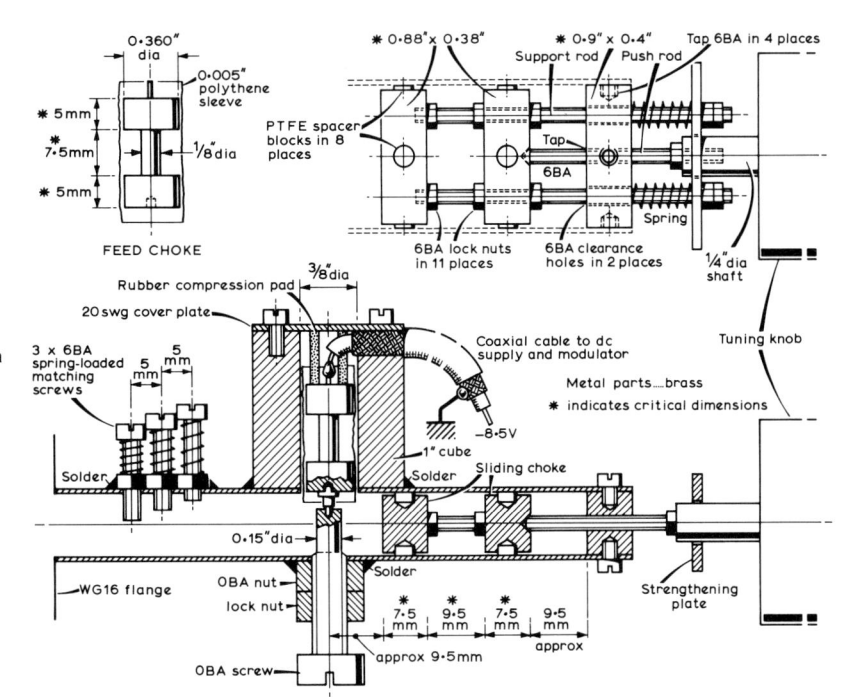

Fig 18.38. Gunn oscillator tuned by a sliding short

Fig 18.37a illustrates a simple rear-cavity (cavity between the Gunn and the end short) design which was originally described [12] by GM3OXX as a self-oscillating mixer for a complete, simple transceiver. The metallic micrometer tuning arrangement can be replaced with a micrometer driven ptfe tuning plunger which will ensure a much smoother tuning characteristic as discussed above. Simple capacitive decoupling is used. Use of the Gunn as a self-oscillating mixer does, however, result in a rather insensitive receiver. Fig 18.37b is another version of the oscillator, this time using choke decoupling instead. Note that it is important to get the cavity length behind the diode accurate. If the frequency is too low, the end plate must be removed and a little ground off the waveguide – the frequency shift is about 300MHz per mm! If too high in frequency, it would be impossible to lengthen the cavity. Fig 18.38 is another rear cavity oscillator using a precision adjustable short circuit for tuning purposes.

The main disadvantage of this type of oscillator is susceptibility to frequency pulling because of the open ended nature of the oscillator, although this is of less consequence when the oscillator works into a constant load.

The next development, due to G8APP, was the front-cavity, iris-coupled oscillator shown in Fig 18.39. Here the cavity is between the Gunn and the output iris. This led in turn to the development of the oscillator shown in Fig

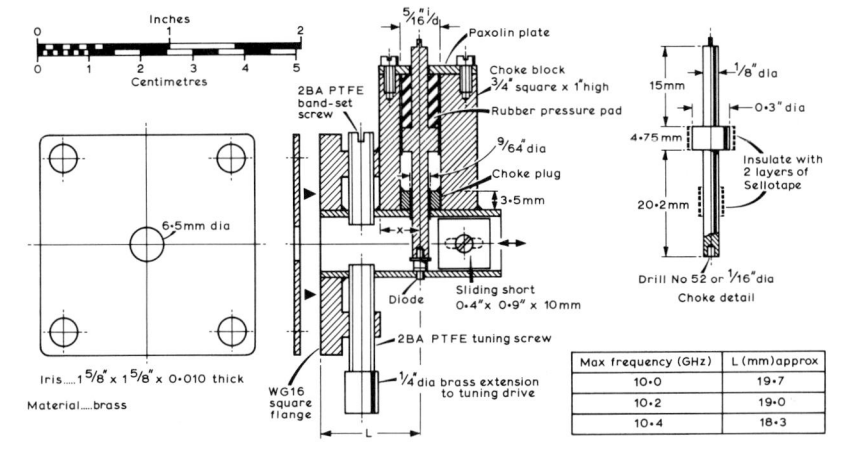

Fig 18.39. Choke-decoupled Gunn oscillator

Max frequency (GHz)	L (mm) approx
10·0	19·7
10·2	19·0
10·4	18·3

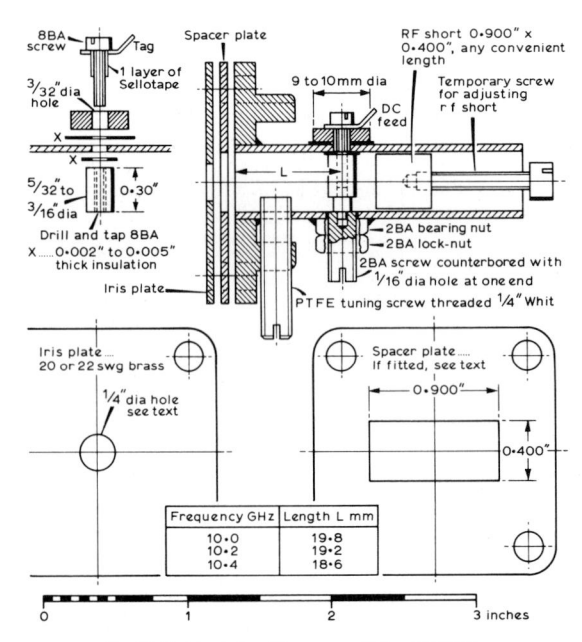

Fig 18.40. Iris-defined Gunn oscillator (G8APP)

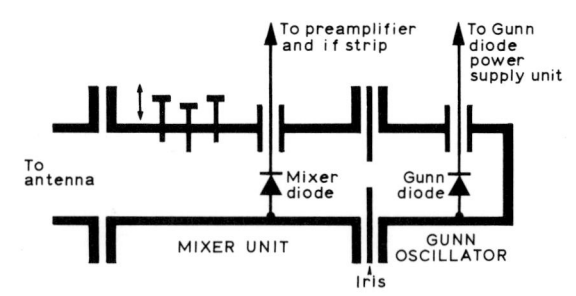

Frequency GHz	Length L mm
10·0	19·8
10·2	19·2
10·4	18·6

Fig 18.41. In-line mixer/oscillator, block diagram (G3RPE/G3WDG)

18.40 (described in [14]). This design was further simplified by using capacitive decoupling and eliminating the sliding short, leading to the oscillator shown in Fig 18.42 as part of the oscillator/mixer assembly described in the next sub-section. Although capacitive decoupling is simpler than the choke, more rf does leak out and this can result in poor frequency stability. This cause can be verified by moving a metal plate around near the decoupling disc and looking for frequency pulling. The design of Fig 18.42, with the choke of Fig 18.38, is probably the best combination. The principal advantages of this design are:

1. Relative simplicity of construction.

2. Reliability and reproducibility.

3. Ease of adjustment of cavity length for coarse frequency setting: if the cavity is made deliberately short, its length can easily be increased to lower the frequency by clamping a thin spacer plate between the flange and the iris plate.

4. Ease of adjustment of oscillator coupling into the external load, by altering the size of the hole in the iris plate.

5. A degree of isolation from the external circuitry making it less susceptible to pulling than the rear cavity types.

Points to be made about construction are these: it is better to drill the 3/32in hole centrally through both top and bottom walls of the guide at one operation, to ensure accurate alignment and then open out the lower hole to take the 2BA thread. It is important that the 1/16in hole in the 2BA screw which takes one end of the diode should be *exactly* central, otherwise the diode may be sheared as the screw is rotated. Care should be taken to remove burrs, especially from the areas in contact with the diode and the insulation. Both the iris plate and the spacer plate (if used) must be flat in order to ensure good contact and maintain the Q of the cavity; these plates are best sawn from sheet rather than cut by tinsmith's shears.

A simple 10GHz Gunn oscillator and mixer

This design was originally described in [15], primarily as a receiver "head" with limited transmit capability. The Gunn oscillator is a particularly reliable and reproducible design and the in-line mixer is described here (rather than in the earlier section on waveguide components or the later section on 10GHz down-converters) because it is a natural companion to the oscillator.

The receiver schematic is shown in Fig 18.41. It consists of a simple mixer assembly which is connected directly to a Gunn oscillator of the type whose cavity is defined by an iris. The mixer uses a length of waveguide into which is fitted the mixer diode, the hot end of which is decoupled and feeds the i.f amplifier in the conventional way. Diodes of the 1N23 type are recommended; those with the later prefixes (E, F) are preferred for their lower noise figures.

The signal-input end of the guide can be any convenient length and is fitted with a matching screw (or screws) to match the mixer diode to the waveguide. The length of the waveguide at the local oscillator end is critical; it needs to be made electrically an odd number of quarter guide wavelengths, i.e. $n \times \lambda_g/4$, where n is 1, 3, 5, 7 etc, as is convenient. This rear cavity is closed by the same iris as that used to define the Gunn oscillator cavity.

Two forms of the mixer assembly which were developed quite independently are shown in Fig 18.42a and b. A third method of mixer decoupling is shown in Fig 18.43 and Fig 18.44, which shows a complete assembly of mixer and oscillator. Also shown is a recommended design of Gunn oscillator which is a modification of the G8APP design [14] with a fixed, rather than adjustable, rf short. A feature of the design given in Fig 18.42a is that it requires a minimum amount of tools in its fabrication. Points that can be made with respect to its construction are:

1. First drill a hole about 3/32in (2.4mm) centrally through the both broad faces of a suitable length of WG16 and open one of them to 0.25in (6.4mm) diameter.

THE 10GHZ ("3CM") BAND

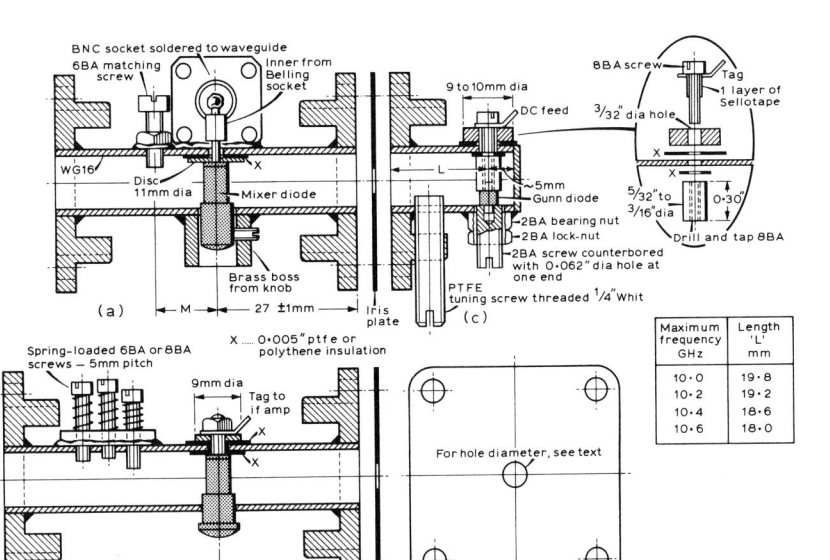

Fig 18.42. Detail of construction of the mixer oscillator of Fig 18.41

Maximum frequency GHz	Length 'L' mm
10.0	19.8
10.2	19.2
10.4	18.6
10.6	18.0

2. Remove the brass centre boss from a knob intended to be used with a 0.25in (6.4mm) shaft by breaking away the surrounding plastic. Fit the two flanges in their positions and solder these and the boss in a single operation. The latter may be jigged using a 0.25in (6.4mm) drill shank. Note that the position of the input flange is not critical in any way, but that at the oscillator end should be within about 1mm of that specified.

3. Drill and tap the holes for the matching screw(s). Remove any excess waveguide projecting from the flanges by sawing, filing and finally by grinding on wet silicon carbide paper backed by a sheet of glass. Carefully remove burrs from the inside of the guide, especially where the insulation is to be fitted.

4. If the only diode available is of the fixed collet type, carefully file away the lip so that the connection is uniformly 0.25in (6.4mm) diameter. If the diode available is of the removable collet type, an earthing block such as that shown in Fig 18.41 can be used instead of the modified fixed collet.

5. Drill the hole in the capacitor plate so that it is a tight fit on the diode pin. When assembling, press the diode against the wall of the guide before tightening the grub screw.

The construction of the design given in Fig 18.42b is similar, but in this case the diode is bolted to the bypass capacitor at one end, while the other end is made a tight fit in the wall of the guide. In mixer diodes that are reversible, it will be found that one pin is solid and, preferably, this pin should be threaded. The pin is undersized for the 8BA thread specified, so forces involved in tapping the thread are small; it can be done while holding the diode with the fingers.

The fabrication of the Gunn oscillator should present few problems. Constructional details are given in [14] if these are required. An alternative to the turned mounting post was suggested by G4DDK [16] who found that the brass bolt from the old type of ceramic-bodied mica compression trimmer capacitor can be used. It is removed by breaking away the ceramic body and consists of a thin disc at the top, suitable for decoupling, and is tapped approximately 8BA for a bolt to pull it to the top of the waveguide. The same tapped hole, enlarged slightly at the end remote from the disc, serves as the diode mounting hole.

Heatsinking in this design is certainly adequate for the low power diodes which generate up to about 20 or 30mW and which dissipate about 1W. It is insufficient for high power devices which dissipate about 10W. Note that for many low power Gunn diodes, the connection with the flange should be made negative.

Alignment can be simply achieved without elaborate test gear if the following method is used:

1. Connect the input of the mixer via a variable attenuator to a suitable rf source, which can conveniently be the Gunn local oscillator to be used. Inject rf at the *signal* frequency and adjust the matching screw(s) to maximise the mixer current while, at the same time, setting the variable attenuator so that this maximum occurs at the optimum value for the particular mixer diode being used. For point contact diodes, a current of 250 to 500μA is suitable. The matching screw(s) should then be locked in position. During this operation the rear end of the mixer cavity should be closed with either the iris to be used or by a blank plate.

2. With the input connected to a matched load and the Gunn oscillator fitted in its normal position, alter the size of the hole in the iris plate until the diode current is the

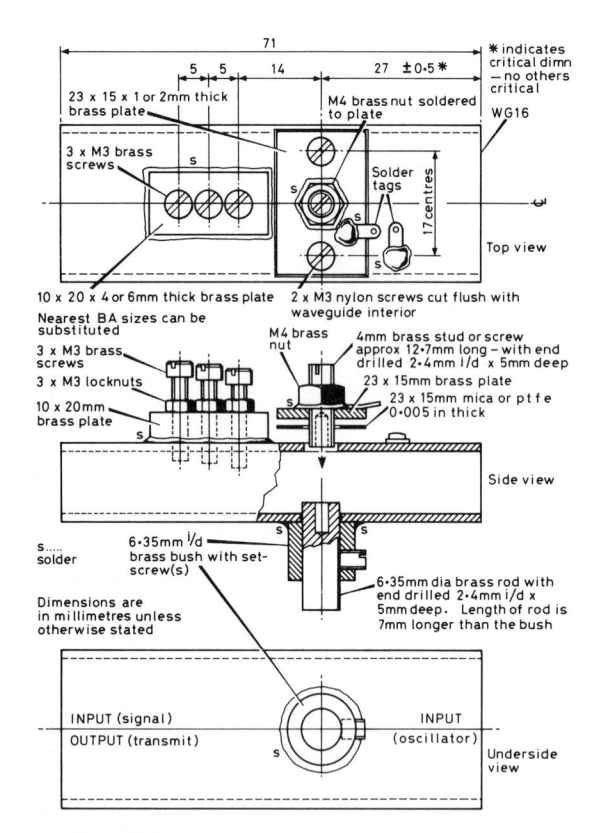

Fig 18.43. Alternative mixer decoupling arrangement (G3PFR)

Fig 18.44. Photograph of the oscillator/mixer constructed as shown in Fig 18.43. (G3PFR). Photo: G6WWM

same as that in 1. Obviously the size of the hole will depend on the output power of the oscillator, but will normally be within the range 3 to 6mm diameter.

An alternative method of alignment used by G3WDG is to attach the receiver to the antenna/waveguide run which is to be used. If this method is used, it is recommended that alignment is carried out outdoors with the antenna pointed at the sky, otherwise reflections from nearby objects (indoor or outdoor) can give rise to confusing results. For initial tests an iris about 3/16in (4mm) is suggested. The matching screw (or screws) is adjusted to set the mixer current to about 250µA. If the current is greater than this, even with no screw penetration into the guide, then the iris should be reduced in diameter. Conversely, if the mixer current obtained with a maximum recommended screw penetration of 3–4mm is still less than the optimum value, then the size of the iris should be increased. If the size exceeds about 6mm, then there is a risk that the stability of the Gunn oscillator might be adversely affected. If the mixer current is still too low, then a fault in construction, a poor mixer diode, a poor Gunn diode or a badly matched antenna should be suspected. The latter can be checked by substituting a large horn (or any other well-matched load) for the antenna in question. If correct operation is obtained, then the matching of the original antenna should be improved using, for example, another set of matching screws fitted adjacent to the antenna. It is also possible that either the Gunn diode or mixer might be at fault; this can be checked by direct substitution of either or both, one at a time.

Alternative configurations of the mixer assembly are possible. The critical dimension of the mixer assembly is the length of the guide between the diode and the iris. This was determined experimentally by fabricating an adjustable iris from 0.02in (0.5mm) thick sheet 0.9in (22.9mm) wide which was bent into the form of a square "U" with base 0.4in (10.16mm) wide. Using the set-up described under alignment, the position of the iris, penetration of the matching screws and the insertion loss of the attenuator were adjusted at signal frequency to peak at the optimum current for the mixer diode. It was found that moving the iris away from its best position by up to 1mm could be compensated for by readjustment of the matching screws. The value given in Fig 18.42, 27mm, represents a compromise length between 10.0 and 10.5GHz. It is somewhat smaller than the values calculated for $3\lambda_g/4$ at these frequencies, namely 29.8 and 27.4mm respectively.

The same procedure is recommended if it is desired to optimise the mixer assembly at another frequency, or to lengthen the cavity by making it $5\lambda_g/4$ or $7\lambda_g/4$ in order to fit a wavemeter. Other Gunn oscillators which employ an iris or its equivalent at the output flange can be substituted directly.

As noted earlier, some of the local oscillator power is radiated from the antenna port and may be used as a low power transmitter. By increasing the size of the hole in the iris plate, the amount radiated may be increased to make the transmitter more effective, although the reduced Q of the oscillator cavity resulting from this change means that the efficiency of the receiver will be impaired. Despite this, the performance of such equipment should be competitive with that of most other wideband transceiver configurations. The size of the iris should not exceed about 6mm diameter, otherwise the stability of the Gunn oscillator may be seriously affected. The in-line receiver is fairly sensitive to antenna match and significant variations in mixer current may be seen with different antennas.

A varactor tuned oscillator

G4EBF and G8EXL [17] suggested mounting a surplus varactor tuning diode in a cavity behind a Gunn diode, rather than alongside it as discussed earlier. Their arrangement is shown in Fig 18.45 and was found to give a change in frequency in excess of 100MHz for a 12V change in tuning bias. The diode will typically withstand a reverse bias of 30V. At this extreme, a change of frequency of the order of 200MHz should be seen. The mounting arrangement for both diodes can be similar to that for Gunn diodes in the oscillators previously described. The dimensions were derived experimentally and some further experimentation may be necessary to optimise the performance of this type of oscillator. As shown, modulation is achieved by applying the modulating voltage to the varactor and the Gunn bias is derived from the varactor tuning supply. With the supplies as shown, the frequency stability will be fairly poor and it may be preferred to use a conventional power supply/modulator for the Gunn and use a separate stabilised supply for varactor tuning. When configured as shown, the exact value of the Gunn series resistor must be selected to suit the particular Gunn diode in use.

Using "surplus" Gunn oscillators

Large numbers of doppler intruder alarm Gunn oscillators and some oscillator/mixers have appeared on the surplus market. Many are dual-cavity side-by-side or "piggy-back" types and are not readily adaptable for amateur use, other than as a source of good mixer and Gunn devices, or for stripping to recover ready made chokes and other mounting/decoupling components.

There are, however, a few types around which can be used directly and effectively. A notable example of an oscillator of excellent design which is still in current production and available from the normal commercial sources, as well as surplus, is the Mullard CL8630 Gunn Oscillator. At first glance the waveguide appears too small for a 10GHz Gunn, but the flange mates with a square WG16 flange and it is intentional that there is a step in waveguide size at this point. This step performs the same function as an iris. The oscillators are normally on a frequency of 10.687GHz, but the tuning screw can be adjusted to pull the frequency down into the amateur band. Usually it is possible to pull them down to 10.1GHz, but there will certainly be no trouble in using them at 10.3 to 10.4GHz. The oscillator is shown in Fig 18.46 and some data for it is also given there. For anyone interested in designing a Gunn mount, some lessons can be learned from this module, as the frequency is remarkably insensitive to the close proximity of objects

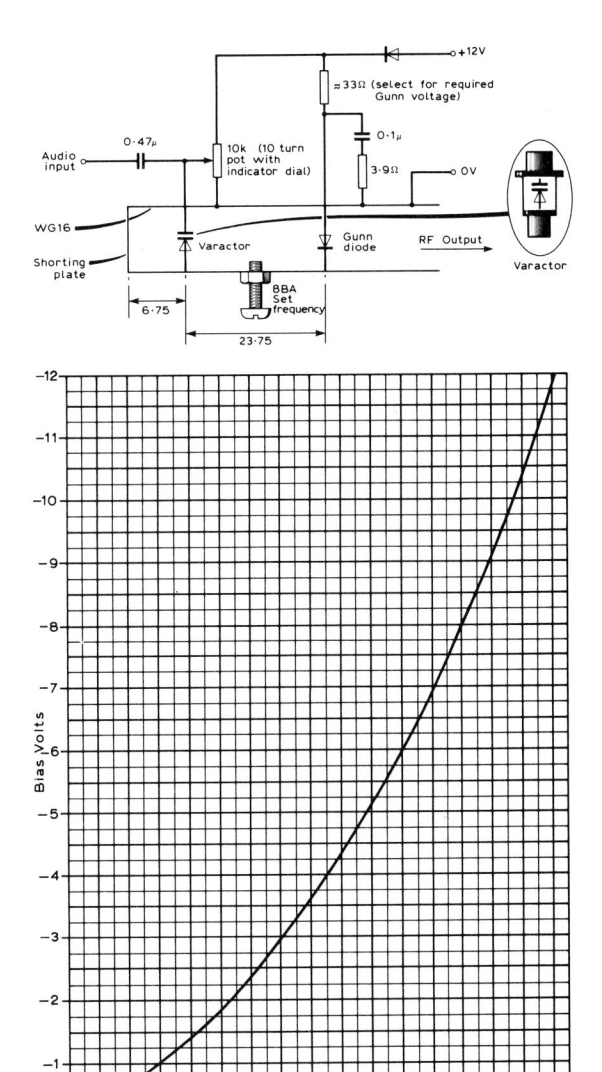

Fig 18.45. (a) Construction of a simple varactor tuned Gunn oscillator (G8EXL) (b) typical characteristics of a varactor tuned Gunn oscillator

(eg. hands) near the open flange, unlike many other designs examined.

Another oscillator which has been found to be satisfactory is made by Solfan in the USA. No technical details are available. However, as will be seen in Fig 18.47, this is a heavy die-cast unit of the iris defined type. It is pretuned by means of a large brass screw and the frequency might lie commonly around 10.68GHz. There are two slightly different types available which differ only in the presence or

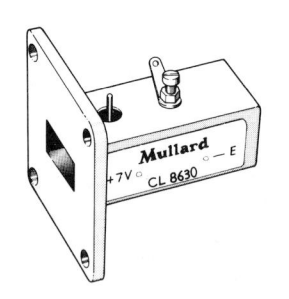

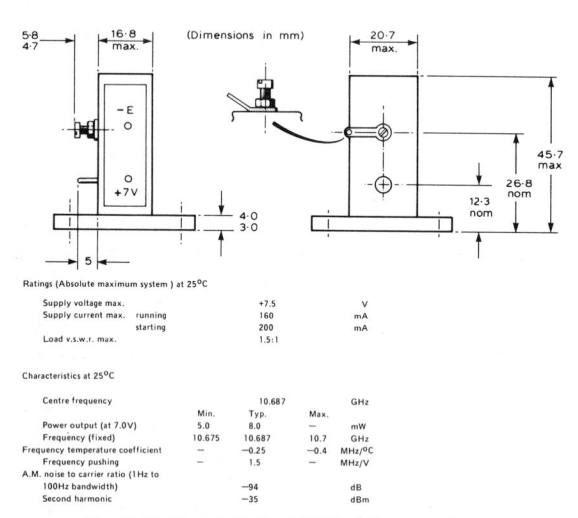

Fig 18.46. CL8630 Mullard (Philips) Gunn oscillator

absence of an additional tuning/matching screw just inside the iris end of the cavity. Either type can be retuned into the amateur band by adjustment of the main tuning screw. The oscillator is easily made mechanically tuneable by the following simple modification. If the oscillator is to be used as a fixed-tuned separate transmit oscillator, these modifications are not needed.

Fig 18.47. Photograph of a Solfan Gunn oscillator (right) and a Solfan in-line oscillator/mixer (left). Photo: G6WWM

Adjacent to the Gunn decoupling disc there is a solder tag, held in place by a screw, used to earth an electrolytic bypass capacitor. The screw and solder tag should first be removed. Next the rear plate, which is held in place by four screws, should be carefully removed and placed on one side. Do *not* remove or damage the piece of black expanded foam plastic which is glued onto this back-plate – it is essential to the correct functioning of the oscillator and *must* be replaced after completing the tuning modification.

The earthing-screw hole is drilled out and tapped 2BA, right through into the cavity. Carefully remove all swarf from the cavity and then replace the back-plate. A 2BA ptfe screw should be fitted into the newly tapped hole to provide the tuning element.

With the tip of the ptfe screw flush with the inside face of the cavity the brass tuning screw should be used to set the *highest* frequency to which the oscillator needs to tune and then be locked in place. Insertion of the ptfe screw will now tune the oscillator down in frequency by at least 300MHz.

A modified oscillator can be used in conjunction with the home-made in-line mixer. The iris of the Solfan oscillator is recessed by about 6.5mm and to use it with the mixer, the distance between the centre-line of the mixer diode and the flange on the oscillator input end of the mixer needs to be reduced from 27mm to 20.5mm or thereabouts. The small screw immediately in front of the iris (outside the cavity) affects the level of oscillator output – in effect it controls the iris size – and it can be used to set the mixer current, although it will cause frequency pulling whilst being adjusted and it may be necessary to readjust the "bandset" (brass) screw to reset the highest operating frequency. Another Solfan module which has been widely used is also illustrated in Fig 18.47 and consists of an oscillator with an integral mixer of the in-line type, but with an offset mixer diode. An identical tuning modification can be made.

A number of problems exist with this unit as obtained. First, the mixer diode is a Schottky barrier diode and is often "under par", possibly from static damage during handling. It should be replaced, preferably by a point contact diode such as the 1N23WE or 1N415E. These are somewhat more robust in terms of static damage. Second, the mixer is usually seriously overdriven with mixer currents in the range 2 to almost 10mA. The mixer current should be reduced to a more suitable level by the means already described. Finally, the mixer decoupling capacitor (disc) is rather too large to allow the use of a high i.f (eg 100 or 144MHz), most of the desired signal being bypassed by the capacitor. The module appears to be more suited to the use of i.fs up to about 50MHz, and is certainly satisfactory below 35MHz.

Using "side-by-side" oscillator/mixers

A modification to adapt a "side-by-side" oscillator/mixer, such as the Philips CL8963 (which is the same as the Mullard or RS Components CL8960), to amateur use was described by VK5ZO [18]. Although this uses a 3dB

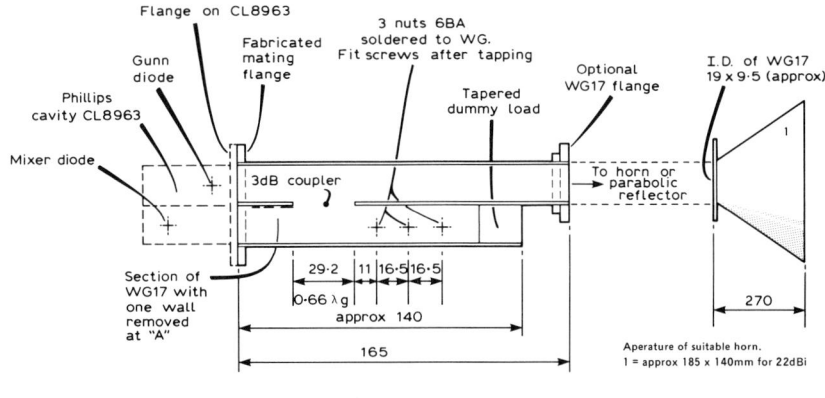

Fig 18.48. Use of a "side-by-side" doppler unit (VK5ZO)

coupler and, therefore, loses half of both the received and transmitted signals, the design is included to illustrate amateur ingenuity. Fig 18.48 is largely self-explanatory in terms of construction. The size of both the Gunn and mixer cavities approximates to the size of WG17. The mating of the cavity to the waveguide was more accurately achieved by completely removing one wall from the shorter piece of WG17 and using this for the mixer section. The longer piece of guide was modified by removing $0.66\lambda_g$ (29.2mm) from the common wall of the main guide, and the unused arm in the mixer section was terminated in a dummy load, in front of which was placed a matching screw tuner. This was used to set the mixer current to about 0.4 to 0.7mA. A flange was fabricated to match the coupler to the doppler module and another to join to the antenna. Even with the large signal losses involved, the modified module should be capable of working over appreciable distances and would certainly be suitable for setting up short distance fixed links.

18.3.3 Dielectric resonator stabilised oscillators

A relatively new device which has become accessible to amateurs, largely as a result of falling prices for "consumer" products, is the GaAs fet dielectric resonator stabilised oscillator, variously abbreviated to drso, dso or dro. The dielectric resonator (dr) has already been briefly mentioned in chapter 6 and the theory will be expanded a little further here.

A dielectric resonator is, simply, an accurately dimensioned disc, cylinder, bar or rod (depending on its particular application) made from a ceramic dielectric possessing a well-defined, very high relative dielectric constant, ε_r, and very low loss. The ε_r value is 10 to 150 (typically 30 to 40) and is much larger than values for other common dielectrics. The electromagnetic fields form standing waves in the dielectric, which acts just like a conventional cavity; the resonances depend on the exact geometric dimensions, the relative dielectric constant, ε_r, the relative permeability constant μ_r and, to a lesser extent, the presence of conducting plates around it.

As with all "cavity" structures, many patterns of standing wave and, thus, resonances, can be set up by the exciting energy. In the case of a dielectric cylinder or disc of diameter D and height H (of approximately D/2), the fundamental or dominant mode of resonance is similar to the TE011 mode of a cylindrical cavity resonator. The field patterns for both a metallic cavity resonator and a dielectric resonator are given in Fig 18.49a and b respectively. In a metallic resonator the fields are limited by the conductive nature of the walls, so that no fields exist outside the cavity and coupling ports must be provided to get energy into or out of it. This makes such cavities difficult to use, especially in conjunction with microstrip techniques, although such use is possible, as will later be seen.

The dielectric resonator does not present this problem. Although the electric field is mainly confined within the disc or cylinder as a result of the high dielectric constant, the permeability (μ_r, usually 1, the same as air) allows the magnetic field to extend beyond the boundary of the material as shown in the figure. This also allows the resonator to be coupled easily to its source of excitation. The resonator is simply fixed in place between the circuit elements to

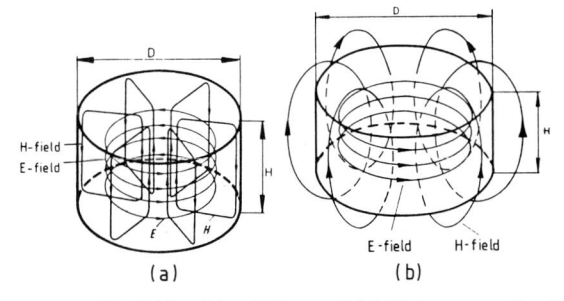

Fig 18.49. The E (electric) and H (magnetic) fields in a conventional cylindrical metallic cavity (a) and a dielectric resonator (b). Note that in the metallic cavity, both the E and H fields are contained by the conductive walls of the cavity whilst in the dr, the H field extends beyond the boundaries of the resonator, allowing direct H-field (magnetic) coupling into and out of the resonator. After DB1NV and DJ3RV, *VHF Communications* 4/1983

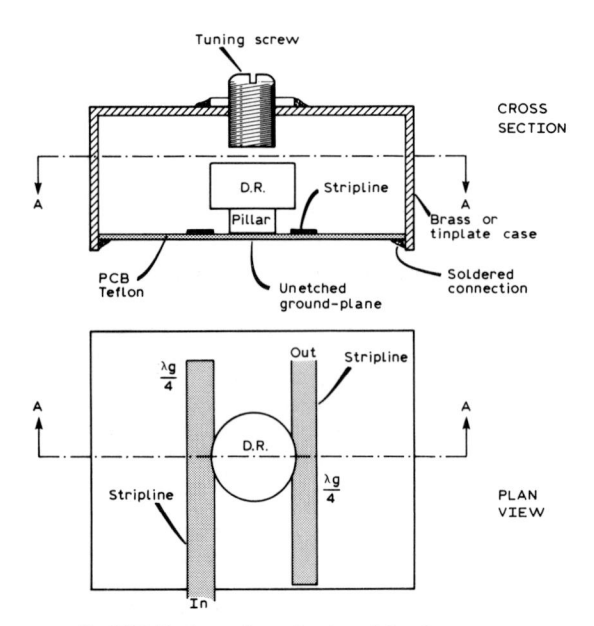

Fig 18.50. Tuning and coupling to a dielectric resonator

be coupled together. These are most conveniently two λ/4 microstrip lines. Such coupling is shown diagrammatically in Fig 18.50, which illustrates the use of a dielectric resonator to couple into two lines and also the method of tuning the resonator.

One of the most important advantages of the dielectric resonator is its small size, which enables it to be used with microstrip. The dimension D (diameter) of an air filled cylindrical resonator is typically one wavelength and its height (H) is half a wavelength. At 10GHz this infers a diameter of approximately 30mm and a height of approximately 15mm. However, in the dielectric resonator, the velocity factor of the electromagnetic waves is reduced by the factor $1/\sqrt{\varepsilon_r}$ and the physical dimensions are reduced by the same factor. Thus, for resonators with a dielectric number in the thirties (a common value) they have a diameter of about 5mm and a height of 2mm.

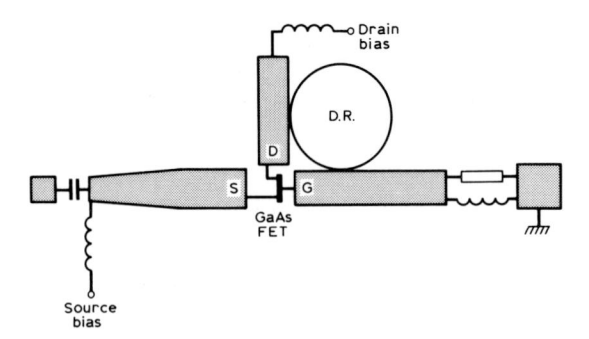

Fig 18.51. Theoretical circuit of a dielectric resonator stabilised oscillator (dro). (Mitsubishi data sheets)

The suitability and usefulness of a particular dielectric resonator is governed mainly by its Q and its frequency versus temperature characteristic. Unloaded Q values of 10,000 are not uncommon in resonators designed for use up to 25GHz, whilst temperature coefficients can be "designed-in" by choice of material, and often values of parts per million (ppm) can be achieved. Both factors are influenced by the degree of coupling into the external circuits. Light coupling will result in preservation of a higher Q and hence a more sharply defined resonance. If too much power is dissipated in the resonator, then heating effects can move the resonant frequency somewhat. However, the material from which the resonator is made can be selected to have a positive, zero or negative temperature coefficient. By this means it is possible to offset the usually positive circuit coefficient with a resonator of negative coefficient, resulting in a virtually temperature-independent oscillator.

Tight coupling is obtained when the resonator is physically close to the coupled lines, usually by being "glued" to the surface of the substrate. This will usually mean considerable loss of Q by reason of the resonator's close proximity to the ground plane of the circuit board.

Lighter coupling and considerably better performance can be obtained by standing the resonator off the substrate by means of a low-loss dielectric pillar as shown in Fig 18.50. Specially dimensioned pillars are available to optimise coupling into the particular resonator in use.

The electrical circuit of a typical dro is given in Fig 18.51. It consists of a conventional drain to gate positive feedback oscillator in which the frequency selective feedback circuit is the high Q dielectric resonator. The complete circuit, other than the fet and resonator but including the decoupling components, is produced in microstrip on a ceramic substrate which in turn is mounted within waveguide. The bias supplies are fed to the circuit via feedthrough capacitors soldered into the walls of the guide. Although the theoretical circuit shows the need for separate source and drain bias, in practice the device is made self biasing, so that only one positive supply is needed.

There is a preset tuning screw which is mounted above the resonator and allows tuning over a range of, typically, more than 500MHz. If the oscillator needs to be continuously tuned, then it should be possible (but not easy!) to replace this screw by a micrometer screw.

Other dros designed as tvro converters, contain additional microstrip circuitry which couples the local oscillator and signal into a Schottky mixer diode. The theoretical circuit is shown in Fig 18.52. These devices will have additional adjustable screws to allow independent adjustment of oscillator drive to the mixer and also to partially suppress unwanted oscillator radiation to the (receive only) antenna. The i.f output is via a feedthrough mounted in the waveguide wall.

A good example of a dro is the NEC MC5808 tvro local oscillator source on 10.678GHz which is stabilised to remain within ±1.0MHz (typically ±0.7MHz) of nominal over the temperature range −40 to +50°C.

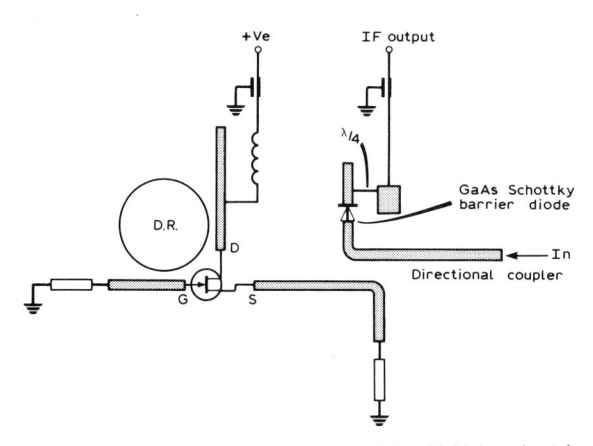

Fig 18.52. Theoretical circuit of a dro/mixer (Mitsubishi data sheets)

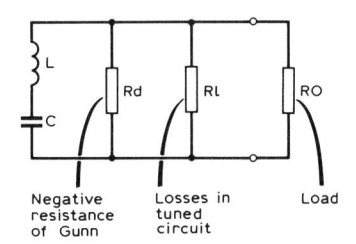

Fig 18.53. Equivalent circuit of a Gunn device

Similar specifications apply to the Mitsubishi high stability oscillator, type FO–1010XS, where the frequency variation with temperature is quoted as ±0.5MHz over the range −20 to +60°C. A less expensive oscillator, the FO–1010X, is quoted as ±5.0MHz over the same temperature range, whilst the oscillator/mixer assembly FO–UP11KF, designed as a tvro "front end", is typically ±10MHz over −30 to +70°C.

Many other modules exist, some designed for doppler use as well as for tvro applications. It should be seen, therefore, that these devices offer very good stability, potentially an order of magnitude better than the traditional Gunn oscillator. Sideband noise is also less than that generated by Gunn oscillators. The dro is considerably more efficient than a Gunn of corresponding power output. For instance, a dro will typically consume 50mA at 6V (0.3W) and produce 10mW output whereas the Gunn might consume three times the current viz 150mA at a slightly higher voltage, 7.5V (1.1W). Until recently, only difficulty of supply and cost have limited their use by amateurs.

Now that dros and dro/mixers are in mass production for satellite tv, the prices have fallen to be comparable to commercially-priced Gunn oscillators and, although still well above "surplus" prices, this disadvantage is outweighed by superior performance. It is expected that these devices will displace Gunns as a means of generating amateur fm signals and lead to the use of narrower bandwidths than previously possible in simple equipment. The implications of this have already been outlined in section 18.1.3.

18.3.4 Power supplies and modulators

Both Gunn oscillators and dros may be frequency modulated by using the effect called "frequency pushing" where the frequency is dependent on the supply voltage.

The frequency of an oscillator is primarily determined by the effective cavity dimensions in the case of a Gunn or the characteristics of the dielectric resonator in a dro. Both can be tuned mechanically. However, in both types of oscillator the frequency can be varied over a small range by varying the dc conditions on the active device.

Any semiconductor junction can be represented by a complex network consisting of both capacitive and inductive reactances and resistance. The simplified equivalent circuit of a Gunn device and its cavity, for instance, is shown in Fig 18.53. These reactances and resistances are all effectively shunted across the resonant cavity and variation in any of them will have an effect on the overall performance. The effect of varying loads is well known − frequency pulling. Assuming that the other parameters (inductive reactance of the Gunn and its mount and the combined negative resistance of the Gunn and the positive resistance of the cavity) remain sensibly constant, the one parameter likely to have the greatest effect on frequency is the capacitive reactance of the device within the cavity. Since the capacitive reactance of the semiconductor junction depends on the applied voltage, it should be apparent that varying the voltage applied to the device is equivalent to varying the capacitor in a tuned circuit. The effects on the resonant frequency of the waveguide cavity and dr are no different to the effects in the familiar "coil and capacitor" combination.

The voltage versus frequency characteristics of two types of Gunn oscillator are shown in Fig 18.54. It will be seen that the low Q coaxial cavity is much more sensitive than a high Q waveguide cavity. That is, the degree of frequency shift with applied voltage change is much higher with a low Q cavity than with a high Q cavity. This frequency pushing effect is, or can be, one of the main sources of fm noise on the carrier. Unless the bias source is remarkably free from noise and spurious oscillation, there is every

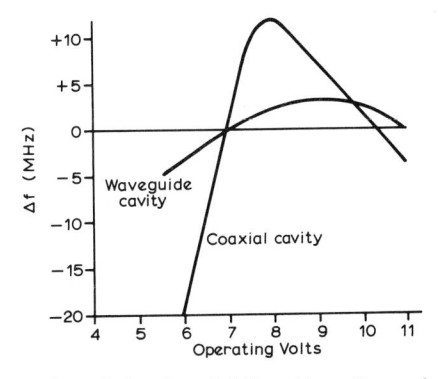

Fig 18.54. Gunn device characteristics – bias voltage vs frequency in high-Q and low-Q cavities

Table 18.4. Pushing factors for various types of Gunn oscillator

Oscillator type	Pushing factor* MHz/V
Home made iris cavity	2 – 10 (depending on iris size)
Home made rear cavity	15
Solfan iris (Osc.only)	13 (in-line mixer type, 5MHz)
Mullard CL8630 step	2
AEI/Pascal	20

*Measured:
1. At 10.38GHz (nominal) with the oscillator under test running through a variable attenuator into a cavity wavemeter/detector assembly.
2. The attenuator was set in all cases to give maximum isolation consistent with measurable detector output.
3. The bias supply was that described in this section (see later).
4. In each case the bias voltage was set midway between the turn-on voltage and the maximum power voltage (see Gunn power characteristic curves, Fig 18.36).

likelihood that the generated carrier will be "dirty" in terms of fm noise sidebands. Even with a well designed and decoupled bias supply, there will still be some fm noise present, arising from random fluctuations in the Gunn device itself, so-called "jitter". There may also be some random a.m noise present, resulting from amplitude variations in the device output and coming from "flicker" noise which is present in all devices.

Thus, it is better to employ a high Q oscillator to minimise these effects and to ensure that the bias supply is clean and free from spurious noise. Further protection against unwanted noise modulation can be gained by attention to proper screening of the entire module – taking the oscillator and its bias supply together as being a module, rather than the oscillator on its own. It is thus a good idea to enclose both the bias supply and the oscillator (or oscillator/mixer, as maybe) in one screened enclosure, rather than feeding the oscillator from a "remote" source via a length of screened cable. This is true where performance must be optimised; elsewhere, for instance in a simple home station installation, it may be preferable to opt for slightly degraded performance in order to avoid the complication of either using a waveguide run or installation of the bias supply and oscillator at masthead. Some pointers were given for this type of installation in chapter 9,

"Beacons and repeaters". Poor stability can also result if the dc feed to the Gunn is not adequately decoupled and the microwave rf leaks out of the cavity via this decoupling. Frequency changes may occur as objects moving around outside the cavity alter the amount of rf that is reflected back into the cavity through the dc feed.

One of the writers recently undertook some simple measurements of the frequency pushing characteristics of several different Gunn oscillators and the findings are shown in Table 18.4.

Practical power supply/modulators for Gunns and dros

A Gunn (or dro) power supply consists of a stabilised supply controlled by a variable voltage regulator. The associated modulator is simply some means of superimposing a few tens or hundreds of millivolts of "audio" voltage (tone or speech) upon this bias. Many circuits have been evolved for the purpose and published in the amateur press.

Most small Gunn devices developing between about 5 and 50mW require a bias supply between about 5 and 9V and at a current of up to roughly 150mA. DROs typically require 5 to 7.5V at 50 to 80mA. Thus, a supply designed to cope with the requirements of a small Gunn oscillator will easily cope with the requirements of a dro and the same supply can be used with either device – with the proviso that the maximum voltage supplied to the dro does not exceed that recommended by the manufacturer. It is advisable to choose the values of the voltage-setting elements of any circuit used to limit the output voltage to a figure slightly below this maximum.

The simplest bias supply would be a zener diode of suitable voltage and rating. However, this does not allow bias voltage optimisation nor does it easily allow modulation facilities. Fig 18.55 shows a simple circuit based upon four transistors and a zener diode. This circuit supplies a fixed voltage bias, chosen to be optimum for many small Gunn diodes but, being fixed, does not allow optimisation for the particular Gunn diode in use. The transistors form a pulsed tone oscillator which a.m modulates the Gunn output, rather than the customary fm. The simplicity of the circuit does not easily allow the user to optimise the modulation level.

Fig 18.56 shows a circuit which allows the user the choice of tone or speech modulation. It consists of a single stage microphone amplifier, a tone oscillator, a 741 op-amp and series pass transistor. The op amp controls the base of the pass transistor. The non-inverting input (pin 3) is referred to a constant zener-derived voltage and the inverting input (pin 2) is referenced to a variable (presettable) voltage derived from the emitter of the pass transistor. The non-inverting input is used to apply modulation to the pass transistor and the stabilised, modulated, variable bias is taken off the emitter of the pass transistor. The tone oscillator is a multivibrator which can be keyed. The deviation level of the tone modulation can be preset and afc (automatic frequency control) can also be applied to the two op amp inputs. A suitable afc control voltage/current can

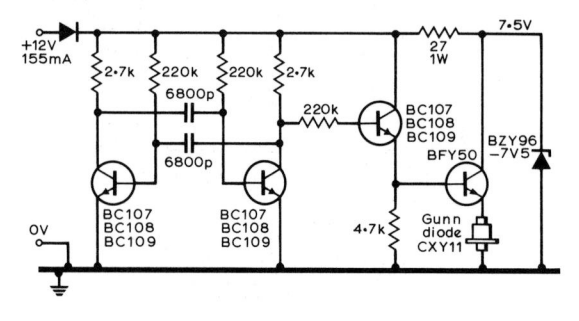

Fig 18.55. Tone (pulse) modulated Gunn supply (Mullard)

THE 10GHZ ("3CM") BAND

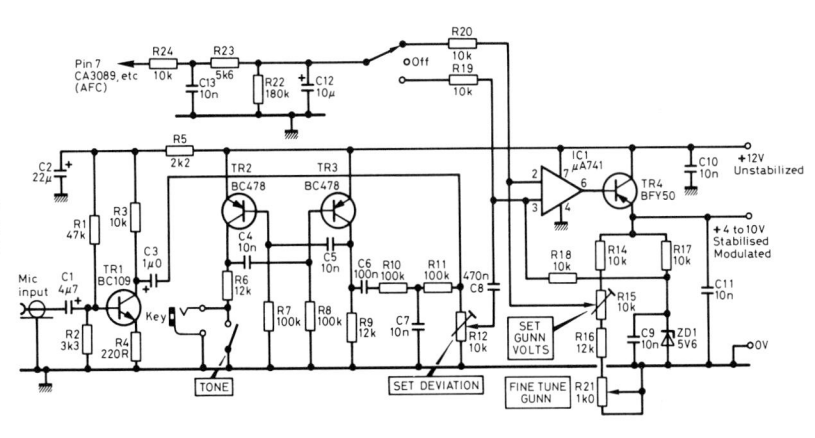

Fig 18.56. A speech and tone modulated Gunn supply, showing automatic frequency control (AFC) circuitry (G3NKL)

be derived from the afc terminals of an integrated i.f signal processing chip such as the CA3089 or CA3189 series. The variable bias available at the output can be adjusted in the range from about 4V to 9V. Replacing the 10k variable with a combination of a smaller value of potentiometer and fixed resistors allows the tuning range to be cut down and also the upper limit to be set; resistor values should be selected on test. The circuit also shows the interfacing needed to apply effective afc, derived from the afc output of a CA3089 chip.

The construction of any of these circuits is not particularly critical, although care should be taken with layout to prevent unwanted coupling, instability or noise pickup. It is wise to provide decoupling both on the Gunn terminal and the power supply. The network shown in Fig 18.59, connected directly across the Gunn, is recommended. Such circuits have been successfully constructed on Veroboard or other prototyping board and even using "bird's nest" construction on small pieces of plain pcb material. A much neater product would result from the constructor designing a small pcb, although this is not really necessary. When completed and tested, the circuit should be checked for spurious oscillation as this type of circuit tends to be prone to high frequency instability, sometimes up to several hundred kHz or even a few MHz. The easiest means of checking is to examine the output on an oscilloscope and, if instability is encountered, the cure is usually the fitting of a small capacitor from the base of the regulator transistor to earth. An alternative to an oscilloscope is the simple test circuit shown in Fig 18.57 (due to GW3PPF, [19]). Any meter reading indicates the presence of spurious oscillation. It may be possible to locate the critical part of the circuit by touching suspect parts of it with the point of a small screwdriver; this may temporarily stop the oscillation and additional decoupling should be added there until the meter reading disappears.

Gunn power supply/modulator with short circuit protection

Fig 18.58 shows the basic circuit of a stabilised variable bias supply which was originally designed for the London 10GHz beacon GB3LBH (now no longer operational). It uses a 7805 positive voltage regulator commonly used in computer equipment to supply a fixed 5V supply to logic circuits. It will supply up to 1A when used with a suitable heat-sink and has in-built short-circuit and thermal protection. Although the regulator chip is designed for a fixed output, by fitting a "pedestal" resistor, between connection C and earth, the output voltage can be increased up to within 2.5V of the input voltage, by an amount equal to the voltage drop across this resistor. The maximum input voltage to the regulator is 25V. The resistor is also a convenient means by which modulation may be applied to the output voltage, as shown.

The use of the pedestal resistor inevitably means some loss in regulation although this is small; in a unit set up to produce 9V from a 12V supply, the output voltage dropped

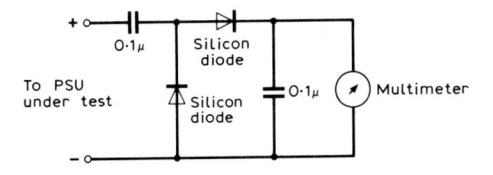

Fig 18.57. Test circuit for detecting supersonic parasitic oscillations in a Gunn power supply regulator (GW3PPF)

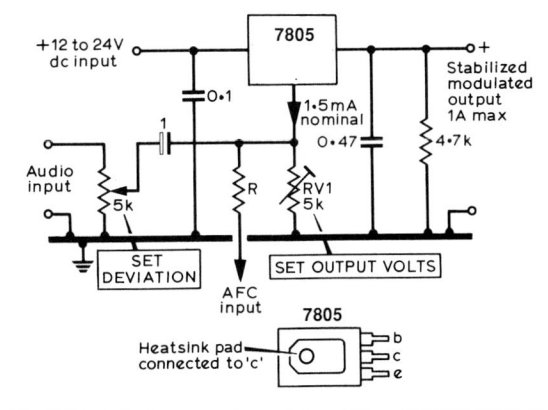

Fig 18.58. Basic Gunn supply regulator using a uA7805 ic (G4ALN)

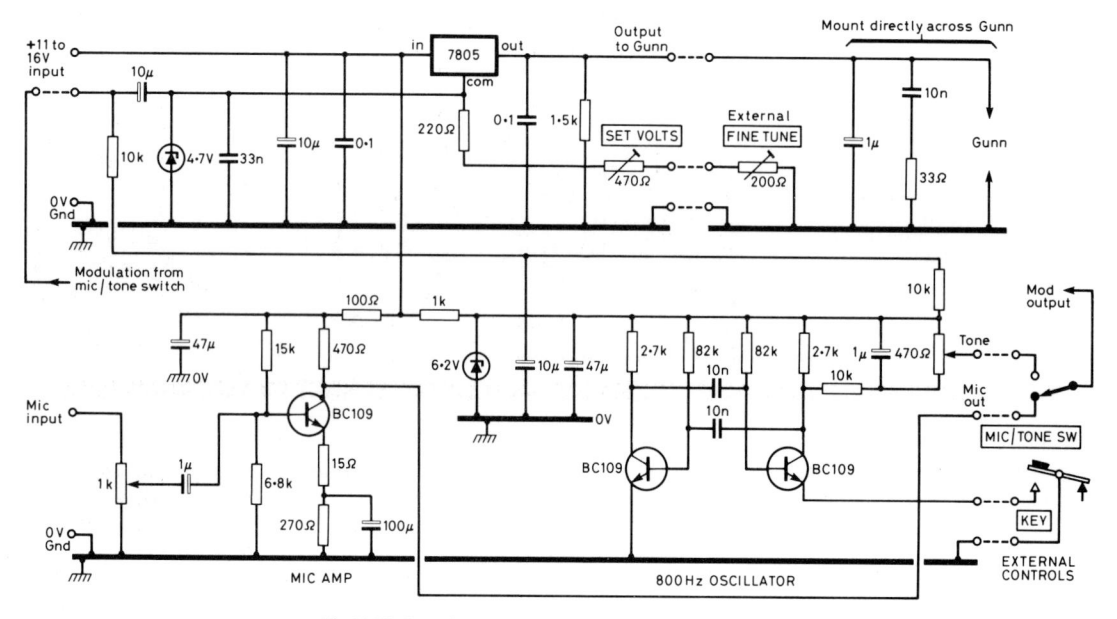

Fig 18.59. Complete tone and speech modulator circuit (G4KNZ)

only by 35mV when a load taking 200mA was connected, and by only about 10mV/V as the input voltage was changed. This is adequate stability for almost all amateur wideband applications and this basic circuit has been used as the basis for the design, due to G4KNZ [20], to be described here.

The circuit is given in Fig 18.59 and the layout of the components is given in Fig 18.60, together with heat-sink details. The pcb track pattern is given in Fig 18.61.

A single stage microphone amplifier provides a gain of up to about ten times and is suitable for use with medium impedance microphones. The tone generator is a two transistor multivibrator which can be keyed on and off for cw. Modulation from the microphone amplifier or tone generator is applied to the common terminal of the regulator ic via a capacitor and the level required depends on the Gunn voltage (and Gunn diode) but is typically from 10 to 100mV peak-to-peak. The output is variable in the range 6 to 9V. Although the ic is protected against short circuit, the heat sink employed is such that it will not withstand this condition indefinitely. For the current normally drawn by a low power Gunn the ic does not actually require a heat sink; however it is better practice to provide limited heat-sinking capacity to guard against short term excess dissipation.

Several decoupling capacitors and resistors have been added to prevent noise on the power supply or tone from the tone generator reaching the microphone amplifier or the Gunn diode. A diode has been included at the power input to protect against accidental supply reversal and

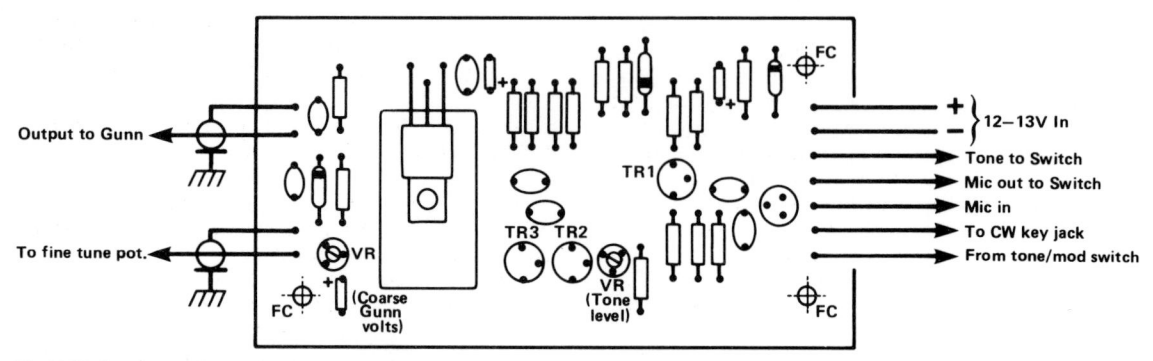

Fig 18.60. Component layout for the circuit of Fig 18.59. The heatsink for the regulator is a piece of 1mm aluminium sheet, about 30mm x 30mm, bent into a U-shape 20mm wide with 10mm sides, to fit the available space on the board. The mounting hole for the ic is 4mm diameter

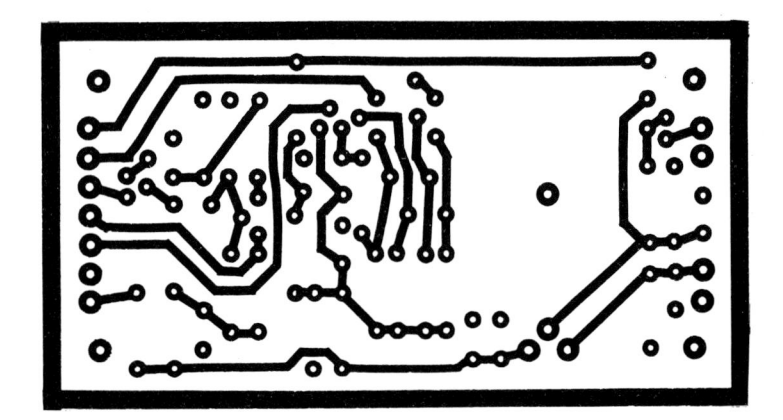

Fig 18.61. PCB track pattern for the circuit of Fig 18.59

every attempt has been made to keep the circuit as simple as possible, consistent with good performance.

The circuit is etched onto one side of a double clad board. The other side remains unetched and all earth connections are made to this ground plane. Where component leads need to pass through the board un-earthed, the simplest procedure for the home constructor – avoiding double etching masks – is to remove a small circle of copper using a drill bit or a Vero-tool of the type designed to remove track from Veroboard, this operation being carried out *after* the component holes have been drilled and the board cleaned up. The regulator ic will need to be electrically isolated from its heat sink if the sink is screwed down to the pcb. A dab of heat sink grease and the usual insulator bush assembly is used to accomplish this task.

Construction is straightforward and any order of assembly can be used. Testing consists of checking the range of output. The "fine tune" potentiometer should be first set to mid travel and then the preset 470R set to give the desired output voltage. The fine-tune control should then give a voltage swing of about ±1V around this preset value. The fine-tune control can be a multi-turn potentiometer with a calibrated drive; this can subsequently be calibrated directly in frequency, thus providing the user with a guide to the oscillator frequency after the Gunn mechanical tuning screw and operating voltage have been set to the required values. After testing in this manner it is a good idea to check the circuit for spurious oscillation (although this has not proved to be a problem with this particular circuit) by examining the output on an oscilloscope or by using the test circuit already described. The completed board is best mounted in a screened box to avoid spurious noise or signal pickup and the incoming power supply may need extra off-board decoupling for the same reason.

Measurements made on the prototype circuit gave the following results: tone frequency 830Hz (this may vary with component tolerance); tone output level 100mV (max); microphone amp 100mV (max) for 10mV input; response −6dB at 100Hz and 25kHz.

Using a Mullard CL8630 Gunn oscillator at a bias voltage of 7V, 150kHz deviation was obtained with a modulation voltage of 30mV p–p at the regulator common terminal.

Setting the Gunn operating voltage

The Gunn diodes' voltage/power characteristics have already been discussed. For best operation the Gunn voltage should be set to about 0.5 to 1 volt above the starting voltage – that is the voltage at which oscillation reliably starts. At this point the oscillator can be tuned smoothly over a wide range using the dielectric tuning screw and its noise sidebands will be quietest. This will be important in a receiver if a low i.f, such as 10.7MHz, is used. Slightly more power will be generated at higher voltages, but the oscillator will then generally be less well behaved, possibly with a tendency to break into spurious modes of oscillation and certainly with a greater tendency to drift due to thermal effects because of the higher device dissipation. The front panel control used to fine-tune the Gunn by frequency pushing should really have a very restricted range, ideally no greater than ±1V. Adjustment beyond this, whilst still tuning the device effectively, may bring the operating point to the non-linear portion of the characteristic curve, so that the deviation will vary greatly as the bias voltage is increased. The optimum operating point should be set by a preset resistor and left at that point!

Adaptation of power supply/modulators for use with dros

The frequency pushing characteristics of currently available dros is not greatly different to those of a Gunn oscillator. For instance, the Mitsubishi FO-UP10KF oscillator/mixer will push at 2MHz/volt, just the same as the Mullard CL8630 Gunn. This means that similar regulator/modulator circuits can be used for either device. However, the dro will normally operate at a nominal 6V, with an absolute maximum of 7.5V which must not be exceeded. The tuning voltage range is thus less than that of the Gunn.

With the circuit of Fig 18.56, the adaptation is simply a matter of changing the values of the resistive divider chain (R14, R15 and R16 in Fig 18.56) to give the required adjustment range, say 4 to 6V. The total value of the chain should be maintained around the nominal 30kΩ value shown, but the effective range of the variable reduced to give the required output adjustment. Replacing the 10kΩ variable directly with a 5kΩ variable in series with a 4.7kΩ

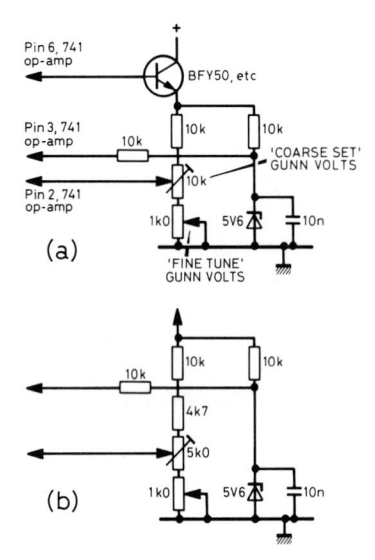

Fig 18.62. Modifications to a "standard" Gunn modulator/regulator to allow use with a dro. (a) original circuit (b) after modification

fixed resistor as shown in Fig 18.62 will cut down the adjustment range to approximately 4 to 6V. An additional safeguard might be to connect a 6.2 or 6.8V zener directly across the output of the unit to protect against any accidental overvoltage.

The 7805-based regulator is a little more difficult to use, in that the output cannot be made to swing below 5V and, at 5V output, the common pin is connected directly to earth, disabling the modulation. However, the common pin pedestal resistor could be chosen to give a range of 6 to 7V. An output of approximately 6V is obtained with a resistor of 200Ω and 7V with approximately 450 to 470Ω. The other alternative would be to replace the 7805 regulator with a variable voltage regulator such as the LM317L (100mA rating) or the LM317M (500mA rating), both of which will supply a regulated output in the range 1.2 to 37V from a supply rail of 4 to 40V. The general circuit for these regulators is given in Fig 18.63a and the graph plots output voltage against R2, the "adjust-pin" resistor. Fig 18.63b gives the circuit and values for either ic, to obtain adjustable voltages in the range of approximately 4 to 6.4V for use with a dro, or 4 to 9.2V for use with a Gunn. A LM317L can supply the current for a dro but an LM317M should be used to cater for the higher current demands of a Gunn.

18.3.5 Narrowband sources for 10GHz

Narrowband signals at 10GHz may be generated by high order single-step multiplication from a lower frequency, multiplication in several steps, by a combination of multiplication and mixing or by phase or injection locking techniques. Whichever method is used, multiplication is invariably involved at some stage and, since this generates harmonics other than the wanted frequency, filtering of the output is essential. Regardless of method, the equipment will be much more complex than the simple wideband sources described so far. The enormous advantages of narrowband over wideband have already been outlined and the amateur already using narrowband equipment will need no reminder of this! Narrowband sources are not only more complex in concept, but are also more demanding in resources (time and money) and technique; this, however, should not deter determined enthusiasts! Another requirement for successful construction of narrowband equipment is a need for test-gear – albeit homemade – for use in the alignment process.

Perhaps the ideal way of approaching the problems of constructing and operating such equipment is first to gain practical experience of 10GHz techniques by building and operating wideband equipment and then to "graduate" to narrowband techniques. One thing is for certain – most, if not all, of the components acquired or built and used in the wideband learning phase will be of equal use later; there is seldom any redundancy of "wideband" components! Many can be used, as later described, in the essential test equipment for narrowband.

At the moment the technique of generating narrowband signals at 10GHz is fairly standard; a high stability, low noise uhf source is multiplied – perhaps directly, perhaps in several stages – to the final, desired frequency. It is then well filtered to remove the unwanted products of the multiplication process.

Alternative techniques are being developed by amateurs as modern, advanced GaAs devices (transistors, fets and microwave integrated circuits – mmics) become available at affordable prices – see the last sections of this chapter. For the moment, however, the techniques to be described are those which are in common use; ideas for alternative techniques are outlined in section 18.9 and for further ideas the reader should refer to such periodic publications as the *RSGB Microwave Newsletter*, *VHF Communications* and *Dubus-Informationen* (the latter two are published in English editions) for the very latest amateur developments in the use of the newest technology, much of which requires pcbs and devices not always accessible to the "ordinary" amateur.

Drive sources

The drive sequence should always start at the highest practicable frequency in order to maximise harmonic spacing and mimimise the multiplication factor. Wide harmonic spacing facilitates filtering and minimum multiplication factor enhances efficiency. Using easily available and reasonably robust crystals, this will usually mean somewhere in the region of 100MHz and it is usually convenient to multiply this into the uhf range (filtering where appropriate) on the driver board. It should also employ an oscillator of good design, in terms of both stability and freedom from noise.

Several uhf drive sources have been described in amateur literature and a couple of designs have been produced commercially (by Microwave Modules Ltd and Wood and Douglas Ltd). One of the most successful high quality sources designed and published was the circuit developed by the RSGB Microwave Committee which was fully described in chapter 8, "Common equipment". This is recommended for use as a drive source suitable for 10GHz narrowband, especially when used in conjunction with the low voltage-drop regulator board described with it.

When completed and checked, it is *essential* to house the source in a robust, die-cast box to provide screening, mechanical rigidity and a degree of thermal protection. All in-going connections should be made via feedthrough capacitors or "filtercons". It is essential to keep the oscillator well screened from the rf output to the multipliers as rf feedback can cause serious frequency pulling. To maximise the stability and minimise the noise content of the output the source, in its own screened enclosure, should be fixed in the same outer rigid casing which houses the subsequent multiplier(s) and filter(s).

Other, later, drive sources are the G4DDK–001 (1.1 to 1.3GHz) or the G4DDK–004 (2.0 to 2.6GHz) which can be substituted for the older design.

Power amplification at uhf

The uhf drive source described will produce between 100 and 400mW output, dependent largely on the particular make of transistors employed in the circuit and the final output frequency. The actual level of output is easily determined with sufficient accuracy using the simple methods of power measurement described in chapter 10, "Test equipment".

In order to drive the subsequent multiplier(s) to obtain a useful level of 10GHz output, it is necessary to amplify the driver output to a level of several watts. Fortunately such an amplifier does not need to be linear and any design for fm use is suitable. Several different amplifier modules have been described in chapter 8, "Common equipment", some of them thick-film hybrid "power gain blocks", e.g. the BGY22, and some constructed from discrete components. One of these designs should be selected, according to the constructor's needs.

Commercially produced fm amplifier kits are also available at reasonable prices, which are easily constructed and highly reliable. Examples are the 432MHz fm amplifiers produced in kit form in the UK by Wood and Douglas; these will tune up at 384MHz and have been used both at 432 and 384MHz in several UK microwave beacons with great success. Similar products are available in most other countries and thus it is not proposed to describe any other designs here.

Most uhf amplifiers are essentially quite broadband devices and should, ideally, be preceded and followed by a bandpass filter: again a suitable, small, filter is available in kit form (from Wood and Douglas in the UK) or any of the simple filter designs described in several VHF/UHF manuals can be used – or designed from the data given in chapter 12, "Filters"!

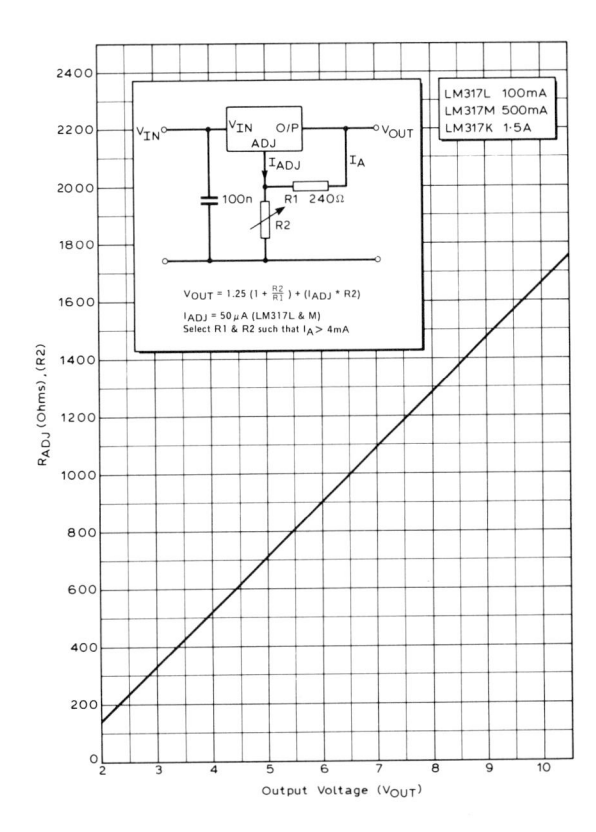

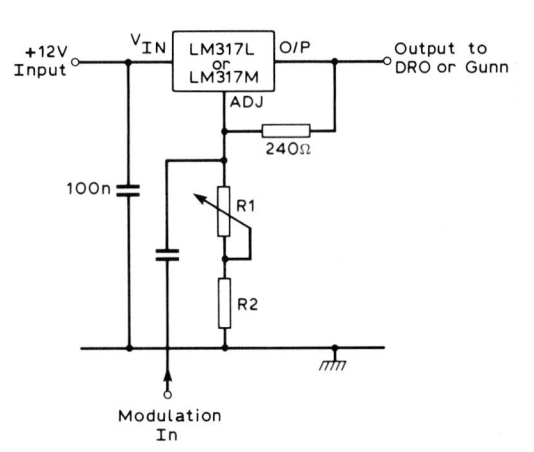

$V_{OUT} = 1.25 (1 + \frac{R2}{R1}) + (I_{ADJ} * R2)$

$I_{ADJ} = 50\mu A$ (LM317L & M)
Select R1 & R2 such that $I_A > 4mA$

LM317L 100mA
LM317M 500mA
LM317K 1·5A

Note:
For approx. 4 to 6.4V output (DRO)
R1 = 470Ω Cermet pot.
R2 = 510Ω 1% high stability metal oxide

For approx. 4 to 9.2V output (Gunn)
R1 = 1.0k   Cermet pot.
R2 = 510Ω 1% high stability metal oxide

For maximum regulation stability the 240Ω resistor should be connected as close as possible to the regulator output pin.

Fig 18.63. Using the LM317-series variable voltage stabiliser ic's

Multipliers

Passive (varactor) multipliers are widely used as a cost-effective means of generating signals in most of the microwave bands. Occasionally, active (valve or semiconductor) multipliers have been used, although designs for these have by no means been common or widely used. Active multipliers are seldom used above about 2GHz because of falling efficiency and rising costs of suitable devices; general purpose GaAs fets can make good doublers, but are not really suitable for higher multiplication factors. So passive multiplication is often the means adopted to reach the higher bands and particularly 10GHz. Whilst passive varactor multipliers are relatively simple and need no power supplies other than the rf energy supplied to them, there are two main problems with them.

The first is their vulnerability to changes in operating conditions; minor variations in drive level or loading can cause the output to break up into a large number of spurii. This effect can be difficult to detect, let alone cure, without the use of a spectrum analyser, to which few amateurs will have ready access. The second problem is that the output, even though crystal-controlled, may have a surprisingly large bandwidth and therefore be far from "narrowband" by the standards of lower frequencies.

The first problem can usually be cured by providing isolation, in the form of resistive attenuation, between the drive source and the multiplier, albeit at the expense of efficiency in terms of drive power.

The second problem is common to any multiplier and is only really solved by maximising the quality of the drive source. Any phase-noise sidebands present in the drive source output will also be multiplied; this is particularly noticeable when using the high order multiplication needed to get to 10GHz. This noise originates in the oscillator and is always present. However, it becomes much more apparent at 10GHz because the bandwidth is proportional to the square of the overall multiplication factor. If the source is to be used as a receiver local oscillator then a considerable improvement can be made by the combined tactic of using a very narrow filter following the multiplier and a high i.f if the narrowband source is used in a receiver or transceiver; for transmit purposes, this noise is of less consequence.

In what follows, reference is made to multiplication from 384MHz to 10.368GHz, ie. directly into the narrowband communication sub-band. If transverting is envisaged, it is common to multiply from 378.6666MHz (or higher) to 10.224GHz and then to use additive mixing with a 144MHz source to give the required output of 10.368GHz. It is possible to multiply directly from 384MHz to 10.368GHz, but since this is a multiplication factor of ×27 in a single stage, the efficiency is low. Another alternative is to triple from 384MHz to 1,152MHz in one stage and then multiply again (×9) to 10.368GHz. Although it is necessary to provide isolation (resistive padding) between the cascaded multipliers to ensure stability and provide higher drive power to compensate for this isolation, this technique results in more output at 10GHz, even though the overall efficiency is still very low. Both types of multiplier are described below. Other multiplication sequences can, of course, be used.

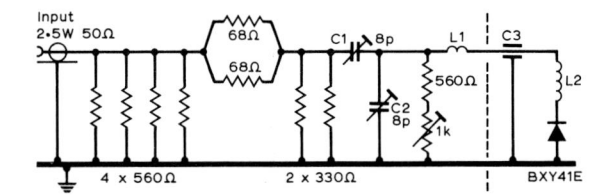

Fig 18.64. Circuit of a 384MHz to 10,368MHz multiplier (G8DEK)

Alignment of multipliers

On the face of it, being simple circuits, it might appear that multiplier alignment would be simple. However, it is not! All the variable circuit elements interact during alignment and the alignment attained at one input level will not hold for other levels of input. Thus, alignment needs to be a methodical and iterative process. It also requires certain items of test equipment – a sensitive multimeter, a directional coupler, a waveguide mounted detector, a variable waveguide attenuator, an absorption wavemeter and a (preferably prealigned) 20MHz-wide waveguide filter, as well as the drive source of suitable frequency and power, together with a fair measure of patience!

The filter is an essential part of the multiplier to final frequency and should be prealigned using a source other than the multiplier, since the chances of correctly aligning the multiplier and filter together are remote; too many harmonics and tuning combinations exist! Methods for prealigning a narrow filter are described in section 18.6. Final alignment must always be done on the combined multiplier and filter.

A 384MHz to 10.368GHz multiplier

This design, due to G8DEK [21], probably represents the simplest means of producing a usable power of a few milliwatts at 10GHz from a source at 384MHz. The circuit is given in Fig 18.64 and, although originally designed for use with a BXY41E multiplier diode, will work equally well with other similar diodes such as the Microwave Associates' MD4901 and MA44150.

The construction of the module should be apparent from Fig 18.65. There are often problems when using thin ptfe or Sellotape as the insulating washer, as it can puncture and short out the diode. An alternative insulator is a mica washer that is normally used for mounting power transistors – these already have holes drilled in them and they can be adjusted in thickness by splitting the layers of mica apart with a fine needle. It is also possible, as shown in the figure, to use the plastic insulating bushes to keep the diode mounting post central and insulated where it passes through the waveguide wall. Note that the diode should be mounted with its heatsink end (i.e. that end remote from the flange) in the 2BA screw.

An important feature of the design is the inclusion of a 6dB resistive attenuator between the varactor circuit proper and the driver in order to reduce the amount of interaction between the two. It is important that the attenuator is mounted on the multiplier itself rather than at the exciter end of a piece of connecting cable. Many of the problems with this type of multiplier appear to be due to instability in the preceding driver stage caused by the widely varying load represented by a varactor.

For a drive level of about 2.5W at 384MHz, about 3 to 5mW of rf at 10GHz should be obtained after filtering. This is sufficient for a receiver local oscillator or as a small transmitter.

As with all multipliers, a filter is *absolutely essential* to remove unwanted harmonics from the output and suitable filters have already been described in section 18.2.10. When using the least noisy driver available at that time, G8DEK measured the sideband noise and found it to be about 10dB worse than a Gunn oscillator at 10.7MHz away from the carrier and about 7dB worse at 30MHz. However, more recent measurements made by G4KNZ and G3YGF on a G3JVL transverter system (see section 18.6) suggests that the oscillator noise, using the combination of modules described in that section, has been reduced to a satisfactory level for use with a receive or transceive i.f of 144MHz. It was shown, using a broad band diode detector, that of the total of 10mW output available directly from the multiplier cavity (which includes all harmonics supportable in WG16), only about 3 to 4mW of the wanted frequency was measured at the output of the filter. A good filter is thus essential both to remove the many harmonics from the output and to attenuate the noise sidebands.

Aligning the 384MHz to 10.368GHz multiplier

Attach the waveguide detector to the output of the multiplier and connect a sensitive multimeter to the detector. Apply about 2W of drive to the input socket. Some detector current should be seen immediately.

Maximise this current by repeated adjustments of the tuning screw, the two trimmer capacitors and the variable bias resistor. When several mA of current have been obtained, the multiplier can be connected to the input of the pre-aligned filter and the detector to the output of the filter.

Again, some detector current should be seen. The output is maximised by repeating all the adjustments already made on the multiplier and by adjusting the filter tuning and input-matching screws. The detector current finally produced should be at least 1mA and possibly as high as 5mA, depending on how carefully the multiplier and filter have been constructed. The presence of excess solder in any cavity, particularly the filter, will cause losses and may lead to lower than expected output. The final check should be repeated switching-on and off of the drive. There should be no sign of instability which might be revealed by varying detector current. If there is any instability, further careful adjustment of the multiplier should eliminate it.

Be careful not to let the 10GHz detector and meter pick up any of the 2W of 384MHz drive, as this can cause confusing results. This is another good reason for keeping the connecting leads at 384MHz short and ensuring that all the circuitry is in screened boxes. The bare coil on the multiplier can radiate quite well!

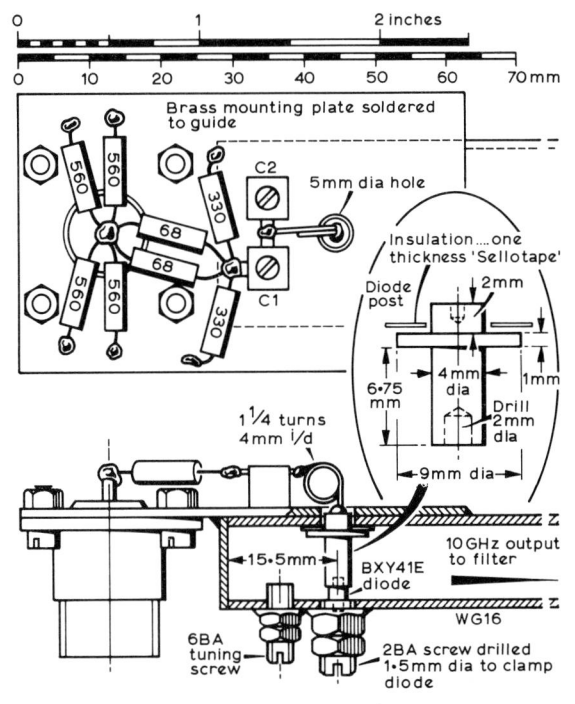

Fig 18.65. Layout and construction of the G8DEK 384/10,368MHz multiplier

384MHz to 1,152MHz multipliers

Complete design and alignment details for a high power tripler were given in chapter 8, "Common equipment". This is suitable for input powers of up to 10W. At this drive level an output of 6W can be expected. A suitable post-multiplier interdigital filter is described in chapter 14, "The 1.3GHz band". This filter will tune successfully at 1,152MHz and give adequate unwanted product rejection.

A lower powered design was described by G4MBS [22] and is included here as an alternative. The circuit is shown in Fig 18.66. For 2W input at 384MHz, 1.2W output at 1,152MHz was obtained from several prototype units. Constructional details are given in Fig 18.67. The heights of the sockets (h) and the diode (l) are determined by the height of the trimmers C6 and C8. The diode is held in position by a 0BA screw which runs in a thread tapped in the side of the die-cast box. Note that the cylindrical block soldered to L4 which holds the diode *must* be mounted so

Table 18.5. Component values for 384 to 1,152MHz tripler

C1	10pF tubular fixed
C2, 4, 5	10pF tubular trimmer
C3	3pF tubular fixed
C6, 8	5pF tubular trimmer
C7	0.5pF fixed
L1, 2	2t 18swg t.c.w. 1/4in diameter, 10mm long with 3mm leads at each end. The varactor end of L2 has a 10mm lead
L3	15mm of straight 18swg t.c.w.
L4	Brass strip 16mm x 8mm
L5	Uncut centre connection of a BNC socket
Diode	BXY27 or similar
R1	Pencil lead rubbed on diode body to give about 10kΩ (see text)
Box	Eddystone 7969P diecast, approximately 38x93x30mm

that it is at right-angles to the wall of the box to prevent the diode shearing when the bolt is tightened. This is *not* the same as making it parallel to the bottom of the box, as the walls of a die-cast box always slope.

The 0BA brass mounting screw has a 1/16in (1.6mm) hole to take one end of the diode. The other end of the diode fits in a similar hole in the 8mm diameter, 5mm long brass rod which is soldered to L4. The axes of the two holes are best aligned using a "dummy" diode turned from a 0BA bolt as shown in the diagram. This can be screwed through the diecast box and into the hole in the other brass rod. When these are well seated, solder L4 to C6 and L3/L2. Then remove the dummy diode and insert the bolt holding the diode proper. Gently tighten the bolt and lock it in place with the locknut.

In the original design, the bias resistor was "fabricated" by rubbing pencil-lead on the diode body to give

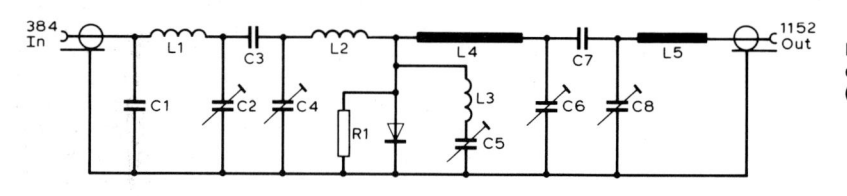

Fig 18.66. 384MHz to 1,152MHz tripler, circuit diagram and general layout (G4MBS). Component values are given in Table 18.5

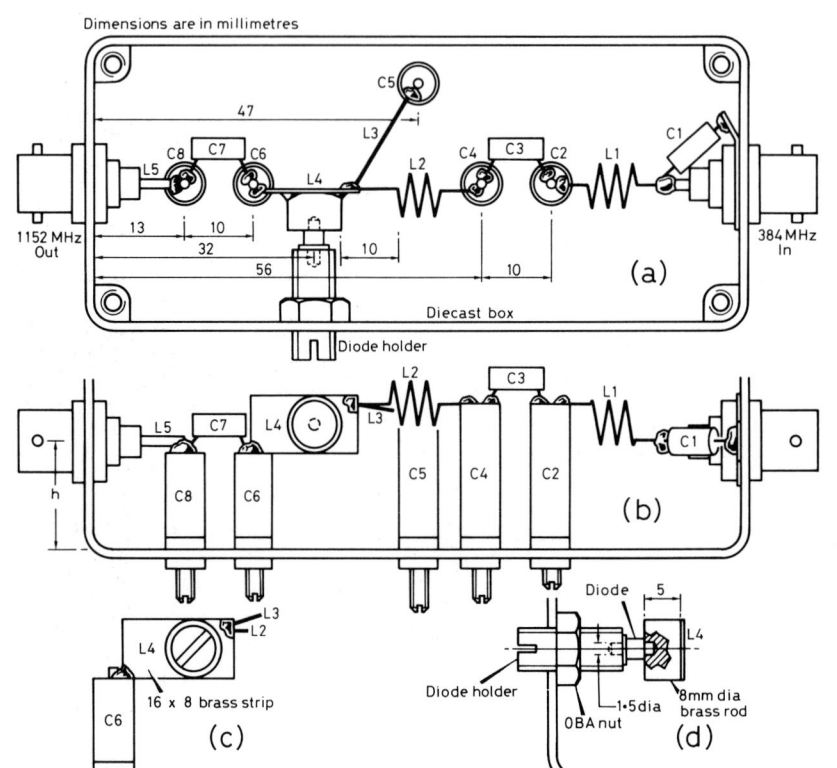

Fig 18.67. Mechanical details of the G4MBS tripler. (a) general layout, top view, (b) general layout, side view, (c) detail of the multplier diode mounting (side view), (d) detail of the multiplier diode mounting (end view)

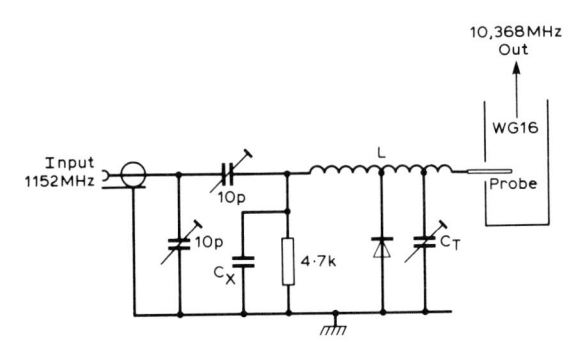

THE 10GHZ ("3CM") BAND

1,152MHz to 10.368GHz multipliers

Two designs for these multipliers are shown in Figs 18.68 through 18.72. The first design, due to DK2VF and DJ1CR and published in *VHF Communications* [23] was based upon a copy of a commercial multiplier which was once available in limited numbers on the surplus market. The schematic diagram is given in Fig 18.68. D1 in the original description was a Microwave Associates' MA44140, L1 and C3 were fabricated as part of the coaxial mounting for the diode, C1 and C2 were JFD trimmers type MVM 0.8 to 10pF or type AT5200 (also 0.8 to 10pF). C4 is a trimmer capacitor formed between the end of a tuning screw and the inner coaxial conductor. The remote (output) end of the coaxial conductor protrudes into the waveguide and acts as a coaxial to waveguide transition. Tuning of the cavity is provided by the sliding short and matching of the output to waveguide is by two tuning screws.

Full constructional details and dimensions are given in Figs 18.69 and 18.70. Note that the former figure includes details of a simple post-type filter built as an integral part of the unit. It would be better, for the reasons given above, to fit a 20MHz-wide iris-coupled filter of the type shown in Fig 18.29, as this is capable of much better suppression of unwanted harmonics and oscillator noise than the post filter. If this is done, then the length of the waveguide can be cut down and a flange fitted close to the coaxial mount. The filter should be provided with a flange and, since it is

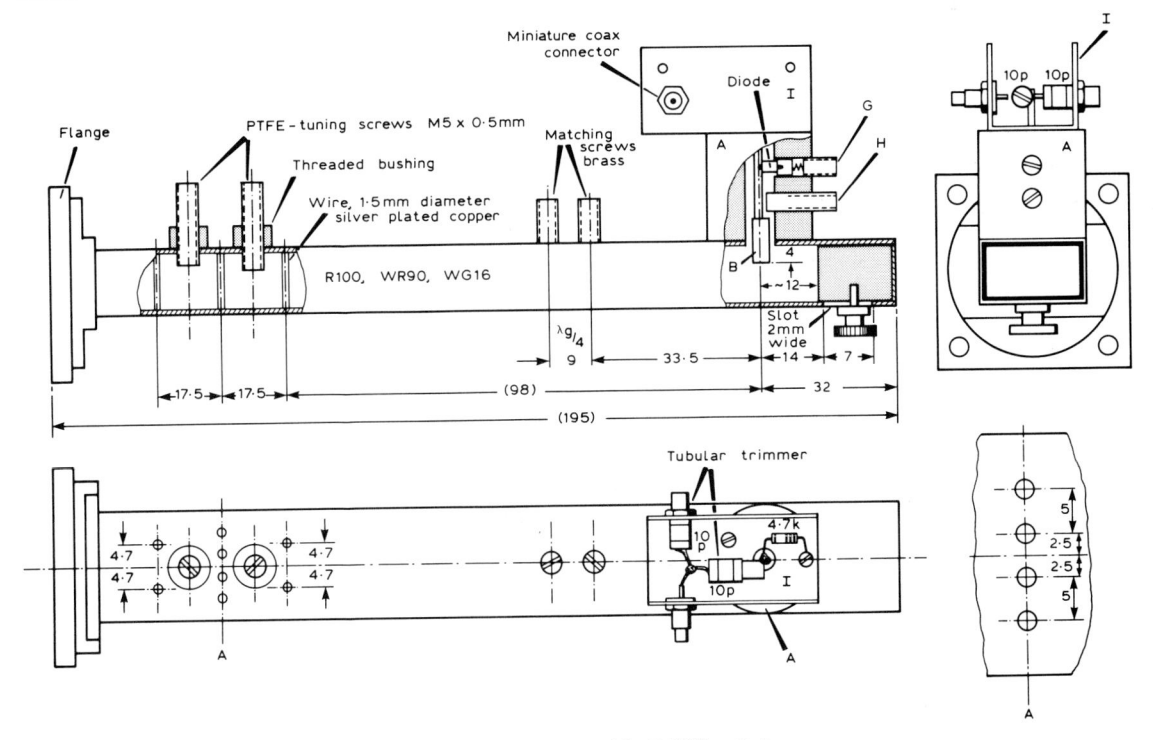

Note;
C_x is a fabricated capacitor consisting of an 11.5mm diameter disc forming part of the coaxial mount.
C_T is a small trimmer capacitor formed between the coaxial element and a tuning screw.
L Represents the inductance of the coaxial inner, into which the varactor and C_T are "tapped".
The probe section of the mount couples the 10GHz components of the output into a waveguide cavity which is tuned by a sliding short circuit.

Fig 18.68. Circuit of a 1,152MHz to 10,368MHz multiplier (DJ7VY)

about 10kΩ. It should be possible (and technically more satisfactory) to fit a miniature preset variable resistor in an accessible position between the junction of L4/L3 and an earthing tag bolted either to the base of the box below L4, or to the box wall close to the diode mounting screw.

Fig 18.69. General layout of the DJ7VY multiplier

MICROWAVE HANDBOOK

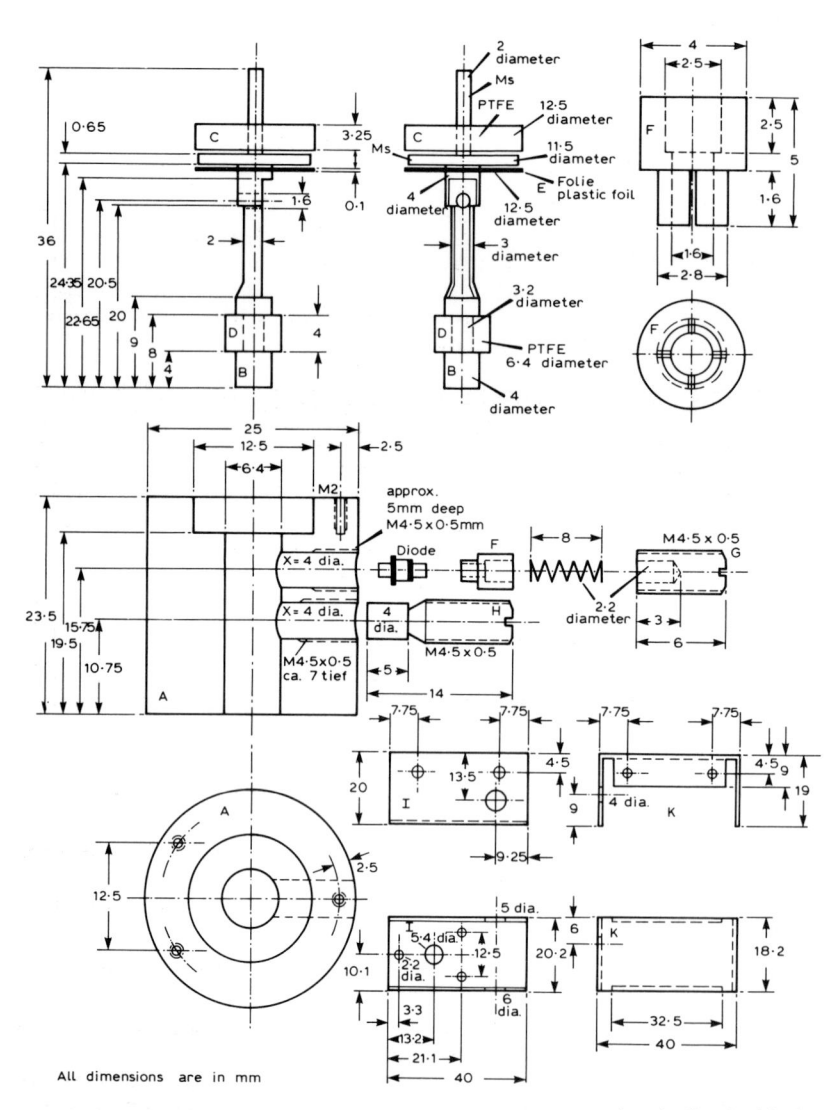

Fig 18.70. Parts detail for the DJ7VY multiplier

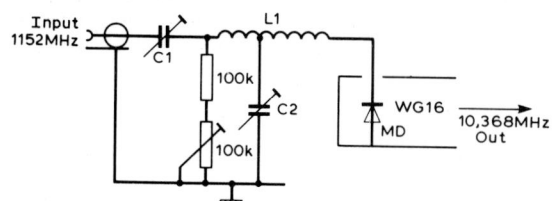

Note:
C1 is a small capacitor formed by a piece of copper foil in proximity with the end of L1. It is adjusted by bending.
C2 is a trimmer formed by the end of a ∅BA bolt in proximity to L1, effectively tapped into L1.
Unlike the previous design, the varactor is waveguide mounted in a cavity also tuned by means of a sliding short circuit.
MD = Multiplier diode

Fig 18.71. Approximate circuit of an alternative 1,152MHz to 10,368MHz multiplier (G8NDJ)

already fitted with three screw tuners at input and output, the two matching screws of the original design can also be eliminated.

The performance of the original multiplier was 15mW output for an input of 100mW and 40mW output for 250mW input.

A simpler design was described by G8NDJ [24]. This can be constructed without the use of a lathe which is clearly needed for fabrication of many of the parts for the first design. The circuit is broadly similar and is shown in Fig 18.71. The input capacitor C1 is an adjustable tab placed close to the inductor L1. L1 is tuned by C2, a 0BA screw. The bias resistor, consisting of a fixed resistor and a preset variable resistor, is tapped onto one end of the inductor. The other end is connected to a copper line which connects to the waveguide mounted varactor. Again cavity

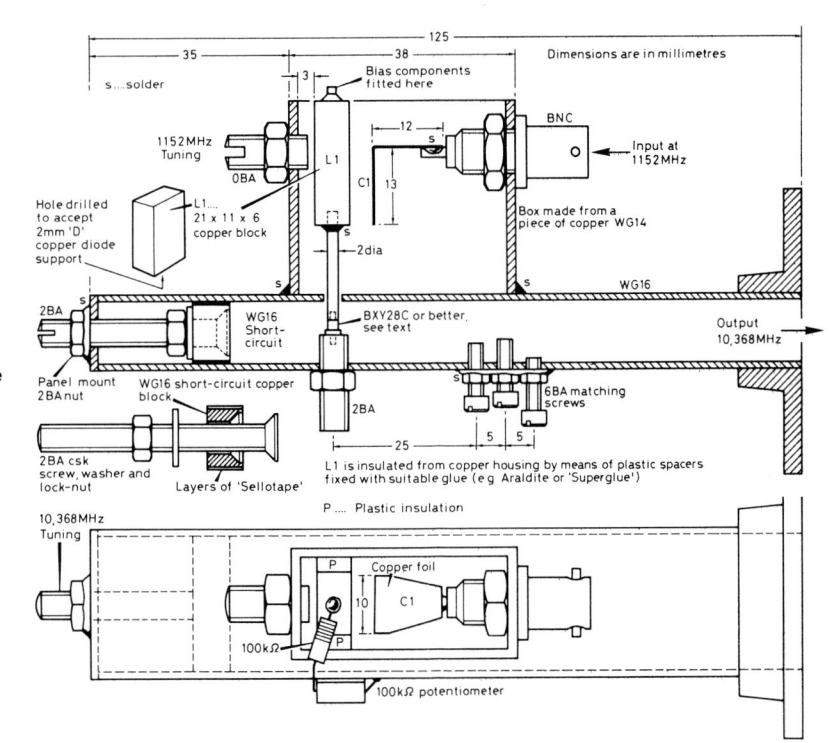

Fig 18.72 Layout and detail for the G8NDJ multiplier

tuning is achieved by use of a waveguide short and output matching by a three screw tuner.

Details and dimensions of the multiplier are given in Fig 18.72. Two prototypes were built by G8NDJ and both gave identical results; 6mW output for 700mW input. However, they were tested using the only microwave varactor available, a BXY28C, designed for use at up to 4GHz. It is expected that, using an MD4901 diode, the efficiency will not be greatly different from that of the more sophisticated and difficult design given earlier.

Aligning 1,152MHz to 10.368GHz multipliers

Alignment of the DK2VF multiplier is made much easier if the design has been modified as suggested. Similar methods can then be used to those described for the G8NDJ multiplier. The order of operations for the latter should be as follows:

1. Connect a WG16 variable attenuator, wavemeter and detector cavity to the terminating flange as shown in Fig 18.73. The attenuator should initially be set to minimum.

2. Attach a multimeter (50µA range) to the junction of the fixed bias resistor and the variable resistor. Set the latter to maximum resistance.

3. Apply the 1,152MHz drive and adjust the 0BA tuning screw for maximum indicated current. Adjust the 1,152MHz source output tuning to maximise current and adjust C1 by moving the foil, again to maximise current. These adjustments all interact and should be repeated several times until no further improvement can be obtained.

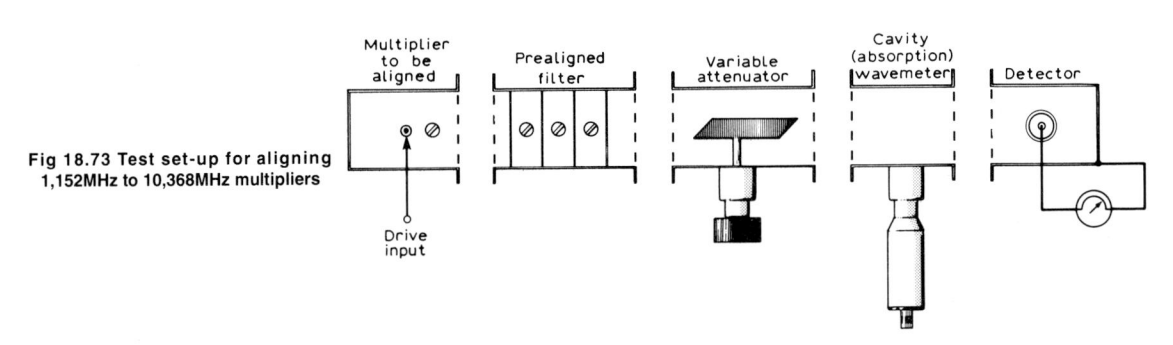

Fig 18.73 Test set-up for aligning 1,152MHz to 10,368MHz multipliers

4. Connect the multimeter to the cavity detector output, with the wavemeter tuned well away from the output frequency, and adjust the waveguide short to maximise the meter reading. Use the wavemeter to determine that output is on the correct frequency and then detune it again.

5. Readjust the 0BA tuning screw, C1 and the 100kΩ variable resistor to maximise the output.

6. Next adjust the three matching screws for maximum output and repeat all adjustments to obtain the maximum possible output at the correct frequency. The variable attenuator should be used from operations 4. onwards to keep the meter reading as low as is possible, consistent with a positive indication of current and the ability to see the wavemeter dip.

7. Switch the 1,152MHz source on and off several times to check that the system is stable. If it is not, continue adjustments until a point is reached where the multiplier is stable; this might mean slight detuning from maximum output, but it is more important to have stable output rather than absolute maximum power output.

8. Next connect the prealigned filter between the multiplier flange and the attenuator and repeat all adjustments, including the filter tuning and matching screws, until best results are obtained. Finally, repeat 7, the stability test, readjusting if necessary.

Very similar methods can be used to align the DK2VF multiplier; in this case the multimeter should be connected in series with a 4k7 fixed resistor before being connected across the bias resistor.

It is recommended that before attempting any multiplier alignment the constructor should build and prealign the output filter to be used with the multiplier. In this way the filter, in effect, becomes an integral part of the design.

18.3.6 Phase locked narrowband sources

In the following chapter, "The 24GHz band", there is an extensive description of a complete phase-locked narrowband source for that band. It is equally feasible to use similar methods for producing narrowband phase-locked 10GHz signals and so it is not proposed to describe suitable equipment here.

18.3.7 Injection locked narrowband sources

If a three-port circulator is available, then it is possible to generate powers of tens of milliwatts of narrowband signal by the simple expedient of injection locking a Gunn oscillator to a low-power crystal controlled narrowband signal. The arrangement is shown schematically in Fig 18.74. It is a much simpler technique than phase locking.

The technique offers a relatively simple method for converting wideband oscillators to crystal-controlled standards and goes a long way towards avoiding the need for a chain of high powered multipliers (and their attendant filters) necessary to generate rf indirectly.

If a low power crystal controlled oscillator is injected into a tuneable oscillator then, as that oscillator is tuned to within the locking range of the system, it will jump in frequency to that of the source and remain locked on to it. Further tuning of the oscillator will have no effect until the locking range is exceeded, when the oscillator will jump in frequency from that of the injection source to that set by its tuning mechanism. This behaviour is similar to the more familiar automatic frequency control (afc) systems.

A second feature of injection locking is that the noise bandwidth of the tuneable oscillator (which may be several hundred kHz at 10GHz) becomes the same as that of the injection source, which may be in the order of a few kHz. In this respect, injection locking is superior to afc systems. If the injection source is frequency modulated, then the tuneable oscillator will follow the modulation. If the injection source is switched off, or the tuneable oscillator is set outside the locking range of the system, then the oscillator can be tuned in the usual manner.

The amount of power needed to lock the oscillator is given by the equation:

$$\frac{P_i}{P_o} = \left(\frac{2 \times Q \times \Delta f}{f}\right)^2$$

where P_i = power of injected signal

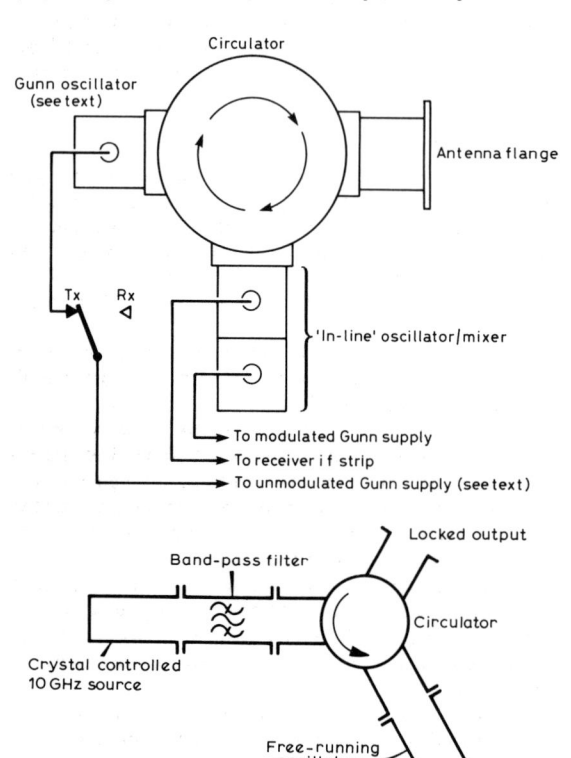

Fig 18.74. Block diagrams of injection locked oscillators using a circulator

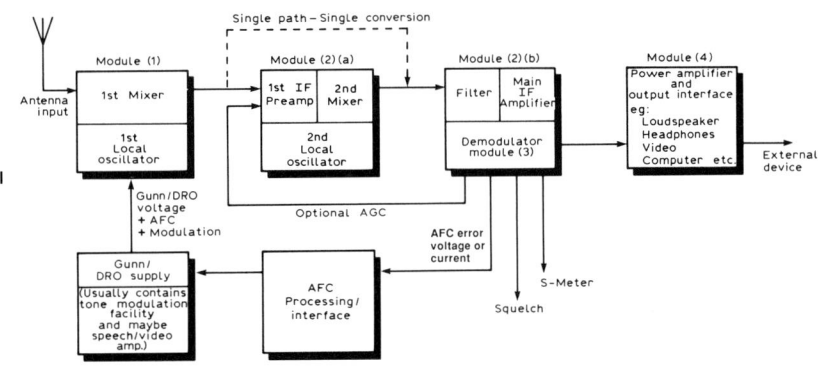

Fig 18.75 Block diagram of a typical 10GHz wideband receiver

P_o = oscillator power
Q = loaded Q of the oscillator cavity
f = oscillator frequency
Δf = locking range

As an example, if the oscillator is assumed to have a loaded Q of 100, then the relative power required to lock over a 5MHz range at 10GHz is 0.01 or –20dB. Thus if the oscillator power is 30mW, then the injection power required is 300μW. Such a system could be described as showing gain, and a feature of this technique which makes it particularly suitable for amateurs is that, for several reasons, it is relatively "fail-safe":

1. The injection source is of relatively low power and not connected directly to the antenna. The level of spurious output does not, therefore, need to be so rigorously controlled.

2. Unlike any conventional amplifier which responds to all signals at its input, the oscillator will only lock on to a harmonic within the locking range of the system and, therefore, only "amplify" inputs very close in frequency to which it is tuned.

3. If the oscillator slips out of lock, then it will drift in frequency by an amount determined by its inherent stability, which is only usually a few MHz.

The implication of the mathematics is that a low Q oscillator should be used, since this will lock at lower injection levels than a high Q oscillator and that for a wider locking range, more locking power will be required. Oscillators most suited to this technique are the open-ended, rear cavity types such as that of Fig 18.37.

The Gunn is simply tuned mechanically to approximately 10.368GHz and the injection source switched on, when the Gunn will automatically lock to the crystal source. With correctly calculated power levels the lock-and-hold range can be made tens of MHz.

Most of the power of the injection source is absorbed in the oscillator although a little may escape with the locked output and the total output power will be less than theoretical, by the forward port-to-port transfer loss of the circulator. Nevertheless, a system gain of around 20dB can be realised by this technique. This represents a very useful increase in power without having to resort to the generation of high levels of 384 or 1,152MHz drive or to the complications of two-stage multiplication.

18.4 GENERAL RECEIVER/TRANSMITTER CONSIDERATIONS AT 10GHZ

So far, several general purpose components and "modules" of 10GHz equipment have been described. In this section, the general features of communications systems based on these modules are discussed. In sections which follow later, complete practical working systems assembled from these modules are described.

18.4.1 Receivers

10GHz receivers invariably utilise the superheterodyne principle. Even a doppler system, which is not used for communications, may be considered as a form of superheterodyne receiver with an audio frequency i.f, more or less equivalent to the direct conversion receivers which are used at lower (short wave) frequencies in simple amateur QRP equipment.

A 10GHz receiver, as at any other frequency, can thus be conveniently split down into a number of "sub-assemblies" or modules as shown in Fig 18.75.

Module 1 is the microwave mixer/oscillator or down-converter assembly which may include filters and preamplification at both signal and intermediate frequency. It is the most crucial assembly in the receiver and presents the amateur constructor with the biggest challenge if a really efficient receiver is to be built.

Module 2 may be a second down-converter, (2a), if double conversion is used, or the main i.f signal processing circuit (2b) if not. The main receiver selectivity is normally provided at this point, as with any receiver, regardless of frequency.

Module 3 is a demodulator of a type suited to the mode in use together with an audio/video amplifier sub-assembly. If an integrated circuit approach is taken, then modules (2b) and (3) will usually be combined on one chip, for instance the CA3189 which provides all the "back-end" facilities for a wideband fm system.

The final module, (4), consists of an output sub-assembly; this might be an "interface" to a loudspeaker/headphones, rtty interface, computer interface or video driver. Again, using ics, the circuit can be compact enough to be combined with the preceding module.

Alternatively, modules 2, 3, and 4 may already be available in the form of an amateur receiver or transceiver (e.g. a 2m transceiver). This would be particularly suitable for a narrowband system.

Factors influencing receiver front-end design

Five factors strongly influence the design of receivers for use at 10GHz at present. These are:

1. RF amplifiers (or lack of them)
2. Oscillator, mixer and signal isolation
3. Oscillator noise
4. Image response and its elimination
5. The choice of i.f and the influence of post-mixer amplification noise

All these factors are, to a great degree, inter-related. For the greatest receiver "potency" all the factors need to be taken into account. This will lead to a complex design best suited to narrowband work where the operator is expecting to handle very weak, but frequency-stable signals. For systems handling wideband signals which are usually much stronger and received over line of sight paths, a simpler approach may be taken and some of the factors can be ignored.

RF amplifiers

RF amplifiers, although slowly becoming practical for the home constructor with the availability of inexpensive GaAs fet devices, are still not in common use and so the performance of the receiver continues to be dominated by that of the mixer. In receivers where it is either not possible or is too costly to use an rf stage, the overall noise figure is set by the sum of the noise figure of the mixer, that of the i.f preamplifier and any coupling losses between the two. For this reason, considerable attention must be paid to the performance of the i.f preamplifier and to minimise the coupling losses it is often mounted as close as possible to the mixer. If the preamplifier has reasonable gain, then its output lead can be of any convenient length and the noise figure of the following stages also becomes less critical.

The effect of an rf preamplifier is quite dramatic; assuming that the mixer noise figure is 16dB, and that of the preamplifier is 3.5dB, then the overall noise figure falls to an estimated 7.7dB. With two such preamplifiers, it would be 4.5dB.

Oscillator, mixer and signal isolation

Since many microwave receivers do not have an rf preamplifier, a low mixer conversion loss is essential for a low noise figure reciever. This loss can be minimized by ensuring that the input signal does not get absorbed in the local oscillator circuitry. This can be done using either tuned circuits on the signal and local oscillator ports or in a double balanced mixer.

The balanced mixer achieves this separation over a very wide frequency range – an example is shown in Fig 18.84. Obtaining the separation using tuned circuits is easier at lower frequencies, but very high Qs are needed at microwave frequencies, particularly when very low i.fs are used. It is possible to use cavity filters for this in fixed frequency receivers, e.g. the G3JVL narrowband design, but this approach is less suitable for wideband designs where wide tuning ranges are needed.

Another fairly broadband method, common at microwave frequencies, is to use a directional coupler to couple a small fraction of the local oscillator power into the mixer. This is inefficient in terms of local oscillator power, but the weak coupling prevents the signal being lost in the local oscillator port. This technique is used in the cross coupler design shown in Fig 18.81a.

The level of local oscillator power needed to drive the mixer to an acceptable level of performance is usually in the range of 0.25 to 1mW. Therefore, if the oscillator develops a power of (say) 30mW, the coupling coefficient into the mixer needs to be in the range 30 to 120:1, i.e. 15 to 22dB. Some means of adjusting this level easily needs to be incorporated into the oscillator/mixer design and means of doing this for wideband equipment is described in section 18.5.1, where a number of practical circuits are described. Using the highest powered oscillator which is available will enable the highest isolation to be used and will bring the reciprocal benefit of minimising signal loss into the oscillator.

Oscillator noise

At these frequencies, local oscillator sideband noise is usually high in both directly generated sources and multiplied sources. Noise in both arises from random fluctuations in frequency and amplitude and is inherent in all circuits. By careful design and choice of components noise can be minimised but never completely eliminated. Noise sidebands can be much worse in crystal derived, multiplied sources, since the level of noise sidebands is proportional to the square of the multiplication factor, which in 10GHz receivers is high – in excess of ×100 when multiplying from a fundamental source around 100MHz.

Two methods can be used to get round this problem:

1. By using the highest i.f possible without degrading the overall noise figure by the increasing noise figure of the i.f preamplifier. The practical limit until comparatively recently was of the order of 200MHz. With the advent of new semiconductors, this limit is now probably in the region of 1,500MHz, with 430MHz (or thereabouts) a practical proposition for the average amateur station and 1,300MHz for the owner and operator of a good 23cm receiver or transceiver. A high i.f also has the advantage that receiver second channel (image) noise is more easily reduced or eliminated than at a lower i.f. As already explained, it is

difficult to obtain sufficiently sharp filtering to remove the close-in noise from a 10GHz oscillator and this has a profound influence on the choice of i.f in the more sophisticated receiver.

2. By using a balanced mixer. In this device the a.m (but not the fm) components of the noise are cancelled. A relatively low i.f can then be used, and a value of around 30 or 40MHz might be typical, although image protection would still be difficult. The only easily realised design for a balanced mixer in waveguide is based upon the use of a hybrid-T, described later.

Image rejection and the choice of i.f

In an unprotected receiver, i.e. one without any selectivity before the mixer, the receiver responds to signals at F(oscillator) ± F(i.f). What the receiver "hears" is the wanted signal plus the noise present in the image channel. If the image channel response can be suppressed by the use of filters, then the receiver has a theoretical gain in sensitivity of 3dB. As already mentioned, it is not easy to make sufficiently "sharp" filters to remove this image if it is close to the wanted signal, i.e. if a low i.f is used. If it is intended to incorporate image recovery techniques, then an i.f of at least 100MHz is advised.

Choice of intermediate frequencies

Whilst there may be a relatively limited number of practical 10GHz down-converter designs available to the amateur, the number of choices open for the "back-end" of the receiver is many. Most amateurs will wish to use either their lower frequency receivers or transceivers or some other, readily available, inexpensive alternative.

For narrowband use the receiver is expected to be capable of extremely good weak signal performance and, for the reasons given above, should employ a high first i.f. 144MHz is widely used since this is sufficiently far removed from the oscillator frequency to allow really effective oscillator and signal image filtering and most amateurs will have a good 144MHz multimode receiver or transceiver available. Other possibilities are 432MHz or 1.3GHz.

The situation on 10GHz wideband is completely different; here the equipment is often deliberately kept as simple as possible to avoid undue constructional problems or higher costs and to make the equipment easily portable and minimise its power consumption. The equipment should be expected to be reasonably efficient over unobstructed paths or where there is perhaps a single obstruction. The difference in performance was outlined very early in the chapter.

Because of the relatively poor stability of wideband equipment based on Gunn oscillators, it is almost impossible to use the narrow bandwidths (typically 2.5 to 3kHz on cw/ssb and 15kHz or less on fm) offered by the receiver of an amateur transceiver. Thus, use has not been made of 144 or 432MHz transceivers for wideband systems. If a dro/mixer is used to replace the Gunn/mixer, then it is possible to consider the use of an amateur transceiver in nbfm mode to receive the more stable signals generated by such modules.

With Gunn-based receive down-converters, i.f bandwidths and transmitter deviations of between 50 and 250kHz are usually used and this is consistent with the bandwidth employed in receivers designed for fm broadcasting in band II (88 to 108MHz in the UK). Thus, one quite popular i.f is (nominally) 106MHz, near the top of band II. This frequency has been chosen by many operators as offering:

1. An i.f sufficiently far removed from the local oscillator to avoid significant noise problems.

2. A frequency at which reasonably low-noise gain can be provided by simple transistor/fet or ic preamplifiers.

3. A relatively "quiet" frequency at which to provide such amplification although, with increasing band occupancy by both broadcasting and public service transmitters, it is becoming increasingly difficult to avoid breakthrough problems, especially during portable hill-top operation.

4. The opportunity to utilise inexpensive pocket-portable fm broadcast receivers as complete, ready built units, often capable of accepting input from the microwave mixer directly with a minimum of modification.

Invariably, if this approach is taken, the receiver becomes double conversion, with a second i.f at the international standard of 10.7MHz. Most pocket portable receivers will benefit greatly by the provision of a good preamplifier which will not only improve sensitivity and noise performance, but may also provide a better match between the microwave mixer and the receiver input, as this is often some peculiar impedance designed to work from a short, non-resonant, retractable whip antenna. Such receivers may also offer limited but useful afc facilities.

The main disadvantage of the band II choice is that it is impossible to define a *precise* frequency due to the presence of (local) strong signals in the band. This makes it difficult to set up full duplex contacts, should the operators wish to do so. It is also an inconvenience where the Gunn oscillator frequency calibration is poor or suspect.

Two other i.fs in fairly common use are 10.7MHz and 30MHz. One disadvantage of single conversion direct to 10.7MHz is that the local oscillator noise can, in a poorly designed or incorrectly set-up Gunn oscillator/mixer, increase the receiver noise level. 10.7MHz, however, has the advantage that it is an international standard frequency and is thus a very good choice of "base-band" i.f for a more elaborate double conversion receiver. Many components such as screened, tuned inductors, ics and inexpensive ceramic or crystal filters are available for 10.7MHz, enabling the constructor to "tailor" receiver performance to individual requirements. Being an international standard, i.f/signal processing strips are available in kit-form or ready made, intended for incorporation into high quality fm broadcast receivers. The use of multifunction ics leads to very compact and versatile signal processing circuits and a complete, basic but comprehensive, wideband transceiver board suitable for use at both 10 and 24GHz is described in section 18.5.6. Whilst it was designed for use as a single conversion receiver, its performance can be improved still further by preceding it with a suitable second down converter and using it as a double conversion receiver. The choice of first i.f is then open to the constructor. The frequency of 30MHz is ideal in many respects and is strongly recommended as a "standard" first i.f for normal wideband speech communication. 50MHz, now in regular amateur use for communications, might also be a good choice of i.f.

If really wideband communication such as full definition fast-scan atv is contemplated, then a different set of i.f standards must be chosen (as described in section 18.5.8) in order to accommodate the necessary bandwidth of the fm-tv signal. The choice of receive i.f also, as will be seen in sub-section 18.4.3, has some bearing on transceivers and transmitter/receivers.

18.4.2 Transmitters

In effect, transmitters have already been discussed in some detail in section 18.3, since any of the sources described there can be used as transmitters.

At the present "state of the art", most amateurs usually work at power levels of a few hundred µW to perhaps a hundred mW. Coupled with high antenna gains on both receive and transmit, these power levels are sufficient to work all line of sight paths and some obstructed paths, especially in narrowband modes.

Almost invariably in simpler equipment, the modulation mode is fm, since this is the easiest mode to implement. Transmitters thus fall into two main categories:

1. Directly generated sources (at 10GHz), typified by Gunn or dro devices. These are used in wideband or "pseudo-narrowband" systems.

2. Indirectly generated sources – by multiplication/filtering, by multiplying/mixing/filtering or by phase or injection locking.

Indirect sources are usually used for narrowband systems, although they can equally well be used for wideband systems, by increasing deviation levels to be compatible with a wideband receiver. However, this would defeat the main objective of operating narrowband, which is the attainment of better receiver sensitivity by utilising narrower bandwidths!

Since, apart from beacon use, transmitters are invariably used with a receiver, there is little else to discuss in this sub-section, other than to stress again the need for adequate precautions to stabilise direct frequency sources and to filter indirect sources when used as transmitters.

18.4.3 Transmitter/receivers and transceivers

With directly generated 10GHz signals the receiver local oscillator is often used also as the transmitter, an approach typified by the in-line mixer/oscillator module of section 18.3.2. This will usually mean, unless common i.fs are in use at both ends of a communications link and full duplex is in operation, that retuning of the oscillator between receive and transmit will be necessary. With a high i.f, this may mean retuning by up to 100MHz or more. This implies both a tuneable oscillator with sensibly constant output over this range and also the ability to set frequency reproducibly and accurately. This is frequently done using Gunn oscillators – and fairly effectively – by many operators. If a dro is used, although it is possible to tune over this wide range, stability may start to deteriorate. A dro/mixer unit is not really suited to transmit use because the module is usually designed to suppress oscillator radiation to such a low level that it becomes an ineffectual transmitter.

The task is made easier in either case by the use of a lower i.f such as 30 or 10.7MHz, but even so still requires a fair degree of operating skill, especially since there will seldom be any image-reception protection! Despite these drawbacks, the in-line Gunn system offers a number of advantages. As a receiver the performance is very good. As a transmitter, the power output is necessarily limited, although the operator may consider the disadvantages to be outweighed by the fact that transceive operation, without switching, is offered.

A better system would be to employ a receive down-converter and add a separate second (possibly fixed-tuned, transmit only) oscillator to the system, together with a suitable changeover switch. In this way, the receiver down converter can be optimised for best receive performance and the transmit oscillator stabilised by the use of an isolator. Such a system should be provided with two independent power supplies. The transmit supply should be capable of both tone and speech (or other) modulation whilst the receive supply will normally only need tone modulation to aid weak signal identification. Both the systems outlined are particularly applicable to simpler wideband equipment although, as will be seen later, the "classic" G3JVL narrowband transverter system which in effect uses in-line mixing, can also be improved in transmit performance by using similar techniques to separate the transmit and receive functions, albeit at the expense of system simplicity.

18.5 WIDEBAND SYSTEMS

Various techniques for combining transmitters and receivers into workable wideband systems are illustrated and described in this section. First, various configurations of the 10GHz down-converter will be described. It is here where a number of the components already described can be configured in a variety of ways to produce workable systems. This is followed by first-i.f preamplifiers and then an optional second down-converter. Next follows a description of a comprehensive but simple 10.7MHz "baseband" i.f signal processing/transmitter module suitable for wbfm voice/cw/rtty communication, and finally a sub-section devoted to fast-scan fm atv at 10GHz.

The problems existing in the design of 10GHz down-converters have been discussed in the preceding section. What remains to be discussed here is the practical realisation of designs which overcome some or most of these problems. In the first two oscillator/mixer systems described, a doppler system and a self-oscillating mixer, it is very difficult to meet these criteria and so the systems are of limited efficiency and utility. The later down converters are based on sounder principles and can be constructed, using components already described, to meet the criteria. As a consequence they are more efficient, even if more complex. Another desirable feature of the microwave converter from the amateur viewpoint, is that it should, if possible, be usable as an efficient receive converter and (with a minimum of complication) as a transmitter also. These two requirements usually conflict, as will be seen!

18.5.1 10GHz down-converters

Doppler radar and a simple doppler system

Whilst not strictly falling into the classification of a down-converter, a doppler module is, nevertheless, an interesting piece of 10GHz technology and the following narrative is offered as a (hopefully interesting and useful) diversion from the more serious intent of this section.

Most readers will be aware of the doppler effect by virtue of noticing the audible effects of a passing police or ambulance siren, or in the amateur radio context, noting the changing offset from nominal frequency on signals received from passing satellites such as the early Oscar series or the experimental UOSAT vehicles. They may also be painfully aware of one practical application of the doppler effect at 10GHz, by being caught exceeding the speed limit in a radar speed trap!

A doppler radar detects the difference between the frequency of a signal transmitted from a stationary source and its echo from a moving target. To detect motion, the source must transmit an unmodulated continuous wave signal and be able to receive an echo signal, reflected from the target, as shown diagramatically in Fig 18.76.

The transmitted wave at any point in space may be described mathematically as a time varying electric field:

$$E = E_o \cos(W \times t)$$

where W (ω or $2\pi F$) denotes the angular frequency and t is the time taken for the signal to travel from the transmitter to the target. This time is equal to D/c, where D is the distance between the transmitter and the target and c is the velocity of the wave (that of light). Since the target is moving, D varies with time, but its velocity, V_d, is small compared with c. The signal arriving at the target may be described as:

$$E = E_o \cos(W \times (t-D)/c)$$

or, rearranged, as:

$$E = E_o \cos(W \times t - W \times D/c)$$

A small proportion of this signal will be reflected from the target and the reflected signal at the receiver may be represented by:

$$E = E_o \cos(W \times t - 2W \times D/c)$$

remembering that a time of D/c has elapsed since the signal left the transmitter and reached the target, and a time D/c is also required for the reflected signal to travel from the target back to the receiver.

Assuming that the target is moving at a constant velocity, the distance D is time-dependent and may be expressed by:

$$D = D_0 \pm (V_d \times (t-t_0))$$

where D_0 is the distance at time t_0, and V_d is the radial velocity of the target. If the target is approaching the transmitter, the term $V_d \times (t-t_0)$ has a negative value and if it is receding, a positive value. Thus the echo signal at the receiver can be described by:

$$E = E_o \cos((W \times t - 2W \times D_0/c) \\ \pm (V_d \times 2W \times t/c) \pm (W_d \times 2W \times t_0/c))$$

Rearranged, this becomes:

$$E = E_o \cos((W \pm W_d) \times t - (2W \times d_0/c \pm W_d \times t_0))$$

$$\text{carrier} \pm \text{doppler} \qquad \text{constant phase shift}$$

From this it can be calculated that $W_d \times 2W \times V_d/c$ is the change in angular frequency caused by the movement of the target. The frequency F_d, where $F_d = W_d/2\pi$ or $2F/c \times V_d$ is known as the "doppler frequency".

To give an example of the magnitude of the doppler shift at 10GHz, the following should be considered:

Source frequency: 10.4GHz
Target velocity: 50km/hr towards the source

Substituting these values in the equation $F_d = 2F/c \times V_d$ will calculate the doppler shift. For example:

$$F_d = \frac{2 \times 10.4 \times 10^9 \times 50 \times 10^3}{3 \times 10^8 \times 3,600} = 962 \text{ Hz}$$

Similarly, if the source frequency remains the same but the target velocity falls to 1km/hour, then the same calculation shows that the doppler frequency falls to 19Hz.

The schematic of a typical doppler velocity measuring system is shown in Fig 18.77. The local oscillator/transmitter and mixer assembly may be either a side-by-side module (such as the Philips CL8963, mentioned earlier,

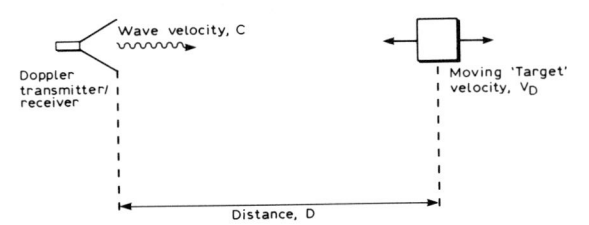

Fig 18.76. Schematic of the principle of doppler speed measurement

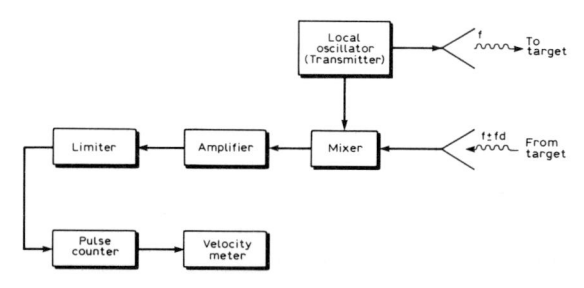

Fig 18.77. Block diagram of a doppler velocity meter

although this type of system is a compromise and only useful with very low gain horns) or an in-line module such as the Solfan unit, also mentioned earlier. A small part of the local oscillator signal is used to drive the mixer, the remainder being the transmitted signal. The return signal (echo) mixes with the local oscillator and the difference (doppler) signal is amplified, limited and counted. The output of the pulse counter gives a voltage which is proportional to F_d which, in turn, is proportional to the velocity. This is applied to a visual indicator such as a meter which displays the measured velocity.

Professionally, such systems are calibrated by using a tuning fork as a target. The doppler shift on the echo from the tuning fork is equal to the audio frequency of the tuning fork and is independent of the exact transmitted frequency.

It may seem a little incongruous to have dealt with a non-communications system in a chapter otherwise devoted to communications. There are several good reasons for so doing. Firstly, it may have given the newcomer an insight into the theory of the doppler devices which he or she might purchase on the surplus market and then use for communications purposes. Secondly, a simple doppler transceiver can be an aid to the fixed station operator in identifying heavy rain and weather fronts. These, and passing aircraft, can enhance signals over an obstructed path by acting as scattering objects. When this happens the rapidly moving object or particles may spread a narrowband cw/ssb signal out over several kilohertz because of the doppler effect, making it sound almost like an auroral signal at vhf.

It is desirable to be able to detect this type of weather by using one's own equipment, without needing a constant signal from another station. Some experiments carried out by G4MBS using a low powered narrowband transmitter and a separate receiver, both employing antennas with fairly broad beamwidths, showed that it was possible to detect such effects, either by listening for increasing signal strength or by listening for the doppler shifted components of the received signal.

It is possible to do exactly the same using a surplus doppler module which, although not as sensitive as a narrowband set-up, is considerably simpler. A doppler amplifier circuit, due to G3YGF, is shown in Fig 18.78. It is simply a very high gain audio amplifier (with an overall gain of about 400,000) capable of driving a small loudspeaker directly. Its low frequency response is probably limited in practice by the loudspeaker or headphones used. A spare doppler module or the oscillator/mixer of an existing wideband transceiver can be used by replacing the i.f amplifier, normally connected to the mixer, with the audio amplifier shown in the figure. The leads from the mixer to the amplifier should be kept short and screened to prevent hum pickup, and the supply may need additional decoupling if it is noisy. Used with a 20dB horn, it proved possible to detect cars at a range of about 1km.

Doppler signals will only be heard if the "target" is moving more or less directly toward or away from the antenna, so that to detect weather phenomena or passing aircraft on a given path, the antenna should be pointed up slightly along that path for best results.

A self-oscillating mixer converter (transceiver)

This design of a down-converter utilises only a Gunn oscillator as the local oscillator source, transmitter and mixer. It is similar in principle to the self-oscillating mixers which have commonly been used for many years in inexpensive long/medium wave and cheaper vhf broadcast receivers. It suffers from the disadvantage that the ratio of

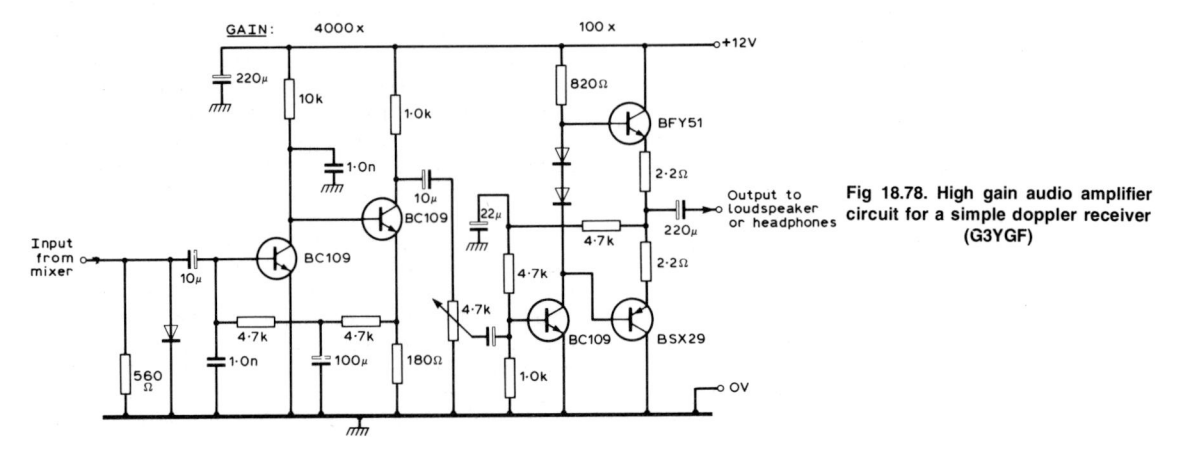

Fig 18.78. High gain audio amplifier circuit for a simple doppler receiver (G3YGF)

THE 10GHZ ("3CM") BAND

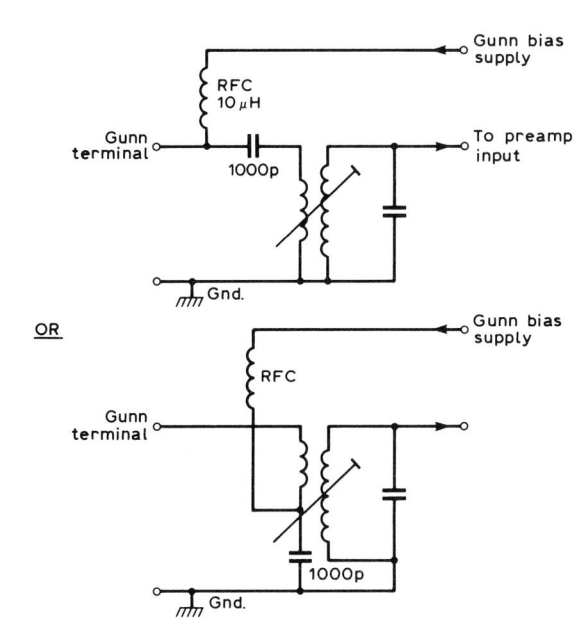

Fig 18.79. Input circuits for a Gunn self-oscillating mixer

local oscillator to signal is not, and cannot be by its very nature, optimised for best conversion efficiency and, as a consequence, tends to have a rather high noise figure, usually of the order of 20 to 25dB.

For unknown reasons, presumably related to random fluctuations in frequency and amplitude of the Gunn output, an occasional sample of Gunn diode seems to exhibit much better mixing characteristics than usual and here the noise figure may be in the region of 15 to 20dB, which is comparable to a badly set-up in-line mixer converter. It is said, although this has not been checked-out, that the performance of a self-oscillating mixer can be significantly improved by the employment of afc, an effect which is presumably related to a reduction in frequency "jitter" in the oscillator, although the theory is not clearly understood.

Self-oscillating mixers are sometimes employed in doppler intruder alarms where ultimate sensitivity is not needed over the short range of operation usually required of such devices. Examples of this type of alarm seen by the writers have employed a variety of oscillators, varying from simple, rear-cavity, open ended oscillators to the Mullard CL8630 and Solfan (oscillator only) units.

Both maximum transmit capability and best signal input transfer to the Gunn are achieved by using a simple, open ended oscillator such as that described in [12] and illustrated in Fig 18.37.

The oscillator is connected into an i.f preamplifier as shown in Fig 18.79. Regardless of the i.f used, the Gunn operating voltage should be optimised at a point 0.5 to 1V above the starting voltage. This is the point where the output is quietest. This type of converter is very simple but is only really suited for use over relatively short distances because of its comparative lack of sensitivity and high noise figure. At 24GHz and above, however, the technique may provide the user with results not greatly different to the results obtained with a separate mixer and is thus worthy of further development at these frequencies. As a transmitter, the open-ended oscillator provides good potential when modulation is applied in the usual way.

The in-line converter (transceiver)

This has already been fully described and discussed in section 18.3.1. It only remains to reiterate that, because of the choice of oscillator and mixer design, it becomes easy to meet most of the criteria outlined earlier. Its use as a transceiver is a compromise between a potentially excellent receiver and an indifferent (in power output terms) transmitter, some receive performance usually being sacrificed by over-driving the mixer to produce more transmit output. Under these circumstances the converter might be expected to yield a noise figure in the region of 14 to 17dB.

If used with a separate transmitter oscillator and suitable switching, the in-line can be optimised as a receiver and may then be expected to be some 3dB or so better in performance. This could be improved still further by the inclusion of an image rejection filter before the mixer, when a further 3dB improvement is theoretically possible. In practice it has been found that a noise figure of around 9dB can be obtained by these means. A straight in-line converter was illustrated in Fig 18.43 and the 20 or 60MHz wide filters of Fig 18.29 could be fitted immediately before the mixer provided that a suitably high i.f (eg 100MHz) is used. Image rejection with a lower i.f becomes less effective because of the relatively wide "skirt" response of either filter.

A cross-coupler converter (transceiver)

Use of a cross coupler enables the criteria to be met quite easily. The cross coupler used can conveniently be the Moreno cross coupler described in chapter 10, "Test equipment", and its coupling factor chosen according to the power available from the Gunn source. The assembly of such a system, which includes a wavemeter to simplify frequency setting, is shown in Fig 18.80. All the required components have been described either in this chapter (oscillator and single ended mixer) or in chapter 10 (wavemeter, cross coupler and matched load). An isolator could, with some further advantage to stability, be placed between the oscillator and the cross coupler.

The microwave head can be used in receive mode, in which case the configuration shown in Fig 18.81a is used, or alternatively in transmit mode whose configuration is shown in Fig 18.81b. To change from receive to transmit it is necessary only to interchange the matched load and antenna as shown in the two diagrams. If ports A and B and the antenna feed are terminated in round flanges, then manual changeover can be accomplished, although this may

Fig 18.80. Photograph of a cross coupler receiver/transmitter head (G3PFR)

be "fiddly" and time consuming. As shown, with a 10mW source, the coupling factor should be of the order of 10dB, since a single ended mixer will usually require about 1mW of drive for optimum performance.

The use of such a coupler, although cumbersome, means that 90% of the received signal reaches the mixer and 90% of the transmitter power reaches the antenna. Both receive and transmit functions are therefore reasonably efficient. Even better performance can be obtained by using a higher powered Gunn; for instance a 20mW Gunn would enable a 13dB coupler to be used, a 40mW Gunn, 16dB and so on.

A circulator converter (transceiver)

A circulator enables a very simple and efficient converter to be realised, which also offers complete transceive facilities without switching.

It consists of a WG16 three port circulator, a single ended mixer and Gunn oscillator, arranged as shown in Fig 18.82 and photograph Fig 18.83. As with the cross coupler converter, an isolator could be added (in this case with little added advantage) between the oscillator and the circulator. The circulator, already described, may possess a forward port to port insertion loss of 0.3 to 0.5dB and a 20 to 30dB reverse isolation. The amount of local oscillator drive

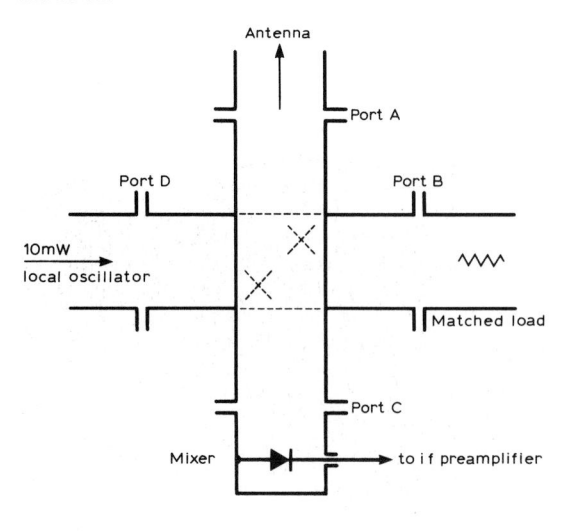

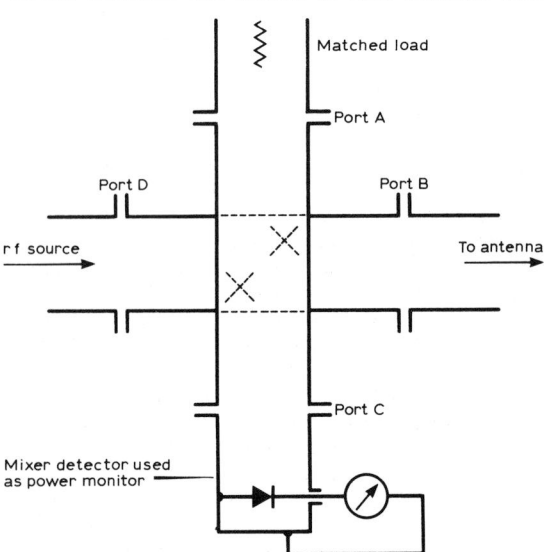

Fig 18.81. (a) Diagram of a cross coupler assembly used as a receiver (b) The same assembly used as a transmitter

THE 10GHZ ("3CM") BAND

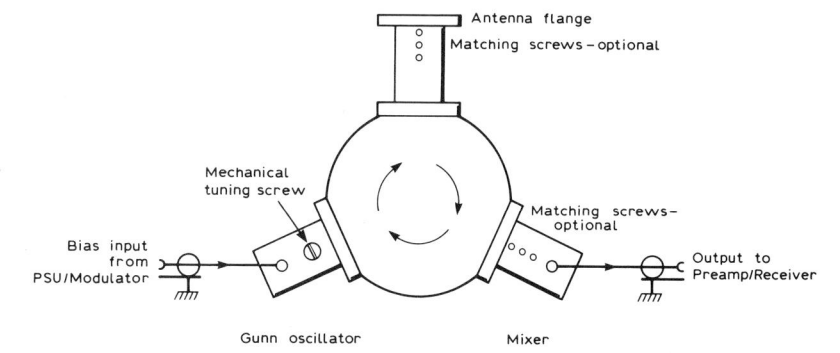

Fig 18.82. Diagram of a circulator transceiver head

transferred into the mixer is partly determined by the match of the load (antenna). If the load were perfectly matched and absorbed all the power transmitted in the forward direction, then the mixer would only receive a signal whose magnitude is determined by the reverse isolation, i.e. 20 to 30dB down. This means that a 20mW Gunn with a 20dB circulator would, under these circumstances, provide a near-ideal drive level.

However, the load is never perfectly matched and this means that additional drive may be received by reflection from the antenna port mismatch and transmission through the circulator to the mixer port. The drive level is thus somewhat unpredictable and should be determined by experiment. If the antenna feed is optimised at the desired receive/transmit frequency by means of a three-screw tuner in the antenna feed waveguide, the mixer drive will be at a minimum at that frequency and the mixer current can next be maximised by adjustment of the three-screw tuner on the mixer. Doing this will optimise the receive signal matching at that frequency also.

If the oscillator is now detuned by the value of the i.f, it will be found that, due to the slight antenna mismatch now existing at the new frequency, more mixer current will be seen, although this may be to some extent offset by the mismatch, at the new frequency, presented by the mixer. In practice, having set up the system as described, it is now possible to check and optimise the mixer drive by the simple expedient of adjusting the Gunn oscillator iris if the drive is not indeed already in the desired range. Using a 20mW iris coupled Gunn and a (nominal) 100MHz i.f with a 20dB circulator, set up as described to receive with the Gunn detuned, it was found that 0.45mA was obtained without the need for iris adjustment, the antenna mismatch in this case being nicely sufficient to supply the right level of drive. Retuning the oscillator to the transmit frequency, the mixer drive fell to 0.1mA, a level sufficient to monitor the Gunn performance. Similar results were obtained with a 30mW Gunn after adjusting the circulator isolation to about 28dB by retuning its matching screws.

Although such a system may need to be set up empirically in the manner described, the circulator transceiver can be made compact and very efficient using the modules already described with the added advantage of needing no additional test equipment to set it up. A circulator is particularly useful with separate transmitters and receivers; for example, the simple mixer could be replaced by an in-line mixer optimised as a receiver, and the tuneable Gunn by a fixed-tuned Gunn as the transmitter for wide-band use. Alternatively an image-protected mixer could be used on the receiver port and a narrowband source on the transmit port for narrowband use.

A commercially available transceive converter using these principles, the Microwave Associates' "Gunnplexer", is available to amateurs and although widely used in the USA, has not been much used in the UK, largely because of its price and availability compared with the home-assembled or surplus in-line units. It would be a recommended unit for those without a circulator who wish to get away from the compromise in-line transceiver approach, or those who are not prepared to experiment with an optimised in-line receiver, separate transmitter and waveguide switch. Full details of the Gunnplexer and its applications are given in [25]. The performance of either the circulator converter or the Gunnplexer should be comparable to a well set-up trasceiver on receive, and very considerably better on transmit.

Fig 18.83. Photograph of a circulator transceiver head (G3PFR). Photo: G6WWM

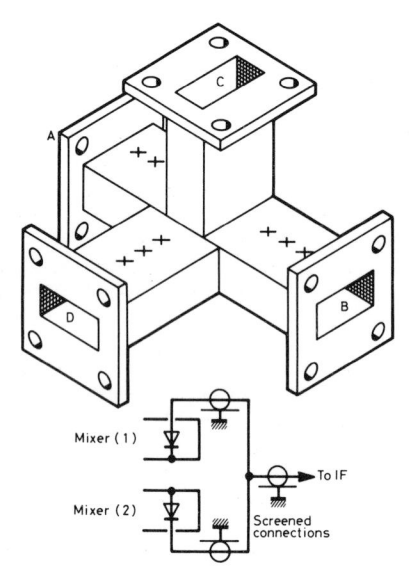

Fig 18.84. Diagram of a Hybrid-T (balanced mixer) receiver head. Identical mixer cavities should be fitted to ports A and B. Note that the diodes should be connected as shown below the main diagram and be of opposite polarity. The Gunn oscillator is fitted to port C and Port D connects to the antenna. Matching screws may be fitted, if needed, at points marked "x"

A hybrid T balanced mixer converter (receiver)

A hybrid T can be used as an efficient balanced mixer by attaching two identical mixer cavities, fitted with matched but opposing polarity diodes, to the A and B ports as shown in Fig 18.84. The antenna is attached to port D and the local oscillator is attached to port C. In a properly matched and terminated hybrid T, there will be little or no oscillator energy appearing at the antenna port and the oscillator power will be equally divided between the two mixers in antiphase. The incoming receive signal will not appear at the oscillator port, but will be equally divided, this time in phase, between the mixers. This means that good isolation exists between the mixer and oscillator and there is little or no received signal loss. There is insufficient local oscillator leakage to the antenna port to allow use as a transmitter.

The advantage to driving the mixer in this fashion is that, being balanced, it will cancel any a.m noise present in the local oscillator signal, giving the mixer some noise performance advantage over the single ended mixers so far used. Additionally, image reception rejection is easy to implement by including a suitable filter in the antenna port.

The hybrid T, configured as shown, makes an extremely good receive down-converter. Matching of the signal into the mixers can be improved by fitting three-screw tuners into the B and C ports. However, this can upset the oscillator injection balance and the mixer match may already be good enough if the cavities are designed and constructed carefully for the frequency in use, are fitted with a pair of carefully matched diodes and are as near identical as the constructor can make them.

Fig 18.85 shows a surplus hybrid T balanced mixer which uses cross-bar coupled matched mixers and cartridge-type diodes. It was found to work satisfactorily without modification when used with an iris coupled Gunn oscillator which was optimised to provide the desired mixer current. Optimisation of the mixer drive must be carried with one diode or other disabled since, with a properly balanced circuit and both diodes in place, there will be no current indicated on a meter connected to the output socket. Better still, both diodes are left in place and one output disconnected at the output junction. The disconnected output should be terminated in a short circuit. Ensure the i.f preamplifier provides a dc return for the other diode. Oscillator injection is now adjusted to the usual range by altering the Gunn iris plate as necessary. If a current check is now made with the disconnected mixer reconnected and the second mixer terminated as described, the current observed should be sensibly the same as before, but of opposite polarity. Ideally it should be identical if the diodes and their cavities are truly matched. On reconnection of both mixers, no current should be seen at the t-junction output. In the unmodified mixer there was slight but acceptable imbalance, amounting to about $10\mu A$ of indicated mixer current. A small degree of imbalance of this order is unlikely to detract greatly from the mixer performance and is almost certainly due to dynamic matching of the diodes. If the constructor has a selection of diodes available for test, a pair showing the smallest possible current should be used.

DRO mixers (receivers)

These recent additions to amateur equipment have already been discussed. They are roughly similar in concept to the cross-coupler receiver by virtue of the fact that they employ pcb couplers between the oscillator and mixer and antenna port (waveguide) and mixer. However, unlike the cross coupler converter, they do not lend themselves to the transmit function because most dro/mixer designs, and the

Fig 18.85. Photograph of a Hybrid-T balanced mixer receiver (G3PFR). For convenience in setting up (see text), the mixer output T-connection is made into a BNC T-adapter. The Gunn local oscillator tuning screw is at the rear. Photo: G6WWM

Mitsubishi FO–UP11KF is typical, incorporate a nulling device to cancel oscillator radiation, so making them unsuitable for transmit purposes. However, they make excellent receivers and exhibit noise figures typically around 7dB. The GaAs fet oscillator is very quiet compared with a Gunn, enabling effective use of a low i.f. The preset tuning screw can be replaced by a fully adjustable tuning screw if so desired, although this is not an easy operation. The stability of the module is such that a "narrowband" (eg 25kHz passband or less) i.f can be used.

The use of a fixed-tuned dro opens up the possibility of using a *tuneable* first i.f rather than a tuneable first oscillator. For this application, a 432MHz multimode or fm receiver or transceiver covering a 10MHz range (or more) would be near-ideal. The dro module's oscillator could be retuned from its nominal 10.4GHz to 10.808GHz. Given that the uhf receiver tunes from 430 to 440MHz, the microwave coverage then becomes, by subtractive mixing, 10,808 – 440MHz = 10,368MHz to 10,808 – 430MHz = 10,378MHz. This would enable the 10GHz receiver to cover the narrowband part of the band (10.368 to 10.370GHz, including narrowband beacons) and the first 8MHz of the upper "wideband" section, from 10.370 to 10.378GHz. The range could be extended slightly, possibly by a couple of MHz, by using the frequency pushing characteristics of the dro as a form of calibrated receiver incremental tuning (rit). Alternatively, if the dro frequency is set to 10.83GHz, coverage of the section 10.38 to 10.4GHz, which includes wideband beacons, could be equally easily provided. "Reverse" tuning produced by subtractive mixing might be regarded by some as an aesthetically undesirable operating inconvenience: to make "forward" tuning possible with this i.f, the dro would need to be set at 9.938GHz or 9.950GHz to enable the receiver to cover the same ranges as before. It is doubtful, although this point has not been checked, that the "standard" dro designed for 10.4GHz will tune so low. There are alternative modules available for lower frequencies, but these are not as easily available as the standard higher frequency modules which are usually available "off the shelf".

If more restricted microwave receive coverage is acceptable, a 144MHz i.f could be used. Amateur 2m receivers and transceivers normally cover 144 to 146MHz, although many modern "synthesised" receivers can be programmed to cover perhaps 144 to 148MHz without too much loss of performance. If the dro frequency is reset to 10.224GHz, then the microwave receive range can be either 10.368 to 10.372GHz (additive mixing, "forward" tuning) or 10.080 to 10.076GHz (subtractive mixing, "reverse" tuning). Again, the tuning range could be extended slightly by frequency pushing. In order to realise the full potential of the dro/mixer, image protection should be provided, whether 430 or 144MHz is chosen as the i.f.

Notwithstanding these limitations, the combination of a dro converter and a high, tuneable first i.f makes an extremely "potent" receiver, better in performance than most existing wideband designs, needing little setting-up, and providing performance close to the more elaborate narrowband designs given in section 18.6. Almost any i.f can, of course, be accomodated if the operator possesses a multimode, wideband "scanner" receiver.

18.5.2 Intermediate frequency preamplifiers

It is important to follow the microwave mixer with a reasonably high gain, low noise, impedance matched intermediate frequency preamplifier. Several designs are given in this section, suitable for use at the most commonly used "standard" frequencies of 144, 100, 30 and 10.7MHz. Some of these are broadband amplifiers which may be usable over the entire range mentioned. Others are "selective" but could be easily adapted to tune to nearby frequencies.

A broadband amplifier is more versatile in terms of being able to be used with a variety of i.fs, although it will, of course, respond to any signals – wanted or unwanted – which are within its gain passband. Preamplifiers designed around the SL560C (RS 560C) ic fulfil requirements between 10 and 150MHz, with usable but lower gain and manageable noise figures up to 300MHz. Such preamplifiers can be used either broadband or with some input (L/C) selectivity.

To minimise signal transfer loss, and avoid pickup of stray signals, it is desirable that the preamplifier is mounted close to the mixer. It is quite an easy matter to arrange for the microwave "head" (receiver or receiver/transmitter) and its associated preamplifier to be mounted in one screened housing suitable for mounting immediately behind the antenna (dish or horn), making a compact assembly which is suitable for either fixed station or portable use. It is also quite easy to arrange for the coaxial cable which carries the i.f signal from the mixer to the main receiver to carry the dc operating voltage from the receiver up to the preamplifier.

With present 10GHz passive mixers, a gain of 20 to 40dB with a noise figure of 3.5dB or better is usually demanded from the preamplifier. Only the circuit diagram and values are given for some of the designs. They are well proven circuits which are largely uncritical in their construction and "ugly" or "bird's-nest" construction is usually quite adequate, with the earthed components soldered to a piece of unetched pcb material and the non-earthed components appropriately suspended between them. The more fastidious constructor might like to prepare his or her own pcb layouts for any of the preamplifiers, although this is not strictly necessary.

BFR34a preamplifier (144MHz)

A typical example of the form of "ugly" construction is the BFR34a preamplifier, used with great success, built into the interdigital converters for 1.3, 2.3 and 3.4GHz (described in the respective band chapters) and onto the 'JVL transverters for 5.7 and 10GHz (section 18.6).

Broadband bipolar transistor preamplifier (70 to 500MHz)

A "lossless negative feedback" amplifier design, by DJ7VY, giving around 19dB gain over a frequency range of 70 to 500MHz was described in *VHF Communications*, volume 10, issue 1, 1978. This preamplifier has

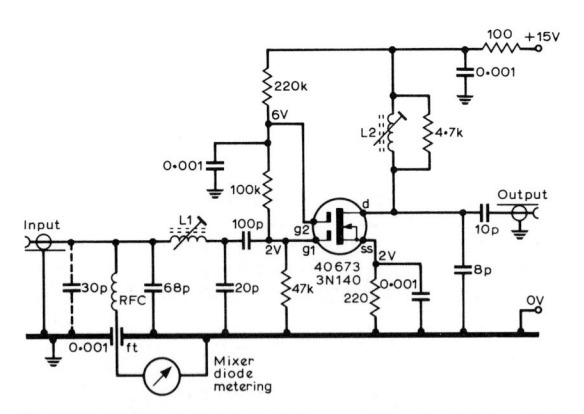

Fig 18.86. 30MHz dual gate mosfet preamplifier circuit. L1 and L2 16t 22swg e.c.w., close wound on 5mm former with "dust" core

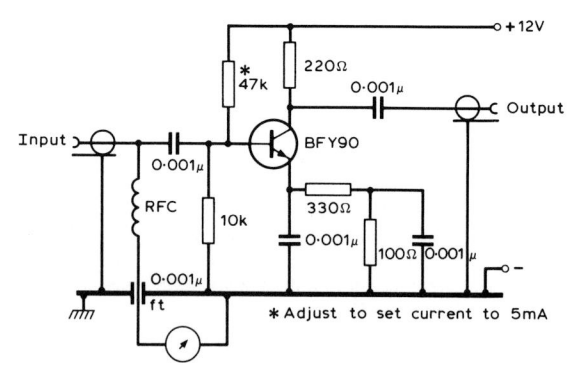

Fig 18.88. Single stage broadband bipolar transistor preamplifier

been used with good performance at both 144 and 430MHz and is recommended as both simple to construct and set up, for there is no alignment as such, other than setting the individual transistor standing currents as described there. It presents a singularly high dynamic range (approximately 102dB) with a noise figure of around 1.4dB at 144MHz and 1.6dB at 440MHz. Since the amplifier is capable of an output of +18dBm (approximately 63mW), it could also be useful as a low powered oscillator-chain amplifier at any frequency within its gain-bandwidth range.

Tuned dual-gate mosfet preamplifiers (30 and 144MHz)

Tuned preamplifiers using 40673 dual-gate mosfets are shown in Figs 18.86 and 18.87. Although designed specifically for 30 and 144MHz respectively, they can easily be modified for nearby frequencies. Any tendency to instability in the lower frequency version (where the gain is highest) can be prevented by unsoldering the Gate 2 lead of the rf amplifier fet and fitting a ferrite bead onto the lead before resoldering it into the circuit.

Untuned bipolar transistor preamplifiers (10 to 200MHz)

Fig 18.88 is the circuit of a single stage untuned amplifier usable from about 30MHz to 200MHz. It has a gain of about 10dB and a noise figure of about 2dB. The mixer current monitoring choke (rfc) should have an inductance of around 300µH at the lower frequencies and 3µH at the upper frequencies. The 47kΩ fixed resistor in the base circuit could be replaced by a miniature skeleton preset potentiometer of similar value. This would assist in easy adjustment of the transistor operating point.

A two-stage amplifier usable from 10 to 200MHz is shown in Fig 18.89, having a gain of about 25dB and a similar noise figure to the previous design. Similar observations about the value of the rfc apply, and again the amplifier performance is set to optimum by adjusting the value of the lower base bias resistor of the input stage (shown as 5.6kΩ). This was found to be fairly critical and in several samples of this amplifier, optimum performance was obtained with a value of 4.7kΩ. It is suggested that the fixed resistor is replaced by a miniature 10kΩ variable which is adjusted, on test, for best results.

Integrated circuit preamplifiers

The designs which follow use the SL560c (RS560c). This chip is very versatile, offering gains of up to a maximum of

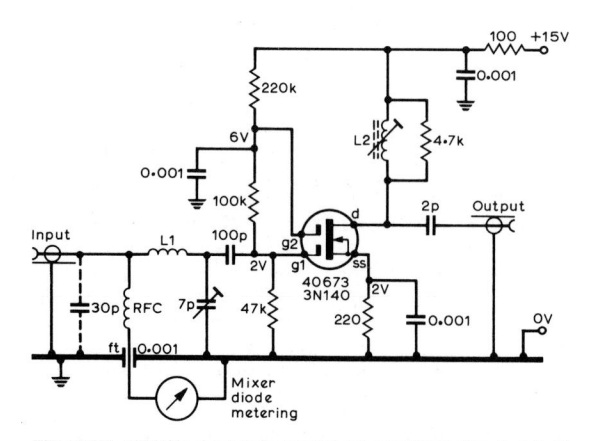

Fig 18.87. 144MHz dual gate mosfet preamplifier circuit. L1, 5t 22swg t.c.w., 6mm long, 5mm former with vhf core. L2, 4t as L1

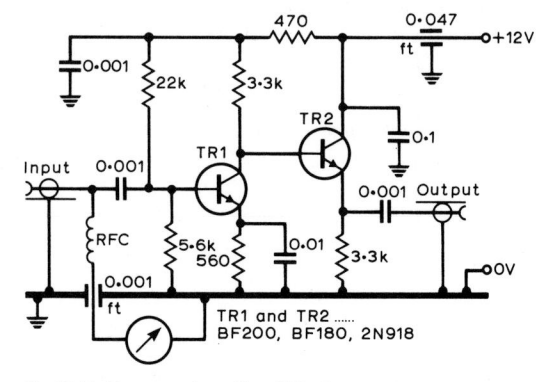

Fig 18.89. Two stage broadband bipolar transistor preamplifier

Table 18.6. SL560c amplifier characteristics: input in common emitter mode

Impedance	Rb	1st stage current	Approx gain high	Approx gain low	Noise Figure
50Ω	10k	4.5mA	40dB	30dB	3.4dB
100Ω	3.9k	2.25mA	35dB	25dB	
150Ω	2.7k	1.5mA	30dB	20dB	
200Ω	2.2k	1.125mA	27dB	17dB	1.8dB
300Ω	1.8k	0.75mA	24dB	14dB	

For an input impedance of 50Ω, the amplifier performance at a supply voltage of 9V (at pin 4 of the ic) is:

Gain, flat from 10 to 60MHz,		approx	36dB
Gain at	100MHz	approx	32dB
	144		28dB
	200		24dB
	300		18dB
Output impedance:			50Ω
Noise figure:			3.4dB

The gain figures can be reduced by 10dB by connecting pin 5 to pin 4.

40dB, a noise figure less than 2dB (when matched into an input source impedance of 200Ω), a bandwidth extending from 10 to 100MHz at high gain (or with roll-off to 300MHz) and an operating voltage range of 2 to 15 volts, depending on how the circuit is configured. The input stage can be operated in either common base or common emitter mode, the latter giving high input impedance, highest gain and lowest noise. Common base mode gives low input impedance which can be adjusted to match 50Ω or less. The gain range is "programmable"; if pin 5 is left open circuit, highest gain is realised and if pin 5 is connected to pin 4, this gain is reduced by 10dB. Full applications notes are available from the device manufacturer and should be consulted for more information than can be given here, especially if the constructor wishes to depart from the circuit configurations given.

One important feature to recognise is that the ic has low impedance voltage output and at high frequencies the inherent 25nH series inductance at the output can give resonance problems when working into a capacitive load above, typically, 200MHz. This problem is resolved by isolating the ic from its load with a series 30Ω resistor. This results in the loss of a little of the available gain, but ensures stable operation over the whole frequency range.

A "programmable" low noise broadband amplifier

The basic circuit of this amplifier is given in Fig 18.90 and a suitable pcb layout in Fig 18.91. One feature is that the input (source) impedance can be "programmed" by operating the ic input stage in common emitter mode and selecting the value of Rb. However, the available gain depends on the current drawn by the first (input) stage of the ic and this current is determined by the value of Rb. The gain is higher with low input impedance than with high. Also the noise figure is dependent upon the input impedance, being at a minimum of 1.8dB with an input impedance of 200Ω, rising to 3.4dB at 50Ω. Values of Rb, first stage current, gain and published noise figures are given in Table 18.6.

A 10.7MHz integrated circuit preamplifier

The circuit shown in Fig 18.92 is a variant of the general purpose SL560c broadband circuit already given. It uses L/C input and output circuits to provide a degree of selectivity and matching from the mixer into the amplifier. The input matches into the few hundred ohms of the microwave mixer and the output into the 330Ω impedance of a low-cost ceramic filter and is, in fact, the preamplifier which

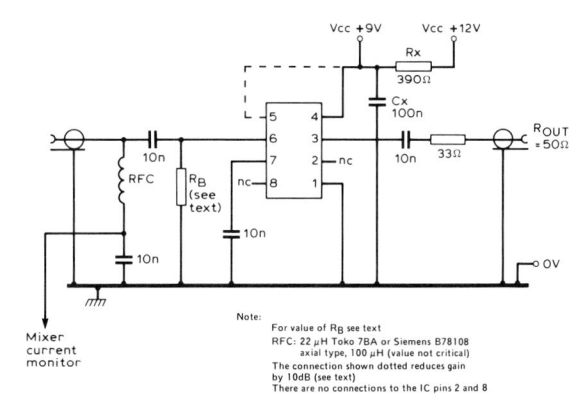

Fig 18.90. General purpose SL560c broadband amplifier

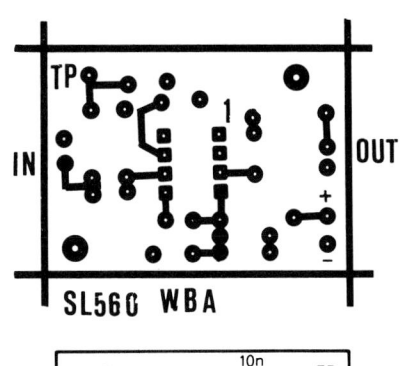

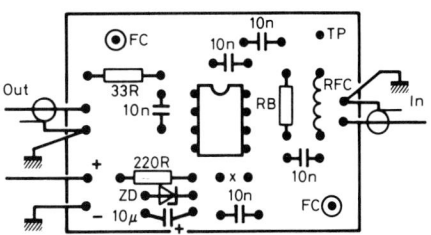

Fig 18.91. (a) track and (b) component layouts for the SL560c amplifier. Note that the isolated pads are earth points, linked to the groundplane with wire or circuit pins and soldered on both sides of the board. Pin 1 is soldered to ground on top groundplane. Pins 2/8 not used. Points "x" are open circuit for maximum gain or linked for 10dB reduction in gain. Performance at different frequencies is given in Table 18.6

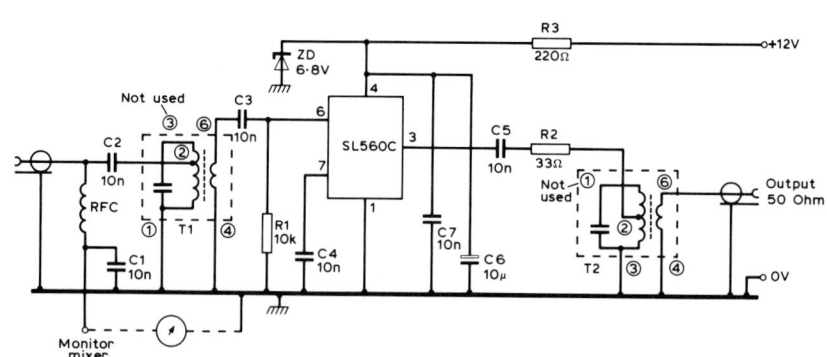

Fig 18.92. 10.7MHz amplifier using the SL560c chip. Component values not shown: Input transformer T1= Toko KAC6184A, Output transformer T2= Toko KAC6184A (50Ω output) or Toko KACSK 3892A (330Ω output). For a 300Ω input source, earth pins 1 and 4 on T1 (3 not used). For a 50Ω input source, earth pins 3 and 4 on T1 (1 not used)

forms an integral part of the 10.7MHz transmitter/receiver board described in section 18.5.4. Suitable ceramic filters are those in the Murata CFSH or SFE series which are available in a variety of bandwidths between 250kHz and 50kHz. By selection of an alternative 10.7MHz i.f transformer, the output matching can be reduced to 50Ω, suitable for driving a coaxial cable, or increased to a higher value to match other filters, according to the constructor's needs.

The circuit and component types and values are given in the figure and a board layout in Fig 18.93. It should be constructed on double clad board with an upper, unetched ground-plane surface. After etching and cleaning the board, the component mounting holes should be drilled and deburred. All earth connections should be made to the ground-plane and, where a component lead needs to go through to board unearthed, a small circle of copper should be removed from the ground-plane side using a larger drill bit or a Vero cutting tool. This method of construction is easier for the home constructor than trying to use a double-masking technique which can cause problems with mask registration. Pre-sensitised fibreglass board is most easily used, together with conventional development and etching.

After assembly, alignment consists simply of adjusting the cores of L1 and L2 for maximum signal. Both input and output tuning are fairly "flat". The design could be equally well used at 30MHz by substituting suitable value of L and C, the inductors being wound on Toko formers, with pin-outs compatible with the pcb tracks.

18.5.3 A 30MHz to 10.7MHz crystal controlled down-converter

A simple, inexpensive, but very effective down-converter from 30MHz to 10.7MHz is given next. Crystal control, incorporating a high frequency crystal without multiplication, is used in order to simplify both construction and alignment. Any converter circuit could be substituted, although this one has been found to be very effective.

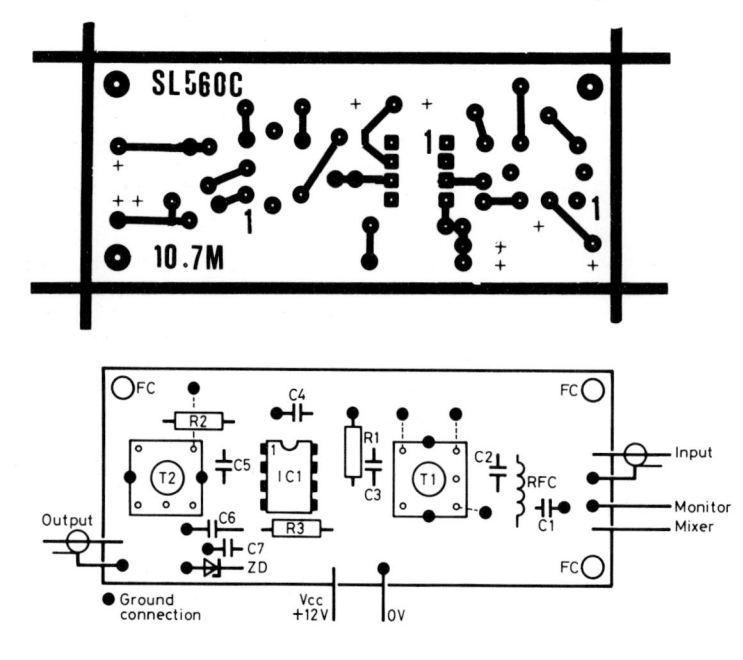

Fig 18.93. (a) PCB track layout for the 10.7MHz amplifier. Small crosses indicate approximate positions of earthing points. Pin 1 of the ic is earthed on the ground-plane only. (b) Component overlay. Components are numbered as in Fig 18.92. T1 and T2 cans are earthed by soldering to the ground-plane. No provision is made for gain reduction – if required (see text), connection is made on the track side of the pcb. See Fig 18.92 for earthing connections on T1 and T2

Fig 18.94. 30MHz to 10.7MHz crystal controlled converter circuit (G3NKL)

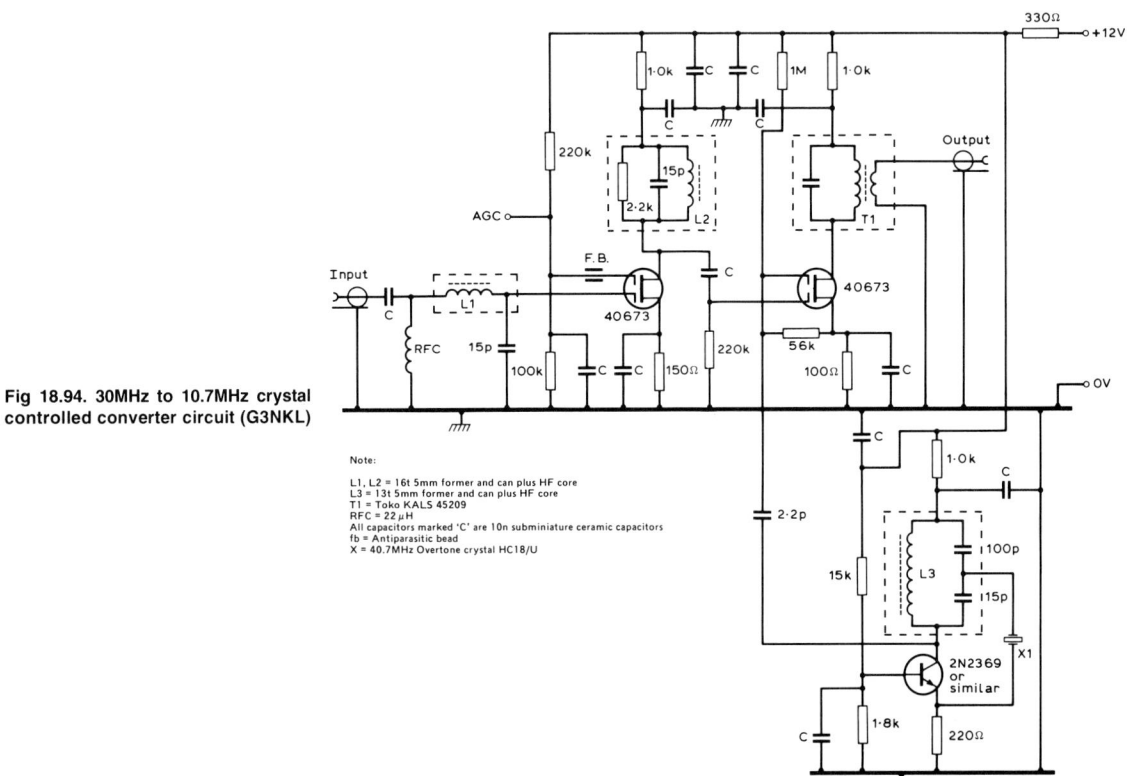

Note:
L1, L2 = 16t 5mm former and can plus HF core
L3 = 13t 5mm former and can plus HF core
T1 = Toko KALS 45209
RFC = 22 µH
All capacitors marked 'C' are 10n subminiature ceramic capacitors
fb = Antiparasitic bead
X = 40.7MHz Overtone crystal HC18/U

The circuit and component values are given in Fig 18.94. Similar construction techniques to those already outlined should be used. Alignment consists of checking and aligning the crystal oscillator which is a "sure-fire" design and should present no problems. The core of L3 is simply adjusted until oscillation is obtained. This will be indicated by a slight rise in receiver noise level when the converter is connected to the input of a 10.7MHz receiver. The input and output circuits are then tuned for maximum noise. If a frequency counter is available, the crystal oscillator can be set to exactly 40.7MHz. The input circuit is, finally, adjusted for best signal to noise ratio with the microwave head connected and tuned to a *weak* signal in the absence of a more "formal" signal generator. Any suitable attenuated 10GHz signal can be used for this purpose – don't forget that house walls can offer suitable attenuation!

18.5.4 A complete 10.7MHz ("baseband") receiver and wideband transmit board

The circuit described (shown in Fig 18.95, with component values in Table 18.7) is a logical extension of some of the "stand alone" designs given earlier. It incorporates an rf preamplifier using an SL560c, described above, and is followed by a CA3089E (or CA3189E, or any other of the pin-compatible derivatives) signal processing chip. This chip provides full i.f amplification, limiting, quadrature discriminator, audio preamplification, agc, afc, squelch and a signal strength meter driver. It is thus an ideal single chip receiver using a minimum external component count. Some of the facilities offered are not used in this application.

Audio power output is provided by an LM380 audio power amplifier ic which will provide up to 2W audio power output to drive a loudspeaker or headphones. The afc output of the CA3089 is a current source/sink capable of providing ±100µA current which is linearly related to degree of detuning of the signal at the discriminator. It was found too difficult to apply this output to the original 7805 regulator/modulator, even when using a conventional op-amp as the interface, so the regulator has been changed to a variable voltage regulator type LM317. This regulator is short-circuit proof and rated for 0.5A output. The output pin follows the voltage on the reference pin, but will be 1.22V higher. In this circuit, it is being used as a power voltage-follower.

A 5V reference is generated by a low-power 78L05 regulator and a variable fraction of this reference is added to the 5V reference in one section of the dual bifet op-amp, whose output can thus be varied over the range 5 to 9.2V by the 470R and 1k presets. This voltage is connected to the reference pin on the LM317 which produces an output of 6.2 to 10.4V at up to 0.5A to bias the Gunn. The op-amp

Table 18.7. Component list for the general purpose wbfm receiver and Gunn supply/modulator of Fig 18.95

Capacitors:
C1–5,7–10,12,13,25,26,33   10nF plate ceramic
C15,17–19, 29,30   100nF plate ceramic
C28   330nF plate ceramic
C21,27,31   1µF electrolytic
C6,11,14,22,32   10µF tantalum
C24   47µF electrolytic
C20,23   100µF electrolytic
C16,34   1000µF electrolytic
All capacitors 10V working, or more

Resistors:
R10	3R3	R4,13	470R	R7, 12	6k8
R5, 9	10R	R17,29	1k0	R1,8, 22,31	10k
R14	15R	R21	2k0		
R2	33R	R18	2k7	R11	15k
R16	100R	R6	3k9	R30	22k
R3,23	220R	R24,26, 27,28	4k7	R25	47k
R15	270R			R19,20	82k

RV3,4   470R 1/4in dia. cermet preset
RV2   1k 1/4in dia. cermet preset
RV5   1k multiturn with turns counting drive (Gunn fine tune)
RV1   10k standard spindle (Audio volume)
RV7,8   47k 1/4in dia. cermet preset
RV6   100k preset carbon
Note: R7 is connected between pins 7/10 of IC2 on track side of board. All fixed resistors 1/4W carbon or metal film

Semiconductors:
IC1	SL560c, RS560c	IC6	LM317N
IC2	CA3189E etc	TR1,2,3	BC109
IC3	LM380N	D1	BZY88C6V8
IC4	78L05	D2	1N4001/4002
IC5	TL072	D3	BZY88C6V2

Inductors and Filter:
RFC1   100µH axial lead choke, Siemens B78108
RFC2   22µH radial lead choke, Toko 283AS–220
IFT1   Toko KAC6184A
IFT2   Toko KACSK3892A
IFT3   Toko KACSK586HM
FL1   280kHz CFSK10.7M1 (Red Spot) or
     180kHz CFSK10.7M3 (Red Spot) or
     50kHz Scan filter UF71N (Maplin) or SFA10.7MF

Miscellaneous:
M1,M2   Miniature edgewise meters, 75/100µA, suitably shunted (adjust on test)
S1   Subminiature SPCO switch
S2   Single pole, centre off switch
J1   Closed circuit jack
SK1   Polarised power socket

is used as a summing junction to add together all the different inputs which can control the output voltage.

AFC from the CA3189E is fed into the other section of the dual op amp which acts as an adjustable gain amplifier. The afc loop-gain can be set by means of the two variable resistors to suit the pushing characteristics of the Gunn in use. Its output is fed into the second section of the op-amp which controls the LM317 output. The appropriate input is selected according to whether the local oscillator is above or below the signal, in order to make the loop lock.

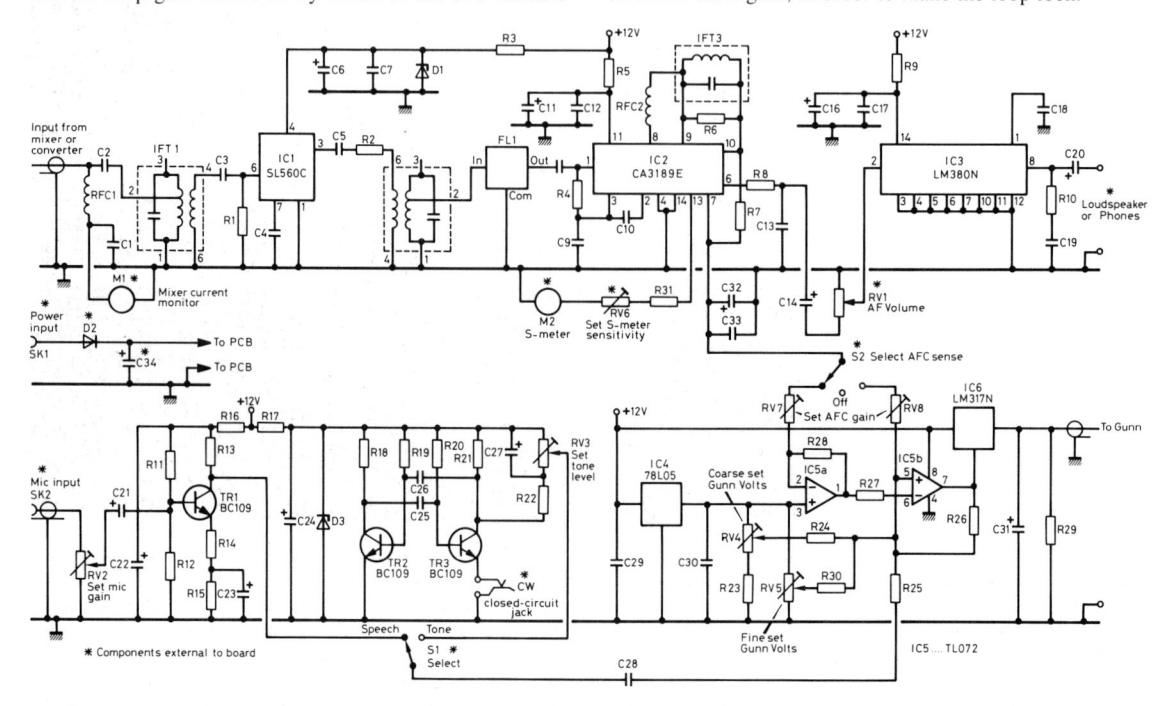

Fig 18.95. 10.7MHz i.f general purpose wbfm Gunn transmitter and receiver circuit, suitable for use at both 10 and 24GHz

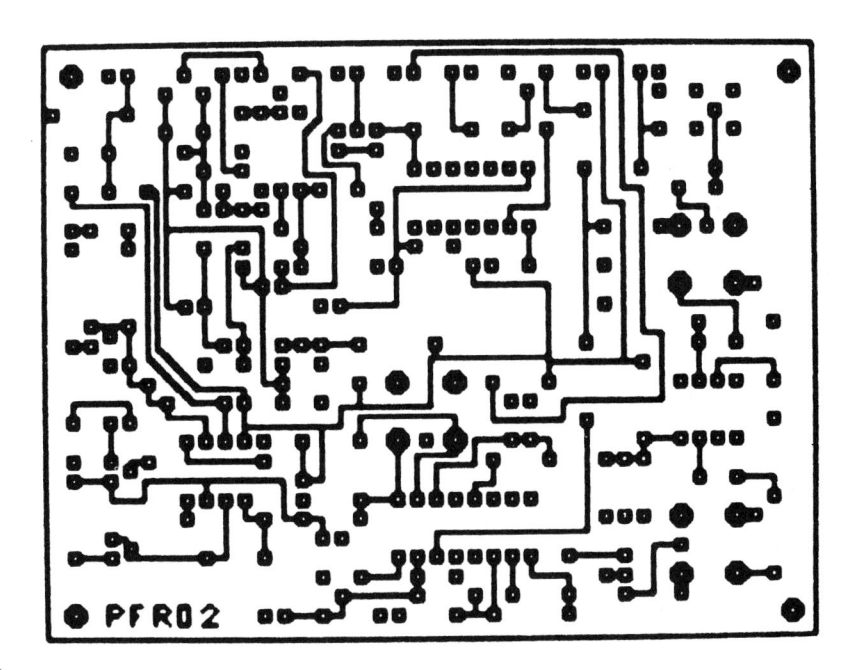

Fig 18.96. PCB track pattern for the general purpose design of Fig 18.95

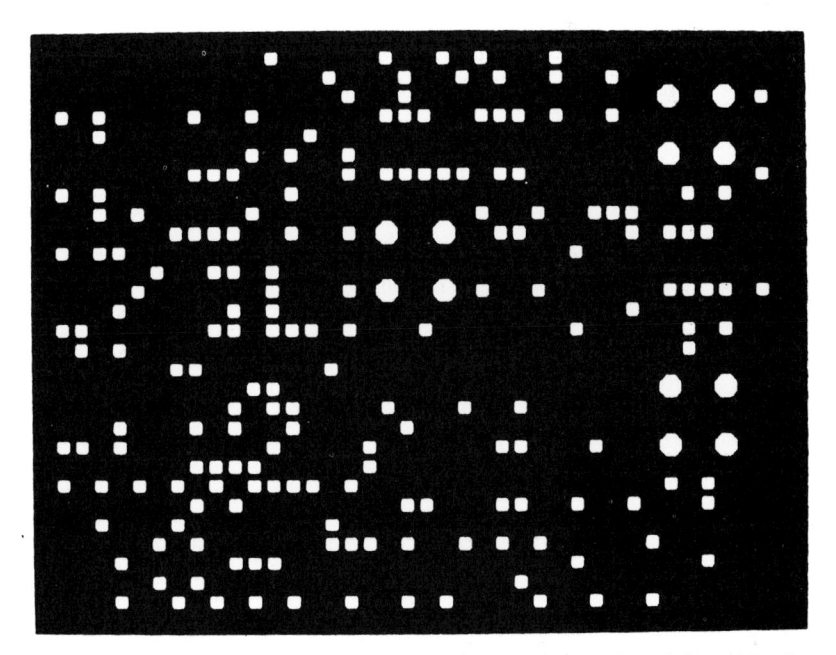

Audio (microphone) gain can be varied by altering the 47k input resistor, and the fine tune range by altering the 22k input resistor. The maximum output voltage can be varied by altering the 220R resistor in series with the 470R pot. The board track layout for the design is given in Fig 18.96 and the component layout in Fig 18.97. There are no particular problems in construction; good practice would be to fit all the passive components first, followed by the ics and transistors.

Alignment is simple but requires a signal either at i.f or, using the Gunn oscillator/mixer, at 10GHz. The input circuits are adjusted for maximum gain (afc off), using the S-meter as a guide and the quadrature coil is adjusted for best signal to noise performance, either by ear or by

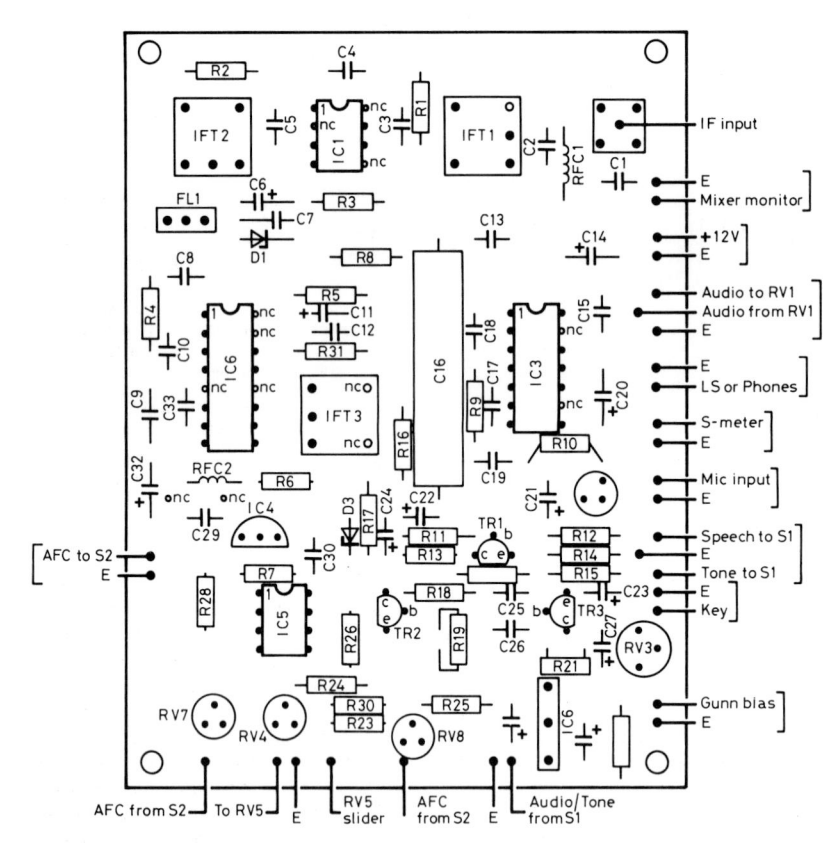

Fig 18.97. Component layout for the circuit of Fig 18.95

using an alignment aid such as that described in chapter 10, "Test equipment". The afc loop gain is adjusted to suit the Gunn in use and should be able to provide lock over several MHz. The microphone gain and tone levels are best adjusted by listening for the signal on another receiver, although it is possible to judge the quality of signal by applying modulation whilst receiving an un-modulated 10GHz carrier. The module can be equally well used with a 24GHz in-line Gunn oscillator/mixer such as the Plessey GDHM32 and a narrower ceramic filter could be fitted instead of the more usual 150kHz filter shown in the circuit design.

18.5.5 "Going to the races": a summary of wideband systems

From what has been described so far, it should be apparent that wideband communications systems of varying complexity and effectiveness can be built up from the various modules.

Assembling and using a wideband system (especially the all-important "front-end" or microwave "head") is a little like going to the races! The best bet is the horse and jockey most suited to the course conditions. In this case the jockey is already decided – the builder/operator. The course should be inspected and, according to its nature (path length and obstructions), the horse (equipment) selected.

The "course" is the nature of the communications paths to be worked – short range fixed-to-fixed paths (home station), longer range fixed or portable operation, contest operation, obstructed paths and so forth. These factors can have considerable bearing on the "quality and breeding" of the horse (equipment). The individual modules can be configured in so many different ways that it is almost impossible to describe them all succinctly and much must be left to the judgement and imagination of the operator. They can vary between a donkey, a mule, an ordinary workhorse or a thoroughbred!

Short range fixed-to-fixed links can be very simple, with the microwave head and intermediate frequency amplifier mounted at mast-head as suggested in chapter 9, "Microwave beacons and repeaters", or Fig 18.98. Suitable units might be a self-oscillating mixer (Fig 18.79), a 3dB coupler ex-doppler unit (Fig 18.48), a home-made in-line unit (Fig 18.42 and 18.43) or a modified ex-doppler in-line unit (Fig 18.47). Whichever microwave head is chosen, it should be followed *immediately* by the first i.f preamplifier and then by the receiver "back end" which can be remotely sited if so desired.

THE 10GHZ ("3CM") BAND   18.63

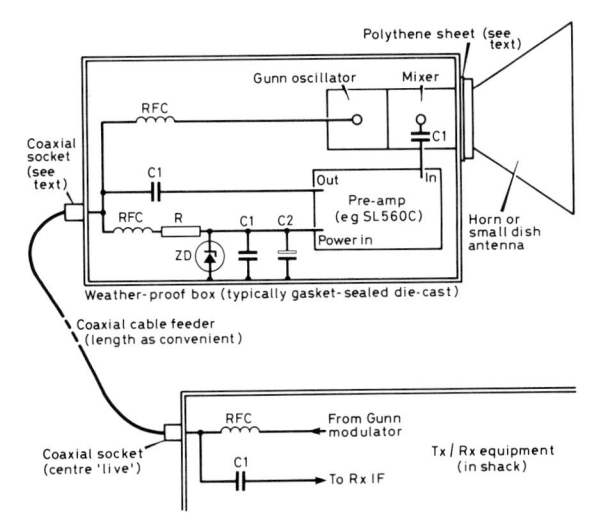

Fig 18.98. Typical masthead installation of wideband (or other) equipment. Note that this configuration allows simultaneous use of the coaxial screened cable to supply dc up to the masthead unit and i.f signals down to the receiver

Table 18.8. Wideband transmit and receive systems summary

Module	Transc've	Tx	Rx	Image Reject	Remarks
Self osc mixer (Gunn)	Y	G	P	N	Noisy RX
3dB coupler (Gunn)	Y	M	P	N	Lose 1/2 signal
In-line Gunn	Y	M	G	N	Compromise
In-line DRO/Mix	N	N	E	Y	No TX
In-line Gunn+ Gunn Tx	Y	VG	VG	Y	Good config
In-line DRO + Gunn Tx	Y	VG	E	Y	V good config
Cross-coupler	Y	E	E	Y	Cumbersome
Circulator	Y	G	G	N	Compact
Hybrid-T	N	N	VG	Y	Bal. mixer
Hybrid-T + Gunn Tx	Y	G	VG	Y	Good config
In-line DRO/mix + DRO Tx	Y	E	E	Y	Excellent config

Key:	Y	=	Yes	P	=	Poor
	N	=	No	M	=	Moderate
				G	=	Good
				VG	=	Very good
				E	=	Excellent

Similar equipment can be used for portable and contest use although here, with the weaker signals to be expected over long line-of-sight paths, only the in-line module (of the simpler configurations) is really suitable. The in-line system performance is greatly improved by using it only as an optimised receiver (and "standby" transmitter) and employing a separate transmitter with suitable changeover arrangements.

Other alternatives for "better" portable use are the cross-coupler transceiver, the circulator transceiver or the hybrid-T receiver with separate transmitter and changeover.

For slightly obstructed paths it is of advantage to increase the "potency" of the system by employing the narrowest practicable bandwidth, for instance by using a dro/mixer for the receiver and incorporating image recovery methods. These measures will squeeze the last few dB out of the wideband system in receiver terms. The transmitter power should be maximised and this is easier to implement by using a separate transmitter than by the simpler but compromise transceive modules.

For very long or badly obstructed paths there is really no alternative but to use true narrowband techniques and employ modes other than wbfm (even nbfm!). Such techniques are discussed in detail in section 18.6.

Some of the many configurations for wideband equipment are summarised in Table 18.8. The simplest system for the beginner could consist of a 20dB horn, an in-line Gunn mixer/oscillator and single conversion down to 10.7MHz where the main signal processing takes place. As the beginner develops the system, the next steps might be a small dish antenna, double conversion receiver, separate transmitter – and so on. Of the simpler "switched" transceivers, the cross coupler, although manual changeover is cumbersome, is particularly recommended.

18.5.6 Fast-scan atv

Using the systems described so far, transmission and reception are limited to "speech frequency" modes; that is, with the modulating signals contained within the audio spectrum from a few tens of hertz to perhaps 20kHz. This includes normal speech, tone modulated cw and tone modulated frequency shift keying (afsk) modes such as slow-scan tv (sstv) and teletype (rtty). In order to transmit and receive full definition, fast-scan tv (atv) the bandwidth of the signal processing circuits, such as the modulator and the receiver back-end must be able to handle bandwidths of several MHz without distortion or undue frequency roll-off which would curtail definition. This requires somewhat different techniques to those so far described. The brief account which follows should be sufficient to allow the experimenter to develop a quite effective atv system on the 10GHz band.

It must be remembered that, due to the increased bandwidth of the tv signal, received signal to noise over a given path will be inferior to narrower band modes. Unless compensated for by either a GaAs fet preamplifier or increased antenna gain at the receiving end and an increase in transmitter power and antenna gain at the transmitting end, the ranges covered are likely to be much less than with speech communication modes. Nevertheless, ranges of up to 100km over line-of-sight paths have been covered with excellent picture quality.

Acknowledgements are made to the British Amateur Television Club (BATC) for most of the ideas and circuits which follow. Further ideas for the construction of fast scan atv equipment on both the 1.3 and 10GHz bands will be found in various BATC publications. With the bandwidths available on the microwave bands, there is less need for slow scan tv, except as a means of improving signal to noise ratio on longer paths using ordinary narrowband transceivers! BATC, however, publish a comprehensive,

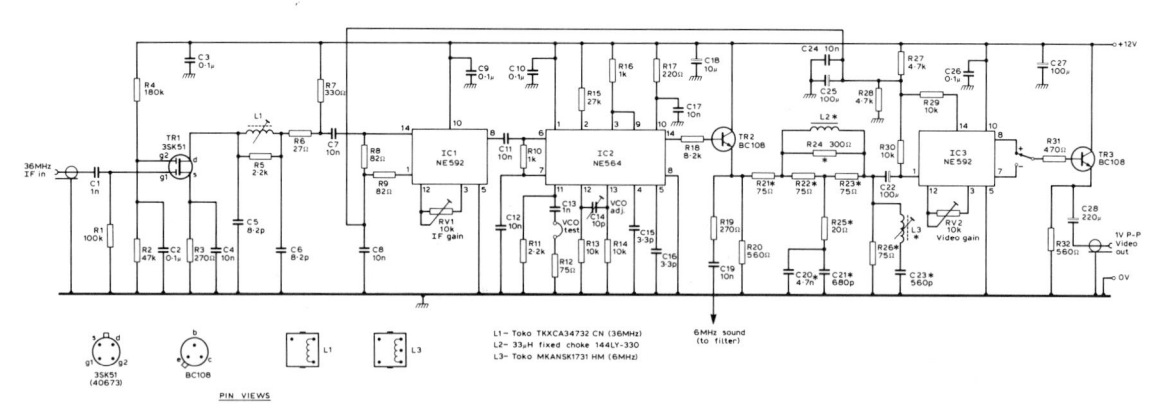

Fig 18.99. Basic fm tv receiver circuit (BATC)

compact handbook entitled *The Slow-Scan Companion*; all the techniques and equipment described there are equally applicable to the microwave bands. The simplest mode to implement on the microwave bands (and particularly on the 10GHz band) is fm.

Television receivers

The system requirements for an fm tv receiver system were enumerated [26] as being:

1. The system needs to have sufficient bandwidth to enable the *whole* of the signal to be demodulated. If the receiver bandwidth is narrower than the signal deviation, video or sync signals will be lost. Conversely, if the receiver bandwidth is too wide, the demodulator will not produce a full amplitude video signal at its output.

2. Front-end gain should be sufficient to cause limiting in the pll demodulator at around the noise threshold. This will ensure that even weak signals will be correctly received.

3. The system should have low impedance input and be capable of being driven, in the case of the lower frequency bands, from a varicap or similar rf tuner. In the case of 10GHz systems, this will usually be interpreted as single conversion, omitting an intermediate tuner.

4. The system should deliver a standard 1V p–p amplitude composite video output suitable for feeding to a monitor or an rf modulator.

5. Variable front-end gain should be provided to cater for different input levels.

6. CCIR standard pre- and de-emphasis should be provided as an option in the transmit and receive designs.

7. Provision should be made to extract an inter-carrier sound signal.

8. The units should be powered from a single 12V power source to enable portable operation.

9. The whole of the circuits should be accommodated on the minimum number of boards.

Almost invariably, video processing strips now employ wideband video ics for both transmit and receive functions, together with pll ics for demodulation. This can lead to very compact printed circuit boards.

The circuit for a basic receiver is given in Fig 18.99. This was originally designed for input from a varicap tuner and intended to be used on the 1.3GHz band. For 10GHz use it is suggested that the input mosfet amplifier is replaced with the broadband SL560C amplifier of Fig 18.90 and 18.91, mounted directly at the mixer of, for instance, an in-line Gunn oscillator/mixer converter.

The component layout for this circuit is given in Fig 18.100. The pcb is double sided with the upper surface used as a ground-plane to ensure circuit stability. Components are mounted with minimum lead lengths. A suitable board is available to members of BATC, through their membership services section.

In the circuit, C14 sets the voltage of the vco to the centre of the i.f frequency and a test point is provided for this purpose. The demodulated video signal is passed to an emitter follower where it is possible to extract the sound signal. The output impedance is suitable for feeding directly into a 6MHz ceramic filter and thence to the sound demodulator which can be based on the CA3189 ic, using a circuit very similar to those already described for wideband speech systems.

The passive network following the emitter follower is a CCIR standard de-emphasis circuit which could be omitted if this is not used in the system. If not fitted, the 6MHz sound trap must also be removed and possibly placed in the base circuit of TR3. The following video amplifier provides either positive or negative going signals at 1V p–p into 75Ω. The use of ic sockets is not recommended.

Subsequent to the publication of this design, it was reported that a number of constructors had found the NE564 pll demodulator to be somewhat critical to align and maintain in alignment. A further article in *CQ-TV* [27] suggested the modifications given in Fig 18.101. These consist of:

1. Removing resistors R12, 13, 18 and 16

THE 10GHZ ("3CM") BAND   18.65

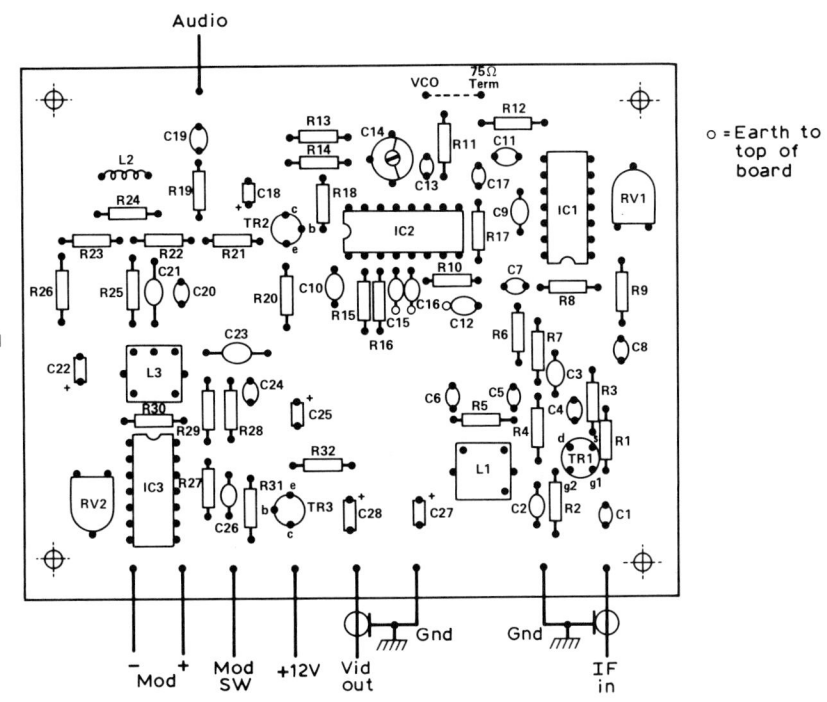

Fig 18.100. Component layout for FM tv receiver (BATC)

2. Removing capacitors C12, 13 and 16.
3. Disconnection of pin 9 from pin 3 by cutting the track and leaving pin 9 disconnected.
4. Replacing C15 with a 10n capacitor.
5. Installing a 1n capacitor between pins 3 and 11 and a 1k resistor between pins 3 and 7, both components fitted underneath the board, directly adjacent to the pins concerned.
6. The circuit has been found to function more reliably at 11V. The constructor is advised to fit a suitable dropping resistor and decoupling in the power supply line to the entire board.
7. It was also found that the network consisting of L1, R5, C5 and C6, which were fitted to "tailor" the bandwidth, are not necessary and removal has negligible effect on circuit operation. Removal will allow the use of other i.fs up to about 70MHz, the approximate upper frequency limit for reliable operation of the pll chip.

Setting the circuit up for correct operation is quite straightforward and consists of connecting a frequency counter to the vco test point and adjusting C14 for a reading equal to the i.f chosen (36MHz in the original design). After alignment, the test point is linked to R12 to provide proper termination. RV1 is turned to maximum and RV2 halfway. RV2 is then adjusted on a received signal to set a satisfactory contrast level. If used, L1 is tuned for best signals, its tuning being very flat. Pins 3 and 9 of IC2 require between 1 and 1.5V on them and this may be adjusted by altering the value of R17; this will, however, alter the demodulator bandwidth which, with the values shown, will be approximately 10MHz.

Two alternative sound demodulators based on phase-locked loop ics were described in [28] and [29].

A tv modulator for Gunn oscillators

The circuit given in Fig 18.102 [30] probably represents one of the simplest means of modulating a Gunn oscillator with a composite video signal. No particular precautions are needed in its construction. The circuit again uses an easily available and inexpensive video amplifier ic which offers either fixed gains of 100 or 400 without external components, or with external adjustable components, gains from unity to 400. The chip contains a two stage differential-output wideband monolithic video amplifier with the input stage designed so that, with external

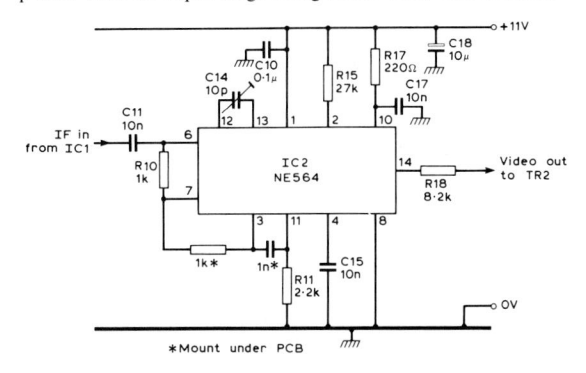

Fig 18.101. Modifications to improve the performance of the circuit of Fig 18.99 (BATC)

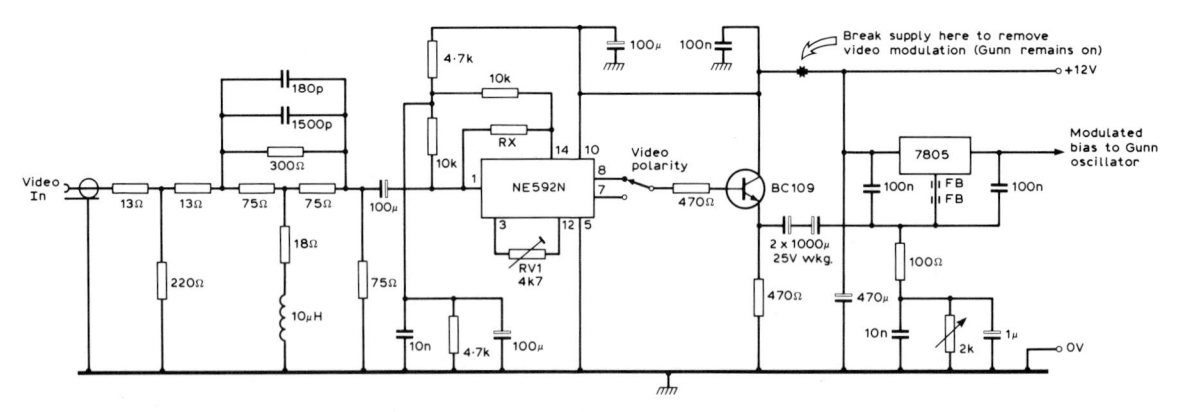

Fig 18.102. Video modulator for a Gunn oscillator (BATC)

components fitted to the gain select pins (3 and 12), the circuit can act as a highpass, lowpass or bandpass filter. In this application these external components are selected to give CCIR standard pre-emphasis. The completed modulator should be housed in the same box as the Gunn oscillator so that lead length between them is kept to a minimum.

The actual modulator is the now-familiar 7805 regulator circuit adapted for this purpose. A suitable pcb is again available to BATC members. About 50mV of video signal, derived from the 1V p–p input signal, should appear on the Gunn bias voltage when examined using an oscilloscope. If this level of signal cannot be obtained by varying the 4k7 gain control, then the value of Cx can be adjusted, on test, until a satisfactory range of control is achieved. If either the receiver or transmitter boards are used with a Gunn in-line mixer/oscillator, then the normal anti-parasitic and mixer resistive load components should be fitted directly on the respective terminals as shown in Fig 18.103.

Fig 18.104 is a block diagram of the interconnection of suitable modules needed for a complete Gunn-based transceiver for 10GHz fm tv.

Alternatives

Just as the advent of satellite direct broadcasting has made dro devices accessible to the amateur, so the first i.f and second i.f/demodulator units with standard composite "baseband" video output have come on to the market at fairly reasonable prices. These would be suitable for those amateurs who want to experiment with 1.3 or 10GHz tv reception without the need to construct all their own equipment, although this might be the more satisfying way to do the job and, in some ways, easier.

Astec, who have made uhf modulators for a number of years, have fairly recently produced two modules which might prove interesting to the amateur. As satellite tv broadcasting increases in popularity, no doubt many other types will appear on the "surplus" market.

The two modules are, respectively, the AT1020 and AT3010. The AT1020 module tunes 950 to 1,450MHz with a nominal 612MHz i.f output, whilst the AT3010 unit accepts the 612MHz output of the tuneable converter and processes it using one of the latest single chip quadrature demodulators. Baseband video output is up to about 3V into 1kΩ impedance.

It is stressed that at the time of writing these units have *not* been tested by the authors. It is almost certain that performance around 1.3GHz will be greatly improved by

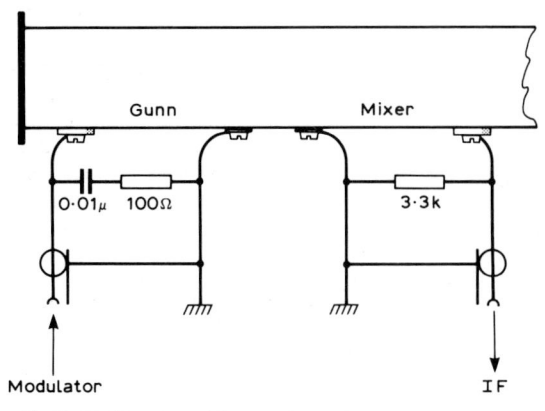

Fig 18.103. Mounting of Gunn anti-parasitic components (BATC)

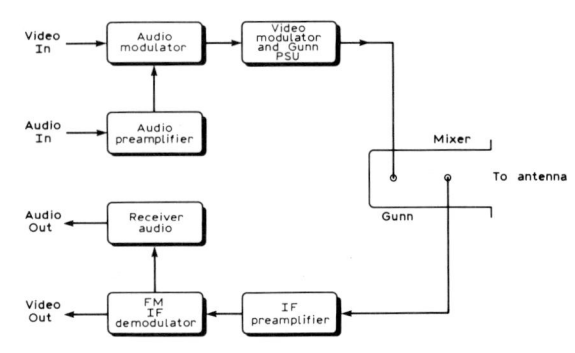

Fig 18.104. Block diagram of a 10GHz atv transceiver (BATC)

the inclusion of a really low-noise preamplifier, which will also enable use for 1.3GHz atv. Although these tuners were designed for use with a microwave converter covering 11 to 12GHz, there is no reason why they should not work equally well with a converter covering the parts of 10GHz recommended for atv, provided that the first oscillator is placed relative to the wanted signal so as to produce an i.f somewhere within the (tuneable) range of the first i.f strip. This would infer a local oscillator somewhere around 9.2GHz to cover the 10.15GHz to 10.35GHz area. Any of the Gunn oscillator designs already given can be adapted to these frequencies by constructing to suitable calculated dimensions. Experience with the iris coupled type of oscillator suggests than an oscillator dimensioned for 9.5GHz will also cover 9.2GHz when tuned by an adjustable dielectric tuning screw. The in-line mixer described elsewhere should be satisfactory without modification. A second alternative might be to follow the mixer with a broadband amplifier at 600MHz and dispense with the first converter, bringing the i.f down to 612MHz. This would then require a local oscillator frequency of 9.43GHz to cover the same portion of the amateur band.

Alternatively, other commercial modules for fm tv are available in the form of ready made or self-assemble kits, notably those of Wood and Douglas who produce the VIDIF 52MHz i.f and demodulator strip, the SCT–2 and SCR–2 transmit and receive sound demodulators boards and pattern generators, to name but a few of the modules available in kit form. Whichever form of construction is used and whatever the i.f chosen, it will be found that the simple techniques outlined in this section can lead to very satisfactory results at 10GHz. Other more elaborate techniques, applicable particularly on the lower microwave bands, are frequently described in the BATC publication *CQ-TV*, which the reader is advised to consult. Very similar techniques to those outlined above can also be applied to the reception of direct satellite broadcasting in the 11 to 12GHz band and could form the basis for experimentation with potentially low-cost antennas and receiver front-ends for that band.

18.6 10GHZ NARROWBAND SYSTEMS

18.6.1 Introduction

Narrowband equipment is inevitably more complicated than the wideband equipment described earlier, in terms of both its construction and also the techniques required in its alignment. The main justification for changing to a narrowband system is to increase the potential of a given size of equipment for working non-optical paths. As has been noted earlier, most wideband equipment will cope with all line of sight paths available, and the extra complexity of narrowband equipment can hardly be justified for this latter type of working.

The potential system gain from reducing the bandwidth and changing the mode from fm to a.m/ssb is highly significant and has already been discussed in section 18.1.3.

Interest in narrowband equipment is comparatively recent in the UK. Nevertheless several approaches have already been investigated – injection locking (section 18.3.7), conventional multiplication (section 18.3.5), phase locking (see chapter 19, "The 24GHz band") and mixing/filtering techniques to be described in this section. Inexpensive generation of narrowband powers above the milliwatt level depends on the development of simple, reliable, economic – and reproducible – amateur designs for GaAs fet amplifiers or readier access to twt amplifiers. Both are happening as we write!

It should be recognised that the design and building of narrowband equipment, at least as it is presently conceived, is difficult for the beginner on 10GHz, both in respect of its comparative complexity, the need for careful and often "fiddly" constructional techniques, and the absolute necessity for test equipment, albeit homemade, for tuning it up. This is the main reason why a lot of space has been devoted to quite detailed descriptions of waveguide components and the simpler wideband techniques early in the chapter: the authors feel that 10GHz wideband is an ideal place to learn simpler waveguide and operating techniques before using this experience to "graduate" to the more difficult narrowband techniques, usually employing precision pcbs, described at the end of the chapter.

18.6.2 The relative performance of am/ssb and wbfm

While signals can be quite strong over line of sight paths, it is often the case that the signals *sound* very strong, but in fact there may only be 10dB in hand. This is a consequence of the wideband fm modulation system, where the audio signal to noise (S/N) ratio can change very rapidly for small changes in the input S/N. Strong signals, e.g. an S9 report, should mean that signals were 40 to 50dB over the noise level (at 6dB per "S" point). If this were the case, then:

1. They would be heard equally well over a wide range of beam headings, as the average sidelobe level on the dish may only be 15 to 30dB down on the main beam.

2. They would also be heard well on open waveguide – typical gain for a 24in (610mm) dish might be 33dBi and for open guide is 6dBi.

In both these cases, there would still be about 20dB in hand. Signals this strong are not very common and, although the *audio* S/N may be 40 to 50dB, signals will usually not be audible unless the dish is pointed in the right direction, implying that there is only 10 to 20dB in hand. We are not usually concerned with having a very high audio S/N. 30dB is probably quite adequate. What we want to know is how much more path loss we could tolerate before losing the signals, either by lengthening the path, or allowing more of an obstruction in it. These will all affect the i.f S/N, and a very high audio S/N does not mean that we can accept that number of dB extra path loss before the signals vanish. The reason for this can be seen in Fig 18.105. At low S/N (above 0dB), the audio S/N increases at the rate of 4dB/dB of input signal. So an increase in signal of 10dB

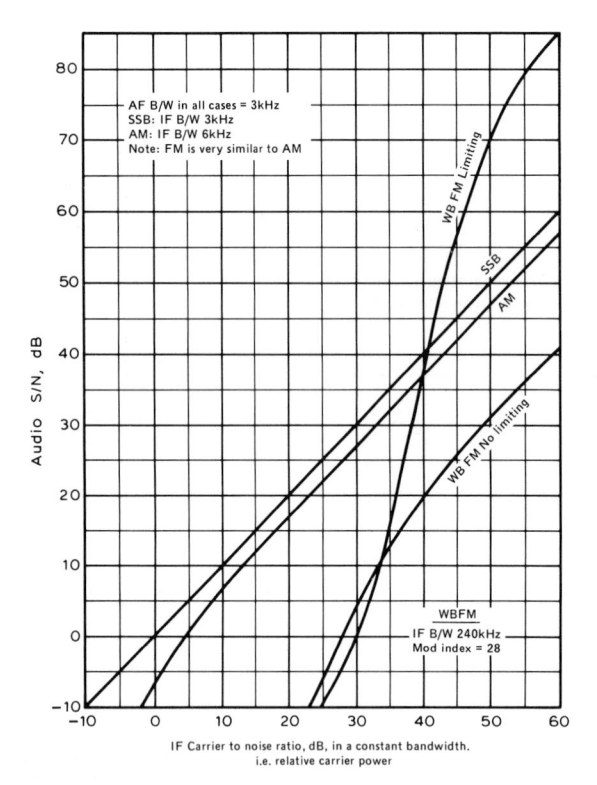

Fig 18.105. Comparison of performance of a.m/ssb/fm modes

(AF B/W in all cases = 3kHz; SSB: IF B/W 3kHz; AM: IF B/W 6kHz; Note: FM is very similar to AM. WBFM: IF B/W 240kHz, Mod index = 28)

will take that signal from being barely detectable (0dB) to apparently S9 (40dB audio S/N). In fact, for very strong signals – a true S9 – wideband fm will give a higher audio S/N than ssb in a 3kHz bandwidth, but for weaker signals ssb or any other narrowband modes are definitely superior.

The wbfm system sacrifices weak signal performance for better performance with strong signals. This is mainly due to the limiting action in the detector and, if no limiter is used, the weak signal performance improves by several dB. It should be possible to operate below limiting by reducing the i.f preamp gain, or placing an attenuator between the preamp and main i.f strip.

In order to obtain maximum range, or contacts over obstructed paths, we are almost certainly working with *very* weak signals and it is in these cases where the superiority of ssb or other narrowband modes will become obvious.

Received signal levels can, therefore, only be properly measured by either a calibrated S-meter or by inserting a calibrated attenuator before the limiting i.f strip – either in waveguide or between the i.f preamp and the limiting i.f strip. The attenuation can be increased until the signal is *only just* detectable and the level expressed as a number of dB over noise – dBN. This is not an absolute unit, as it depends on the receiver sensitivity, but is very meaningful and is constant for any one receiver; the noise figure can be measured at some later time!

18.6.3 The G3JVL image-recovery mixer/transmit converter for 10GHz – original design ("Mark I")

The unit consists of a mixer diode mounted between two filters, one on the local oscillator frequency (10.224GHz), the other on the signal frequency (10.368GHz). The local oscillator filter is designed for maximum rejection of local oscillator noise and thus has a very narrow bandwidth while the signal filter is designed primarily for low loss and is therefore somewhat broader. The construction of the filters was described in detail in section 18.2.10 and so is not reproduced here in full detail. The relevant dimensions, i.e. the spacing and diameter of the iris holes are given in Fig 18.106.

Both the multiplier and mixer diodes are mounted using Gunn type posts and discs, as described in Fig 18.42, (but reproduced again in Fig 18.107 for convenience). The matching into the multiplier diode can be the same as that described in Fig 18.65. The diode requires, typically, 0.5 to 1W drive at 384MHz. A suitable drive source and amplifiers were described in chapter 8, "Common equipment".

The mixer diode is a GaAs Schottky barrier type, typically a DC1501E, although devices of the same general type by any manufacturer will work satisfactorily. The 144MHz i.f output is fed by a suitable matching network to a low noise preamplifier. The matching will depend on the exact diode current used and is best set up using an automatic noise figure meter, such as that described in chapter 10, "Test equipment".

When optimised, the noise figure should be around 7dB, allowing for a 1.5dB noise figure i.f preamplifier. This is a very good performance indeed and will only be bettered by the use of a low-noise GaAs fet rf preamplifier. As a transmit converter, feeding in a few milliwatts of 144MHz ssb into the mixer diode will produce up to about 1mW pep output at 10.368GHz. This is adequate for local contacts or longer distance contacts over unobstructed paths, or is sufficient to drive a twt amplifier to several watts output.

In use, few problems have been found with this design. Like the wideband in-line mixer described earlier, this mixer makes an excellent receiver, but many constructors are disappointed at the output power level when used in the transmit mode. Part of this may be due to poorly made filters which introduce unexpected losses, or perhaps due to poor alignment, although it is recognised that alignment for optimum receive performance does not necessarily correspond to best transmit performance or vice versa. One thing which the design specifically dislikes is water finding its way into the cavities, so operation in wet weather requires careful attention to be paid to waterproofing the unit. Most constructors mount the mixer, together with the changeover system and exciter in a large, diecast box. Also, some detuning of the cavities can occur over extremes of temperature, usually resulting in low mixer current. This is usually a result of its effect on the performance on the varactor multiplier. The first screw to adjust in this event is the local oscillator filter tuning screw nearest to the mixer. Do not attempt to move two screws at once, or the mixer may well go irrecoverably out of alignment!

THE 10GHZ ("3CM") BAND

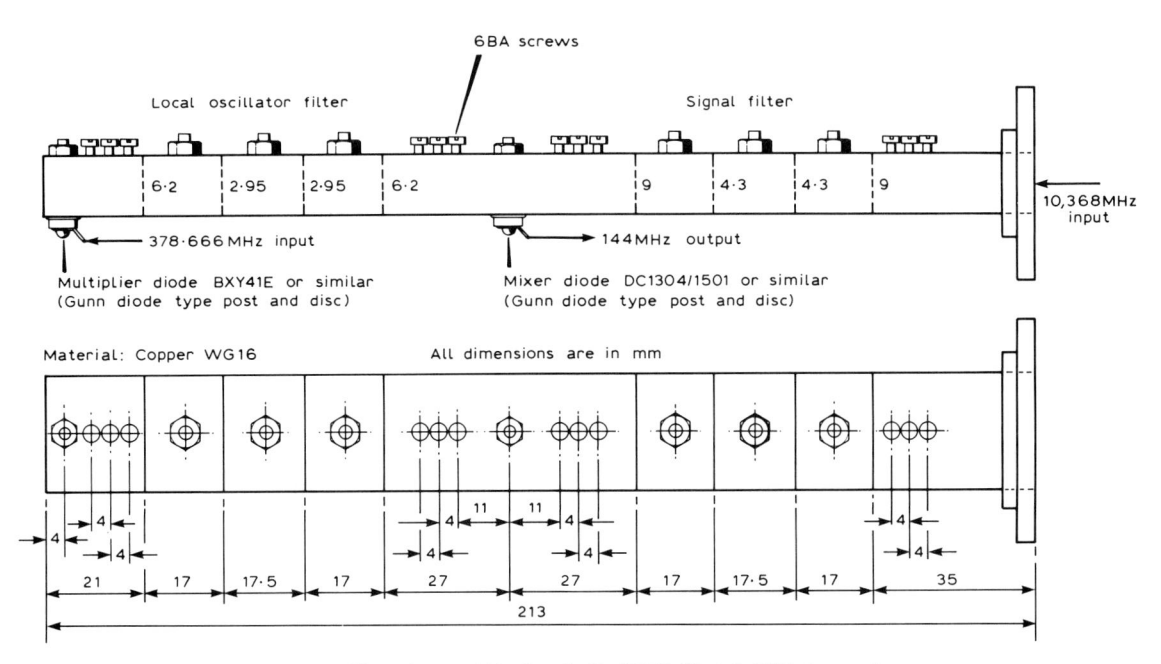

Fig 18.106. Dimensions and iris sizes for the G3JVL (Mark 1) 10GHz transverter

18.6.4 Construction of the mixer

The construction of the mixer can be very straightforward if a careful procedure is followed. There are several "irreversible" stages which must be done in the correct order to avoid later trouble.

First saw off a length of (copper) waveguide and roughly square off the ends by filing. Mark out the positions of the iris slots carefully (aim for ±0.1mm accuracy). The slots may then be cut out using a junior hacksaw (with a new blade!), starting at each corner and working inwards towards the centre to maintain the accuracy of each cut. All holes are then marked out and drilled 2.3mm (6BA tapping). The centre hole must be drilled carefully, using a vertical drill, through both walls of the waveguide. The upper hole is then opened out to 3/32in and the bottom hole to 4mm (2BA tapping). This hole is tapped 2BA and the 18 holes for the tuning and matching screws tapped 6BA.

Next the whole assembly is carefully deburred by running a fine file through the waveguide. This tends to push metal back into the holes and slots, which are then cleared using a hacksaw blade and appropriate drills or taps. Further filing and cleaning out should be repeated until all traces of burrs and loose metal are removed. The final deburring operation is to run a 4mm drill through the 2BA tapped hole *just* into the 3/32in hole and rotate slowly by hand to cut a *slight* chamfer on the inside wall of the waveguide. This ensures a flat surface for the diode post to be mounted against, with no burrs remaining to puncture the insulation.

Details of the mixer mount are given in Fig 18.107. After manufacturing the component pieces of this, deburr the surfaces which are to be mounted against the walls of the waveguide, by rubbing them on emery paper placed on a flat surface. The diode mount should then be assembled and checked for insulation and alignment using a scrap diode of the same outline as that to be used (eg a "blown" Gunn diode). For this operation the 2BA nut should be screwed up against the waveguide wall by hand. Should the diode not quite fit, the upper hole may be filed slightly so that the assembly may be moved to the correct position. A very useful tool for checking the alignment of diode mounts of this type can consist of a 2BA screw with a 1.8mm diameter pin turned on one end. This is used in place of the 2BA screw and scrap diode referred-to above, and will positively assure that the mount is correctly aligned. The importance of this cannot be too highly stressed, as a

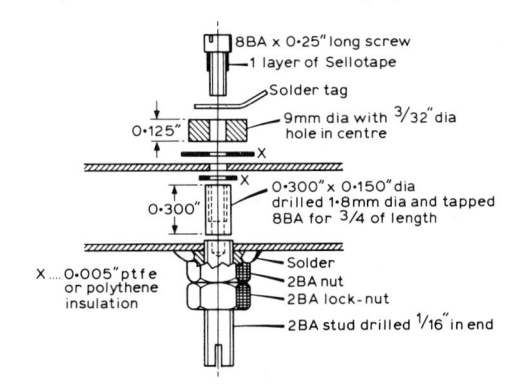

Fig 18.107. Gunn-type mounting for the G3JVL mixer

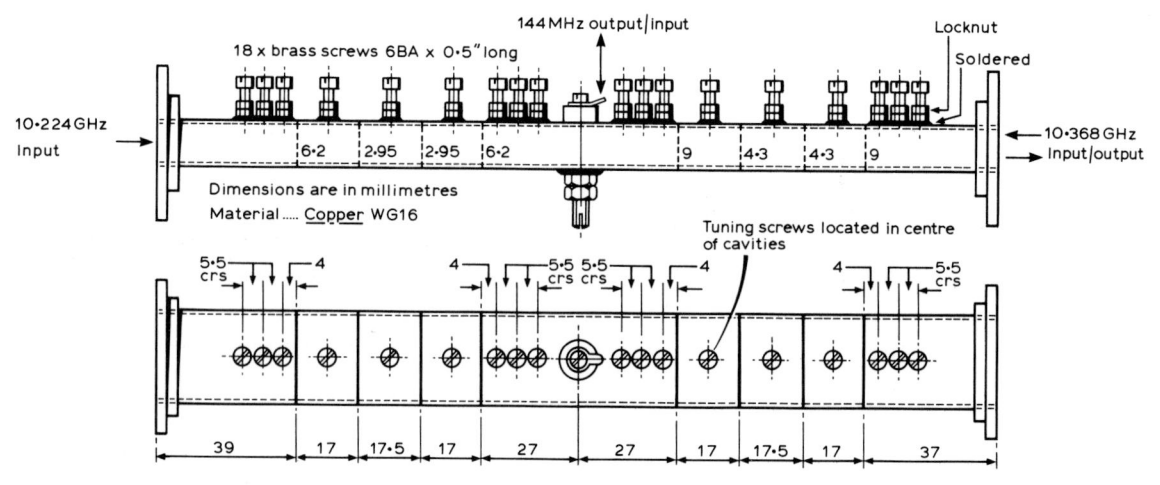

Fig 18.108. Dimensions for the G3JVL (Mark 2) transverter

mount which is out of alignment will result in broken diodes (an expensive experience!). When the diode mount has been checked it should be dismantled. In order to obviate problems later on, ensure that the post will pass freely through the 2BA nut and the 2BA tapped hole in the waveguide. If not, open these out *slightly* with a suitable drill.

The next items to be made are the iris plates. These are manufactured from 0.9in by 0.5in pieces of 0.024in copper sheet and are carefully filed to size; the 0.5in dimension not being as critical as the 0.9in dimension, which should be in the range 0.895 to 0.900in. The iris holes are drilled centrally in the plates; the diameters are given in Fig 18.106. After deburring and flattening the plates between two smooth pieces of metal in a vice, they may be assembled into the waveguide. The 6BA nuts which are to be soldered to the waveguide are jigged into place using non-solderable screws.

The assembly is then soldered using a hotplate to provide most of the heat – direct heating with a flame is much less satisfactory. A suitable hotplate could consist of a piece of 1/8 or 1/4in aluminium plate on a domestic cooker ring. If the temperature of the assembly is held just below the melting point of the solder, local additional heat can be applied by using a soldering iron or fine gas flame to "run" the solder at each point in turn. Perhaps the most critical part of the construction of the mixer is this soldering stage, as too much solder or flux inside the cavities can introduce excessive losses and/or detuning. Thus cleaning of all parts to a bright finish before soldering is necessary. If all parts are clean and fit together well, it should only require the application of a *very small* amount of solder to each side of the irises to solder them. However, be careful that no gaps remain after soldering. 18swg multicore solder or solder paste/paint have both been found satisfactory for this purpose. Speed is essential or the copper oxidises and becomes more difficult to solder. In this context the temperature of the hotplate should be only just high enough to allow the solder to run freely. This is why it may be more effective to keep the hotplate temperature just below the melting point and to use local flame heating.

When all the 6BA nuts and upper sides of the irises have been soldered, the waveguide is turned on its side and the bottom iris joints soldered, applying solder to the tops of the joints and allowing it to flow along the length of the joints. The 2BA nut is jigged in position and soldered. Next the flange is fitted in position, having been preheated on the hotplate, and soldered. A small amount of waveguide should be left sticking out of the flange, to be removed later by filing and rubbing on a sheet of emery paper laid on a flat surface, which then leaves a well-finished surface. Remove the main source of heat from the assembly and, when it has cooled to just below solder melting point, solder several solder tags to the top and bottom walls of the waveguide, using a soldering iron. These are to act as anchoring points for the i.f preamplifier housing. When the unit has fully cooled, remove all the jigging screws and run a 2BA plug tap through the nut and into the waveguide. The tap should be inserted vertically upwards and withdrawn vertically downwards to reduce the chance of metal dust falling into the cavity. This tapping operation ensures that the 2BA diode-mounting screw runs freely in the thread. Next the diode mount may be reassembled. This *must* be done in the following way to avoid losing pieces inside the waveguide, which are very difficult to retrieve (as the authors know to their cost!).

Pass a 1in (25mm) long 8BA screw through the top hole and out through the 2BA nut. Place the small insulating washer over the screw and screw on the post hand-tight. Then pull this piece through the 2BA nut up against the inside wall of the waveguide and insert a 2BA screw, or preferably the alignment tool mentioned above, and screw in until the post is pushed firmly against the wall. The 8BA screw is then removed and the top part of the diode mount assembled. The 2BA screw is next removed and replaced

THE 10GHZ ("3CM") BAND 18.71

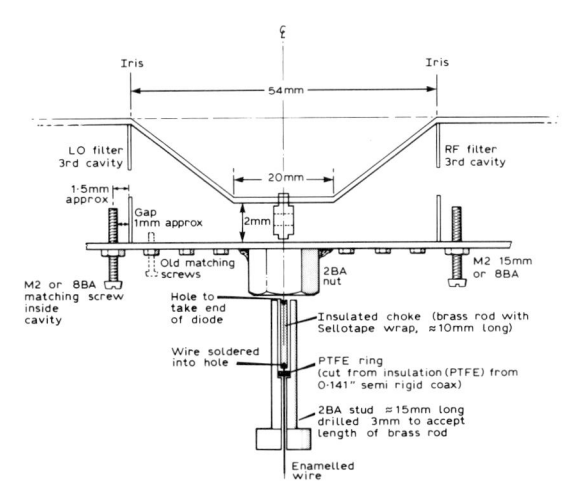

Fig 18.109. The G3JVL (Mark 3) 10GHz transverter. The main differences are: (a) the use of a taper-match section in the mixer and (b) the positioning of a single matching screw inside the third local oscillator and signal cavities, replacing the two three-screw tuners in the mixer cavity

by the 2BA diode-mounting screw complete with diode. In order to avoid dropping the diode inside the cavity, the screw and diode are held with the diode uppermost and then fitted vertically upwards into the waveguide. The screw is tightened by hand so that the diode is held in place, and the locknut tightened up. Diode removal is achieved by slackening off the locknut, unscrewing the 2BA screw one turn, tapping the screw gently sideways to release the diode from the upper post and then unscrewing downwards. The mount itself can be dismantled by reversing the assembly sequence. Other mechanical and soldering methods suitable for assembling this type of filter and mixer are suggested in chapter 7, "Constructional techniques".

18.6.5 Improved versions of the G3JVL mixer

The "Mark II" converter/transverter

The Mark I was designed for single-piece construction which has proved difficult to construct and align. Two modifications to the "Mark I" design can simplify the construction and tuning up procedure considerably. These are the construction of the mixer and step-recovery diode multiplier as two separate units and a minor modification to the arrangement of the matching screws. These modifications are shown in Fig 18.108. With this arrangement, the 384MHz/10.224GHz multiplier described in Fig 18.65 can be conveniently used. The overall length of the multiplier from the front of the flange to the rear short circuit was 29mm in the authors' versions.

The "Mark III" converter/transverter

Both the original and the Mark II versions were rather critical to align and had many interactive matching screws. Details of the "Mark III" version are given in Figs 18.109 and 110. In this version the mixer diode is mounted in reduced height waveguide that matches the diode's impedance better. A tapered transition between the reduced height guide and the filters makes the matching much less critical.

The coupling from each filter to the mixer cavity is made adjustable by a new screw, just inside the last filter cavity, which effectively alters the size of the coupling hole in the iris plate. Since this screw is inside one filter cavity, it only affects the matching at that filter frequency; it does not affect the matching at the frequency of the other filter. It effectively removes most of the interaction problems encountered during alignment of the original versions of the mixer.

This design has given results as good as the original design, but without needing the six matching screws in the mixer cavity. It may be worth leaving provision for some matching screws there, as it may enable a dB or so improvement to be "won". Several different ways of making the tapered section are suggested in Fig 18.111.

GM4HFM suggested bolting a tapered block of brass or copper in place to form the tapered section, as shown in Fig 18.112. This makes it possible to use a socket for the mixer diode connection. Choke decoupling rather than the capacitive disc is used. The block is inserted through a hole in the upper face of the guide and held in good contact with the walls using 6BA bolts in the sidewalls. It is then very

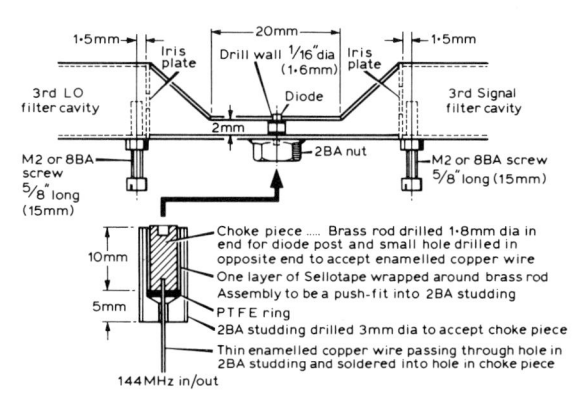

Fig 18.110. Further details of the Mark 3 transverter

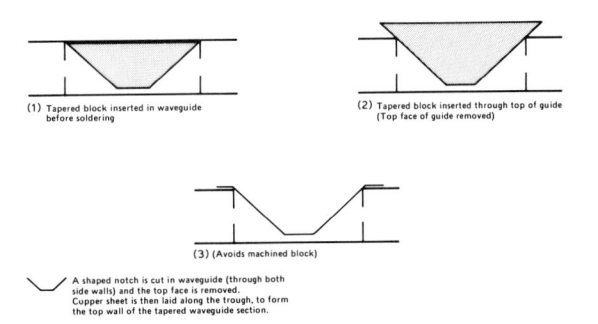

Fig 18.111. Alternative methods to make the tapered matching section for the Mark 3 transverter

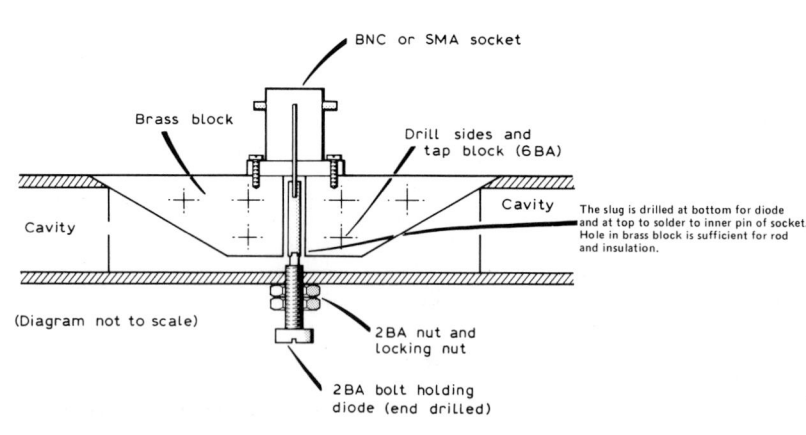

Fig 18.112. Further modifications to the taper-match section, using a machined, doubly-tapered block

The slug is drilled at bottom for diode and at top to solder to inner pin of socket. Hole in brass block is sufficient for rod and insulation.

easy to drill and tap the block to accept the matching screws. The advantage of using a socket mounted on the guide is that it can be used to support the choke and diode connection. In some of the other designs, either the assembly is difficult or the insulation is likely to puncture after a while. Either a BNC or an SMA socket could be used. The choke is made from a section of low impedance transmission line formed by the large diameter inner rod, insulated by a thin dielectric layer. The diameter of the rod is not critical, somewhere between 5 and 8mm, but the gap between the rod and the hole in the block should only be sufficient to allow the rod to be wrapped in one or two layers of insulant, for instance "Sellotape" or thin ptfe tape. The characteristic impedance of this rod can be calculated from the following expression:

$$Z_0 = \frac{\sqrt{(\mu_0/(\varepsilon_0 \times \varepsilon_r))} \times \ln(b/a)}{2\pi} \text{ Ohm}$$

where μ_0 = permeability of free space, 4×10^{-7} H/m
ε_0 = permittivity of free space, 8.85×10^{-12} F/m
a = diameter of rod
b = diameter of hole
ε_r = dielectric constant of the insulator

For 1/4in (6.15mm) rod with 0.25mm gap around it, filled with $\varepsilon_r = 2$, then Z_0 works out at 3.2Ω.

Further information on the G3JVL converter/transverter

The original version of the G3JVL converter/transverter specified a BXY41E diode for the multiplier and an AEI DC1304 GaAs Schottky diode for the mixer. The former device is no longer readily available, while the latter is somewhat expensive. Some cheaper, more readily available diodes manufactured by Microwave Associates have been tested with very satisfactory results indeed. Suitable replacements for the BXY41E are the MA44150 or the MD4901, whilst the MA40150, MA4E024 and DC1501E perform very well as mixers.

A further modification to the design was described by G8AGN in [31]. This is shown in Fig 18.113 and consists of an offset mixer using a common, very inexpensive diode of the 1N23 series. The performance of the modified mixer is not markedly inferior to that of the "Mark II" using a much more expensive diode. Although definitive measurements have not been undertaken, it is doubtful whether the performance as a receiver is more than a dB or so worse. In terms of transmit output, the two were indistinguishable.

Little has been said concerning the 378.666MHz (or 384MHz for direct frequency generation) exciter. This is, of course, a crucial part of the system and careful attention must be paid to its design. The most important point concerns the noise generated by the crystal oscillator. A noisy oscillator can seriously degrade the overall noise figure; this problem beset the early experimenters on 10GHz. Also, the oscillator must be extremely stable, implying the use of high quality crystals as well as a good circuit. The exciter described in chapter 8, "Common equipment", is especially recommended for the converter.

F6DLH suggested that the local oscillator in all narrow-band equipment should be set *slightly low* of its nominal frequency, so that the receiver covers from about 10,367.9 to 10,368.3MHz or so (using the usual 144MHz transceiver for the i.f), so that signals on 10,368MHz, generated from

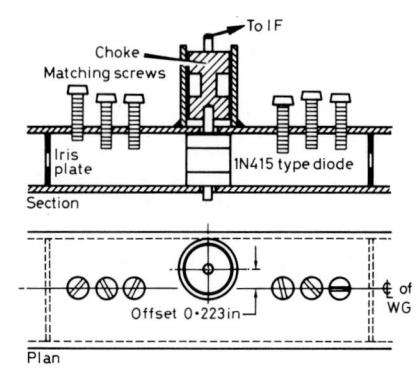

Fig 18.113. The G3JVL transverter mixer using an offset 1N23-type diode (G8AGN). Section, above, plan below. The choke diameter is 0.360in and similar in construction to that in Fig 18.38. It is insulated by a layer of "Sellotape" or other self-adhesive tape

cw transmitters using 96.000MHz crystals, can be found even in the worst case when both equipments drift the wrong way. He also warned that it is possible to listen to a spurious signal, rather than the main one, when listening to a 96MHz-derived 10.368GHz transmitter on a receiver with a 144MHz i.f. The receiver normally "listens" on 10.368GHz and, with 144MHz i.f, has an image response at 10.080GHz. A harmonic of the 96MHz crystal comes out on both these frequencies (96×105 = 10,080, 96×108 = 10,368). The same is true for a transverter on transmit – some power will also be produced on 10.080GHz as well as 10.368GHz. So, particularly when testing over short paths, it is possible to hear the signal on the image frequency despite the very large attenuation that the 10.368GHz filter provides. Be careful not to align the equipment on the image by mistake!

18.6.6 Preliminary alignment of the 'JVL transverter

To undertake the alignment of the G3JVL mixer/filter assembly requires a minimum of the waveguide components shown assembled, as a test rig, in Fig 18.114. The same set-up can be used for aligning any filters on their own using a second detector after the filter, which is needed if such a "stand alone" filter, not containing a mixer, is to be aligned. Such filters might be destined for use with one of the multipliers described in section 18.2 where it was stressed that the possession of a pre-aligned filter made the setting-up of the multiplier much easier. If the constructor has an isolator this can, to great advantage to the Gunn oscillator frequency stability, be fitted between the oscillator and the cross coupler to provide better isolation against frequency pulling during alignment.

The directional cross-coupler, ideally 6 to 10dB, is orientated so that detector current is a monitor of the reflected power, i.e. the current should be very small or zero with a matched load connected in place of the 'JVL mixer, assuming that it is off tune. Set the Gunn oscillator to the alignment frequency, which would be 10.224GHz for a local oscillator filter, or 10.368GHz for either the 'JVL signal filter or a filter for direct frequency generation. Allow the Gunn to stabilise for some time and recheck and reset the frequency as necessary. This operation is best carried out in a thermally stable environment!

Adjust the tuning screw of the input cavity (the one nearest the directional coupler) until a sharp change is seen in the level of reflected power received by the detector. This change indicates resonance. Leave the tuning screw at the position of resonance and adjust the second cavity tuning screw for a similar effect. Adjust the third cavity tuning screw likewise. Depending on just how the test bench is set up, the first cavity may give *a dip* in meter reading, the second *a peak* and the third *a further dip*. See chapter 12 for further information on the alignment of bandpass filters (Dishal's method).

Repetitively adjust all three screws until little or no reflected power is seen on the meter of the detector. Recheck constantly during the alignment that the Gunn oscillator is still on frequency as it is possible that, despite the

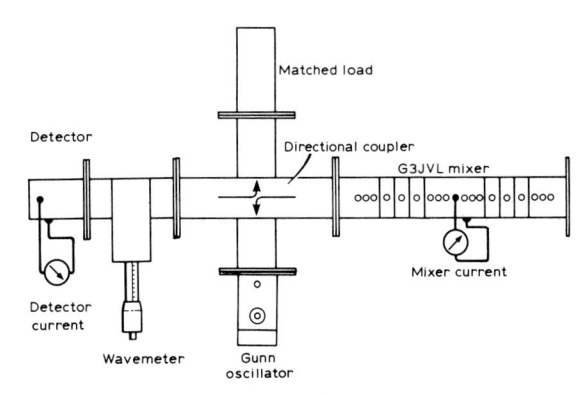

Fig 18.114. Alignment test-rig for narrowband filters

precautions outlined above, the frequency may be pulled by the tuning of the cavities.

At this point there should be an indication of a few µA (referred to as "mixer current") on the meter connected to the mixer diode, indicating some transmission of power through the filter. If no current is seen, slowly insert a matching screw into the hole nearest to the first cavity. As the reflected power is seen to rise, readjust the first cavity tuning screw for minimum reflected power. As this procedure is continued with increasing penetration of the matching screw, at some point some mixer current should be seen. When some mixer current has been obtained, readjust the filter tuning screws and insert matching screws on either side of the filter and experiment with their settings until maximum mixer current is obtained. As all the screws interact with each other to some extent, it is necessary to go several times round the "loop" until optimum settings are found.

A mixer current of several mA should be achieved using a 5 to 10mW Gunn oscillator. It is emphasised that sharp resonance will *not* be obtained unless the screws are very clean and a tight fit into the bearing nuts. This can be assured by keeping a gentle torque on the locknut (to overcome contact problems) whilst the screw is being adjusted. Contact problems can be overcome by using plastic screws, although it was found that by using nylon screws the loss of the filter was increased from about 0.8dB to about 3.2dB for the 60MHz wide iris coupled filter. PTFE may well be less lossy, although this point has not been checked. The best remedy would seem to be to keep slight torque on the screws, using the locknuts.

The screws should be locked in place carefully, avoiding any screw movement whilst tightening the locknuts. If a 'JVL mixer is being aligned, then the mixer assembly should be turned round and the whole process described above repeated for the second filter.

G3YJH suggested an alternative method for approximate prealignment of the filters, particularly useful for the "Mark I" 'JVL. By inserting a short piece of insulated wire, which acts as an antenna, into the tuning screw holes in turn, the signal can be tuned in on a wideband receiver tuned

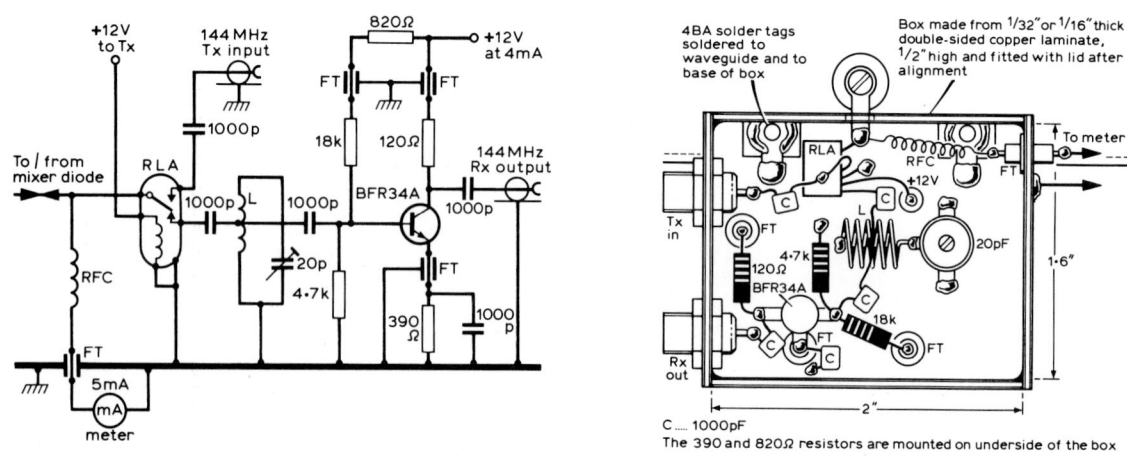

Fig 18.115. (a) Circuit of a low-noise 144MHz preamplifier for the G3JVL transverter. (b) Schematic layout of the preamplifier

to 10.224GHz (the local oscillator frequency). When the tuning screw of the preceding cavity is tuned through resonance, a "blip" will be heard in the receiver as the signal peaks. By working through the cavities in this fashion several times, it is possible to peak the signal quite easily until sufficient mixer current is seen to allow the meter to be used for more precise alignment. G3JVL has used this method with a simple diode detector mounted on semi-rigid coax, with the probe formed by removing the outer of the coax for the last 6mm or so. The probe is then put into the hole for the tuning screw one cavity beyond the cavity being aligned. With a few mW input into the filter, there should be plenty of signal to operate the detector.

18.6.7 A low-noise i.f preamplifier and changeover systems for the 'JVL transverter

Final alignment of the 'JVL mixer cannot be carried out without a suitable i.f preamplifier and the subsequent receiver connected. Since the noise figure of the mixer is partly dependent on the noise figure of the subsequent receiver, a low-noise preamplifier is vital. For optimum performance it should be mounted directly on the waveguide assembly. A recommended 144MHz preamplifier is that shown in Figs 18.115a and 115b; the reader should be familiar with this one, for it has been used for several receivers and transceivers for the other bands already!

Fig 18.116. Photograph of a complete G3JVL transverter system (G3YGF)

THE 10GHZ ("3CM") BAND

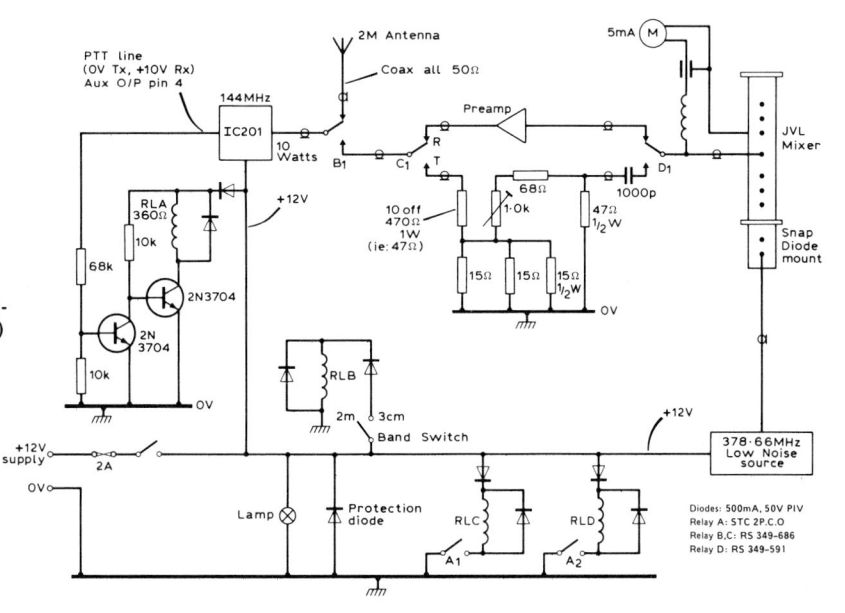

Fig 18.117. General purpose transverter changeover system (G3AYJ)

Note that for receive-only operatione, the relay may be omitted. Fig 18.116 is a photograph of the complete transverter.

If full transceive operation is intended, it will be necessary to add a coaxial or other miniature screened relay, switched by the transceiver ptt line, to switch the transceiver from the 10GHz receiver output socket to the transmitter input. The value of attenuation required in the transmit line should be such as not to overdrive the mixer. Suitable starting values are 20dB for a 3W output 144MHz exciter and 25dB for a 10W exciter. Extra attenuation should be inserted until the 10GHz output *just* starts to decrease; the attenuator can then be rebuilt to that value. It is important to ensure that the resistors comprising the attenuator will handle the power safely, since any failures could result in a large drop in attenuation with the consequent risk of damage to the mixer diode.

A complete changeover system used by G3AYJ is shown in Fig 18.117. Besides the changeover from receive to transmit on the mixer, the facility is provided to switch the 144MHz transceiver directly to the 2m antenna for talkback. While it may still be necessary to retune the transceiver when doing this (although this can be eliminated by the use of "memory", if available on the transceiver in use), the operation of unplugging and plugging-in leads is eliminated.

On transmit, a power attenuator is switched into position when operating on 10GHz, and a 2m preamp is used on receive on 10GHz. Note that the power attenuator will also be switched into position when operating directly on 2m, and also in the event of the ptt line being accidentally disconnected. It is not possible to have 2m power output available without a supply voltage to the relay control circuits. The use of transient suppression diodes across all the relay coils is essential.

Whilst t/r control of the transverter is usually taken from the 2m transceiver ptt line, a number of transceivers do put a dc voltage on their antenna sockets which can be used as a very simple and elegant means of control – the rf lead is the only one required between the two equipments. This dc voltage usually results from the use of a pin diode switch in the transceiver. The transverter senses the dc level on the coax via a high value resistor in the transverter. Unfortunately the polarity of this voltage varies from transceiver to transceiver – the IC202 is 0V on transmit and 10V on receive, whereas the FT290 is 0V on receive and 10V on transmit – so it has to be used with caution. An alternative switching system for the IC202 transceiver is shown in Fig 18.118 and for the FT290 transceiver in Fig 18.119. Both use the dc voltage present on the antenna socket to activate

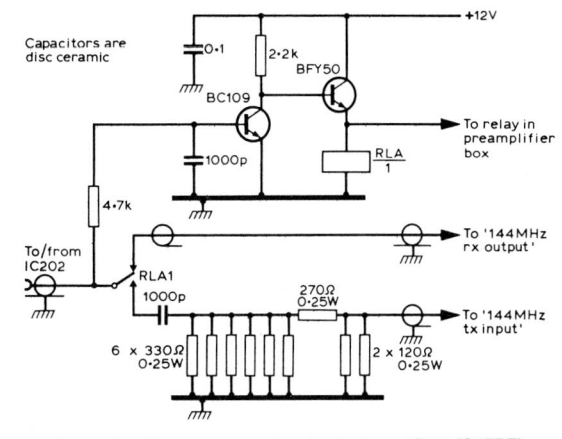

Fig 18.118. Changeover system for the Icom IC202 (G4FRE)

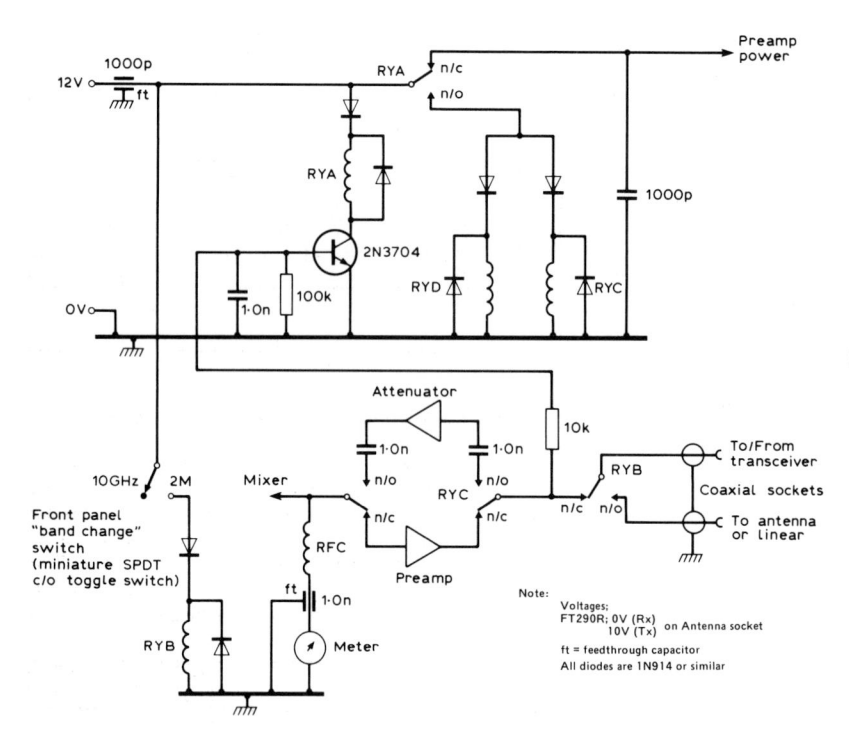

Fig 18.119. Changeover system for the Yaesu FT290R (G3PFR)

the relay switching circuits, which are otherwise similar to G3AYJ's circuit.

18.6.8 Final alignment of the 'JVL transverter

Final alignment of the narrow filters used for either a receiver local oscillator or for the direct generation of 10GHz signals should be carried out using the crystal-controlled source with which the filter or filter/mixer assembly is to be used. Since the filter has already been brought into approximate alignment, this is simply a matter of slightly re-peaking some of the matching screws and possibly the cavity tuning screws. Even before final alignment, the filter will be well-enough aligned to select the correct harmonic from the comb produced by the multiplier, and mixer or detector current should be seen immediately on connecting the multiplier. The input tuning and output matching screws on the multiplier should be adjusted for maximum mixer current before attempting to peak either the filter input matching screws or the cavity tuning screws.

As far as optimising the signal filter and matching the mixer output into the i.f low-noise preamplifier is concerned, this can only be carried out when the i.f preamplifier and a suitable 144MHz receiver is connected. The alignment of the mixer for transceive purposes is a compromise between transmit and receive efficiency. The optimum receive performance does *not* correspond to the maximum transmit power output. It is usually better to optimise the receive performance and accept a somewhat lower power output on transmit; better an efficient receiver with a fraction less transmit output rather than a less sensitive receiver with a little more transmit power. The construction and use of an automatic alignment aid was described in chapter 10, "Test equipment".

18.6.9 Some measurements of the performance of the 'JVL transverter

The following summary of the performance of the transverter was derived from measurements made by G4KNZ and G3YGF on a "Mark II" (non-tapered) version. The exciter used the high quality uhf driver described earlier, without temperature compensation.

1. Variations in supply voltage

The effect of the oscillator power supply voltage on the oscillator frequency was measured and found to be approximately 10kHz over a range of 11.5 to 14V. A low voltage drop regulator pcb was then fitted. The output from this was set to 11.1V to allow regulation down to a supply voltage of 11.5V. The change in the output voltage of this pcb measured over the same input voltage range as before was found to be less than 10mV. Thus the effect on the frequency should be less than 50Hz, provided that the supply voltage is adequate for the regulator to regulate.

2. Variations in temperature

There were two effects due to temperature. First, the oscillator frequency drifted by approximately 2.2kHz/°C over the normal operating range of −5 to +25°C (measured at 10GHz). Second, the waveguide (plus other components such as matching screws) expands or contracts with changes of temperature and so affects the matching between the varactor and the mixer diode. The electrical properties of the varactor diode also change with temperature. The net result was that the output fell by about 50% with a 10°C temperature change. The frequency change as a function of the temperature is shown in Fig 18.120.

3. Noise sidebands

Both the "close-in" noise sidebands and the ultimate noise floor of the 10.224GHz local oscillator were measured. The results of these two measurements are combined in Fig 18.121. The measurement technique is described below.

The main reason for interest in the level of the noise sidebands in the region 1 to 500MHz from the carrier is that it can cause receiver desensitisation even if no other signals are present. This noise is fed straight into the mixer and if it is above the thermal noise at a frequency spaced from the carrier by the i.f, then it will be indistinguishable from the signal as it is on the signal frequency, as shown in Fig 18.122.

The problem in trying to measure the level of these noise sidebands (which may be around the thermal noise level,

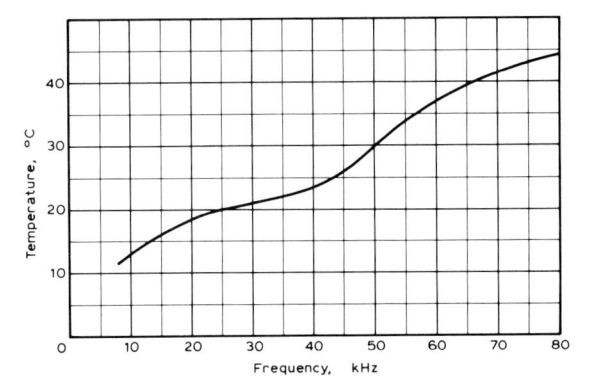

Fig 18.120. The G3JVL transverter – temperature effects

i.e. –140dBm in a 3kHz bandwidth), is the presence of the carrier whose level is around +10dBm – a total dynamic range of 150dB. This cannot be measured directly, but the following technique can be used to enable these measurements to be performed with relatively simple test gear. The dynamic range involved is reduced by using a narrow notch circuit to attenuate the carrier and its very close-in

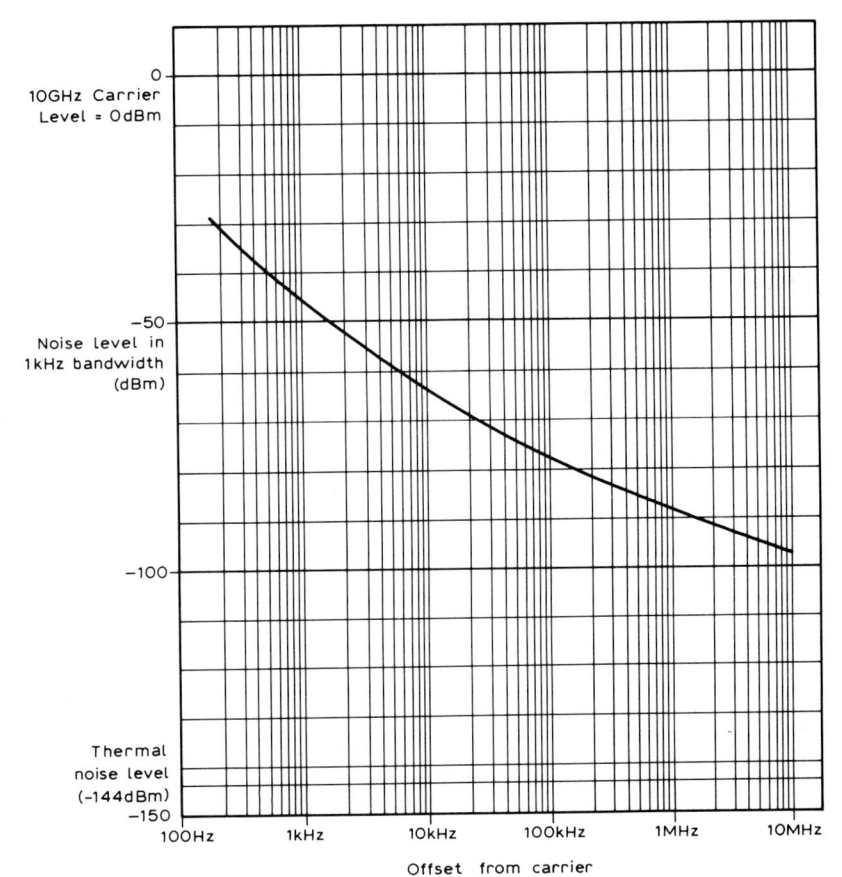

Fig 18.121. The G3JVL transverter – noise sidebands

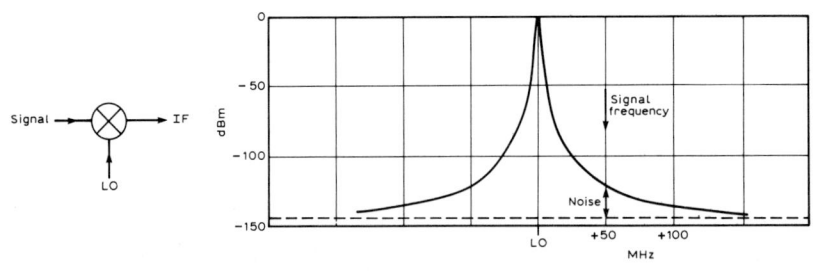

Fig 18.122. The G3JVL transverter – noise sidebands

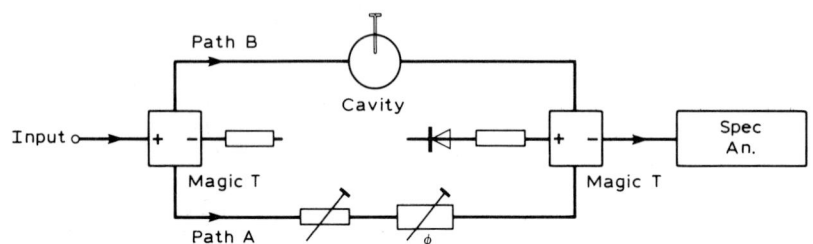

Fig 18.123. Test circuit for measurements on the G3JVL transverter

sidebands, by about 60dB, so as to bring the dynamic range of the signals to be measured down to below 90dB – a much more manageable figure. The circuit to do this is shown in Fig 18.123. The signal to be measured is split into two parts by a hybrid-T (3dB coupler). One part travels through a transmission cavity wavemeter and the other through a variable attenuator and a phase shifter. They are then both recombined in another hybrid-T coupler which produces both the sum and difference of the two signals applied to it. The attenuation and phase shift through the two arms are shown in Fig 18.124.

The attenuator and phase shifter are adjusted to make the phase shift and attenuation through them identical to that of the cavity at its centre frequency so that, at the centre frequency, identical signals arrive at the two ports of the hybrid-T, producing no output from the difference port and all the output from the sum port.

For signals off the centre frequency, the cavity introduces a phase shift and amplitude imbalance which in turn results in some power appearing at the difference port. A long way off the centre frequency, all the power will appear at the difference port. The overall attenuation through this network, as a function of frequency, is shown in Fig 18.125. The spectrum of a typical signal is also shown in Fig 18.126 and the resulting output in Fig 18.127. By adding the value of the attenuation of the network to the observed signal levels, a plot of the actual signal noise levels can be obtained well into the notch and, thus, to within less than 1MHz of the carrier.

It is useful to monitor the transmitted power by mounting a diode detector on the sum port. This can be maximised as the first step, by setting the attenuator to a few dB and tuning the wavemeter to the frequency of the signal. After that the residual signal at the difference port can be monitored on a receiver or spectrum analyser and minimised by adjusting the phase shifter and attenuator alternately for minimum signal. A practical limit will be reached when the notch is about 60dB deep, as the adjustments will then be very critical and leakage outside the guide will become a problem.

The width of the notch is determined by the Q of the cavity wavemeter – a Q of 5,000 might be typical, giving a bandwidth of 2MHz at 10GHz. Thus it will be possible to "see" the sidebands unattenuated to within about 1MHz of

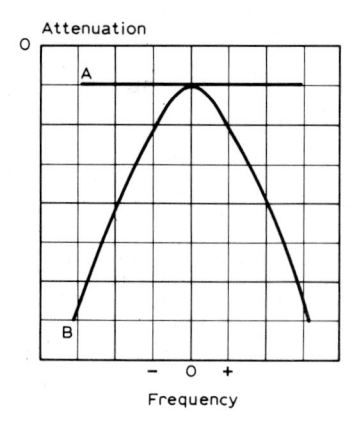

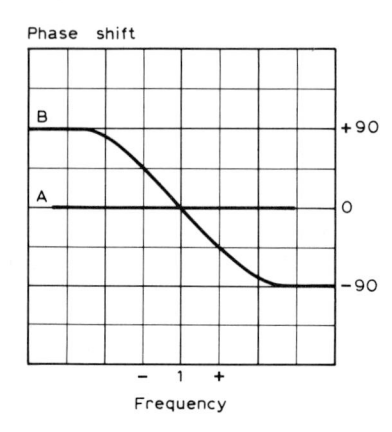

Fig 18.124. Attenuation and phase shift in the test rig

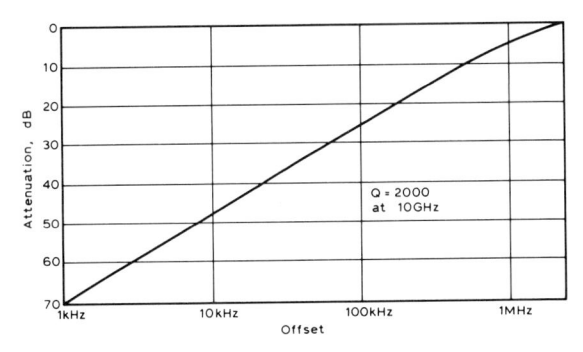

Fig 18.125. Overall attenuation through the test rig

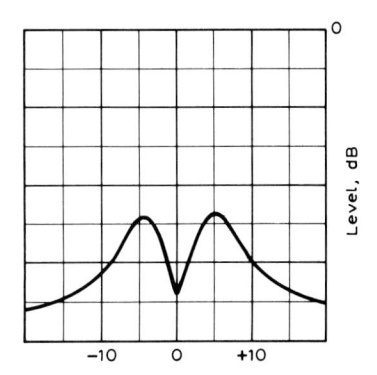

Fig 18.127. Output from the test rig

the carrier. The circuit also converts phase modulation of the signal into amplitude modulation and can thus be used to measure the phase noise on the signal.

4. Power output and noise figure

A broadband diode detector mounted in WG16 indicated a total of 10mW output directly from the multiplier cavity with 0.5W of input at 384MHz. Not all this output is on 10.244GHz; 3.25mW was measured at the output of the 10.224GHz filter. On transmit 30mW at 144MHz was injected into the mixer diode. When optimised, the power output on 10.368GHz was measured as 420μW. The overall noise figure, when optimised thus, was measured to be 12dB by comparison with a 10GHz receiver of known noise figure.

Off air frequency standards

Very accurate frequency setting is required when using narrowband modes such as ssb and cw at 10GHz to ensure that you are looking in the right place for the signal – an accuracy of 100Hz at 10GHz is 1 part in 10^8. Although carefully built crystal oscillators can have short term stabilities of this order, they will drift over a period of time and with temperature variations. Whilst ovened crystals can just attain this sort of accuracy, the most practical solution for the amateur is to build an oscillator which is phase locked to Droitwich (198kHz) or Rugby (60kHz), or some other high accuracy off-air standard, and use it to calibrate the transceiver from time to time. Both of these stations can be heard over most of the UK, although Droitwich which was on 200kHz has recently moved to 198kHz. The block diagram of a simple off-air reference is given in Fig 18.128. Other circuits and ideas may be found in the RSGB publication *Test Equipment for the Radio Amateur*.

A conventional 96MHz crystal oscillator is multiplied up to 10GHz to generate a comb of frequencies every 96MHz.

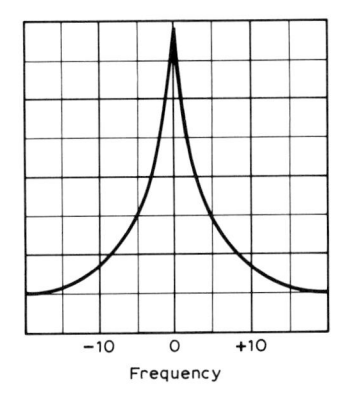

Fig 18.126. Spectrum of a typical signal from the G3JVL transverter

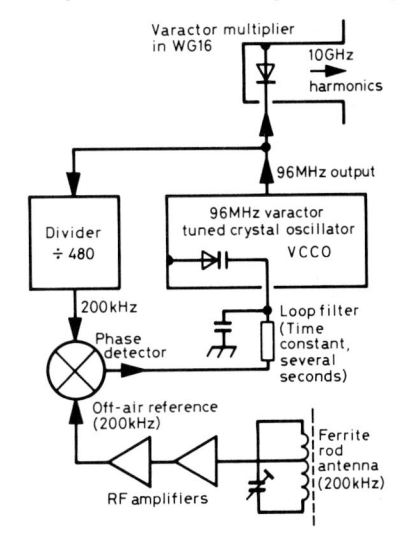

Fig 18.128. Block diagram of an off-air frequency standard. Originally designed to be locked to BBC Radio 4, Droitwich, on 200kHz, the circuit is adaptable to the new Droitwich frequency of 198kHz by altering the division ratio to 485 or the vcco to 95.04MHz with the same division ratio. The advantage of using exactly 96MHz is that the 108th harmonic falls at 10,368MHz providing a strong marker at the bottom of the international narrowband section

The crystal oscillator can be pulled a few hundred Hz at 96MHz by means of a varicap diode in the oscillator, thus forming the vco.

A ferrite rod antenna and several gain stages receive and amplify the 198kHz signal, and feed it to a mixer where its phase is compared with that of a 198kHz signal divided down from the crystal vco. The dc output from this mixer is used to control the vco frequency. It is important to keep the 198kHz from the crystal well screened from the receiver, otherwise it may lock to itself.

The loop filter needs to have a very low cutoff frequency because of the a.m on Droitwich for the Radio 4 transmission, and also some phase shift keying at a few tens of Hz. The psk is very difficult to remove completely and is audible as a few hundred Hz fm on the 10GHz signal. In view of this, Rugby is probably the better choice as a reference for the future.

18.7 10GHZ ANTENNAS AND FEEDS

An account of microwave antennas was given in chapter 4, "Antennas". Most 10GHz systems will use either parabolic dishes or optimum gain pyramidal horns, although beacons or repeater/beacons might use stacked, slotted waveguide antennas for omnidirectional, horizontally polarised radiation or sectoral horns for wide angle, semi-directional coverage.

The theory and general design considerations have already been discussed. What is given here, therefore, is a number of different practical designs for 10GHz, together with some ideas to aid their construction. Where appropriate, a little of the theory is restated to clarify the design objectives.

18.7.1 Horn antennas

Large pyramidal horns can be an attractive form of antenna for use at 10GHz and above. They are fundamentally broadband devices showing virtually perfect match over a wide range of frequencies, certainly over the amateur band. They are simple to design, tolerant of dimensional inaccuracies during construction and need no adjustment. Horns are particularly suitable for use with transmitters and receivers employing free-running oscillators, the frequency of which can be very dependent on the match of their load (antenna). Another advantage is that their gain can be predicted within a dB or so (by simple measurement of the size of the aperture and length) which makes them useful for both the initial checking of the performance of systems and as references against which other antennas can be judged. Their main disadvantage is that they are bulky compared with other antennas having the same gain.

Large (long) horns, such as that illustrated in Fig 18.129, result in an emerging wave which is nearly planar and the gain of the horn is close to the theoretical value of $2\pi AB/\lambda^2$, where A and B are the dimensions of the aperture. For horns which are shorter than optimum for a given aperture, the field near the edge lags in relation to the field along the centre line of the horn and causes a loss in gain.

Fig 18.129. Large 10GHz horn

For very short horns, this leads to the production of large minor lobes in the radiation pattern. Such short horns can, however, be used quite effectively as feeds for a dish.

The dimensions for an optimum horn for 10GHz can be calculated from the information given in Fig 18.130 and, for a 20dB horn, are typically:

A = 5.19in (132mm)
B = 4.25in (108mm)
L = 7.67in (195mm)

There is, inevitably, a trade-off between gain and physical size of the horn. At 10GHz this is in the region of 20dB or perhaps slightly higher. Beyond this point it is better to use a small dish. For instance, a 27dB horn at 10GHz would have an aperture of 11.8in (300 mm) by 8.3in (210mm) and a length of 40.1in (1,019mm) compared to a focal plane dish which would be 12 inches (305mm) in diameter and have a "length" of 3in (76mm) for the same gain.

A computer program for the design of both optimum and sectoral horns has been given in the antenna chapter. Either manual calculation or the program can be used.

18.7.2 Construction of 10GHz horns

Horns are usually fabricated from solid sheet metal such as brass, copper or tinplate. There is no reason why they should not be made from perforated or expanded metal mesh, provided that the size and spacing of the holes is kept below about $\lambda/10$. Construction is simplified if the thickness of the sheet metal is close to the wall thickness of WG16, i.e. 0.05in or approximately 1.3mm. This simplifies construction of the transition from the waveguide into the horn. The geometry of the horn is not quite as simple as appears at first sight since it involves a taper from an aspect ratio of about 1:0.8 at the aperture to approximately 2:1 at the waveguide transition. For a superficially rectangular object, a horn contains few right angles, as shown in Fig 18.131 which is an approximately quarter-scale template

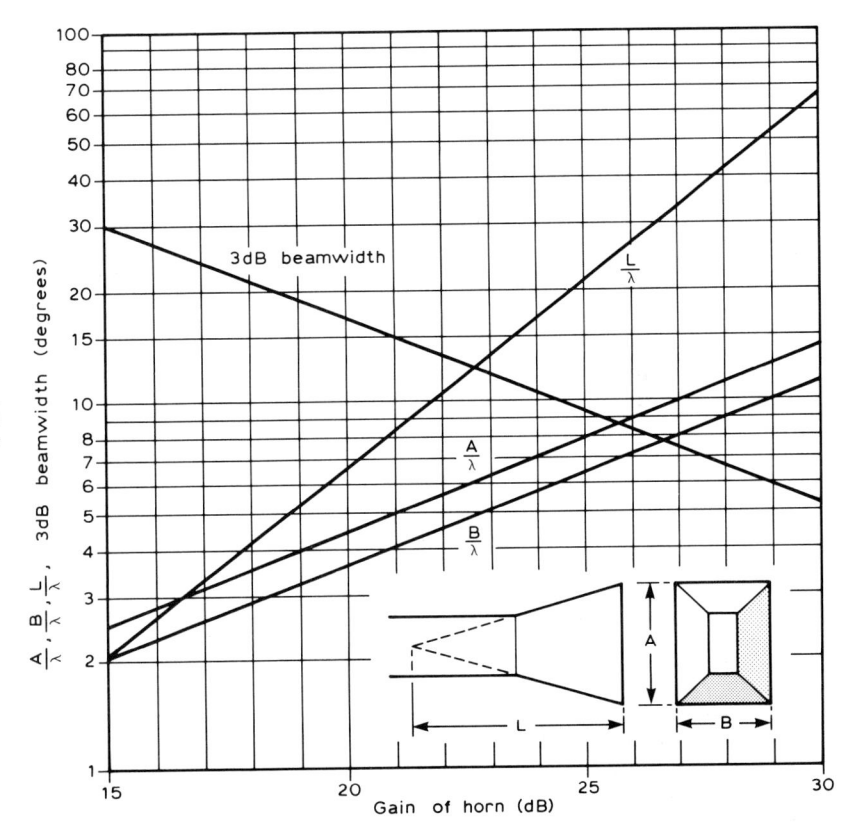

Fig 18.130. Horn design chart. Optimum gain horns can also be designed using the BASIC program in chapter 4, "Microwave antennas"

for a nominal 20dB horn at 10.4GHz. If the constructor opts to use the "one piece" cut and fold method suggested by this figure, then it is strongly recommended that a full-sized template be drafted on stiff card which can be lightly scored to facilitate bending to final form. This will give the opportunity to correct errors in measurement before transfer onto sheet metal and to prove to the constructor that, on folding, a pyramidal horn *is* formed!

The sheet is best sawn (or guillotined) rather than cut with tin-snips, so that the metal remains flat and undistorted. If the constructor has difficulty in folding sheet metal, then the horn can be made in two or more pieces, although this will introduce more soldered seams which may need jigging during assembly and also strengthening by means of externally soldered angle pieces running along the length of each seam. Alternative methods of construction are suggested in Fig 18.132.

It is worth paying attention to the transition point which should present a smooth, stepless profile. The junction should also be mechanically strong, since this is the point where the mechanical stresses are greatest. For all but the smallest horns, some form of strengthening is necessary. One simple method of mounting is to take a short length of Old English (OE) waveguide, which has internal dimensions matching the external dimensions of WG16, and slitting each corner for about half the length of the piece. The

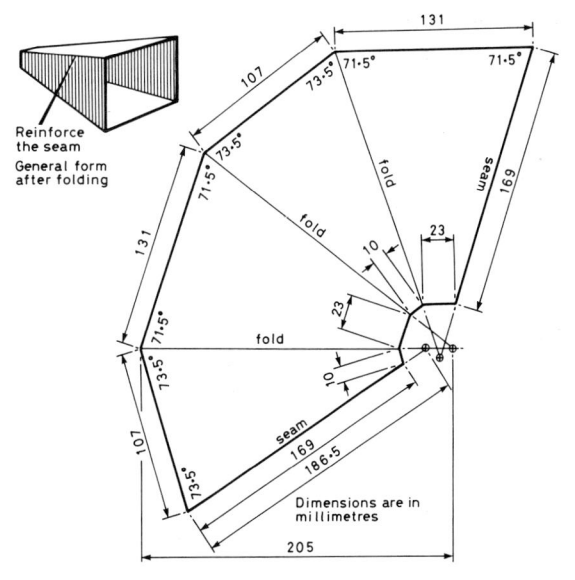

Fig 18.131. Dimensioned template for single-piece construction of a 20dB horn

(a), (b)

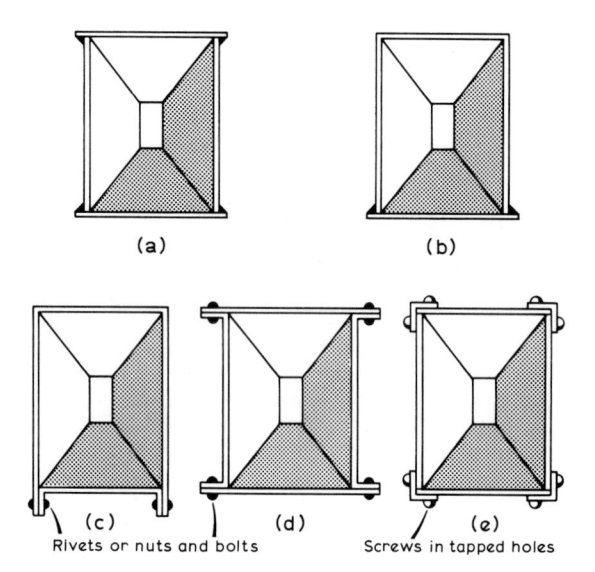

(c) Rivets or nuts and bolts   **(d)**   **(e)** Screws in tapped holes

Fig 18.132. Alternative construction methods for horns

sides can then be bent (flared) out to suit the angles of the horn and soldered in place after carefully positioning the OE guide over the WG16 and inserting the horn in the flares. One single soldering operation will then fix both in place. After soldering, any excess of solder appearing inside the waveguide or "throat" of the horn should be carefully removed by filing or scraping. The whole assembly can be given a protective coat of paint.

An alternative method would be to omit the WG16 section and to mount the horn directly into a modified WG16 flange. In this case the thickness of the horn material should be a close match with that of WG16 wall thickness and the flange modified by filing a taper of suitable profile into the flange.

Whichever method of fabrication and assembly is used, good metallic contact at the corners is essential. Soldered joints are very satisfactory provided that the amount of solder in the horn is minimised. If sections of the horn are bolted or rivetted together, then it is essential that many, close-spaced bolts or rivets are used to ensure such contact. Spacing between adjacent fixing points should be less than a wavelength, i.e. less than 30mm.

10GHz horns from oil-cans and other materials

Empty 5 litre oil cans are an excellent source of tinplate for the purposes of building horn antennas. Select cans with no rust visible either inside or outside. Cut off the top and bottom of the can with tin-snips or a hacksaw, then cut along the length of the four corners and flatten out the four pieces of tinplate thus obtained.

Although a horn of any gain could be built, two values suggest themselves from the sizes of tinplate available: a 130 by 160mm horn of 22dBi gain requiring two cans (it uses four of the larger sheets), or an 80 by 80mm horn of 16dBi gain made by cutting down two of the larger sheets (or using four smaller sheets) which need only be about half the length.

Remove any sharp edges or slivers of metal, then wash and degrease the sheets. It is not necessary to remove most of the paint, only where you want to solder. This can be done with wire wool or paint stripper. Mark out the centre lines along each piece, then the dimensions of the open end of the horn, and the appropriate outside dimensions of the waveguide, separated by the length of the horn. Join up the ends of these lines and extend them about 20mm past the waveguide end line. See Fig 18.133a.

Cut out one pair of opposite sides *exactly* along the marked lines; on the other pair, leave about 3mm extra along the sides, see Fig 18.133b. Then fold the tabs (a) back about 10 to 15°. Hold two adjacent sides in position and make several spot solder joins along the seam. Repeat this process until all four sides are joined up. Make sure that the narrow end is a snug fit over the waveguide by holding the waveguide in place while doing the spot soldering. Then extend the solder joints all along the four seams, both inside and outside, for rigidity. All this soldering can be done with a medium powered electric soldering iron.

Then insert the waveguide into the tabs, making sure that they grip it firmly. A gas flame will probably be needed to help solder the waveguide in place. A suggestion is to prop the horn horizontally on a gas stove, with the guide a few centimetres above the flame. The flame should only be used to keep the metal just below the melting point of the solder, and the soldering iron used to melt the solder locally where required. This avoids the whole structure falling apart when the all the solder melts at once! Also solder the end of the waveguide to the inside of the horn, making a nice smooth joint at the throat of the horn. This process is made much easier if all the mating surfaces around the guide and tabs are tinned before assembly; if all the surfaces are very clean and a separate flux is available, this

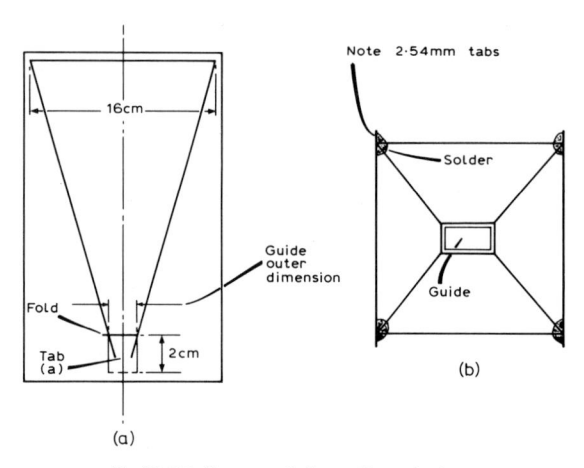

Fig 18.133. Horns made from oil-can tinplate

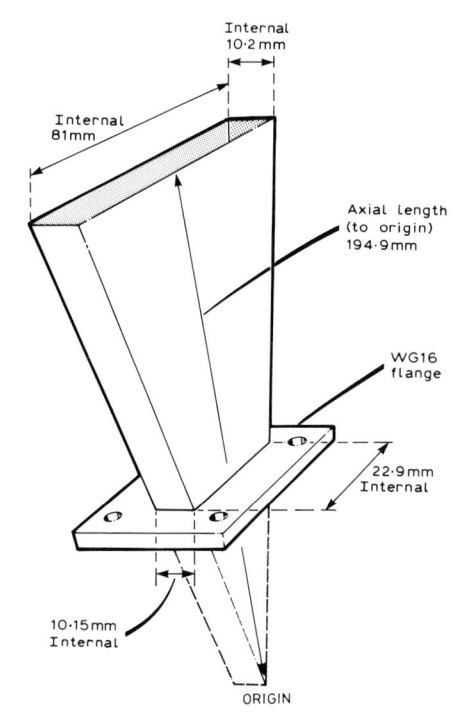

Fig 18.134. Typical sectoral horn dimensions. Both E-plane and H-plane sectoral horns can be easily designed using the BASIC program in chapter 4, "Microwave antennas"

precaution may not be necessary. While the waveguide is hot, the flange can also be soldered into place.

When the horn has cooled down, scrape off any remaining flux and give it several coats of paint, both inside and outside, to prevent it rusting. A red-brown epoxy based paint made for car repairs is recommended for this and other outdoor items, as it seem to adhere to surfaces very tenaciously.

Having said earlier that excess solder should be removed from the inside of the horn, it is apparent that this precaution has not been (and could not be) taken with this particular design. The outcome is simply some loss of gain which proved acceptable for the purposes to which the horn was to be put. If the horn is to be constructed as a reference antenna, then such precautions *must* be taken.

Double-clad fibreglass pcb material has also been used successfully to make light-weight horns of the type described here. These are eminently suitable for occasional, "light" use and can be easily and very cheaply made by adapting the methods outlined above. Soldering with a very small soldering iron is possible, although some ingenuity is needed in making the horn to waveguide transition!

Sectoral horns

The most useful type of sectoral horn is that with an H-plane flare. With the broad faces of the guide vertical a horizontally-polarised field is produced which has an azimuth pattern of nearly 180° but a vertical pattern which is compressed into a few degrees (depending on the gain), thus making it useful for a beacon where semi-omnidirectonal coverage is needed. Construction is similar to that described above and the dimensions of a nominal 10dBi horn are given in Fig 18.134. The design of either E-plane or H-plane sectoral horns of other gains is possible using the computer program mentioned earlier.

18.7.3 Dish antennas and feeds at 10GHz

The geometry and general points concerning paraboloidal reflectors has been fully discussed elsewhere. Their main advantage is that they offer high gain for their size, can be designed for any gain and can operate at any frequency. Their disadvantages are that they are not easy to make accurately, they can be difficult to mount and feed and may have a high windage. Nevertheless, they are widely used by amateurs and some practical dishes and feeds are discussed next. The basic geometry of a dish antenna is given in Fig 18.135, together with the relationship between diameter, depth and focus. From a practical point of view the important relationship is the ratio of f to D (focus to diameter), for this will determine the optimum feed design.

A dish can be illuminated by a front feed, where the radiating element is fixed in front of the reflector (dish) by means of a tripod arrangement, or a rear feed, where the radiating element is mounted at the end of a length of feeder (waveguide) which goes through the centre of the dish and acts as the support for the radiator. The rear feed is usually more convenient and can take several forms, according to the f/D ratio of the dish.

Direct feed dishes

The geometry of the front feed is illustrated in Fig 18.136. The phase centre of the feed is placed at the focus of the dish and power radiated by the feed as a spherical wave is

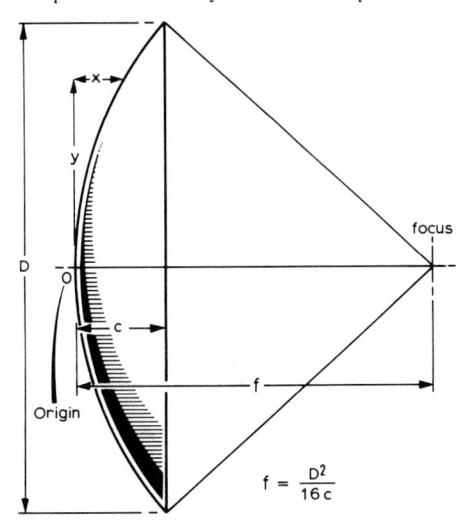

$$f = \frac{D^2}{16c}$$

Fig 18.135. Basic parabolic dish geometry

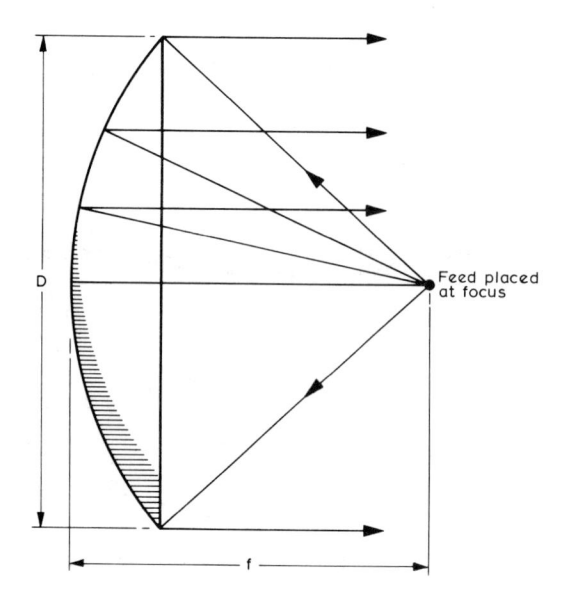

Fig 18.136. Geometry of the front feed

converted by reflection at the paraboloid into a plane wave. Some years ago it was found that an ordinary galvanised dust-bin lid of the smoothly rounded type was a sufficiently accurate paraboloid to be usable as a dish reflector at 10GHz. Most of the lids examined have a relatively long focus, with an f/D ratio in the range 0.7 to 0.9. A lid with the smallest ratio will make a more compact antenna. Lids which have dents or "wrinkles" greater in depth than about $\lambda/10$ at the design frequency should be avoided.

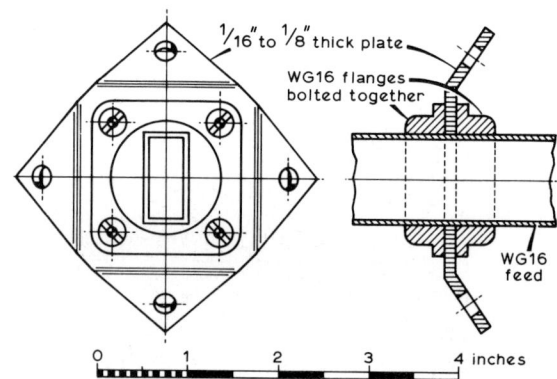

Fig 18.138. Dust-bin lid – feed mount (G3RPE)

It is well worth spending some time in measuring the lid to make sure that it is a reasonable parabola. The mathematical methods for this are described elsewhere. Suffice it to say that the 19in (483mm) lid chosen for the purpose only deviated from a true parabola by $\lambda/17$ at the centre and by $\lambda/50$ (at 10GHz) elsewhere.

The front feed most suitable is a small horn and the means of supporting the feed is illustrated in Figs 18.137 and 138. Fig 18.139 is a photograph of the completed antenna. Fuller details of the construction of such an antenna will be found in [32].

Fig 18.140 gives the dimensions of a small feed horn as a function of the f/D ratio of the dish. The horn used should be designed using this data once the f/D ratio of the lid is determined. As an example, if a paraboloid with an f/D ratio around 0.53 is available, the dimensions of a suitable

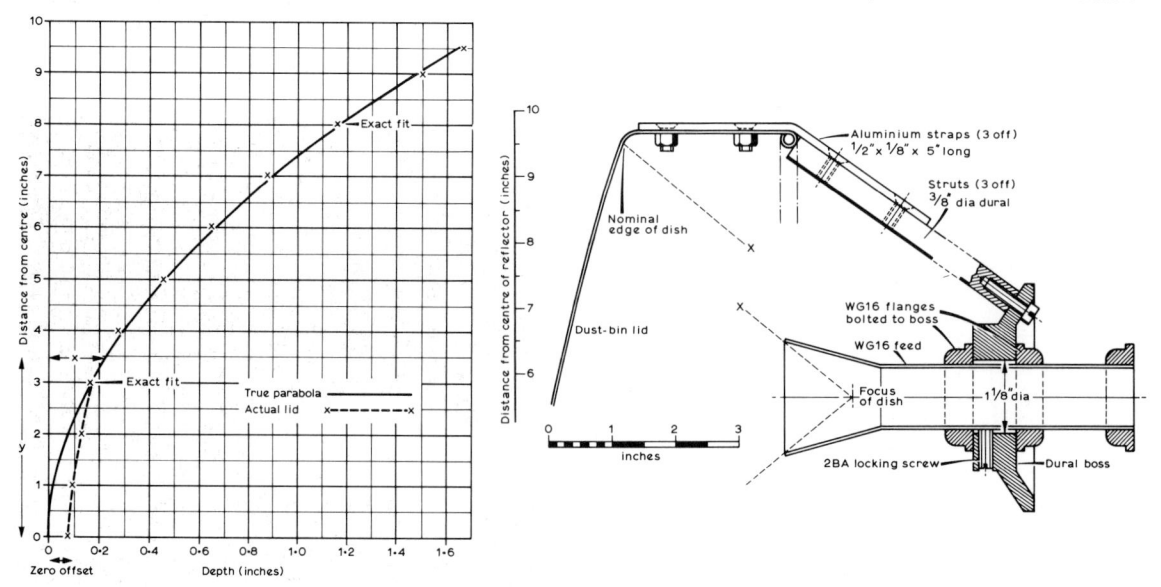

Fig 18.137. "Dust-bin lid" antenna – (a) the profile of the lid, as measured, compared to a true parabola (b) the feed method (G3RPE)

THE 10GHZ ("3CM") BAND   18.85

Fig 18.139. Photograph of a complete dust-bin lid antenna (G3RPE)

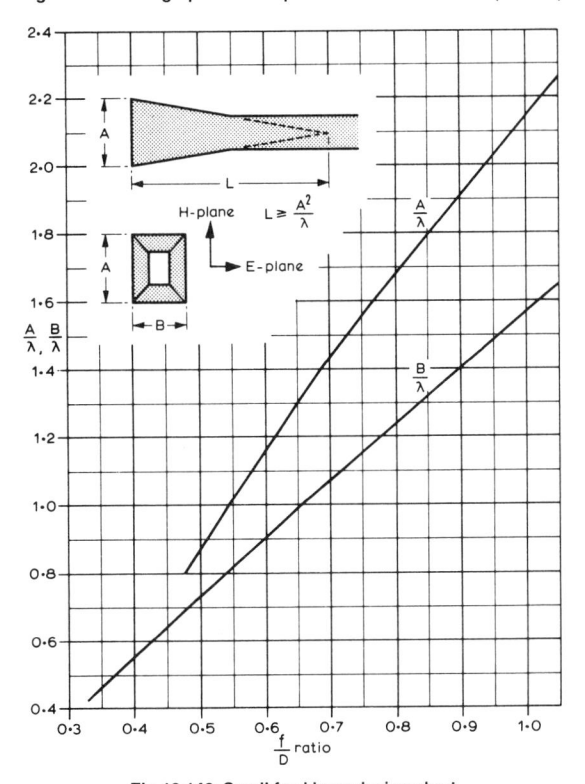

Fig 18.140. Small feed horn design chart

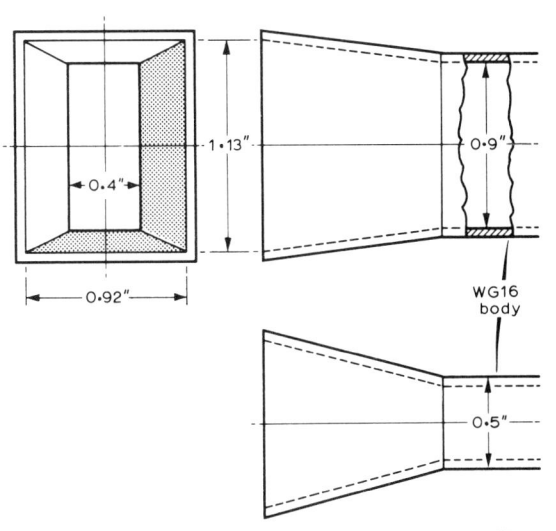

Fig 18.141. Dimensions of a feed horn for a dish with 0.53 f/D

feedhorn are shown in Fig 18.141. A dish of this type is shown, complete with flexible waveguide feed, in Fig 18.142.

A simple modified Cutler feed, which has come to be known as the "penny feed" because an old (pre-decimalisation) penny was just about the right size for the end disc, is shown in Fig 18.143. This was originally described by G4ALN as being suitable for dishes with an f/D ratio in the range 0.25 to 0.3. Whilst it will work reasonably well with a focal plane dish (f/D = 0.25), there is a significant degree of under illumination and consequential loss of gain. However, the constructor may be prepared to trade this loss of efficiency for the simplicity of construction and relative lack of criticality of the feed dimensions.

Fig 18.142. Photograph of a complete front fed dish (G3RPE)

18.86 MICROWAVE HANDBOOK

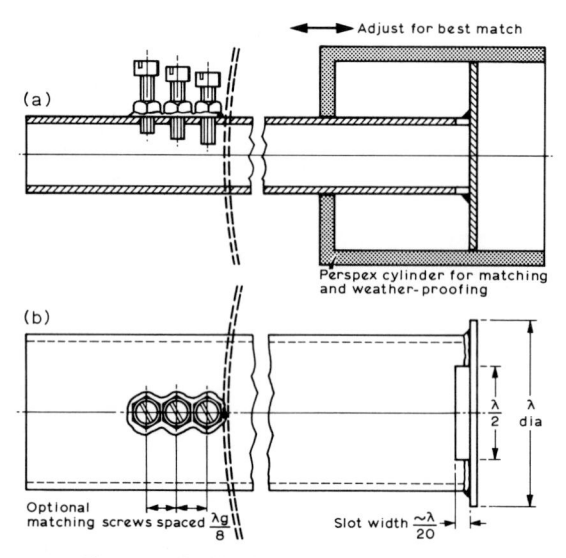

Fig 18.143. Modified Cutler ("Penny") feed (G4ALN)

Table 18.9. The "Penny" or modified Cutler feed

Measurement:	Professional	G4ALN	Suggested Dimensions
Frequency:	(11.15GHz)	–	(10.38GHz)
Disc diameter	1.25λ (33.36)	1.0λ (28.9)	1.25λ (35.83)
Slot length	0.5λ (13.35)	0.5λ (14.45)	0.5λ (14.34)
Slot width	0.085λ (2.25)	0.05λ (1.45)	0.085λ (2.42)
Scatter pins:			
Diameter	0.08λ (2.1)	None	0.08λ (2.3)
Length	0.1λ (2.8)	None	0.1λ (3.0)

The feed is made by cutting two appropriately dimensioned grooves in the end of a length of WG16 and soldering on a circular end disc (the "penny"). The length of the slot formed and the diameter of the disc are thought to be not critical within a few percent, and the width of the slots even less so. Signals with standard horizontal polarisation are produced with the broad walls of the waveguide vertical. The feed can be used without any attempt to improve the match – the vswr is typically about 1.5:1. The match may be improved by conventional matching screws fitted behind the dish as shown in the drawing or by means of the Perspex protective and matching sleeve also illustrated there. A professional 610mm (24 inch) dish complete with WG16 feed, designed for 11.1 to 11.2GHz, was acquired. The feed was a version of the feed just described. In the knowledge that this had been designed for use at (nominal) 11,150MHz, on the assumption that a professional designer would not have used this kind of feed had it not been acceptably efficient and in light of reports that this type of feed was not very efficient with certain types of dish, it was decided to make some careful physical measurements of both the dish and feed. The results of these measurements, when compared with the original version, are quite interesting and may help to shed some light on the apparent inefficiency of the modified Cutler feed.

First the dish was measured and it "weighed in" with the following characteristics:

Diameter (D) = 610mm
Depth (C) = 115mm
Focus (f) = 202.23mm
f/D ratio = 0.332

The measurements of C and calculations for f were confirmed by measuring the distance between the detachable feed mounting plate and the *centre of the slots*. Next the feed was examined and a number of interesting points emerged. The feed is illustrated in the dimensioned diagrams (Fig 18.144) and summarised and compared with the original G4ALN dimensions in the following table. Suggested (calculated) dimensions for 10.380GHz are also given (all in mm) in Table 18.9. The most significant differences between the professional feed and G4ALN's version are:

1. The disc is much thicker, about 0.185λ (4.9mm) and this may be significant. "Scaled" for 10.380GHz, this would be 5.26mm.

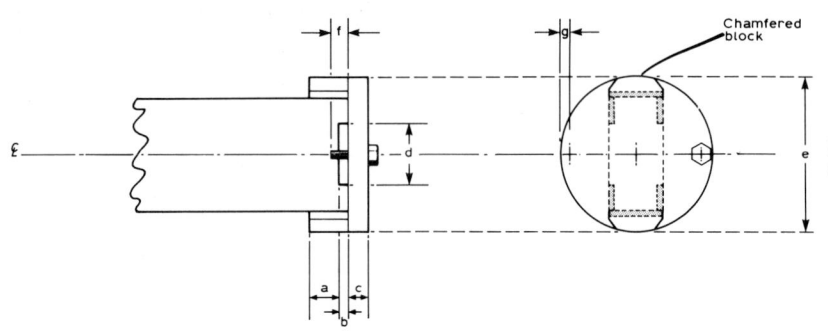

Fig 18.144. Modified "Penny" feed (G3PFR). Dimensions are given in Table 18.9

2. The disc is backed by two chamfered blocks, as shown in the diagrams. It is probable that these, too, have a significant effect on dish illumination.

3. The width of the slots is greater in the professional feed.

4. The presence of two "scatter" pins mounted near the rim of the disc and lying above the midline of the broad face of the guide. Since they were pieces of studding fitted into tapped holes and locked with a lock-nut, they were presumably "tunable" at some stage of assembly.

5. The f/D ratio of the dish is significantly greater than that described in the original text.

A dipole and reflector feed

The problem of feeding short focal length dishes, and particularly focal-plane dishes which are often used by amateurs simply because they are available, is highlighted by the last feed design.

A form of feed known to be very effective with a focal-plane dish is the dipole and reflector shown in Fig 18.145. It is not an easy feed to make but, constructed properly and accurately (particularly with regard to dipole dimensions), it is capable of very good results with dishes of f/D = 0.25. The method of construction is self-evident, except that the dipole and reflector are best made longer than specified and trimmed to length after assembly. Care must be taken to remove excess solder at the base of the dipole and at the seams between the web and waveguide.

The E-plane and H-plane phase centres are not coincident, but their separation can be neglected for most purposes. The mean phase centre is located approximately one third of the distance between the dipole and reflector, nearer to the dipole. This point should lie at the focus of the dish and the exact position determined by experiment. If the feed waveguide passes through a sliding bush in the dish, then the whole assembly can be easily adjusted and may be locked in position after optimisation, either by set-screws or by soldering. Again, a professional focal-plane dish and dipole/reflector feed was obtained and measurements taken. The figures confirmed the dimensions given in the diagram, the only difference being that the reflector was a disc 29mm (1.14in) in diameter, rather than the rod reflector shown. This disc was also used to support a thin plastic (polythene) sleeve which, fitted over the entire length of the feed and clamped to the mounting boss inside the dish, provided weather protection. As in the previous design, it is possible to improve the feed match by fitting conventional matching screws immediately behind the dish.

Indirect feeds

The basic geometry of the sub-reflector feed is shown in Fig 18.146, where the sub-reflector is a plane disc. The disc area should not exceed 30% of the dish area if losses, due to aperture blockage, are not to exceed 1dB. Once the size of the disc is chosen, its position is automatically fixed – it must *just* intercept lines drawn from the real focus of the dish (Fr) to its rim. The position of the feed is also fixed

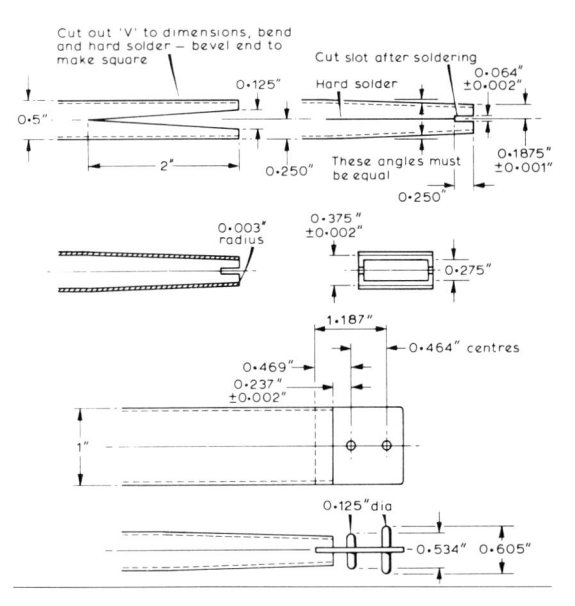

Fig 18.145. Dipole and reflector feed for short focal-length dishes

and its distance from the sub-reflector must be equal to the distance of the disc from the real focus (Fr), the distance m in the figure.

With this type of feed it is usual to adopt a Cassegrain configuration in which the plane reflector is replaced by a hyperboloid and the geometry of this system is illustrated in Fig 18.147. The main result of this change is that the virtual dish "seen" by the feed has a longer focal length than the real dish and thus a dish of short focal length, which can be difficult to illuminate efficiently, can be converted into

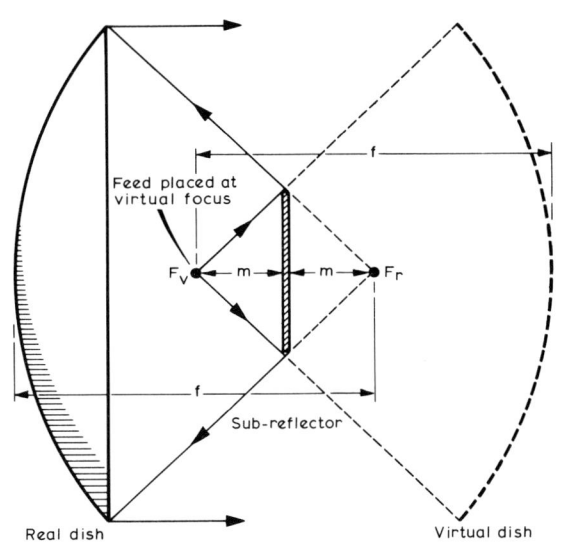

Fig 18.146. Geometry of the plane sub-reflector feed

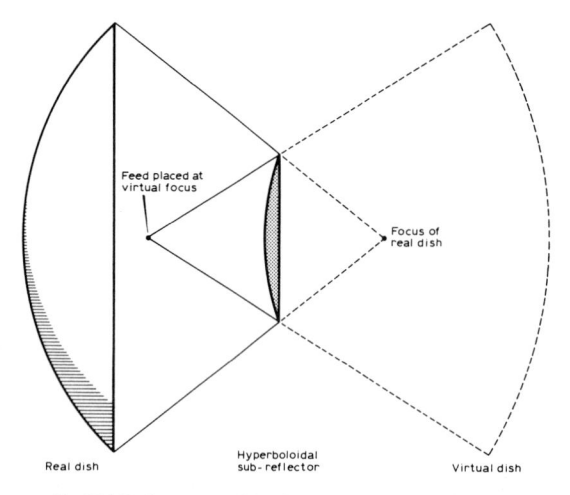

Fig 18.147. Geometry of the Cassegrain sub-reflector feed

one of longer apparent focal length. The methods for determining the form of the hyperbolic sub-reflector are given in the "Antennas" chapter, together with a computer program which will also calculate the required profile.

Making the sub-reflector is not easy for amateurs without a lathe. However it is possible, using either annealed copper or soft alloy, to carefully "beat" a disc into shape with repeated *light* hammer blows, with the disc supported on a firm but resilient backing (for example stiff foam plastic) or on a hard wood profile block. If the former method is used, then progress is best judged by applying a template cut from stiff card or another piece of metal. Gentle beating should continue until the profile is reached. The convex surface of the hyperboloid can then be smoothed with fine emery paper to remove the hammer blemishes (which should be few, if the hammering has been sufficiently light!). A good enough profile for amateur purposes can be achieved with patience.

The sub-reflector, when complete, is supported in the dish as shown in Fig 18.148 and the feed horn (designed from the data given in Fig 18.130) fitted to complete the assembly.

An alternative method of construction, for those possessing the necessary equipment and skills, might be to make a profile from two hardwood blocks, one "male", one "female", and to use a press to form the metal sheet. However, this is probably beyond the means of most amateur constructors and the slow, patient "manual" approach may be followed.

18.8 TEST "BENCHES" AND METHODS

Many meaningful measurements can be made at 10GHz by means of a "test bench" which is simply an assembly of waveguide components and is the microwave analogue of the optical test benches with which some people may be familiar from their school physics!

Fig 18.148. Photograph of a Cassegrain fed dish showing the sub-reflector mounting and feed horn (G3RPE)

The most useful items of equipment which will enable quite a range of measurements to be made are as follows:
- A waveguide mixer/detector (Fig 18.15)
- A calibrated absorption wavemeter (Fig 10.44)
- A calibrated variable attenuator (Fig 10.18)
- A Gunn oscillator (Fig 18.42) *or* a Gunn in-line assembly (Fig 18.42)
- A modulated Gunn supply (Fig 18.59) *or* (better) a complete wbfm board (Fig 18.95)
- A calibrated cross coupler and matched load (Fig 10.38 and 10.15)
- A sensitive multimeter of known characteristics
- An oscilloscope, with timebase output, calibrated input and graticule.

(Figs 10.15, 18, 38 and 44 are to be found in chapter 10, "Test equipment").

Some of the types of measurement which can be meaningfully undertaken are:

Frequency and power measurement
Gunn performance measurement
Simple spectrum analysis
Filter alignment
Component and antenna matching

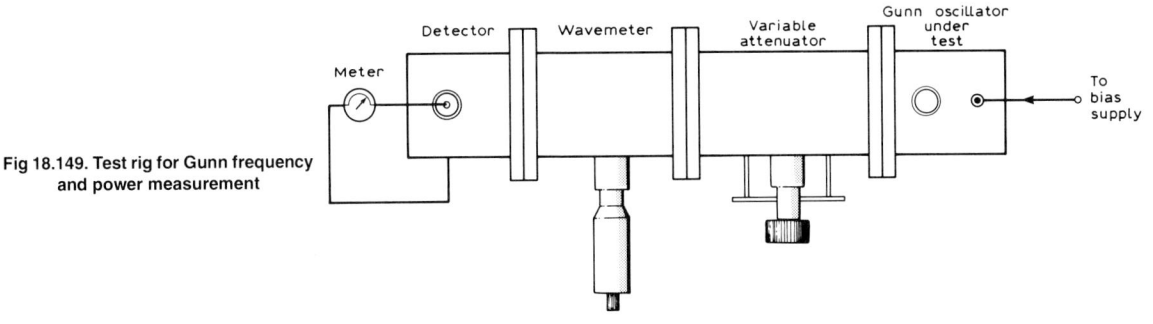

Fig 18.149. Test rig for Gunn frequency and power measurement

Simultaneous frequency and power measurement

With a calibrated absorption wavemeter at 10GHz the frequency resolution obtainable might vary from 10–20MHz with a relatively low Q wavemeter to about 2MHz with a high Q wavemeter. One difficulty in making measurements to this degree of resolution when using free-running oscillators is that, unless isolation is provided between the oscillator and the wavemeter/detector, the oscillator frequency will be pulled by the wavemeter and lead to false measurements. If the oscillator is very load sensitive such a precise measurement is of little practical value unless it is performed in the system it is used in. The simplest set-up to ensure that measurements will be as accurate as the system resolution allows is that shown in Fig 18.149. The same system can be used for simultaneous power measurement.

The test method is simple: set the attenuator to about 10dB, detune the wavemeter well away from the expected frequency and set the multimeter to the 10mA range.

Start the Gunn oscillator – if this has an output in the range 10 to 20mW, then 1 to 2mA current should be seen on the multimeter. Next increase the attenuation until, say, 25µA is seen on the meter and then switch the multimeter down to the 100µA range. Increase attenuation until a reading of, say, 10µA is obtained. This infers an isolation of 30 to 33dB between the oscillator and wavemeter. Now tune the wavemeter slowly for dip, read the calibration scale and record the frequency. Detune the wavemeter again.

Calibration charts of power versus detector current for both point contact and Schottky diodes are given in chapter 10. Depending on the type of detector diode in use, the current can be translated directly into power using one of the two graphs. Better indications of low power levels at the detector are given by the use of a point contact diode for power measurement purposes. The current should be noted and the reading translated into incident power. The setting of the attenuator should also be noted. If the indicated power is 0.01mW (current about 15µA for a point contact diode with a multimeter of resistance 1kΩ on the 100µA range) and the reading on the calibrated attenuator is 30dB, then the oscillator power output is 30dB up on 0.01mW, i.e. 10mW. The accuracy of the measurement will depend on the calibration curve for the diode but should be accurate enough to ensure that the constructor does not make any gross errors of calculation in designing the rest of the equipment or in estimating path capability.

At the same time as making these measurements it is possible to determine several other Gunn parameters. By attaching a voltmeter to the output of the Gunn supply to measure the Gunn voltage, it is possible to measure the starting voltage, the power output at various bias voltages, the bias voltage for maximum power output and the maximum (turn-off) voltage. The optimum Gunn bias voltage for quietest operation can thus be measured. Measurements of frequency at each of the voltage settings will enable the voltage pushing factor of the Gunn to be determined, along with an estimate of the optimum voltage pushing range to be set by the "fine tune" potentiometer on the Gunn supply. It is sometimes possible that the Gunn tuning is not as smooth as expected and it may frequency jump some tens of MHz and then continue tuning normally, due to the oscillator changing modes. This may be very hard to detect as the power output may not change much. These jumps can be the cause of unexplained inability to hear another station – the oscillator may jump past the other stations' frequency! The best way to check for such effects is to mechanically tune the Gunn across its frequency range whilst either following it continuously with a wavemeter or examining it on a spectrum analyser or panoramic receiver and watching for sudden jumps. This process should be repeated at several different Gunn bias voltages within its electrical tuning range.

A Gunn diode and oscillator test rig

The test rig described here will give a very full and rapid display of the characteristics of a Gunn oscillator, measuring most of the parameters (instantaneously) which were done manually with the last test bench. It can be used to optimise the tuning and coupling to give the most reliable performance and show what operating voltage should be used.

The test rig was first devised by G3LQC and then modified to suit the Solartron CD 1400 oscilloscope by G6XM [33]. It can be used for the following tests:

1. Ascertaining the frequency at which the Gunn oscillator is operating.

2. Selecting the best Gunn diode for maximum output.

3. Checking that there is no spurious output and that the diode is operating correctly.

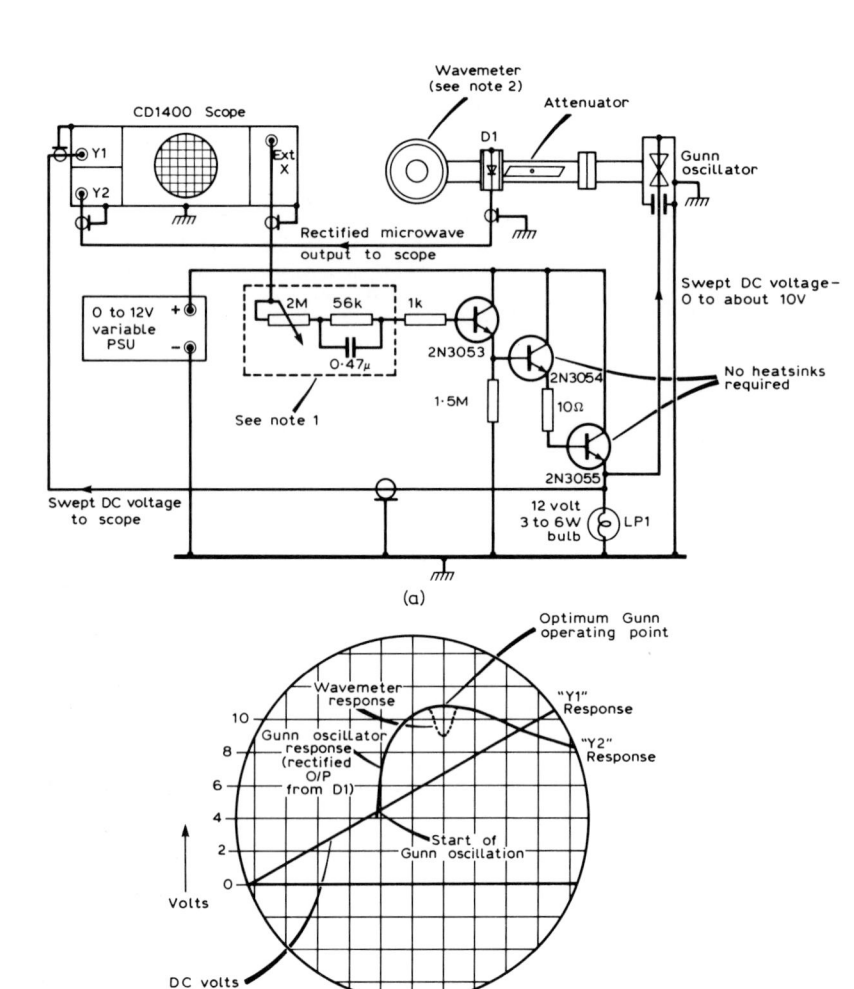

Fig 18.150. (a) The circuit of a test rig for swept frequency Gunn measurements (G3LQC). (b) Typical oscilloscope display

4. Finding the optimum operating voltage.

There is nothing critical in its construction except that the leads to the "Y1" and "Y2" plates of the 'scope must be screened to avoid external pick-up. The power supply should be well smoothed also. Refer to Fig 18.150a and then proceed through the following steps:

1. Connect a 1 to 2 amp power supply to the test rig and set it to about 10V. This level is not critical, but should not be less than 9V.

2. Connect the "X" sweep voltage from the 'scope as shown.

3. Switch the 'scope on and set the "Y1" sensitivity to 2V/cm and the "Y2" sensitivity to 0.2V/cm (this may be altered to suit the 'scope used).

4. Set the timebase of the 'scope to about 20Hz. Observe that the lamp LP1 flickers and that a "saw-tooth" waveform sweeping from zero to about 10V is shown on the "Y1" trace.

5. Check the Gunn diode polarity is correct *before* connecting!

6. Connect the Gunn oscillator and see if it is oscillating. A trace similar to that shown in Fig 18.150b should be seen after adjusting the waveguide variable attenuator. The optimum voltage for best output from the Gunn oscillator can be seen by inspection of the two 'scope traces.

7. The output from the Gunn oscillator should be smooth as shown. Any spurious output will be seen immediately on the "Y2" trace.

8. The wavemeter can now be tuned to ascertain the frequency at which the Gunn oscillator is operating. Reduce the wavemeter coupling to the minimum required to show a positive "suck-out" point similar to that shown in Fig 18.150b.

9. The 'scope timebase speed may be varied but it has been found that the best speed is below 50Hz.

The following notes may be found of assistance in setting up the rig:

1. The X sweep output of the Solartron CD 1400 'scope is high impedance and the values shown within the dotted line gave the best results. For a different type of 'scope, these might have to be changed.

2. The wavemeter shown terminates the end of the waveguide. An "in-line" wavemeter should be connected between the attenuator and the detector diode, D1.

3. Other 'scopes with different "X" sweep outputs may need an inverter or gain stage, using an op-amp to provide a suitable sweep signal. This is determined by experiment.

Simple 10GHz spectrum analysis – a "panoramic adapter"

It is very useful and informative to "see what is going on" during equipment alignment and testing. It is also much more meaningful to be able to see what is happening over even a small range of frequencies, rather than taking readings from a meter attached to a detector in the hope that what is indicated *is* happening at the frequency concerned! A mechanical analogue meter is relatively slow to respond to changes, because of its inertia, and will not readily reveal small and subtle changes which might be significant. By contrast, an oscilloscope beam is inertia-less and can be made extremely sensitive to small changes. Swept frequency displays are particularly good for revealing these small and subtle changes.

A spectrum analyser is a complex piece of equipment which provides many facilities: wide frequency coverage, a narrow passband (high resolution), high dynamic range and sensitivity, an "X"-axis which is linear with frequency and a "Y"-axis which is logarithmic with respect to amplitude (power). It is, in effect, a high quality "panoramic" receiver.

With such a comprehensive specification, an analyser is a very expensive instrument and thus beyond the reach of most amateurs. If amateurs are prepared to accept considerable compromise in return for some meaningful and useful measurement facilities, then it is possible to arrange 10GHz wideband equipment and a fairly basic oscilloscope to provide a simple form of frequency analysis. Such a test bench will provide relative (qualitative) rather than absolute (quantitative) information which, nevertheless, is useful. The compromises are, briefly:

1. Limited sweep bandwidth of perhaps 10 to 50MHz, depending on the oscillator used in the rig, and which may not be quite linear because of curvature of the Gunn characteristics.

2. Frequency resolution which is determined by the relatively wide filter employed in a typical 10GHz wbfm receiver.

3. Both receiver and oscilloscope amplifiers are linear, which would mean that the amplitude display is linear. However, the S-meter driver in the fm chip is approximately logarithmic between 0.2 and 3V and use is made of this.

4. Relatively low dynamic range, although this can be controlled by the manual setting of a variable input attenuator.

5. Frequency measurement which is based on a relatively "broad" absorption wavemeter included in the test rig.

Nevertheless, it is felt that the system to be described (it has picowatt sensitivity) will provide the experimenter with a number of useful "visual aids" facilities based upon the frequency pushing characteristics of a Gunn oscillator. The items of equipment needed for the test rig are:

- A complete wbfm receiver or transceiver strip based upon a CA3189E (or similar) ic.
- A microwave down converter, either in-line or (preferably) a cross coupler, to match the receiver/transceiver strip.
- A high Q absorption wavemeter.
- A variable (preferably but not necessarily calibrated) waveguide attenuator.
- Preferably, an isolator.
- An oscilloscope with high impedance "X"-sweep output.

The test bench can take several forms, two of which are shown in Figs 18.151 and 18.152. The first is based upon a "standard" Gunn in-line oscillator/mixer and the second on a cross coupler receiver. Both are set up as shown in the appropriate diagram.

A simple (and vintage) Heathkit OS2 service oscilloscope which had been modified to provide a sawtooth "X"-scan output at high impedance was used. This output was fed into the sweep circuit described in the preceding section. The same observations on the need for gain stages or

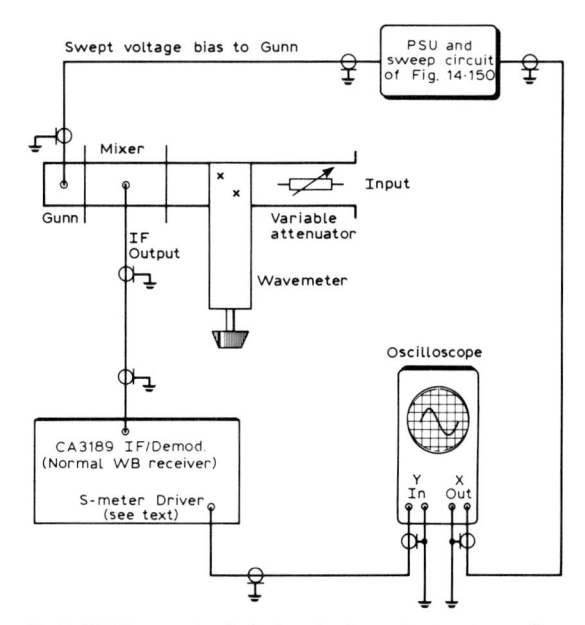

Fig 18.151. "Panoramic adapter" or simple spectrum analyser using a Gunn in-line receiver and the sweep circuit of Fig 18.150. The input transistor of the sweep circuit will need to be biased to a point where supply to the Gunn diode produces continuous output, i.e. just above the turn-on voltage for the diode used

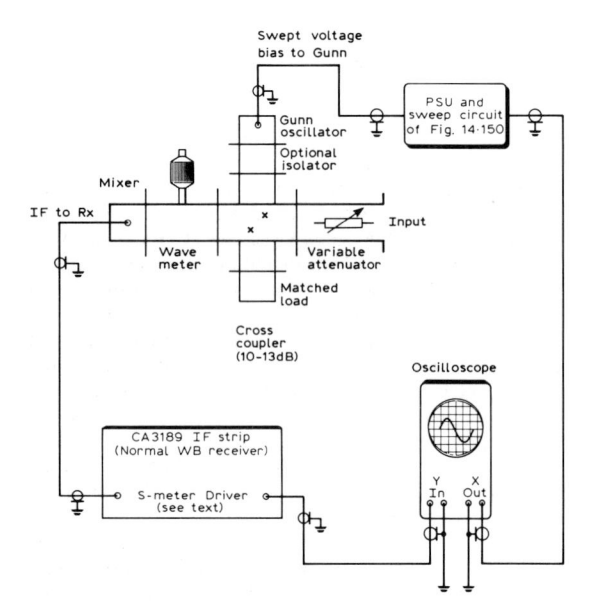

Fig 18.152. Simple analyser using a cross coupler receiver

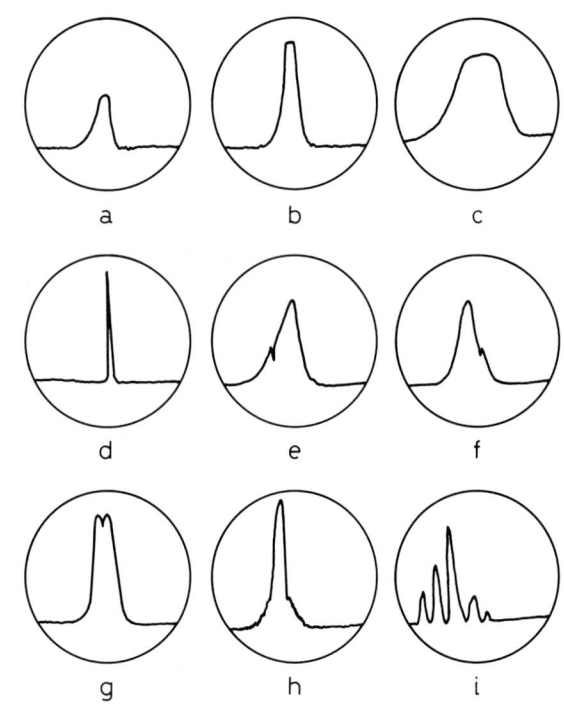

Fig 18.153. (a) to (i) Typical oscilloscope traces from the simple analyser

inverters apply to this configuration, depending on the oscilloscope used. The main difference between this test rig and the preceding rig was that the Gunn oscillator of the 10GHz down converter was swept with the voltage derived from the external swept power supply which replaced the normal modulated Gunn supply built into the wideband receiver. The adjustable components of the test rig "modulator" (those enclosed within the dotted lines of Fig 18.150a) are adjusted to sweep the Gunn voltage from a point somewhere between the "turn-on" voltage and the optimum voltage, up to the maximum output voltage – that is, somewhere in the range from about 6 or 7 to 9V. All connections between the sweep circuit, the oscilloscope, the Gunn oscillator and the receiver were made in screened cable to avoid noise and hum pickup.

The main receiver was that described in section 18.5.4, used as a single conversion (10.7MHz i.f) receiver. The only other modifications made were, first, to fit a 50kHz filter (Murata type SFE10.7MF) in place of the 150kHz filter usually used (this will benefit receiver performance in any case) and, second, to disconnect the S-meter drive from the S-meter and reconnect it to a convenient output socket in turn connected to the "Y" input to the oscilloscope. Further refinement might be to fit an even narrower filter and to use double conversion with a first i.f of 30MHz. The use of a 10.7MHz i.f with a sweep of greater than 20MHz can display both the sum and difference products of any incoming signal, although it is quite easy to determine which is which by using the in-built wavemeter.

Set the input attenuator to minimum attenuation. The "S-meter" output is connected to the vertical amplifier input, whose gain should be initially set to about 0.25V/cm sensitivity.

The "analyser" is now ready to receive and display signals. The output of another Gunn oscillator can be used as the signal source for adjustment of the test rig.

Mechanically tune either the receiver Gunn or the test oscillator until a "blip" is seen on the oscilloscope screen, looking something like Fig 18.153a. If the blip is too small, increase the vertical amplifier gain until the blip is of satisfactory size. If it extends beyond the top of the screen, decrease the vertical amplifier gain until a satisfactory trace is obtained. If the blip is flat-topped as in Fig 18.153b, then the received signal is so strong that one or more of the amplifier stages is overloaded. If this is the case, then the input attenuator should be adjusted to get rid of the flat-topping, and the height of the trace readjusted by means of the vertical gain control. It should, by judicious use of the two controls, be possible to obtain a display like that shown in (a).

The effect of adjusting the "X"-sweep output voltage applied to the modulator can now be examined. With a small voltage applied, the blip will be wide, as the frequency sweep of the receiver oscillator will be small (Fig 18.153c). As the sweep voltage is increased, so the swept frequency range will increase and the blip get narrower, as in Fig 18.153d.

Once these effects have been explored, try tuning the wavemeter. As the wavemeter frequency passes through the signal frequency, a display such as that in Fig 18.153e or f should be seen, with the wavemeter "dip" sliding up and down the frequency response curve. By tuning the wavemeter to give the pattern shown in Fig 18.153g, the frequency of the signal is determined. A high Q wavemeter will give a narrow, deep dip, whereas a low Q meter will give a shallower, broader dip in the displayed response. It should also be possible to see the oscillator noise sidebands as a broadening of the "skirt" of the response, as shown in Fig 18.153h. "Shot noise" appears on the trace as random scintillations which may occur anywhere on the trace. Small random fluctuations in Gunn output appear as a slight "jittering" of the height of the peak displayed whilst frequency "jitter", if really bad, may make the display appear slightly defocussed. The effects of modulation can be demonstrated if a tone is applied to the test oscillator. Sidebands are produced and might give a display something like Fig 18.153i, although under some conditions the modulation may be unsymmetrical, with sidebands appearing only on one side of the carrier peak, or more on one side than the other.

At this point the experimenter is advised to play around with the various controls available, to "get the feel" of what can be done to achieve a satisfactory display for the purpose in mind: it might even be the optimisation of the wbfm rig and its microwave head, using a remote test oscillator (very weak signal)!

Once optimum settings have been found and the operator is thoroughly familiar with the effects of the various controls, the test rig can be used for many of the measurements normally associated with a more complex analyser and may be particularly useful for the final alignment and setting-up of narrowband equipment. If the input attenuator is calibrated then some estimate of the test rig's sensitivity can be obtained and also a sensible estimate of how far down on carrier level any unwanted products may be. Another use is to determine the relative levels of local oscillator and i.f (144MHz) signals required to produce maximum clean and undistorted output from a G3JVL mixer in the transmit mode. Many other uses will occur to the reader exploring these simple techniques. As an "aside", the Astec tv modules mentioned in section 18.5.6 could be candidates for the front end of a wide coverage (950MHz to 1.45GHz) spectrum analyser using these simple principles.

Filter alignment and component/antenna matching

The cross-coupler test bench needed to align 10GHz filters and its use has already been fully described in section 18.6. By connecting the mixer diode output to an oscilloscope instead of a milliammeter and sweeping the Gunn, you have a simple form of network analyser which shows how the reflected power varies with frequency. This is much more helpful than a single frequency measurement when aligning filters. Once aligned sufficiently to allow good transmission through the filter, the spectrum analyser could be used for final "visual" optimisation of a received signal with the mixer attached to the antenna with which it is to be used. In this way the antenna matching could be adjusted at the same time as the filter input-matching screws.

Antenna matching (or any other component for that matter) can be accomplished by a variety of means. The same cross-coupler test bench can be used to monitor the reflected power from an antenna feed. This may be considered sufficiently well matched when the matching screws have been adjusted for no, or very low, reflected power. Earlier, a slotted-line vswr detector was described in outline: this too, could be used in conjunction with any oscillator to adjust the matching screws of the antenna feed (or other component, which should be terminated either in a matched load or attached to the system with which it is to be used) for minimum reflected power. Be careful when using this method to match antennas that false readings are not obtained as a result of reflections from surrounding objects. The best way to ensure that this does not happen is to point the dish up to a clear area of the sky. One of the writers spent quite some time working out why the antenna match could not be improved until he realised that a strong reflection was being picked up from a wooden fence, 30 metres away, at the end of the garden!

Alternatively, the "spectrum analyser" receiver could be used to maximise a received signal from a remote oscillator. It is then a matter of making adjustments to the various matching screws to give the maximum input to the analyser – that is, the largest obtainable oscilloscope display response.

18.9 RECENT DEVELOPMENTS IN MICROWAVE TECHNOLOGY

As a direct outcome of the consumer electronics industry development work which supports direct satellite tv broadcasting on both C-band and Ku-band, there are a number of components, devices and modules (some of them still under development at the time of writing) which are likely to be of use and significance to the microwave amateur, particularly in the 1.5 to 4GHz region and again in the 10.5 to 12GHz region. Devices and modules for the 600 to 1,500MHz region also should not be overlooked.

The first and main aim of this section is to outline briefly some of the techniques becoming available and to indicate how these might be of use to the amateur, with particular reference to 10GHz. Although many devices are intended for application at uhf or in the lower part of the microwave spectrum, it should be obvious that they can also be used at higher frequencies by applying them, in conjunction with either mixing and filtering or multiplication and filtering, to the design and construction of amateur equipment for 10GHz. The relationships between L-band, S-band, C-band and X-band are favourable in terms of efficient multiplication from any of the lower bands into the 10GHz band. Fortunately many of the devices which are designed for frequencies that are *not* harmonically related to the 10GHz band will still work extremely well at nearby frequencies which are.

The second aim, having outlined some of the techniques and devices available, is to try to stimulate the UK amateur (in particular) into thinking about the use of these newer design concepts, in order to get away from the "classical" techniques which have become established over many years and which have sometimes led to a rather stereotyped approach to amateur design and construction. Much of what follows is speculative, though by no means all, and it was felt appropriate to include a section in this chapter rather than elsewhere in the handbook, largely for the reason that the builders and operators of equipment for the higher bands are more inclined to depart from the conventional because few "standard" designs exist!

Only devices and modules likely to become readily accessible to the constructor are considered. This means consumer devices which are usually moderately priced. They fall into a number of categories of interest, as follows:

    PCB materials
    GaAs fets and amplifiers
    DRs and dros
    Phase locked loop ics and vcos
    Wideband modular amplifiers (mmics)
    Divider ics and frequency counting

PCB techniques – general

Many of the users of the 10GHz band will be accustomed to building equipment using WG16 components. These are certainly robust, relatively uncritical in dimensions, and quite easy to build and use as already described, although they do impose constraints on the user.

An alternative now is the use of pcb techniques. The average amateur will almost certainly have used copper clad epoxy-glassfibre board at hf, vhf and possibly uhf and thus be reasonably familiar with its characteristics and use. Such material is usable up to about 1.5 or perhaps 2GHz without incurring unduly serious losses, and is sufficiently dimensionally accurate to allow such use. However, at higher frequencies the losses become disproportionate and unpredictable. The manufacturing method (and materials) does not allow a degree of control sufficient to guarantee predictable results at higher frequencies. Thus it is necessary to use a more precisely dimensioned and controlled medium based, in professionally produced equipment, on a copper clad ceramic (alumina or "sapphire") substrate or, more appropriate to amateur use, glass-loaded ptfe. The former substrates are the more satisfactory but cannot be "worked" without extremely specialised equipment to which the amateur will not normally have access. The glass-ptfe material is workable with ordinary tools and a little patience and care, and so is suited to home construction. It is also suitable for the fabrication of small value, precise capacitors, an example of which is given in the 3.4GHz bipolar preamplifier design in chapter 16.

The most suitable materials are a series of double clad boards of various thicknesses typified by "RT–Duroid" and 3M "CuClad", although other manufacturers market similar products. These materials have precisely controlled dimensions; not only must the substrate thickness be defined, but also its dielectric constant, ε_r, and the copper thickness (defined in ounces per square foot). Precisely defined materials such as these are made in smaller pieces than common fibreglass/epoxy board and are far more expensive than conventional media.

Conventional means of masking and etching can be used, but the constructor should be aware of the need for fairly precise line dimensioning when designing microstrip circuitry. Professionally produced microstrip circuits must be precise for mass manufacture; for amateur use this high precision is not necessary, since the amateur will be prepared to "tweak" a one-off circuit for best performance. Thus, the professional circuit might be drafted, with due attention to precision, at several times the final size and reduced photographically to its *exact* mask size. The home constructor should be careful to avoid over-etching which may lead to "undercutting" of the copper track and a narrower line than intended! For very narrow lines, eg chokes, it is better to use a piece of thin wire rather than to try to etch a very thin line. Simple circuits consisting only of lines can be made by the means described in the constructional chapter, i.e. by the use of a straight-edge, scalpel and soldering iron.

It will be appreciated that whilst the design procedures are fairly simple, actually translating the desired dimensions into practice can be quite difficult for the home constructor. At higher frequencies such as 10GHz, the dimensions of the microstrip become increasingly critical for optimum performance. Fortunately the amateur, although striving for perfection, is often prepared to accept *slightly* inferior performance and to try to make up for this in some other way – perhaps by accepting less than optimum gain or a slightly degraded noise figure and making this up by additional stages of amplification or by more antenna gain, for instance. The dimensioning of pcb resonant structures, such as "tuned circuits" and filters, is no longer so critical since the arrival of dielectric resonators and some of their speculative uses are described later in this section. Provided that the constructor is *reasonably* accurate in terms of electrical length and impedance, then the fabrication of simple microstrip pcb is feasible for the home constructor with limited facilities.

PCB design – matching networks for amplifiers

Knowing all the board parameters enables relatively simple design techniques to be applied to define line lengths and widths so that they present both the desired impedance and electrical characteristics in terms of wavelength or fractional wavelength. A simple design program has already been presented in chapter 7 which should enable easy design of simpler microstrip circuits. However, do not rely too heavily on manufacturer's S-parameter data, as different device mounting methods used in amateur designs can make significant differences to device input and output impedances at higher frequencies. Be prepared to optimise the matching by the means described later.

In its original form the program used two of the four scattering coefficients (S-parameters) normally quoted in

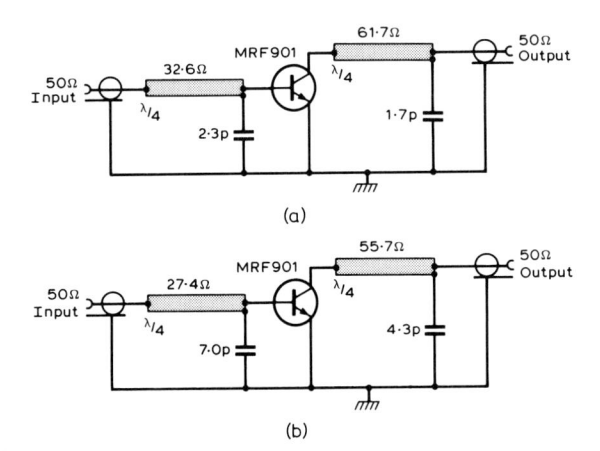

Fig 18.154. (a) and (b) 1.3GHz bipolar amplifier circuit designs produced as a check on the microstrip design program

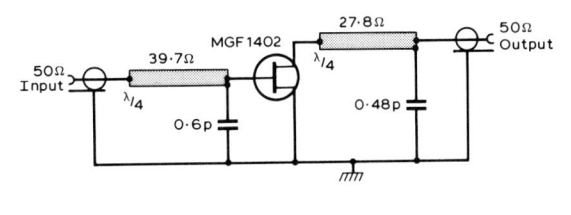

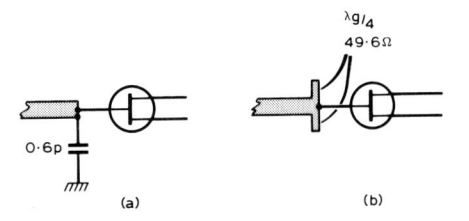

Fig 18.155. (a) and (b) A maximum gain MGF1402 GaAs fet amplifier for 10GHz designed by the program

the device manufacturers' data sheets, namely S11 and S22, to calculate approximate input and output matching networks, based on quarter-wave lines and shunt capacitors. This approach is valid provided that the ratio between the other two scattering coefficients, S12 and S21, is very small. A procedure (PROCline) was added to calculate the actual dimensions of the microstrip lines for a given Z_0 and type of substrate material. There is also an option to represent the shunt capacitive matching elements by 1/8 wavelength open-circuit sections of line. In practice the length of these would have to be "adjusted on trial" since the program does *not* account for "fringing" capacitance at the open circuit ends of the lines. The program prompts for frequency (MHz), whether a maximum gain or low-noise amplifier is required and then asks for substrate dielectric constant and substrate thickness (1oz cladding is assumed). The S-parameters are then prompted in magnitude and angle.

Examples of designs produced using this program are given below:

Example 1

As a check, the program was first used to design a maximum gain amplifier stage at 1,296MHz, using an MRF901 bipolar transistor, whose S-parameters at that frequency are typically:

S11 = 0.47, angle 161°   S21 = 3.10, angle 63°
S12 = 0.08, angle 64°   S22 = 0.43, angle −41°

S12/S21 is 0.026 which is small, as required for the technique to work. Using the program resulted in matching networks as shown in Fig 18.154a. For comparison a similar amplifier was designed using another more exact program which used all four S-parameters and this resulted in slightly different values for the matching network elements as shown in Fig 18.154b. In practice the "approximate" design given by this program resulted in an amplifier whose theoretical gain was only 0.25dB less than that predicted by the more exact program (13.2dB). Biasing networks are not shown in any of the diagrams, but could take the form of rf chokes made from quarterwave sections of Z_0 = 100Ω, connected to the base (gate) and collector (drain) of the device used.

Example 2

A single stage, maximum gain amplifier for 10.4GHz using an MGF1402 GaAs fet. At 10.4GHz, Id = 30mA

S11 = 0.64, angle 144.1°
S22 = 0.58, angle −134°

This produces the circuit values shown in Fig 18.155a and the capacitors can be replaced by eighthwave lines having Z_0 equal to the calculated capacitive reactance, as shown in Fig 18.155b. As noted above, the open eighthwave sections would need to be trimmed. The predicted gain is about 10 or 11dB.

Example 3

This is a single stage, low-noise amplifier for 10.4GHz, again using an MGF1402. At 10.4GHz, Id = 10mA and the MGF1402 has, typically, S22 = 0.599, angle −138° and needs to "see" a source impedance of about 0.407, angle −160° for lowest noise. Hence S11 is entered as 0.407, angle −160°, resulting in the circuit of Fig 18.156. The gain

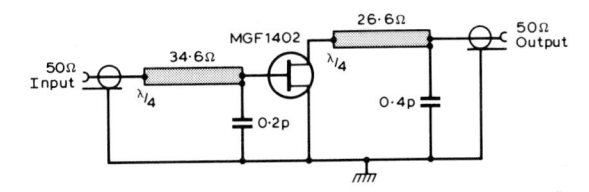

Fig 18.156. Low noise version of the circuit of Fig 18.155; (a) physical layout, (b) circuit and values

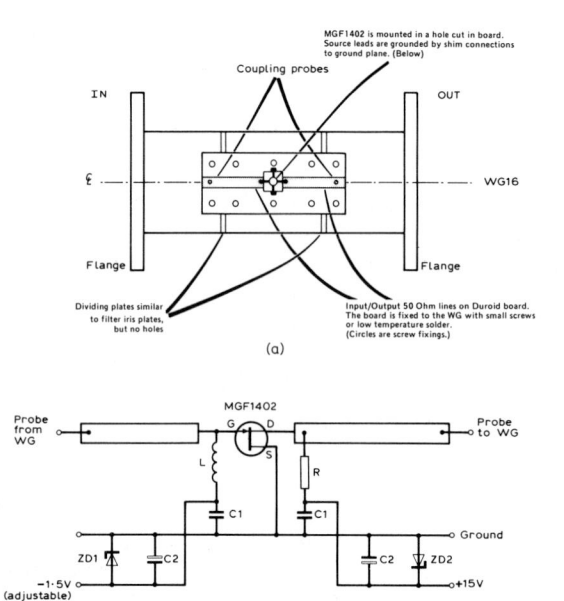

Fig 18.157. Ideas for a waveguide-mounted GaAs fet amplifier. (a) Physical layout. (b) Typical circuit (G3WDG)

is now about 8 or 9dB because of the noise mismatch at the device input.

Ideas for practical GaAs fet amplifiers at 10GHz

The theoretical designs given above need translation into practical circuits! If the constructor uses coaxial input and output, then the techniques outlined in the 3.4GHz chapter can be used, with the substitution of SMA connectors for the N-type used in the bipolar preamplifier described there. Special SMA sockets with "launcher" spills are available and should be used.

"Interfacing" into WG16 is a little more difficult and could use WG to SMA transitions such as those described earlier in this chapter. An alternative idea, due to G3WDG (Fig 18.157a), is to mount the microstrip pcb on the outside of the waveguide and to couple to and from the waveguide by means of two probes. The dimensions of the microstrip can be calculated by the microcomputer program given in chapter 7. The board can be mounted by means of small screws fitted into tapped holes in the waveguide. The size of the probes and their spacing from the short-circuits can be based on the information given for the design of waveguide/coaxial transitions. Input and output isolation in the waveguide is provided by the short-circuits which divide the guide into two electrically isolated cavities, thereby preventing the input from "seeing" the amplified output.

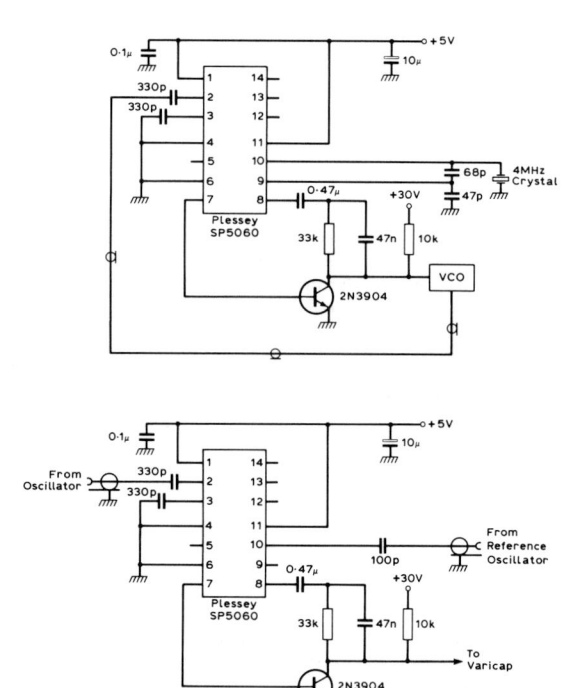

Fig 18.158. Basic circuit of an SP5060 ic frequency synthesiser. The chip will synthesise frequencies up to 1.8GHz and is referenced to either an on-chip 4MHz oscillator (top) or to an external crystal reference (bottom). The power required is 50mA at +5V. (Plessey Semiconductor application notes)

The circuit of the amplifier can follow standard practices and a suitable circuit is shown in Fig 18.157b. All the other components including the microwave trapezoidal or disc decoupling capacitors and earth tags can be mounted directly on the waveguide surface. The remaining components can be suspended between these points as needed. The GaAs fet is mounted in a small hole cut into the board.

Built in this way, the circuit needs tuning. The simplest way to do this is to cut some small pieces of thin copper shim and, with power applied to the amplifier, move these around on or alongside (and touching) both gate and drain lines (using an insulated tool) until gain is optimised. They can then be fixed in position by a "dab" of solder but remember to power down before doing so! It should be noted that the saturated output (1dB compression) of an MGF1402 amplifier is in the order of 50mW, so that a suitably designed multi-stage amplifier would not only improve the performance of a receiver, but could also be used to "boost" the low power transmit output usually obtained from the G3JVL mixer. Maximum gain is obtained with a standing drain current around 20mA, whilst minimum noise requires about half this value.

Other design approaches have been described in [34] and [35]. No easy, reliable "definitive" design for 10GHz has yet appeared in amateur literature (but see section 18.10).

Fig 18.159. Basic circuit of an SP5052 ic frequency synthesiser. This chip is similar in function to the SP5060 but extends coverage up to 2.3GHz and requires a dedicated microprocessor to program frequency information using 14 bits of a 16 bit serial data word. For a synthesiser tuning 1.5 to 2.3GHz in 200kHz steps, the crystal reference would be 6.4MHz. (Plessey Semiconductor application notes)

Frequency generation

The use of dros has already been described in some detail. Other alternatives may be based on phase-locked loops and voltage controlled oscillators (vcos). A circuit for phase locking a 24GHz oscillator is described in detail in the next chapter and, as already indicated, similar techniques and circuits can be used at 10GHz.

However, there are now a number of synthesiser and vco chips becoming available which will operate directly in the lower GHz region, making it possible to generate signals directly at around 1,152MHz. For instance, the writers are aware of two frequency synthesiser devices in the Plessey range. The preliminary information following is given by kind permission of Plessey Semiconductors Ltd and is intended only as a guide to the potential of such devices.

The first device is the SP5060 1.8GHz fixed modulus frequency synthesiser which will, when used with a suitable vco and reference frequency (crystal), generate any frequency within the range 300MHz to 1.8GHz. The operating supply needed is 5V at 50mA. The crystal reference used is typically in the 4.5MHz region. A basic applications circuit is given in Fig 18.158.

The second is the SP5052 which is a single chip 2.3GHz synthesiser, again operating from a 5V supply rail at 85mA. This device is designed for satellite tv down-converters and requires a suitable microprocessor to feed in the frequency information to set the output of the device to the frequency required for subtractive mixing (oscillator higher in frequency than the signal in a down-converter configuration) in a tv receiver "head". The basic circuit is given in Fig 18.159.

The introduction of such devices has created a requirement for an economic vco design covering the range 1.5 to 2.2GHz. This is, of course, possible using hybrid oscillator designs produced commercially, although these are likely to be too expensive for amateur use. An alternative design using easily obtained components is illustrated in Fig 18.160 and 18.161. The frequency of operation can probably be lowered to around 1,152MHz by experimenting with the length of the "inductor" shown as a "0.6cm lead" in Fig 18.160. The output level of the vco could be brought up to the 10dBm (or more) region by using one or more mmics at the output of the vco. Great care would have to be taken in the construction and housing of the vco if "jitter"

Fig 18.160. Circuit of a vco suitable for use with synthesiser circuits. The oscillator specification is: operating voltage +9 to +14V; frequency range 1.5GHz to 2.2GHz; varicap voltage range 0V to +30V. (Plessey Semiconductor application notes)

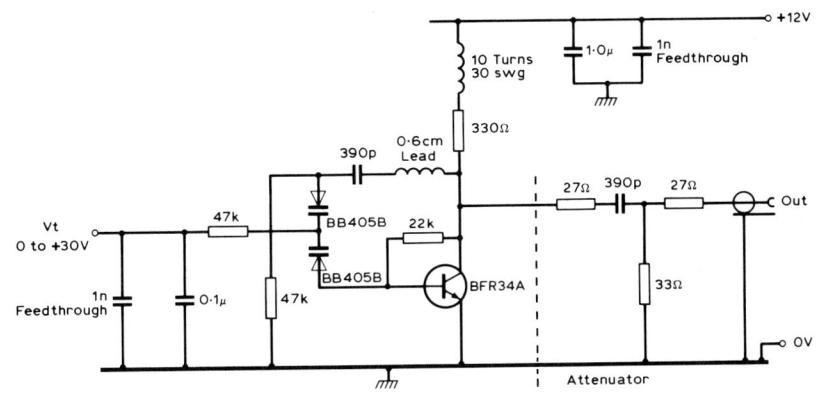

is to be held to acceptable levels and it may be appropriate to consider the use of a λ/4 coaxial cavity to achieve this end. Such jitter might not be of too much consequence in a truly wideband (fm tv) system, but would be of high significance in amateur narrowband equipments. It is almost certain that, despite elaborate precautions to avoid jitter, the stability of the signal would be insufficient to support real narrowband working, but would be sufficient to support nbfm or fsk modes in a bandwidth of 25 or 50kHz.

One speculative use for these modules, suggested by G4DDK, is the direct generation of vco controlled, crystal locked signals for the 3.4GHz band. The block diagram for such a system might look like that in Fig 18.162, where the pll/vco generates output at 1,728MHz which could be used with a sub-harmonic (antiparallel diode) mixer for receive purposes or amplified and multiplied as indicated to produce quite respectable output levels for transmit purposes on that band, or for further multiplication to the 10GHz band (at 10.368GHz).

Frequency measurement and counting at 10GHz

If the operator already has a frequency counter with, say, a capability up to 600MHz then two approaches to frequency measurement at 10GHz are possible. The first (indirect method) uses only the counter and perhaps an external clock reference as the counter timebase; the other (direct method) uses readily available and relatively inexpensive modules as down converters into the counter range.

Frequency measurement for narrowband equipment is comparatively simple by the indirect method; it is usually sufficient to measure the source crystal frequency (around 90 to 100MHz) *accurately* and then to extrapolate the final frequency by the multiplication factor, remembering that a 1kHz inaccuracy at 96MHz with a multiplication factor of ×108 will result in an inaccuracy of 108kHz at 10.368GHz. This measurement is an order of magnitude better than the use of an absorption wavemeter such as might be used in a wideband system, but still not good enough for a narrowband system. If the primary measurement is to an accuracy of 100Hz, then the final calculation will be 1.8kHz out – better, but still not good enough!.

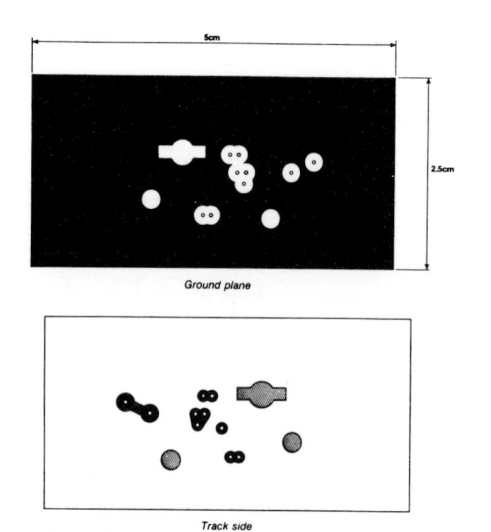

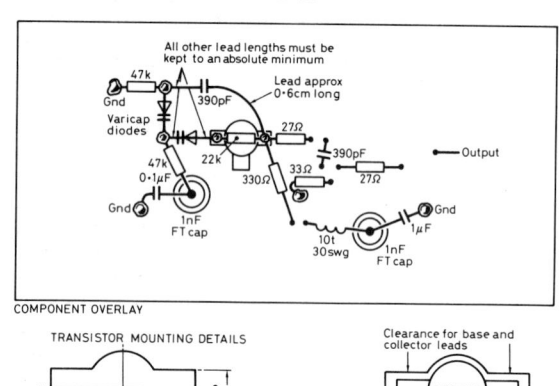

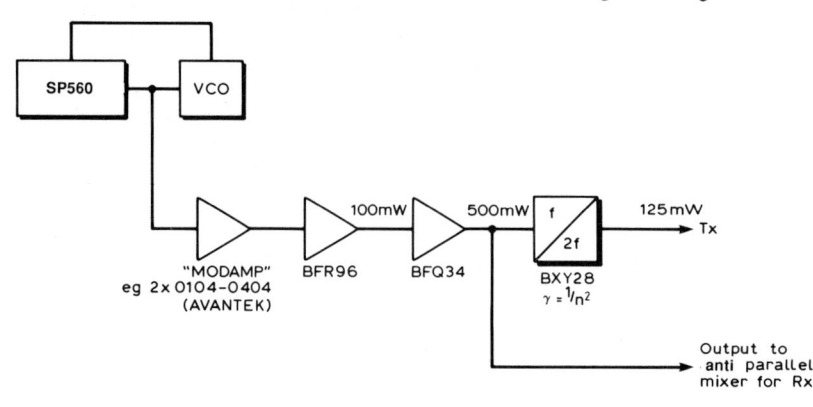

Fig 18.161. Layout of the vco. (a) Ground-plane pattern for the pcb. (b) Track pattern for the pcb. (c) Component overlay. (d) Transistor mounting details. In order to reduce the transistor lead lengths to a minimum, a hole is cut into the board as shown and the transistor mounted flush on the component side. Note that the artworks are actual size. (Plessey Semiconductors application notes)

Fig 18.162. Ideas for a pll/vco with output at 3.4GHz (G4DDK). In addition to the circuit of Fig 18.160, some simple vco designs were given in chapter 10, "Test equipment"

It is usually possible, using a reliable clock standard in the frequency counter, to measure with meaningful accuracy to 10Hz, so that the final frequency can be calculated to an accuracy of 180Hz! This, of course, assumes that both the source crystal and the counter reference have a negligible temperature coefficient; means of achieving this for the source crystal were discussed in chapter 8, "Common equipment".

It is still desirable to verify that the correct harmonic has been selected during the alignment process. This can be done using a well calibrated, high-Q absorption wavemeter with sufficient accuracy to differentiate between harmonics which are typically nearly 100MHz apart.

Direct frequency counting can be achieved using one of the high stability dro modules mentioned earlier in the chapter. Such an oscillator might possess a stability better then ±0.5MHz across the temperature range of –20 to +60°C. This infers a frequency stability of 12.5kHz/°C and, if the oscillator is used under thermally stable conditions, then this stability is sufficient to allow frequency counting to an accuracy good enough for most amateur purposes, assuming that the oscillator can be set accurately to a known frequency and that load pulling and temperature drift of the dro is sufficiently small – this should be verified. Care will have to be taken with the design of the power supply, for this also determines the stability of the oscillator. Frequency pushing may be of the order of 1.5MHz per volt, so to maintain stability of the same order as the thermal stability would require a supply stability of better than ±0.5mV. This, also, is not too difficult to achieve.

The recommended configuration for a high stability down converter suitable for use as a counter "prescaler" is shown in Fig 18.163. It consists of a high stability dro, fed via an isolator into a 10 to 13dB cross-coupler and thence into a terminating load. The coupled output is fed into a conventional single-ended mixer together with a sample of the signal to be counted. The difference product (around 400 to 600MHz) is fed to the input of the counter. Without rearranging the counter timebase, some mental agility will be needed to interpret the display, although this should not be too much of a problem! Some counters may need amplification of the product signal; this could be provided by any broadband amplifier module, for example the mmic amplifiers described in chapter 10, "Test equipment". The use of prescalers to extend the range of counters to about 3GHz is also described in that chapter.

Wideband modular amplifiers

The use of these devices is discussed in chapters 8 and 10. Although the usable gain range of the commoner and less expensive types (for instance the Avantek "Modamps") does not at present extend much beyond 3GHz they are, nevertheless, potentially useful devices when incorporated within 10GHz equipment. For instance, their use in a local oscillator chain for 3.4GHz has already been mentioned (Fig 18.162). It would be possible to take the output of that

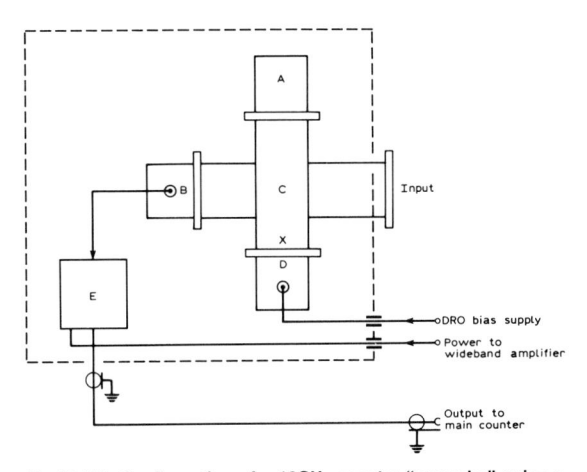

Fig 18.163. Configuration of a 10GHz counter "prescaler" using a high stability dro. A= waveguide matched load; B= mixer cavity; C= 10dB (nominal) cross coupler; D= high stability dro; E= wideband preamplifier; X= position for isolator, if fitted. Note that all leads should be screened and decoupled where appropriate. The output is coaxial. The dotted enclosure provides both screening and thermal stability. (G3PFR)

oscillator chain and multiply the 125mW available to give, theoretically, about 14mW output at 10.368GHz (3.456GHz ×3). Even allowing for circuit losses and some isolation between cascaded multipliers, there would still be plenty of power in hand, after filtering, to drive a mixer for a receiver or to act as a low power transmitter. Because of the high drive frequency, the filtering requirements are much less stringent than with the close-spaced harmonics which are generated from a lower frequency source. Sufficient selectivity would almost certainly result from the use of a single-pole dr filter such as that outlined in the next section.

Dielectric resonators

The dielectric resonators described as the frequency determining elements in dros are available in a number of forms. For instance, the Murata "Resomics" range presents them in disc (puck) and hollow cylinder format as well as rod, bar and sheet for special purposes. The frequency range at present covered is from about 1.5GHz to 25GHz. At 10GHz their dimensions are small enough to enable them to be very effectively used as part of a pcb design. Since they are high-Q devices, it is not too difficult to realise quite narrow, tunable filters based on a dr (or possibly several drs) coupled into microstip lines simply by being sited between the lines as in Fig 18.164. Whilst experimenting with such filters, it is sufficient to affix the dr to the pcb by means of a small piece of "BluTack" until the optimum position is found. It may then be fixed more permanently by means of "Superglue" or epoxy adhesive. If the output line is terminated in a detector, then this configuration could even be used as a wavemeter if the tuning screw is calibrated!

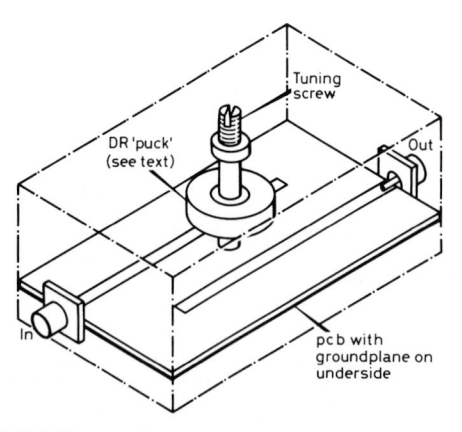

Fig 18.164. Diagram of a dr filter suitable for inclusion in a 10GHz transmitter or receiver strip. The microstrip pcb is mounted within a small, rigid, brass, copper or tin-plate box with the groundplane underneath. The dr may be attached with adhesive, or may be pillar mounted for higher Q (see text). A tuning screw and locknut (or micrometer screw) is mounted above the dr to tune it. Input and output sockets are SMA. If a detector is attached to the output socket and the tuning screw penetration is calibrated, the filter could also be used as a wavemeter

A comprehensive review of dielectric resonators may be found in [37]. By combining printed circuit board techniques, GaAs fets and dielectric resonators, it is possible to visualise a complete, very compact 10GHz receiver and transmitter. An "outline" layout, due to LA6LCA [36], is shown in Fig 18.165. Other, similar approaches to the problems of compact, pcb based narrowband equipment may be found in [34] and [35]. These designs use relatively inexpensive GaAs fets, active mixers and copper pipe cavities soldered to the pcb and coupled by means of probes, rather than the dielectric resonators of LA6LCA's design. [38] describes to construction of the copper filter cavities. Either of the transverter designs may be regarded as typical of current European practice.

These various techniques are put forward in the hope that it will encourage the developmetnt and the use of the newer technologies. It is confidently predicted that there will be a revolution in 10GHz equipment design in the very near future: indeed, many of the devices briefly discussed in this section are bound to have a profound effect on the lower microwave amateur bands also.

18.10 PRACTICAL 10GHZ EQUIPMENT USING THE NEW TECHNOLOGY

Equipment using GaAs fets in microstrip circuits offers higher power levels, better receiver performance and is arguably easier to build than the traditional amateur waveguide-based equipment. Construction techniques follow lower frequency practice rather than "plumbing", although care and attention to detail are still required if good results are to be obtained.

A number of GaAs fet-based designs for 10GHz have appeared in amateur literature, for example [34], [35]. These have tended to use relatively expensive GaAs fets and the cost of construction has been too high for many amateurs. Recently a large quantity of GaAs fets became available on the surplus market in the UK at a fraction of the cost of "new" devices. The GaAs fets in question were manufactured by the Plessey 3–5 Group for use in 11GHz satellite tv lnbs and have excellent performance. The low cost of these led the to development of a number of new UK designs for 10GHz.

18.10.1 The G3WDG–001 2.5 to 10GHz multiplier/amplifier

The G3WDG–001 [40] is a ×4 multiplier/amplifier chain which can provide 50 to 100mW output anywhere in the 10 to 10.5GHz band. It can be used as the basis of a simple cw/fm narrowband transmitter, a beacon or personal signal source, as an atv transmitter or as a packet radio link transmitter.

This design is intended for home construction without the need for either difficult construction methods or elaborate test equipment. It has been duplicated with relatively little difficulty by a number of independent constructors. Wherever possible, low cost components have been identified and designed-in, but in some cases it has been necessary to use more expensive components.

The specified parts *must* be used throughout or the hard work put in by the designers to make the designs reproducible will have been wasted! All the specialised components with the exception of the GaAs fets are available, as a "mini-kit", from the Microwave Committee Components Service. Some problems were encountered during the design-proving phase when some constructors had not used the correct grade of GaAs fets in some locations. The

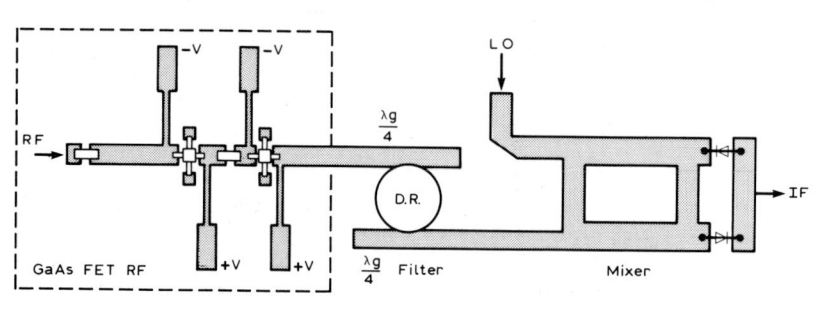

Fig 18.165. Schematic pcb 10GHz receiver front-end using GaAs fets (LA6LCA)

THE 10GHZ ("3CM") BAND

Table 18.10. Components list for the G3WDG–001 2.5 to 10GHz multiplier/amplifier

Capacitors
C1,2,3,4,5,6,7,8,9 — 220pF smd, 0805 size
C10,11,12 — 10µF tantalum bead or miniature electrolytic, 10V
C13, 14 — 2.2pF ATC chip capacitor, series 100 or 130

Resistors
R1 — 39R 1/4W carbon film
R2,3,4,5,6 — 47R smd, 0805 size
R7 — 220R (may need to be adjusted on test) smd, size 0805. Alternatively, use a wire-ended 1/4W carbon or metal film resistor, stood on end
RV1,2,3 — 2k2 miniature horizontal pre-set with legs bent out at 90 degrees and cut to fit board. These are fitted after all other parts except the GaAs fets. This is necessary due to other parts being located below them. Connect to the points marked "X" on the layout diagram. No tracking is provided between the bias connection points. This should be done with fine, insulated wire

Inductors
L1 — 8 turns of 0.315mm diameter enamelled copper wire (e.c.w.) close wound and self-supporting.
L2 — 16mm length of 0.315mm diameter e.c.w. formed into a hairpin shape and laid flat on the board. 1mm at each end used for soldered termination.
L3 — As L2, but 19.5mm long.
L4,5 — Straight length of 0.315mm e.c.w. between stub and track as shown in Fig. 18.167
L6,7,8,9 — Straight length of 0.2mm silver or tinned copper wire between radial stub and track as shown in Fig. 18.170. A single strand of braid from a scrap length of RG214 cable is suitable. Mount flat to pcb.

Semiconductors
F1,2,3 — P35–1108 GaAs fet (Birkett *black* spot)

Miscellaneous
FL1 — Brass cavity resonator, details in Fig. 18.170. Optionally silver plated. Tuning by means of an M4 screw with lock-nut. Two probes each 4.7mm overall length as shown.
PCB pins — RS Components 433–864. 1mm dia., 1.5mm head dia. (approx)
2 off SMA sockets
Tin-plate box type 7754 (37 x 111 x 30mm) from Piper Communications
Solder feedthrough capacitors 1nF to 10nF or Filtercons
Positive and negative regulated supplies eg. +ve from 7808 ic, -ve from ICL7660 voltage converter, see Figs 18.168 and 18.169.

---

different surplus GaAs fets available are *not* interchangeable!

It is a good idea, for those not yet skilled in the art of microwave pcb construction, to build a simple design like this before attempting something more complex, such as a receiver or transverter. Many of the construction techniques are described in detail as they are probably unfamiliar to most amateur constructors.

The design requires drive input of 10mW or more in the 2.5 to 2.6GHz region for full output at 10GHz. The exact frequency depends on the particular application. A suitable oscillator/multiplier, the G4DDK–004 design, has already been described in chapter 8, "Common equipment". This gives the required level of output in this range, and pcbs for the design are available through the Microwave Committee Components Service (G4DDK–004 pcb). If the output level of the DDK–004 board is measured at 10mW or greater, then it will be possible to omit the Avantek MSA0504 mmic amplifier stage in the design.

The circuit is shown in Fig 18.166, the layout of the board and components in Fig 18.167 and their values in Table 18.10.

General circuit features and components

The design is built on ptfe-glass board and, in the main, surface-mount "chip" (smd) devices are used, although some more familiar "ordinary" components are also used. The "heart" of the unit is the use of GaAs fets as active multipliers and amplifiers. Microstrip circuitry is used to provide the correct operating impedances for the GaAs fets and the circuit has been designed to cover the whole of the 10GHz band from 10.0 to 10.5GHz. A reliable method for

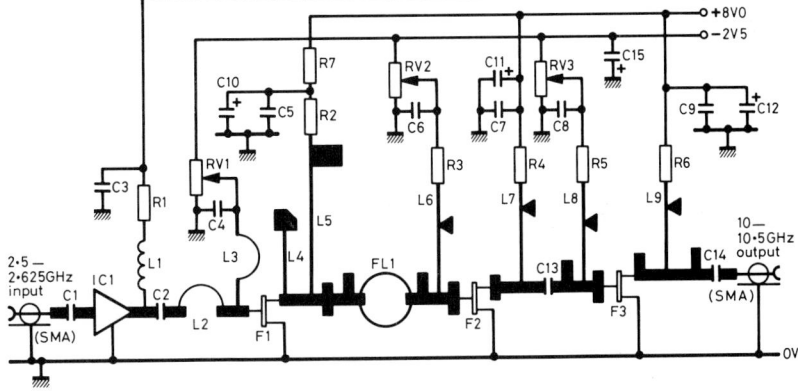

Fig 18.166. Circuit of the G3WDG–001 2.5 to 10GHz multiplier/amplifier Component values are given in Table 18.10

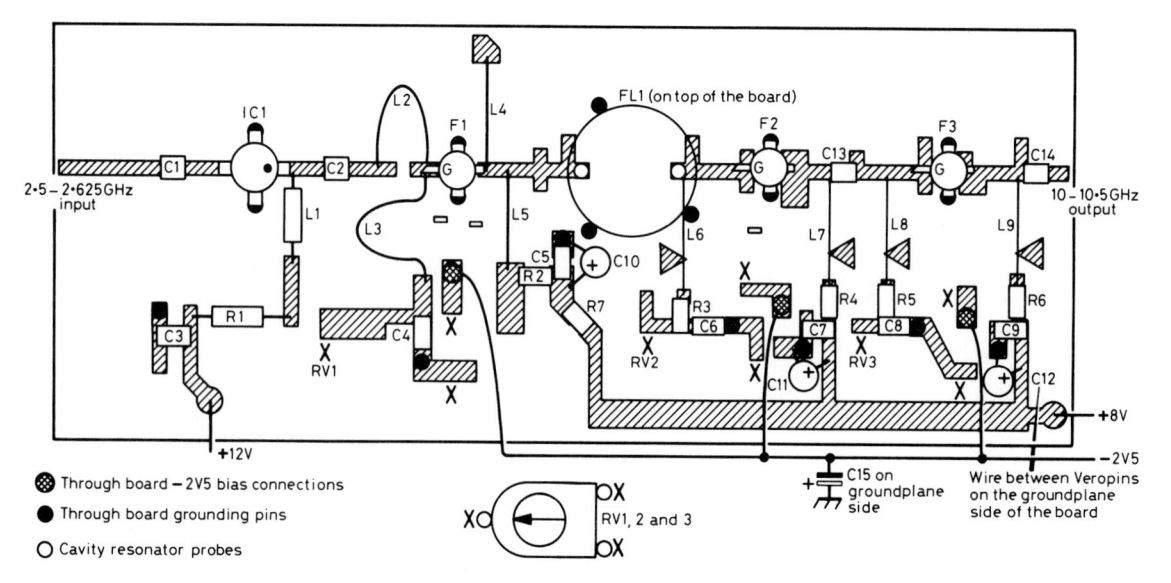

Fig 18.167. Layout of the G3WDG-001 circuit

grounding the source leads of the GaAs fets was developed to ensure that the designs would be reproducible. Earlier attempts using copper foil "wrap-arounds" failed because the inductance of such connections was too variable.

Where high selectivity is required, to discriminate between harmonics, this is provided by the use of a small "pill-box" tuned cavity resonator soldered to the board. Coupling from the microstrip lines into and out of the resonator is accomplished by the use of probes. This technique has been common in German amateur microwave designs for some time, for example [35], and avoids the use of critically dimensioned and spaced printed microstrip filters which are almost impossible to make with enough accuracy. It will be noted that a high drive oscillator frequency has been chosen (around 2.5GHz), also in order to minimise the stringency of filtering. The drive source chosen gives output at the required level, with all unwanted products at least −40dBc or better: this minimises the filtering requirement at the final signal frequency and makes the use of a single, simple cavity resonator possible wherever such selectivity is needed.

Each GaAs fet amplifier stage provides a gain of about 10dB. Matched input and output circuits are realised by the use of microstrip lines. Rather than attempt to etch very narrow (high impedance) microstrip lines, where these are necessary, easier construction results from the use of short lengths of thin wire soldered to the lines and pads on the surface of the board.

It is recommended that the pcb available from the RSGB Microwave Committee's Components Service, as part of a "mini-kit", be used for this design. Both the board material and the dimensions of the microstrip lines are critical to the success of this type of circuit. The other critical components, such as the ceramic chip capacitors, resistors and the resonators, are also included in the kit. Virtually all the other components are available from easily accessible amateur sources.

Only the recommended components should be used and only first-grade *known* components employed – substitution from the "junk-box" or components "salvaged" from other microwave equipment is likely to cause problems!

It is strongly recommended that the pcb is installed in a tin-plate box or an alternative, specially made sheet-metal (brass or copper) enclosure of similar size and form. By so-doing, not only is the somewhat flexible board housed rigidly, but is also well screened and thermally insulated to some degree. The finished, boxed unit(s) should be housed in a rigid outer case to provide further mechanical and thermal stability – the "boxes within boxes" approach which has long been advocated for high performance microwave equipment, almost regardless of frequency.

The use of other than SMA connectors for input and output is not recommended. The 12V power supply (or any other ingoing supplies) should be well decoupled by 1nF to 10nF solder-in feedthrough capacitors or filtercons. The power supplies must be stabilised to the voltages given in the circuit diagrams: if these voltages are exceeded, or the gate bias voltage fails, the GaAs fets can be damaged, if not instantaneously destroyed. By incorporating resistors in the drain circuits a degree of current-limiting protection is afforded. Nevertheless, it is well worth spending time on this aspect of the circuits, using only generously rated and reliable components in the bias circuits.

Care and attention to detail is essential *and* your soldering techniques must be good! Components should be mounted in the order given and the GaAs fet devices should always be the last components to be soldered into place, taking the usual precaution of grounding together the constructor, the body of the soldering iron and the case/groundplane of the pcb whilst soldering them in place. In this way the risk of damage by static discharge is minimised or eliminated.

Drive at 2,556MHz (for output at 10,224MHz) or 2,592MHz (for output at 10,368MHz) is required at a level of 10mW or greater. This is supplied by a suitably crystalled G4DDK–004 oscillator multiplier strip, the crystal being a fifth overtone HC18/U type in the range 106 to 108MHz. To accommodate the varying output levels of the drive source (variations being caused by individual component tolerances and the accuracy of construction and alignment), provision is made for the inclusion of a broadband mmic amplifier, using an Avantek MSA0504 device. This is capable of providing more than sufficient gain to drive the GaAs fet multiplier, F1 (Fig 18.166), adequately at the chosen input frequency. If the output of the G4DDK–004 driver is measured at 10mW or more, this mmic amplifier stage can be omitted and the gap in the microstrip line occupied by the mmic "patched" with a small piece of copper foil soldered across the gap. L1, R1 and C3 may be omitted, but C1 and C2 must be fitted. If the mmic is used, it will be necessary to make a hole large enough to take the body of the mmic so that it can be mounted on the board with minimum lead length – in a manner similar to the BFR91/6 transistor package used in the driver board G4DDK–004.

The output from the mmic is matched to the input impedance of the multiplier fet by a "lumped element" network consisting of L2/L3. L3 is also used to feed the negative gate bias to the fet from RV1 which sets the optimum operating bias for the multiplier. The output circuitry at the drain of the fet consists of a series resonant circuit at 2.5GHz followed by some microstrip matching elements at 10GHz. A series resonant circuit is formed by L4 and a printed capacitor (identified by the cut corner). The function of this circuit is to "short circuit" the input frequency to ground, which improves multiplier efficiency considerably. The drain bias circuit consists of a quarter-wave choke, L5, connected to a quarter-wave stub. The tip of the stub is at very low impedance at its resonant frequency (10GHz). This is transformed by L5 to a very high impedance. In this way there is virtually no disturbance to the signals on the microstrip line from the bias network. This type of bias network is fine at its operating frequency but needs additional decoupling at lower frequencies. This is accomplished primarily by R2/C5 which load the drain of the fet resistively at low frequencies, giving broadband stability. C10 is used for further decoupling at very low frequencies. This bias configuration is used throughout the 10GHz designs, except that in many locations the quarter-wave stub is replaced by a triangular element. This has the same properties as a stub, i.e. the tip of the element is at very low impedance but is very much smaller, resulting in a more compact layout. R7 is used to set the drain voltage to an optimum value for best multiplier performance.

The output from the multiplier consisting of several harmonics, but principally the wanted fourth harmonic of the

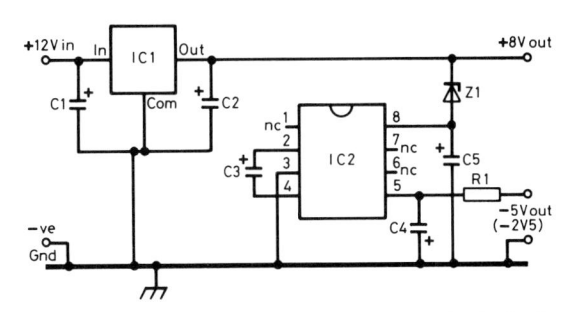

Fig 18.168. Positive and negative voltage regulator circuits for the G3WDG–001 module (G4FRE). Component values: IC1, uA7808; IC2, ICL7660PCA; Z1, 3V0 or 3V3, 400mW zener diode; R1 3k3 1/4W metal film; C1 1µF tantalum bead, 16V wkg; C2 0.1µF tantalum bead, 10V wkg; C3, C4, 22µF tantalum bead, 10V wkg; C5 10µF tantalum bead, 10V wkg; pcb, G4FRE–023

drive frequency, is fed into FL1, the resonator filter, by means of a probe. Filtered output is coupled to the gate of the first amplifier, F2, by means of a second probe. The resonator is tuned by means of an M4 screw, locked in place with a lock-nut after tuning to resonance. The resonator body and tuning screw can, with advantage, be silver plated. Not only does this increase the Q of the cavity significantly, but also makes the cavity much easier to solder to the board.

GaAs fets F2 and F3 form a cascaded two stage amplifier yielding about 20dB gain, each with gate bias and drain bias arrangements similar to those of the multiplier stage. DC blocking and rf coupling in the drain circuits is accomplished by means of 2.2pF chip capacitors which *must* be designed specifically for such frequencies, e.g. ATC series 100 or 130. The ordinary chip-C's used for lower frequency coupling or decoupling are definitely not suitable in these two positions.

The drain supply of the three GaAs fet stages must not exceed 8V measured at the supply rail, otherwise their ratings may be exceeded and they will become expensive, fast-acting fuses! Similarly a gate bias rail at –2.5V is needed to set their operating points. Two simple ic regulator circuits can be built onto a small piece of fibreglass board housed in the groundplane compartment of the box. A suggested circuit is shown in Fig 18.168 and layout for the in-built regulators is shown in Fig 18.169. Note that the pcb layout shown for the regulator board requires the components to be mounted on the track side, as if the components were surface mount devices.

Construction

It is strongly recommended that the following procedures are followed in detail and in the order described:

1. If the mmic amplifier is to be used, carefully drill a hole, 3.7mm diameter, in the middle of the gap in the microstrip lines identified from Fig 18.167 as the position of the mmic. It is suggested that a small pilot hole is drilled from the track side which is then carefully opened to the required size by drilling from the groundplane side. Care will

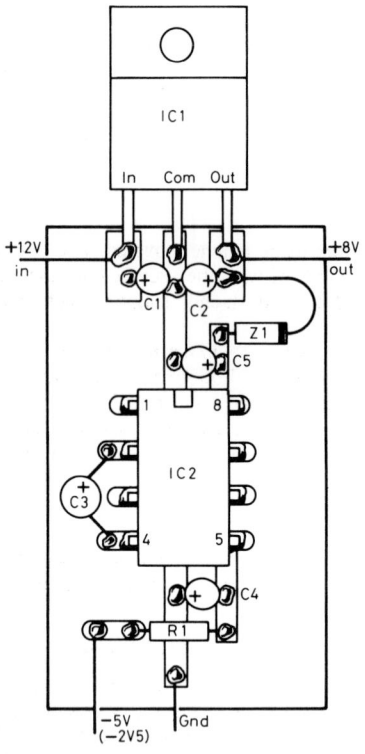

Fig 18.169. Layout for the G3WDG–001 regulator circuits (G4FRE). All components are mounted on the track side of the pcb. Board size approximately 32x13mm

be needed to avoid tearing the copper of the groundplane. If this amplifier stage is not to be used, then simply bridge the gap with a piece of copper foil, the same width as the lines, soldered across the gap.

2. Ensure that the copper of the tracks and the groundplane is clean and bright. Various means of achieving this are described later in connection with local cleaning just before critical soldering operations. Where light tinning is required, use a small soldering iron and very fine solder. If too much solder is applied accidentally, remove excess using a solder sucker – before trying to solder the small chip components in place!

3. Fit grounding pcb pins and filter locating pins (see later) and solder in place. Lightly tin around the edge of the groundplane.

4. Solder the filter into position (again, see later).

5. Locate the pcb into its box and trim to a neat fit if needed, particularly in the corners of the box where there are joints. The pcb material will cut quite easily with a sharp scalpel blade and straight-edge. Locate the groundplane 17mm from the top of the open box and mark its position. Locate and mark the SMA socket centre-pin clearance holes. Drill the holes and de-burr. Locate, drill and de-burr holes for any feedthrough components needed for power supplies. Tack-solder the corner seams of the box and make sure that the lids are a neat fit. Adjust as necessary. Check also that the board will fit neatly. When satisfied, solder the corner seams fully. Solder the SMA connectors and the feedthroughs in position.

6. Relocate the pcb so that the input and output tracks touch their respective socket spills, tack-solder the pcb in place and, when satisfied that it is correctly located, solder all round the groundplane and solder the SMA socket spills to their respective tracks. Note that the +8V rail is very close to the edge of the board and care must be taken to avoid an accidental short circuit to the wall of the box whilst soldering the board in place. This completes the mechanical construction of the module.

7. Fit inductors L1 – L9, as specified in the parts list (Table 18.9), into position, ensuring inductors L2 – L9 lie flat to the board.

8. Fit all chip components using the mounting techniques described later. You will need a pair of fine-pointed tweezers to handle these small devices and, maybe, the assistance of a magnifier!

9. Fit all components which have leads, ensuring that static-sensitive devices (ics, fets) are put on the board last of all to minimise the risk of damage to the devices.

Note: it is best to apply the supply voltage to the board *before* fitting IC1 and the fets, to check that both the +8V and –2.5V voltages are present and correct on the respective tracks/pins. On completion of this test, disconnect power and solder in the devices only if everything checks out correctly.

Individual "build" techniques

The following sections cover specific techniques required in the construction of the board:

PCB-pin grounds

Clean the groundplane on the rear face of the pcb using an RS Components pcb eraser (part number 555–303), or similar mildly abrasive pad – a really clean surface enhances solder flow. Tin around the hole on the groundplane side. RS Components pcb pins, part number 433–854, 1mm diameter with a head diameter of 1.5mm, are used wherever there is a need for through-board grounding. Other makes of pin, e.g. Veropins, could be used provided that their size is the same.

Place the pcb-pin in the hole with the pin head on the track side and the body of the pin sticking up through the groundplane, with the exception of the three filter-locating pins (see below). Place the head of the pin on something hard and flat and press the board until the head butts up against the track side of the board.

Solder by starting with the iron at the top of the pin; tin the pin generously and, while applying more solder to the joint, flow the solder down the pin and onto the

THE 10GHZ ("3CM") BAND   18.105

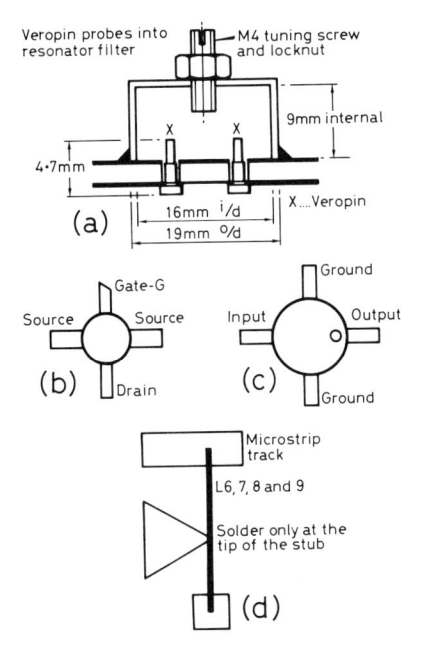

Fig 18.170. (a) Resonator filter and probe pin details. Resonator material – Brass rod, 19mm diameter, silver plated if possible, but not absolutely necessary. (b) FET package outline. (c) MMIC package outline. (d) Wiring of decoupling stubs

groundplane to ensure good pin to groundplane contact. Trim the pin back using flush-cut cutters.

Fitting the filter

This is potentially the most difficult soldering operation on the board! Note that cleanliness is a *must* for this operation and the rear face of the board ie. the groundplane, should be "shining clean". The surface should be cleaned carefully using the RS Components pcb eraser or other similar mildly abrasive pad and, if possible, degreased with an aerosol cleaner, e.g. RS Components part number 567–660 or 554–838. Details of the cavity are given in Fig 18.170(a), together with the dimensions of the coupling probe pins. First prepare and clean the pcb, then fit the three pcb-pins which mark the filter cavity position. These pins are fitted from the groundplane side through to the track side. Solder the pins to the pads provided on the track side and cut off excess pin length after soldering. The board is now ready to take the cavity filter.

Pre-heat the cavity with its tuning screw and locknut assembly in position. Heat it by placing on a hot plate (e.g. a 3 to 6mm thick sheet of aluminium placed over a gas ring) and heat until 60/40 tin/lead solder melts easily on touching it to the cavity wall near the base (open end). Quickly transfer the hot cavity, using pliers to grasp the tuning screw, to the board, position it between the three guide-pin heads on the groundplane side of the board and apply fine (22 swg or finer) solder at the junction of the board and filter to fix the cavity in place. Ensure the cavity does not jump outside the guide pins whilst soldering. Ensure a continuous small fillet of solder all round the cavity, but do not apply too much solder. Allow the cavity to cool without disturbing it. When it has cooled fully, fit the two pcb-pins which probe through the board and into the cavity, having pre-cut them to the length shown in Fig 18.170(a).

Inductors

For inductors using enamel covered wire (e.c.w.), cut the required length of wire then scrape/chip the enamel from the last 1–2mm of each end using a scalpel blade. Tin each end.

For inductors using one strand of a standard multi-strand wire, tin one end and fit to the board as shown in Fig 18.170(d): solder first at position 1, then at position 2 as close as possible to the apex of the triangle, then at position 3. If any excess wire remains at 1 or 3, trim off carefully with a scalpel blade.

Chip components

To fit chip components across two circuit tracks or pads, adopt the following procedure for best results: (see Fig 18.171)

a. Lightly tin one of the tracks or pads.

b. Fit component and reflow solder to make a solder fillet at the tinned side – the tip of the tweezers may be used to hold the chip in position whilst the solder solidifies. Use as little solder as possible to form a very small fillet.

c. The component should now be secure: tin the other track and make a solder fillet on the second side of the chip component to complete the mounting.

This procedure ensures that the components are flat to the board, good contact is made and the best circuit performance is achieved.

Static-sensitive components

Components such as ics and fets should always be fitted last to minimise the risk of static damage. The GaAs fets have the gate lead bevelled for identification, as shown in Fig 18.170(b). Grounding of the two fet source leads is via pcb-pin ground "pads" fitted as shown in Fig 18.172. Cut the source leads to minimum length, but note that before handling static-sensitive devices, you should make sure that you and the handling implements (e.g. tweezers) are grounded together: it is often a good idea to work on a

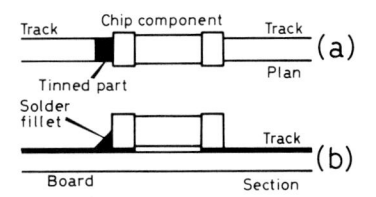

Fig 18.171. Fitting surface mount chip components

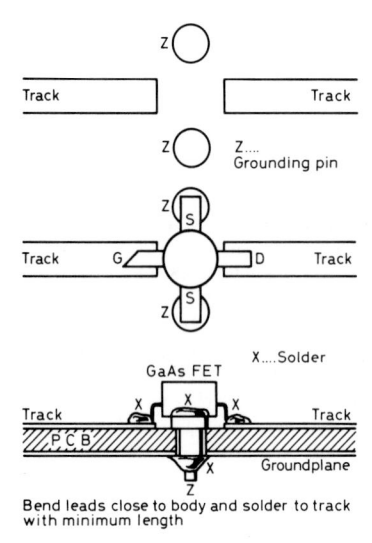

Fig 18.172. Fitting the GaAs fets

grounded sheet of aluminium foil spread on the work surface, resting the wrists on the foil, with the board and implements also on the foil when not in use. You may find, if using surplus GaAs fets, that one source lead is already trimmed short. Cut the other to a similar length.

Lightly tin the source grounding-pins and the ends of the lines – the same remarks about cleanliness and minimum amounts of solder apply here also! Place the trimmed fet as shown in Fig 18.172 and reflow the solder on one of the source leads, then the other. Push the gate and drain leads down flat onto the board as close to the fet as possible, cut off excess lead length carefully with a sharp scalpel blade, then solder them down to the respective tracks, making sure that the device is orientated the right way round!

Alignment with simple test-gear

Once completed, the pcb should be carefully examined for poor joints, accidental solder bridges and other forms of short circuit. Once satisfied that all is well, the alignment procedure may begin. You should already have checked before mounting the ic (if used) and fets that the correct supply voltages will appear on the positive and negative supply rails when a 12V supply is connected to the input feedthrough capacitor.

1. Turn the bias potentiometers RV1, RV2 and RV3 fully clockwise so that full (cutoff) bias will be applied to the gates of the three GaAs fets when power is applied.

2. Insert a multimeter in series with the +8V supply between the regulator output and the +8V rail and set it to, say, 500mA full scale deflection.

3. Attach a suitable load/power indicator to the output socket. This might consist of an SMA to waveguide 16 transition, variable attenuator, wavemeter and detector connected as shown in Fig 18.173. Here's where some of your older waveguide components come into their own!

4. With no oscillator drive applied to the input socket, apply +12V to the power input feedthrough. Current consumption should be no more than a few microamps of "leakage" current. Switch the range of the multimeter as needed. If considerably more current is measured, look for short circuits or mis-connected or damaged components.

5. If all is well, adjust RV1 so that the current consumption rises to about 2 or 3mA.

6. Adjust RV2 for a further rise of 20mA in total current.

7. Repeat for RV3, again looking for a rise in current of a further 20mA. At this stage the current consumption should be of the order of 42 to 45mA.

8. Set the variable attenuator in the load/power indicator to minimum and detune the wavemeter well away from the expected frequency.

9. Apply drive to the input socket at a level of 10mW or more. Some output may be seen on the power detector meter and the current consumption may rise slightly.

10. Starting with the tuning screw just entering the cavity, slowly adjust the filter tuning screw inwards until the output begins to rise. Continue adjusting for rising power output whilst increasing the attenuation to prevent excess current in the detector diode. When adjusted for maximum output, check that the correct harmonic has been selected by using the wavemeter. The tuning range of the cavity is such that at least two, and probably three, harmonics of the 2.5GHz drive frequency can be peaked by the tuning screw – hence the need to check that the right harmonic has been chosen. Continue adjusting the tuning screw until you are certain that the correct harmonic has been selected and peaked. Lock the screw in position with the locknut.

11. Adjust the gate bias settings slightly to optimise output, starting with the multiplier fet, F1, and finishing with the output fet, F3. Readjust and lock the cavity tuning screw as necessary.

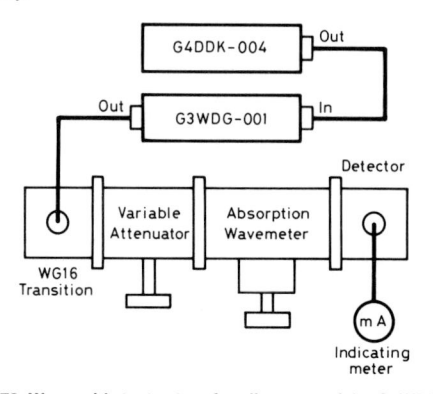

Fig 18.173. Waveguide test set-up for alignment of the G3WDG–001 unit

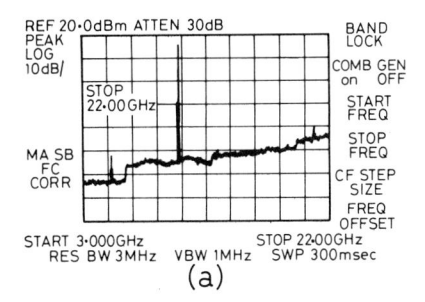

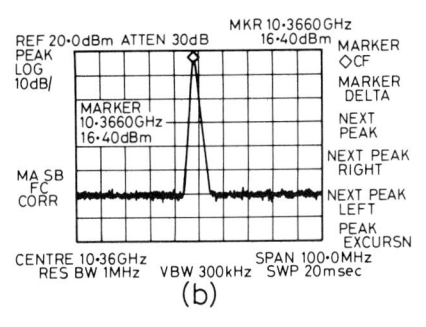

Fig 18.174. Spectrum analyser traces of the output from the G3WDG–001 multiplier/amplifier. (a) The output between 3GHz and 22GHz, showing the remarkably low level of sub-harmonics and harmonics. The rising noise level is due to the spectrum analyser, not the multiplier! The drive source was the G4DDK–004 board at a level of +10dBm. (b) Close-in view of the output spectrum showing the clean output obtainable with this design

12. There is some spread of characteristics of the completed units, so the bias current settings given in 5, 6 and 7 above should be regarded as starting figures only. However, no stage should be allowed to pass more than about 50mA and the output stage should always work into a well matched load. Some current limiting protection is afforded by the 47R resistors in the drain feeds.

13. This completes the alignment: the power output should lie between a minimum of 50mW and a maximum of 100mW. Spectrum analyser traces of the output are shown in Fig 18.174. It is not advisable to try to squeeze more power output than this, especially if the load is not too well matched although, if you can guarantee a good match, more output can be obtained by changing R6 for a 10R resistor. However, this is rather "caning" the final amplifier and excess dissipation in any stage may lead to rapid destruction of the device concerned. You have been warned!

14. Some constructors have been able to obtain more power by tuning the output circuit of the final amplifier. This was done by sliding a small piece of copper foil (about 3mm × 3mm) up and down the microstrip line and soldering in position when the best point had been found. Don't forget to power-down before soldering the copper foil in place! In one case this increased the power output from 75 to 125mW.

Modulation and stability

The frequency stability of the drive source oscillator is of paramount importance since the crystal frequency is being multiplied by a total factor of 96. The drive source design G4DDK–004 was evolved with this purpose in mind and, with a good quality crystal (10ppm or better), housed as suggested, it will be found to give good stability at the 10GHz output frequency. However, stability can be improved still further by fitting a crystal heater, such as the Murata "Posistor" type PTH507B01BM500N016 which simply clips onto the crystal casing, is supplied with 13.5V nominal and, after a short warm-up, maintains a steady crystal temperature regardless of ambient conditions. The initial in-rush (switch-on) current for a "cold" heater is of the order of 0.5A but requires only 25mA maintenance current once at operating temperature.

With regard to modulation, the simplest form of modulation is fm or fsk achieved most simply using the varicap modulator described in chapter 8, "Common equipment".

If either crystal heater or modulator are to be fitted to the drive source, make sure that their supplies (dc or audio) are well decoupled, where they enter the box, by means of additional solder-in feedthrough capacitors or filtercons. Additionally, make sure that the modulator components are fitted directly across the oscillator inductor with minimum lead length and maximum rigidity! Unwanted stray coupling, either rf or hum, must be avoided if maximum stability and minimum sideband noise on the 10GHz carrier are to be achieved.

18.10.2 The G3WDG–002 10GHz to 144MHz receive converter

Like the G3WDG–001 multiplier/amplifier described the preceding section, this module [41] also requires a drive input of approximately 10mW in the 2.5 to 2.6GHz region.

This design, a down-converter (receiver), G3WDG–002, to 144 – 146MHz incorporates a ×4 multiplier chain to generate the local oscillator signal, a dual-diode mixer and two stages of low-noise pre-amplification before the mixer. The design also incorporates a low-noise post-mixer amplifier at the intermediate frequency. The front-end noise figure of several prototypes has been measured at less than 3dB. It is possible to improve this figure by using an external pre-amplifier.

This design is also available as a "mini-kit", pcbs and all critical components, except for the GaAs fets, from the Microwave Committee Components Service.

Circuit description and operation

The circuit is shown in Fig 18.175, the layout of the board and components in Fig 18.176 and their values in Table 18.11.

Referring to the circuit diagram, Fig 18.175, the 2,556MHz local oscillator input signal is fed to F1 which acts as a ×4 frequency multiplier producing a few milliwatts at 10,224MHz. The multiplier circuit is identical to that used in the G3WDG–001 module. The input signal is

Table 18.11. Components list for the G3WDG–002 10GHz to 144MHz receive converter

Capacitors
C1,2,3,6, 7,8,9	220pF smd, 0805 size
C13,14,16 17,18	1000pF smd, 0805 size
C4,5,11	2p2 ATC chip capacitor, 100 or 130 series
C12	22 to 47μF tantalum bead, 10VW
C10,19	2μ2 to 10μF tantalum bead, 10VW
C15	30pF trimmer, 5mm diameter. eg. Murata TZ03Z300 (green).

Resistors
R1,3,4,5,6	47R smd, 0805 size
R2	220R smd, 1206 size, (or 1/4W leaded)
R7	18k smd, 0805 size
R8	4k7 smd, 0805 size
R9	270R smd, 0805 size
R10	100R smd, 0805 size
R11	560R smd, 0805 size
RV1,2,3	2k2 horizontal preset, eg. Allen-Bradley 90H, Bournes VA05H or Philips OCP10H etc.

Inductors
L1	16mm length of 0.315mm diameter e.c.w., formed into a hairpin. 1mm each end to be tinned and soldered to tracks as shown in layout
L2	As L1, but 19.5mm long
L3	Straight length of 0.315mm diameter e.c.w., tinned 1mm each end, as above. The wire should be soldered to the edge of the stub and as close to the drain connection of F1 as possible
L4	Straight length of 0.315mm diameter e.c.w., tinned 1mm each end, as above. Solder between the stub edge and the probe connection to FL1
L5,6,7,8	Straight length of 0.2mm dia. (not too critical) tinned or silver plated copper wire. Solder between track, stub point and terminal pad as shown in layout
L9	10mm length of 0.2mm diameter wire, as L5. Bend to fit between earth pin and mixer connection
L10 + L11	Single 20mm length of 0.2mm tinned or silver plated copper wire, as L5. Solder between mixer centre, stub and C13, as shown in layout
L12	4 turns of 0.6mm diameter tinned or silver plated copper wire. Wound to 5mm inside diameter, turns spaced 1/2 wire diameter. Centre-tapped. Mount 1mm above the board

Semiconductors
F1, 3	P35–1108 GaAs fet (Birkett *black* spot)
F2	P35–1145 GaAs fet (Birkett *red* spot) or P35–1108 GaAs fet (Birkett *black* spot) – see text
D1	Alpha series dual diode (DMF 3909–99)
T1	BFR90/91

Miscellaneous
FL1, 2	Cavity resonators as for G3WDG–001. See text and diagrams for details
PCB pins	approx 33 off RS Components 433–864. 1mm dia., 1.5mm head dia. (approx)

SMA sockets
Tin-plate box 3 off type 7754 (37 x 111 x 30mm) from Piper Communications
Feedthrough capacitors – Solder-in, 1nF to 10nF, or filtercons
+ve and –ve regulated supplies eg. +ve from 7805ic, –ve from ICL7660 voltage converter, see Figs 18.176 and 18.177

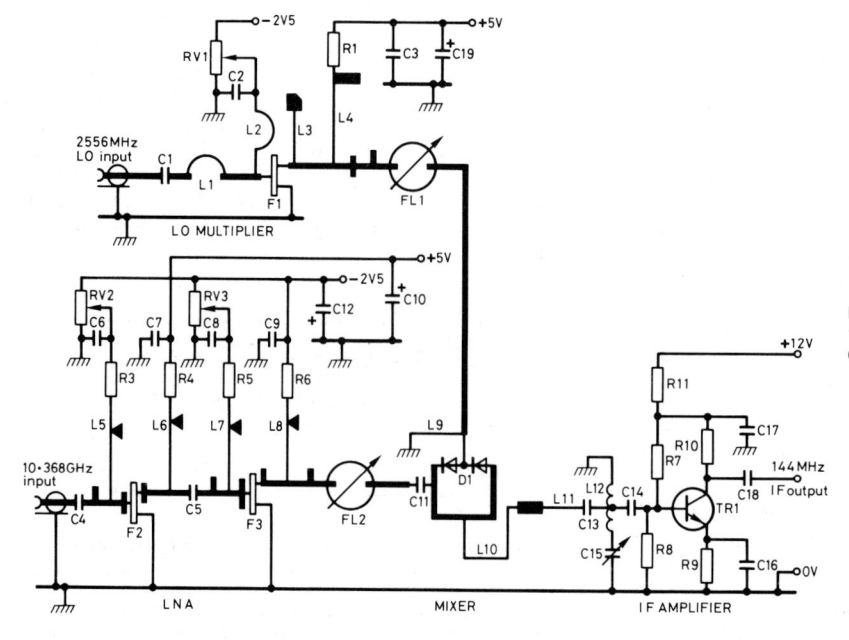

Fig 18.175. Circuit of the G3WDG–002 10GHz to 144MHz receive converter. Component values are given in Table 18.11

THE 10GHZ ("3CM") BAND   18.109

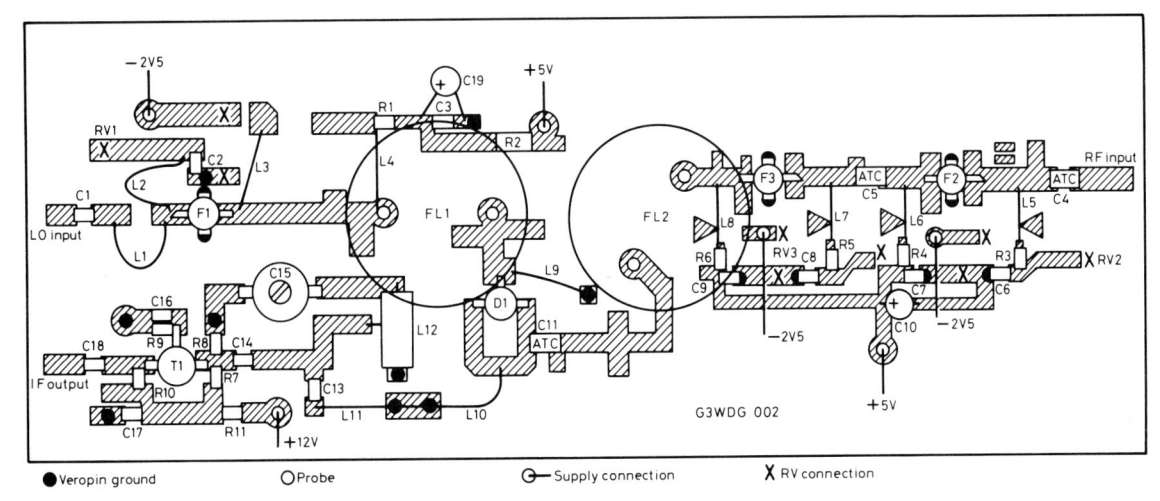

Fig 18.176. Layout of the G3WDG-002 circuit

fed to the gate of the fet via a lumped element matching network L1/L2. The cold end of L2 is decoupled via C2 and negative gate bias is applied via L2. The output of the fet is matched to 50Ω via microstrip elements, and drain bias is fed via a quarter-wave line, L4, which is decoupled at 10GHz by a low impedance quarter-wave stub. L3, and the chamfered element (a microstrip shunt capacitor) form a series-resonant circuit at the input frequency (2,556MHz) to improve the efficiency of the multiplier. Wideband stability is provided by decoupling elements R1, C3 and C4. A number of harmonics are present in the output from the fet. The wanted fourth harmonic is selected by the cavity filter FL1 and passed to the local oscillator port of the hybrid ring mixer via a microstrip matching network.

The mixer uses a series diode pair, D1, connected between the ends of a folded three-quarter wave line. This configuration gives good rejection of the local oscillator signal at the rf port and vice-versa. L9 is a shorted quarter-wave line to provide the required low impedance i.f return path, while having no effect at the local oscillator frequency. The conversion loss of the mixer including the matching networks is about 6–7dB.

A two stage low noise rf amplifier (lna) is provided to reduce the noise figure of the unit to a more acceptable level (less than 3dB). The amplifier is of conventional design and uses "Birkett surplus" (Plessey) fets. It is very similar to the power amplifier used in the G3WDG-001 design, except that a low noise fet can be used in the input stage as an option. F2 can be either a red or black-spot surplus GaAs fet according to the level of performance required. Overall noise figures of 2.6dB with a red-spot fet and 3.2dB with a black-spot fet are typical, but see the comments on stability later. The input circuit of F2 has an "optional" stub which, in some cases, when connected, can reduce the noise figure by a small amount. It is usually not required. The overall gain of the lna is about 19–20dB and its own inherent noise figure in the region of 1.9 to 2.5dB.

The output from the lna goes to the rf-port of the mixer via a microstrip matching network and a high-pass filter (C11) after passing through filter FL2 which provides about 20dB of image rejection (with a 144MHz i.f). Note that FL2 can easily be tuned to the wrong image (10,080MHz), so care should be taken when tuning up (see later). The i.f output from the mixer is fed to a low noise amplifier via a low-pass filter consisting of L10, a quarter-wave line, and L11, to prevent 10GHz energy from reaching the i.f amplifier. The i.f amplifier is a well-proven design from another application [39], constructed in surface mount form to save space.

The negative bias generator used to supply the gate bias for the fets uses the same pcb as that used in the G3WDG-001 module (G4FRE-023). Two modifications have been made for this application – the use of a 7805 regulator and the omission of the zener diode. The circuit is shown in Fig 18.177 and the board layout in Fig 18.178. Note that C12, the negative rail decoupling capacitor, is not shown as it is fitted on the reverse side of the board. It can been seen in Fig 18.179, which shows how the power supply board is fitted.

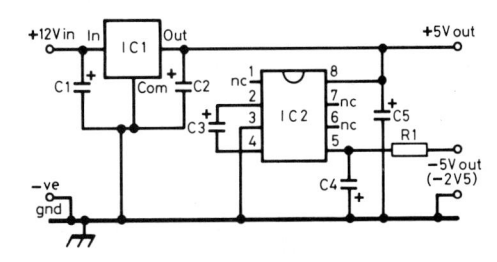

Fig 18.177. Positive and negative voltage regulator circuits for the G3WDG-002 module (G4FRE). Components values: IC1, uA7805; IC2, ICL7660PCA; R1, 680R 1/4W metal film; C1 1μF tantalum bead, 16V wkg; C2 0.1μF tantalum bead, 10V wkg; C3, C4, 22μF tantalum bead, 10V wkg; C5 10μF tantalum bead, 10V wkg; pcb G4FRE-023

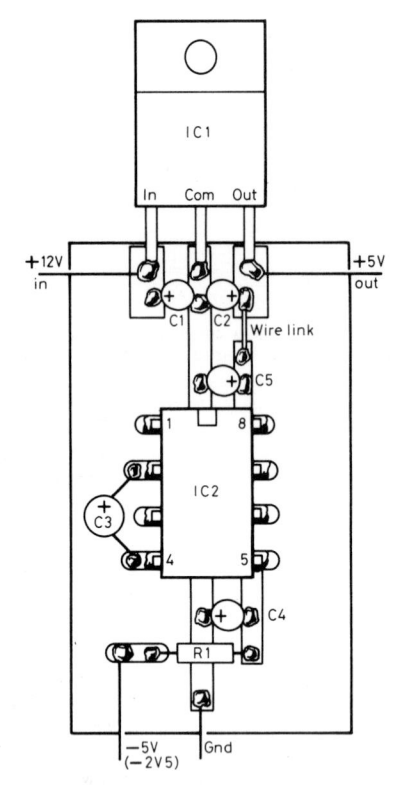

Fig 18.178. Layout for the G3WDG–002 regulator circuits (G4FRE). As in Fig 18.169, all components are mounted on the track side, using the same pcb (G4FRE–023)

Construction of the G3WDG–002 receive converter

This module uses the same construction methods detailed for the G3WDG–001 module and it is strongly recommended that the procedures described are used as follows:

1. Fit grounding pcb pins and filter locating pins and solder in place. Lightly tin around the edge of the groundplane.

2. Solder both filters into position. Leave the tuning screws and lock nuts in position to avoid unwanted debris accidentally falling into the cavities.

3. Locate the pcb into its box and trim to a neat fit if needed, particularly in the corners of the box where there are joints. The pcb material will cut quite easily with a sharp scalpel blade and straight-edge. Locate the groundplane 17mm from the top of the open box and mark its position. Locate and mark the SMA socket centre-pin clearance holes. Drill the holes and de-burr. Locate, drill and de-burr holes for any feedthrough components needed for power supplies. Tack-solder the corner seams of the box and make sure that the lids are a neat fit. Adjust as necessary. Check also that the board will fit neatly. When satisfied, solder the corner seams fully. Solder the SMA connectors and the feedthroughs in position.

4. Relocate the pcb so that the input and output tracks touch their respective socket spills, tack-solder the pcb in place and, when satisfied that it is correctly located, solder all round the groundplane and solder the SMA socket spills to their respective tracks. This completes the mechanical construction of the module.

5. Fit inductors L1 – L11, as specified in the parts list (Table 18.11), into position, ensuring that the wires lie flat to the board.

6. Fit all chip components using the mounting techniques described earlier.

7. Fit all components which have leads, ensuring that static-sensitive devices (ics, fets, etc) are put on the board last of all to minimise the risk of damage to the devices.

Note: it is best to apply the supply voltage to the board *before* fitting the fets, to check that both the +5V and –2.5V voltages are present and correct on the respective tracks/pins. On completion of this test, disconnect power and solder in the devices only if everything checks out correctly.

Individual "build" techniques are exactly as described for the G3WDG–001 module and need not be described further. In this design there are two "pill-box" filters to fit and note that the lengths of the filter probe-pins are significantly different for the local oscillator and signal filters: make sure you fit the right pins in the right places!

Alignment with simple test-gear

As before, once completed, the pcb should be carefully examined for poor joints, accidental solder bridges and other forms of short circuit. Once satisfied that all is well, the alignment procedure may begin. You should already have checked before mounting the fets and other semiconductors that the correct supply voltages will appear on the positive and negative supply rails when a 12V supply is connected to the input feedthrough capacitor.

1. Preset the FL1 and FL2 tuning screws as shown in Fig 18.180. Note that the 7.5mm dimension shown is the length of screw protruding from the locknut.

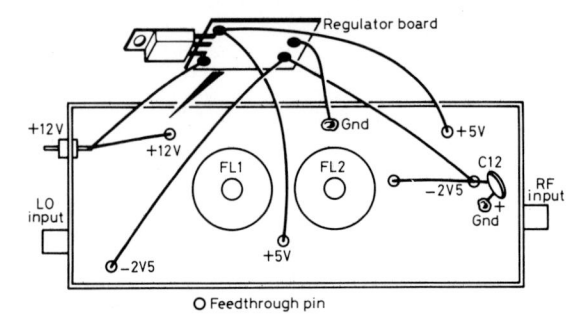

Fig 18.179. Positioning the regulator circuit and decoupling capacitor within the module enclosure (ground-plane side)

2. Turn the bias potentiometers RV1, RV2 and RV3 so that full negative bias will be applied to the gates of the three GaAs fets when power is applied.

3. Insert a multimeter in series with the +5V supply between the regulator output and the +5V rail and set initially to, say, 500mA full scale deflection.

4. Connect some form of matched load to the 10,368MHz input socket, such as a 10GHz-rated termination, attenuator or SMA-to-waveguide transition with a horn or similar well matched (low vswr) antenna connected.

5. With no oscillator drive applied, apply +12V to the power input feedthrough. The indicated current should be no more than a few microamps. Switch the range of the multimeter as necessary. If considerably more current is measured look for short-circuits or misconnected components.

6. Adjust RV1 to give an indicated current of about 1mA on the meter.

7. Apply the local oscillator signal (approx. 10mW at 2,556MHz). The indicated current should rise to approximately 7mA. The absolute value is not critical. If this order of value is not reached, the input drive may be too low, in which case retune the local oscillator drive source and/or change the lead length between the local oscillator source and the module.

8. Connect a 144MHz ssb receiver to the i.f output socket and tune C15 for a noise peak. This should be quite a significant peak if the i.f amplifier is working correctly.

9. Carefully adjust the FL1 tuning screw a turn or so either side of the preset position. A clear drop in the 144MHz noise level should be heard at the correct tuning point. Lock the tuning screw in this position.

10. Adjust RV2 to cause the indicated current to rise by approximately 12mA, and RV3 to cause a further increase of 15mA. The total current shown should now be in the order of 34mA. The lna is now powered up at approximately the optimium bias currents for the two fets.

11. Preliminary alignment is completed by tuning FL2 for a peak in the noise level. Careful tuning will show that two peaks can be heard. The correct one is with the tuning screw at the smaller penetration. Lock the screw in this position.

12. If a signal source is available (maybe your, or your friends', G3WDG–001 personal beacon?), check that this can be heard satisfactorily. It is worth reconfirming that FL2 is set to the correct image by tuning it for maximum signal. If no such signal source is available, try listening for other local signals on the band, perhaps harmonics from lower frequency equipment.

13. Remove the test meter and make good the connection.

14. Final alignment can be done using a noise figure meter or weak signal to optimise the performance. Adjust RV1,

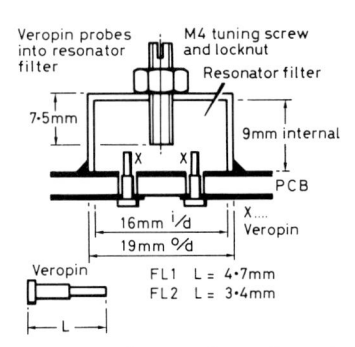

Fig 18.180. Details of the filter coupling probes and approximate filter tuning-screw positions for the G3WDG–002 module

RV2, RV3, FL1, FL2 and C15 for best results. If a noise figure meter has been used, recheck with a signal source or generator that FL2 is on the correct frequency, as the noise figure meter will not tell you if you are tuned to the wrong image!

After alignment, the converter should have a noise figure below 3.5dB. Prototypes have varied from 2.4 dB to 3.3dB. The overall gain should be in the region of 27 – 30dB.

Stability

During the initial tuning-up phase, the module should be operated with both top and bottom lids off. Under these conditions the module should be perfectly stable with either a red or black-spot fet in the front end. However, problems with stability were encountered with some of the prototypes when the lid was put onto the component/microstrip side of the box. Note that the module works perfectly well without a lid and can be operated like this with no problems, unless mounted close to another metal surface. In one of the prototypes, the impedance connected to the rf input socket had an effect too. Quality of construction, particularly how well the groundplane side of the board is soldered to the box, may also have an effect. Prototypes using black-spot fets seemed to be more stable then those using red-spot fets.

A common cure for lid-induced oscillations (often used in commercial lnbs) is to mount a piece of lossy material in the lid above the lna section of the unit. The choice of material is quite important and a lossy rubber, the same as that used professionally in many microwave applications, should be used. This is known as r.a.m. (radar absorptive material). Black anti-static ic foam was tried initially, but was not nearly as effective as the proper material. The rubber sheet should be glued on the inside surface of the lid, as flat as possible, above the two stage amplifier. Note that the lossy material on the lid does not degrade the performance: indeed the designers have often seen an improvement of up to 0.2dB with the lid in place!

Other Applications

As described above, the application is for narrow-band use at 10,368MHz. However, the design has been made sufficiently wideband so that the unit can be used anywhere in the 10,000–10,500MHz band with virtually the same performance. All that has to be done is to choose the appropriate local oscillator frequency and tune FL1 and FL2 to the desired local oscillator and rf frequencies. Two particular applications might be in the 10,450–10,500MHz band for receiving future amateur satellites, and, lower down in the band, for atv.

A large amount of flexibility also exists with the choice of the i.f frequency, limited only by the tuning range of the "144MHz" tuned circuit. However, i.fs below 144MHz are not recommended for high performance applications as the image rejection will be insufficient and the noise performance will suffer. (You will not see this on a noise figure meter, though, so beware!) Higher i.f frequencies should be possible by modifying the i.f amplifier, although this has not been tried by the authors. However, the mixer on its own has good performance with i.fs up to at least 1.3GHz. For atv use, it should be possible to accommodate a standard amateur fm tv signal within the FL2 bandwidth, but the i.f bandwidth might be too narrow. A damping resistor across L12 should increase the bandwidth, but this has not been tried. A better solution would be to use a higher i.f, e.g. 480, 612 or 1,240MHz with a modified i.f amplifier to suit. If a higher i.f frequency is used, the bandwidth of FL2 could be increased by using longer probes. The authors would be pleased to hear of work done to use the G3WDG-002 for 10GHz atv.

18.10.3 The G3WDG–003 144MHz to 10GHz transmit converter

Circuit description and operation

The G3WDG–003 is the final module needed to complete a high-performance 10GHz amateur band transverter. When combined with a 144MHz transceiver, 10GHz antenna and suitable changeover relays, the system is capable of outstanding performance either from a fixed, home location or in portable use. Even better results may be obtained by adding a HEMT receiver preamplifier and a power GaAs fet pa, modules for which were being developed at the time of having to close the computer files used for printing this handbook!

This design is a linear up-converter (transmitter), from 144 – 146MHz to any 2MHz segment in the 10GHz band. It incorporates a ×4 multiplier chain, a GaAs fet mixer and four amplifier stages to reach an output power of at least 50mW. By providing the G3WDG–002 and –003 modules, described earlier, with a common 2.556GHz local oscillator source and suitable transmit/receive changeover arrangement, the result is a complete "state-of-the-art" linear transverter.

The circuit is shown in Fig 18.181, the layout of the board and components in Fig 18.182 and their values in Table 18.12. Referring to the circuit of the G3WDG–003 transmit converter, Fig 18.181, the 2,556MHz local oscillator signal is fed to a mmic amplifier IC1 which provides

Table 18.12. Components list for the G3WDG–003 144MHz to 10GHz transmit converter

Capacitors				
C1–5,7,11, 13,14, 16–19, 21,22	220pF chip capacitors, size 0805 smd		L4,5	Straight length of 0.135mm dia e.c.w., tinned 1mm each and and soldered flat to pcb between the tracks as shown
C6	Printed on PCB		L6,8–16	Straight lengths of 0.2mm dia (approx.) silver plated or t.c.w., soldered between the tracks and the radial stubs as shown. A single strand from a miniature coaxial cable such as RG174/U is suitable
C8,9	10nF, size 0805 smd			
C10	22pF, size 0805 smd		L7	4.5t of 0.135mm dia e.c.w. wound through the centre of an FX1115 ferrite bead. Alternatively, use a 10µH radial lead miniature choke such as Toko type 348LS100.
C12,15	10µF, tantalum bead, 10V Wkg			
C20,23	2.2pF ATC porcelain 100 or 130 series (0.050in size)			
C24	1pF, size 0805 smd			
Resistors			**Semiconductors**	
R1	100R, axial lead 1/2W rating		IC1	MSA 1104, MAV11 mmic
R2	100R, size 0805 smd		F1, 2, 3 4, 5, 6*	P35–1108 GaAs fet (Birkett *black* spot)
R3,5–13	47R, size 0805 smd			
R4	220R, size 1206 smd			
R14	47R or 10R, size 0805 smd(see text)		**Miscellaneous**	
VR1,2,4–7	10k skeleton cermet potentiometer, horizontal mounting. Suitable types: Allen Bradley 90H, Bournes VA05H or 3309, Philips Components OCP10H or similar		FL1–4	Cavity resonators as for G3WDG–001. See text and diagrams for details. Overall probe length 4.7mm for FL1 and 3.4mm for FL2 to 4
			PCB pins	approx 58 off RS Components 433–864. 1mm dia., 1.5mm head dia. (approx)
VR3	2k2, type as above.		SMA sockets.	4 off, 2-hole mounting, or pcb types with lugs removed. IF output could be SMB or SMC (Conhex)
Inductors			Tin-plate box	Type 7760 (74 x 111 x 30mm) from Piper Communications
L1	6t of 0.315mm dia e.c.w., close wound, 2mm dia, self supporting, 1mm above the pcb. Lead length 2mm			Feedthrough capacitor – solder-in, 1nF to 10nF, or filtercon.
				+ve and –ve regulated supplies eg. +ve from 7808ic, –ve from ICL7660 voltage converter, see Figs 18.168, 18.169 and 18.183
L2	16mm length of 0.135mm dia e.c.w., tinned 1mm each end, soldered flat to PCB as shown			
L3	As L2, but 19.5mm long			

THE 10GHZ ("3CM") BAND 18.113

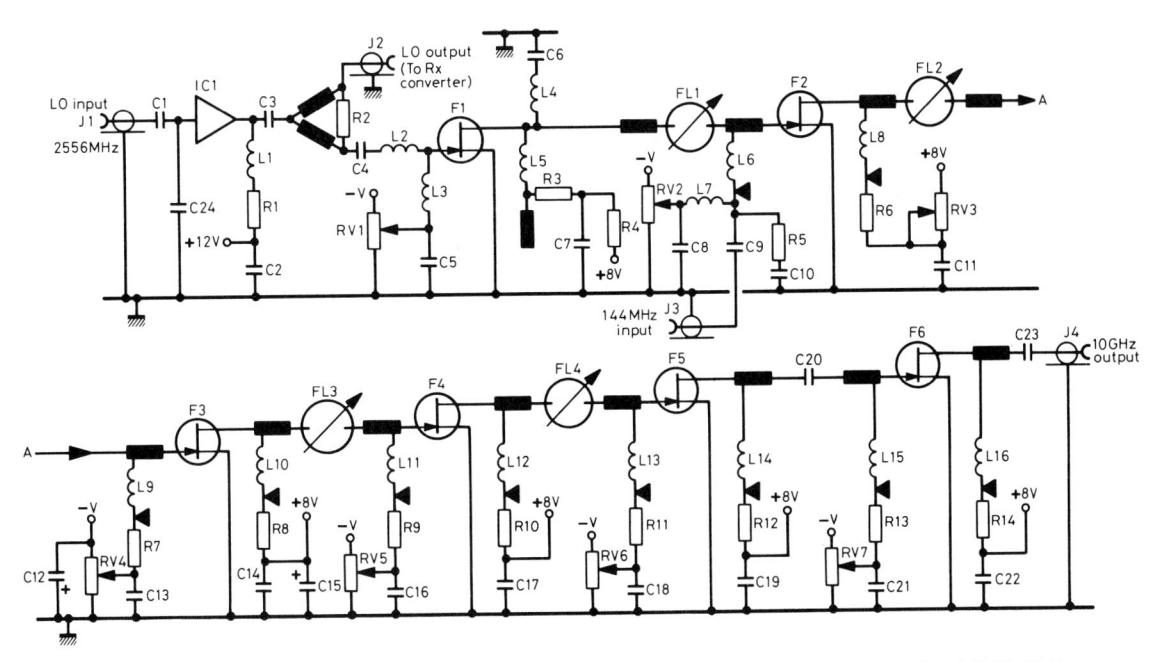

Fig 18.181. Circuit of the G3WDG-003 144MHz to 10GHz transmit converter. Component values are given in Table 18.12

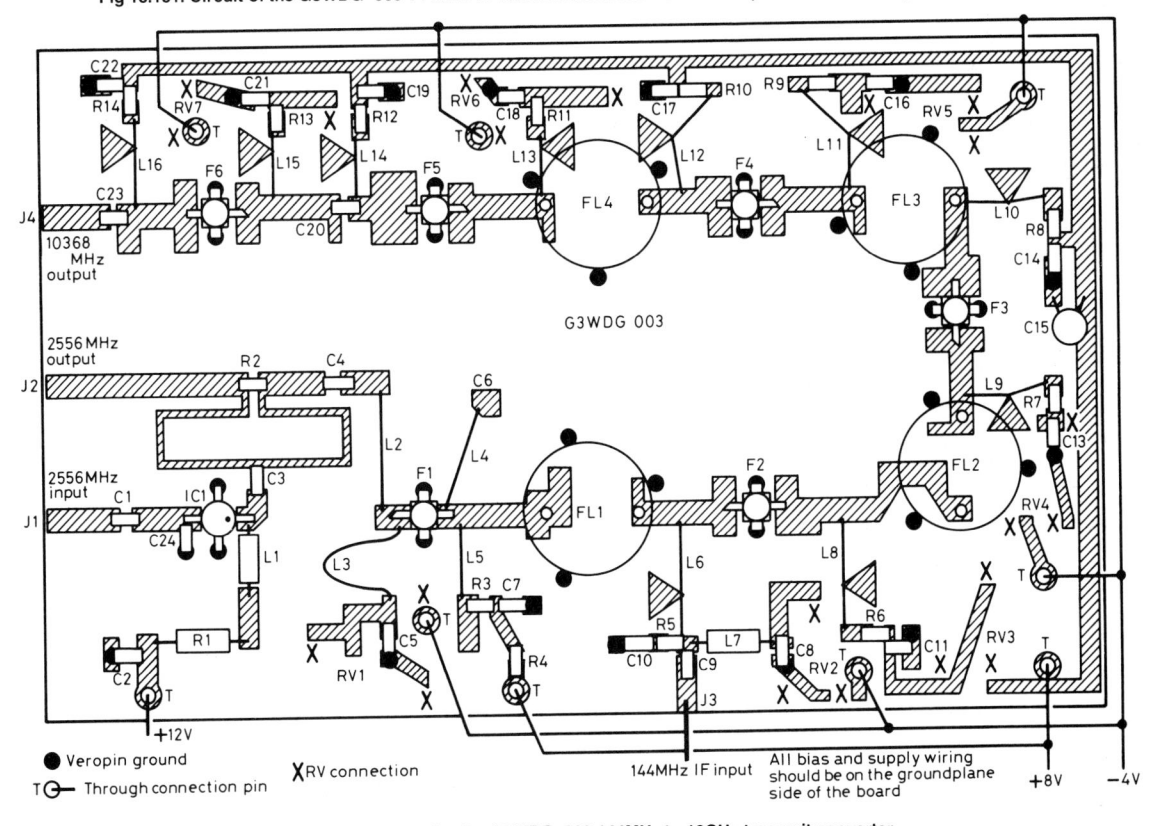

Fig 18.182. Layout for the G3WDG-003 144MHz to 10GHz transmit converter

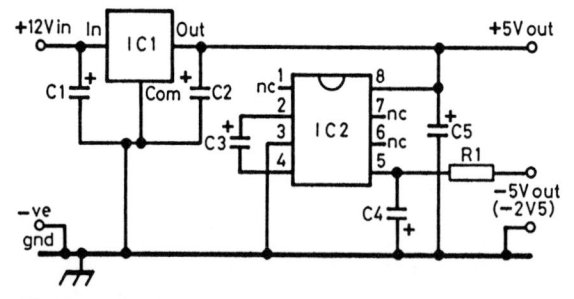

Fig 18.183. Positive and negative voltage regulator circuits for the G3WDG-003 module (G4FRE). The circuit is the same that in Fig 18.168, although some component values are different. The layout is exactly the same as in Fig 18.169. Component values: IC1, uA7808; IC2, ICL7660PCA; Z1, 3V0 or 3V3, 400mW zener diode; R1 1k5 1/4W metal film; C1 1µF tantalum bead, 16V wkg; C2 0.1µF tantalum bead, 10V wkg; C3, C4, 22µF tantalum bead, 10V Wkg; C5 10µF tantalum bead, 10V wkg; pcb, G4FRE-023

about 5.5dB gain. The output from this goes into a printed Wilkinson divider, which splits the signal into two equal, well isolated, outputs. One output is connected to J2, which is intended to provide the local oscillator signal for the G3WDG-002 receive converter. The gain from J1 to J2 is about 2.5dB, allowing proper operation of the unit with local oscillator powers in the range 5–15mW. The input circuit to the mmic is modified from that used in the G3WDG-001, in that a 1pF shunt capacitor has been added. This improves the input vswr from about 5:1 to less than 1.5:1, and is a useful modification for existing G3WDG-001 units. It decreases 2.5GHz drive requirement by about 2.5dB and stops resonant cable length effects from detuning the G4DDK-004 driver output circuit).

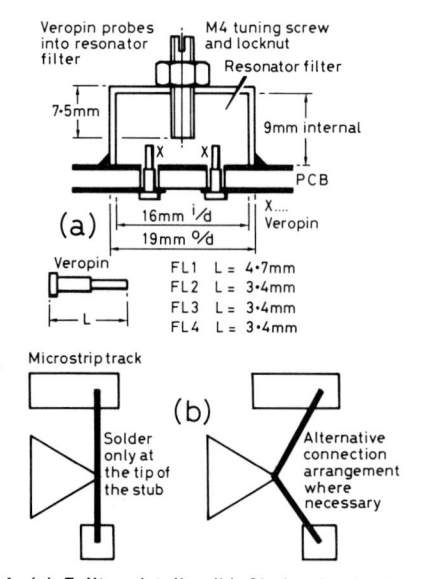

Fig 18.184. (a) Filter details. (b) Choke details for the G3WDG-003

The other output from the Wilkinson is fed to the ×4 multiplier stage F1, via a matching network L2/L3. The multiplier is identical to that used in the G3WDG-001 and 002 modules, and details may be found in earlier sections.

The 10,224MHz output from FL1 is fed to the gate of F2 together with the 144MHz signal from J3 via C9/L6. F2 acts as an up-converter, and is provided with both variable gate bias (RV2) and drain bias (RV3) to optimise the conversion efficiency. These adjustments interact to some extent (see later). The output from F2 contains three main signals, the local oscillator at 10,224MHz, the wanted 10,368MHz signal and the image at 10,080MHz. FL2 selects the wanted output.

The remaining circuitry is a four stage amplifier containing further bandpass filtering (FL3 and FL4) to clean up the output signal. The value of the drain resistor R14 is optional, 47 or 10Ω. The use of 10Ω may increase power output by a small amount in some cases.

The board layout is shown in Fig 18.182. Note that C12, the negative rail decoupling capacitor, is not shown as it is fitted on the reverse side of the board. The negative bias generator used to supply the gate bias for the fets uses the same pcb as that used in the G3WDG-001 module (G4FRE-023). The circuit has already been given in Fig 18.168 and the board layout is given again in Fig 18.183, together with the appropriate components values, some of which are different in value to the earlier circuit.

Construction of the G3WDG-003 transmit converter

Construction of the G3WDG-003 transmit converter uses the same techniques and order of construction as those already described for the other modules, and reference should be made to preceding sections for the "mechanical" construction of the module. Once the grounding pins and filter cavities have been fitted and the pcb soldered into its box, the remainder of the components may be fitted by the methods already described. All parts are specified in the parts list of Table 18.12. Fitting should, briefly, be in the order:

1. Wire inductors L1 – L11, as specified in the parts list (Table 18.12), into position, ensuring that the wires lie flat to the board.

2. Fit all chip components using the mounting techniques described earlier. You will need a pair of fine-pointed tweezers to handle these small devices and, maybe, the assistance of a magnifier!

3. Fit all components which have leads, ensuring that static-sensitive devices (ic, fets) are put on the board last of all to minimise the risk of damage to the devices.

Individual "build" techniques have already been described in some detail – see earlier sections. Filter details are given in Fig 18.184 (a), the fet pin-out in Fig 18.184 (b) and the choke decoupling details in Fig 18.184 (c). For inductors using enamel covered wire (e.c.w.), cut the required length of wire then scrape/chip the enamel from the last 1–2mm of each end using a scalpel blade. Tin each end. For inductors using one strand of a standard multi-strand wire, tin one end and fit to the board as shown in Fig 18.184(c): solder first at position 1, then at position 2 as close as possible to the apex of the triangle, then at position 3. If any excess wire remains at 1 or 3, trim off carefully with a scalpel blade. Fig 18.185 shows the positioning of the regulator pcb and associated decoupling capacitor.

Alignment without laboratory equipment

The recommended way of tuning up the module without resorting to laboratory test equipment is to align a given stage while monitoring the drain current of the following stage, which acts as a power indicator. The gate bias of the monitor fet should be set to give a drain current in that fet of approximately 1mA. The procedure is to make temporary connections to the drain resistor of the following stage using thin flexible wires, which are then connected to an *analogue* multimeter. Digital multimeters are much more difficult to read and should be used *only* as a last resort, since they do not operate in real-time! For example, when optimising the setting of RV1 and FL1 while aligning the multiplier stage, the multimeter is connected across R6 and RV2 is set to give a drain current of 1mA in F2. Adjustments are then made to R6 and FL1 for maximum current. In some cases the increase in current may be small and the adjustments need to be made carefully, especially the tuning screws. It helps if all tuning screws are preset to approximately the right penetration, i.e. with 7.5mm of thread below the bottom of the locknut. When adjusting the tuning screws, particularly FL2–4, the locknut should be kept reasonably tight with a spanner while turning the screw with a screwdriver (like setting the tappets in a car engine!), or the filter loss may be very high. The final tuning may be done using the locknut tightness as a fine adjustment, but take care that the correct tuning point is with the locknut quite tight, or again the filter loss may be higher than normal.

1. Connect a 50Ω load or 10dB or larger attenuator to J2, and a 10GHz power indicator to J4 (eg coax-waveguide transition and waveguide diode detector).

2. Set the gate bias on all stages to cause all fets to be pinched off (zero drain current).

3. Set gate bias of multiplier to 1mA using RV1.

4. Apply 2.5GHz drive to J1 (5mW minimum). The drain current of F1 should increase.

5. Using F2 as a power meter as described above, adjust FL1 and then optimise RV1.

6. Temporarily remove 2.5GHz drive and dc power, and transfer the test leads to R8. Set RV3 to its centre position and, after restoring dc power, set the current through Q2 to 0.5mA with RV2. Reconnect 2.5GHz drive and apply 1mW of 144MHz to J3. The mixer current should increase incrementally on application of the drive signals. Next, using

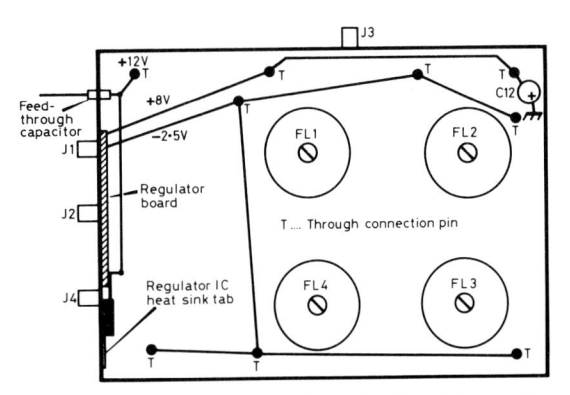

Fig 18.185. Positioning the regulator circuit and decoupling capacitor within the module enclosure (ground-plane side)

F3 as the power indicator, adjust FL2. This is the most difficult stage in the alignment, as there are three peaks since the filter resonates to 10,080, 10,224 and 10,368MHz. Finding the correct one may take a little time! The correct setting is with the tuning screw at the smallest penetration, and the power should disappear when 144 drive is removed. If the 10,080MHz peak is selected by mistake, it is possible to carry on with the alignment and all will appear normal except that no-one will hear you (unless their '002 is also on the wrong image!).

7. Next, optimise RV2 and RV3 for maximum power. The adjustments interact, so it is necessary to "loop" through the adjustments until the best combination is found.

8. The same procedure should be followed to align the rest of the stages. Before starting optimisation, drain currents of the amplifier stages should be set to 15mA. After optimisation, the current can remain at whatever value resulted, with no risk of damage to the device. Finally, some power should be seen at the output, and a wavemeter should be used to confirm that the output is on 10,368MHz as desired.

9. The final stage in the alignment is to go around all the adjustments again for maximum output power at J4. In case the unit is operating at significant compression with 1mW 144MHz drive, adjustments will be easier if the 144MHz drive power is reduced. Prototype power outputs have been greater than 50mW, but it may be possible to achieve even more by tuning the microstrip lines with small pieces of copper or brass foil. If attempting this, start at the output and work backwards, as most will be gained (if any) around F6. Remove dc power before soldering the foil in position.

10. Check the level of output from J2. This should be 2–2.5dB higher than the drive power applied (below 10mW). At higher drive powers, IC1 will go into compression but this does not matter, as plenty of drive will be available for the receive converter. Spectrum analyser traces of the output from the G3WDG–003 module are

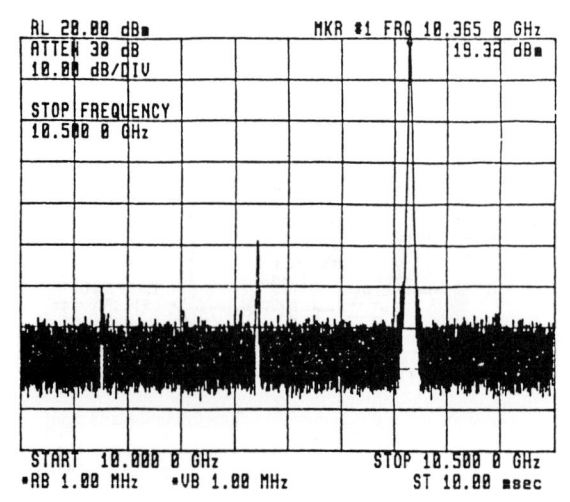

Fig 18.186. Output spectrum of the G3WDG-003 module from 10.0 to 10.5GHz, showing the wanted signal at 10,368MHz, the local oscillator signal at 10,224MHz and the image signal at 10,008MHz

shown in Figs 18.186 and 18.187. Fig 18.188 plots power output from the module as a function of the level of 144MHz drive, clearly indicating the optimum drive level to be about 1mW.

18.10.4 Switching microwave transverters

The circuits given in this section were originally developed as a "universal" interface and transmit/receive sequence switching circuits for the "surplus", ex-Mercury "White Box" 10GHz transverter units made by M/A-COM (Microwave Associates) which have been available in limited numbers on the UK market. However the sequencer, in particular, has many other applications. It was developed by G3SEK [43] from a design by N6CA in the *ARRL Handbook*.

144MHz i.f Interface

The first unit, Fig 18.189, is a low noise 144MHz i.f head pre-amplifier with good broadband 50Ω match, together with tx/rx switching and a power attenuator suitable for 144MHz powers of 1 to 3W. There is also an optional connection for amateur 144MHz transceivers such as the Icom IC202 and Yaesu FT290R, both of which apply a dc "push to talk" (ptt) signal through the antenna socket.

The circuit is loosely based around various muTek transverters. The receive gain is about 15dB and the nf about 2dB. The maximum rf power output to a tx mixer is about 10mW for 3W input. With a minimum capacitance of 2pF for VC1, the output is adjustable down to the levels required to drive the G3WDG-003 transmit converter circuit, if the power input to the interface is 3W. Component values are given in Table 18.12. The whole unit must be tightly screened to prevent leakage of 144MHz signals in or out.

Input levels much higher than 3W would cause a problem with heat dissipation. For higher powered transceivers, it would be better to use an external power attenuator which could probably be left in-line on receive as well as transmit, as most microwave transverters could do with a bit less rf gain!

TX-RX Sequencer

Even with the price of GaAs fets coming down, transmitter powers are going up, so we need to be increasingly careful about the proper sequencing of tx/rx changeover. The correct sequence for switching from rx to tx is:

1. Turn rx power off (but leave the local oscillator running).
2. Energise the coaxial relay and wait for the contacts to settle.
3. Enable the pa and let it reach its quiescent operating point.

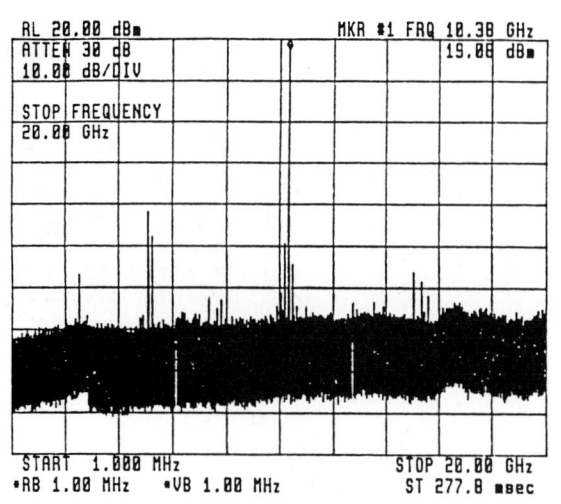

Fig 18.187. Wideband spectrum analyser trace of the G3WDG-003 module, from 1MHz to 20GHz

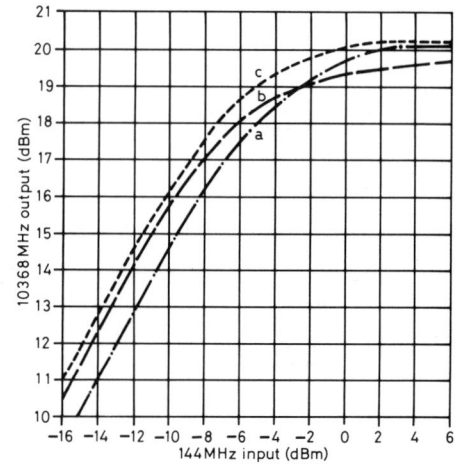

Fig 18.188. Power output of the G3WDG-003 at 10,368MHz vs. drive power at 144MHz, for three prototypes

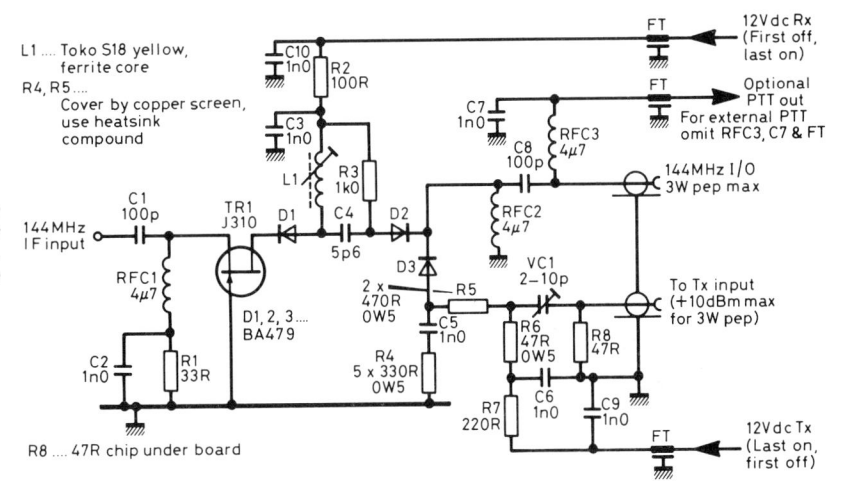

Fig 18.189. Circuit of a transverter interface for use with low power 144MHz transceivers (G3SEK). Component values are given in Table 18.13

4. Enable tx exciter/driver.

The sequence in going from tx to rx is exactly the opposite.

The sequencer circuit shown in Fig 18.190 is an adaptation of the design by N6CA in the *ARRL Handbook*. TR1 and TR2 provide a ptt interface suitable for relay-switched 0V–rx, 12V–tx or alternatively the dc levels provided by the IC202 or FT290R (see earlier). Switching between rx and tx causes the voltage across C1 to ramp up or down, taking about 200ms. Comparators IC1A to IC1D have progressively increasing reference voltages, so they will switch over in sequence.

The only major change from the N6CA design is to bring the coaxial relay and pa-enable supplies out to separate connections to allow the use of voltages other than 12V. For instance, commonly available SMA microwave relays often operate at 28V. Component values are given in Table 18.13.

Construction

PCB track layout for the 144MHz interface is given in Fig 18.191 and the component overlay in Fig 18.192. The track layout for the sequencer is given in Fig 18.193 and the component overlay in Fig 18.194. The track layouts are full sized, viewed from the track side. The i.f interface board has a copperclad groundplane on the upper side and is designed to fit into a Piper Communications size 2A box for screening, whereas the sequencer board is single sided and should require no special screening. The prototype was mounted on self-adhesive plastic pillars in a convenient corner of the transverter case. All circuit values are given in Table 18.13.

Fig 18.190. Circuit of a switching sequencer suitable for use with the G3WDG transverter and interface of Fig 18.189 (G3SEK after N6CA, ARRL Handbook). Input connections on J1:
1   + to tx (FT290R)
2   Ground
3   0V to tx (IC202)
Output connections on J2:
1   RX stages, 12V rx, 0V tx
2   +12V supply
3   Coaxial relay, 0V rx, 12–28V tx
4   +12 to 28V supply to 3
5   Ground
6   +12 to 28V supply to 7
7   TWT (or other pa) enable, 0V rx, 12 to 28V tx
8   Ground
9   +12V supply to 10
10   TX stages, 0V rx, 12V tx

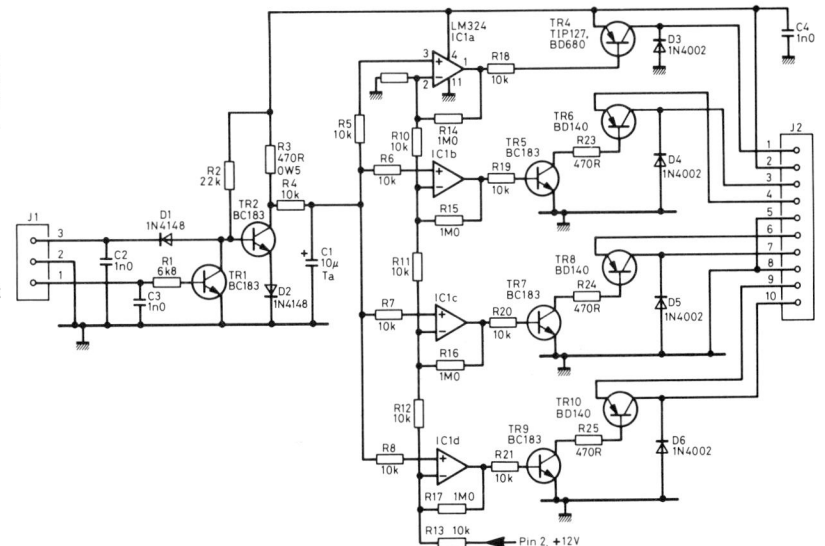

Table 18.13. Component values for the G3SEK 144MHz interface and tx/rx switching sequencer (after N6CA, *ARRL Handbook*)

144MHz interface		TX/RX Sequencer (numbering corresponds to the original N6CA circuit)	
C1	100p		
C2,3	1n0		
C4	5p6		
C5,6,7	1n0	C1	10µF Tantalum
C8	100p	C2 3 4	1n0
C9, 10	1n0	D1, 2	1N4148
D1,2,3	BA479	D3–6	1N4002
R1	33R	TR1, 2	BC183
R2	100R	TR3	(not used)
R3	1k0	TR4	TIP127/BD680
R4	66R 2.5W (5× 330R, 0.5W)	TR5,7,9	BC183
		TR6 8,10	BD140
R5	235R 1W (2× 470R, 0.5W)	R1	6k8
		R2	22k
R6	47R, 0.5W	R3	470R, 0.5W
R7	220R, 0.5W	R4–13	10k
R8	47R Chip (under board)	R14–17	1M0
		R18–21	10k
L1	Toko S18 yellow (ferrite core)	R22	(not used)
		R23–25	470R *
TR1	J310	IC1	LM324
RFC1–3	4µ7H		
VC1	2–10p, SKY		

* Note: R23 should be 470R 0.5W. With 12V supplies to pin4 and/or pin 6, R24 and R25 should be 470R 0.5W. When using 24–28V supplies, change R24 and/or R25 to 1k5, 0.5W.

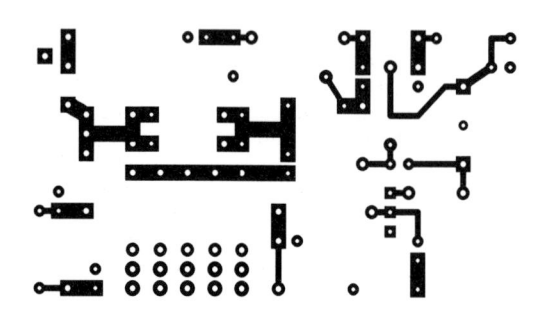

Fig 18.191. PCB track layout for transverter interface (G3SEK)

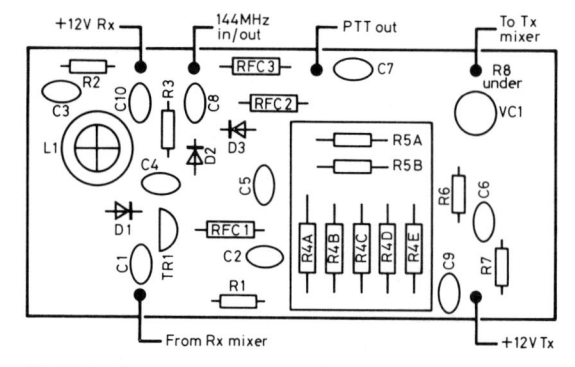

Fig 18.192. Component overlay for the transverter interface (G3SEK)

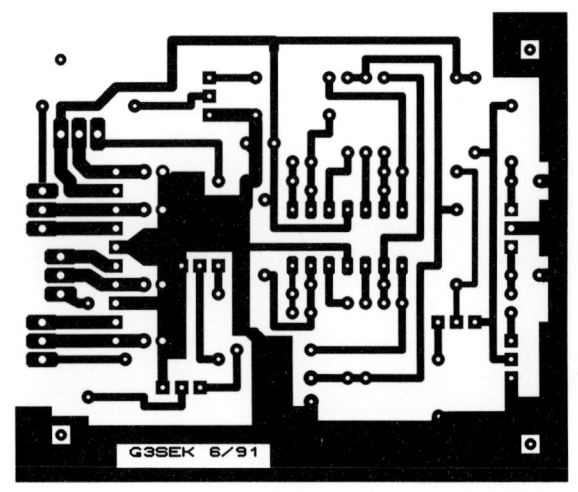

Fig 18.193. PCB track layout for switching sequencer (G3SEK)

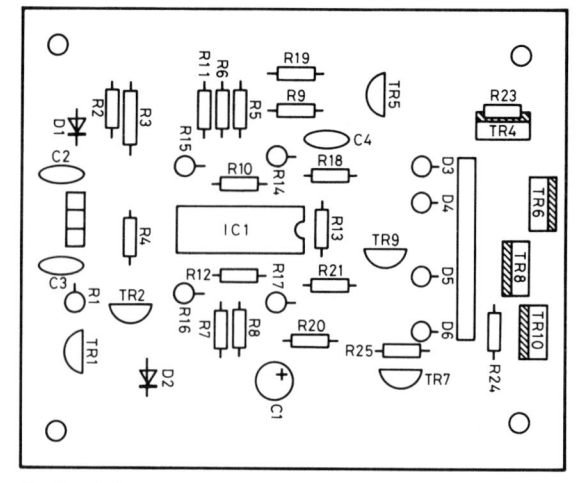

Fig 18.194. Component overlay for the switching sequencer (G3SEK)

18.11 TRAVELLING WAVE TUBE AMPLIFIERS

18.11.1 Introduction

The theory of travelling wave tubes (twts) was discussed in chapter 6, "Microwave semiconductors and valves". Recently, numbers of "second user" twts, withdrawn from professional service, have become available to amateurs in the UK and elsewhere. This has been more-or-less (in the UK) coincidental with the development of the technically advanced G3WDG 10GHz narrowband designs described in the last section. It seemed, therefore, appropriate to include an account of some of the findings and experiences in the use of these devices recently gained by amateurs in the UK. The use of such high-gain devices (typically 30 – 40dB) to generate output powers of up to 10 or 20W should not be undertaken lightly, for a number of reasons:

1. High and potentially lethal voltages (several thousand volts, albeit at relatively low current) are required to operate twts.

2. High density rf fields, which are potentially hazardous, are generated at some localities within the system equipment, e.g. in open waveguide, leaky waveguide flanges, dish feeds and antenna near-fields, at the power levels generated by twtas.

3. TWTs are relatively delicate, high-gain devices which are quite easily destroyed by incorrect operation. Excess power dissipated in any of the tube electrodes, especially the helix, can result in burn-out. The tube should *always* be operated into a matched load: power reflected from an un-terminated output can rapidly destroy the helix.

The safety aspects of the use of high voltages and the potential hazards of rf radiation were fully discussed in chapter 11, "Safety". It should be remembered that it is not only common sense to follow good practices, but that it is a condition of the amateur licence that it is the operators' responsibility to ensure that the (rf) power flux density, in areas where people have access, does not exceed the limits recommended by the competent authorities. Misused or poorly engineered systems using twtas can certainly infringe this requirement. Be sure you know what you are doing before you try to do it!

18.11.2 Using "surplus" twts

The following hints and tips were derived from unpublished notes by Mike Walters, G3JVL, and are offered as suggestions to those who wish to try to use "retired" twts!

Most twts acquired by amateurs are at the end of their professional service life ie. they are officially "dead". This state is recognisable by the following characteristics:

- The helix current has risen from the normal level of 1mA (no drive) or 2mA (at saturation drive level) to perhaps 3 or 4mA.
- Power output is reduced from the rated 10W to around 2W (or less).
- The signal drive level required to saturate the tube has risen from 1mW to anywhere between 2 and 6mW, i.e. tube gain is reduced drastically.
- Higher than rated anode voltage is required to attain the rated cathode current. This may be coupled with a requirement for higher than normal collector and helix voltages in order to produce power output.

All these characteristics are indicative of failing emission from the cathode emissive coating. The emissive properties of the cathode can often be restored to more normal levels by being run at a higher temperature than usual. This may be achieved by the means which was often used to prolong the useful life of "low emission" cathode ray tubes used in early tv receivers in the 1950s – by raising the heater voltage above the normal, rated 6.3V. The twt performance may respond to raising the heater supply voltage to 7V or even 8V, although some tubes will not respond. Where a tube responds, the helix current will drop, the cathode current increase and the power output rise towards, or even back to, rated values. The tube can last for many hours of active amateur use, although it tends to have a "short but merry" life – how long it survives lies in the lap of the Gods! It should be noted that once the heater has been run at higher than normal voltage, reducing the supply back to 6.3V will result in the tube producing less power than it did before. To keep on producing power, the temperature must be increased further and further until, in the end, the heater burns out.

Several different mains-driven twt power supplies have also become available on the surplus market and many of these may be adaptable to amateur use. In general, the raising of the heater supply voltage and other electrode voltages, as outlined above, may be simply a matter of adjusting transformer primary tappings. For instance, in the UK, the mains transformers will normally be tapped for voltage inputs between 200 and 240V, often in 10V steps. Thus, it may be a simple matter of re-setting the primary tap(s) from the normal 240V setting down to 200 or 210V in order to achieve the needed increases in secondary voltages. If, however, a separate heater transformer is preferred or required, the user must install a transformer with adequate insulation – usually at least 5kV. Other changes may need to be made, such as adjustments to the various protective circuits (eg. helix current trip and switching delay circuits) and the user should try to be familiar with the circuit of the psu before attempting modifications. Remember, in particular, that high and potentially lethal voltages are present and may remain present (eg. on high voltage capacitors) long after the unit is switched off and disconnected. As a general principle the heater and helix voltages, in particular, must be well stabilised and free from noise and hum which will otherwise modulate the electron beam. Collector, grid and anode voltages are not so critical. These points will be emphasised in the next section.

18.11.3 A practical twt power supply

The following power supply, which was first described in [44], was designed by Andy Talbot, G4JNT, specifically for operating twts such as the EEV N10012 and STC W3MC/11A, both capable, when new, of 10W output at 10GHz. The power supply requirements are:

Helix	4,000 – 5,000V at up to 3mA, very stable and "clean"
Collector 1	1,450V at 15mA, stabilised to 10% or better
Collector 2	650V at 15mA, stabilised to 10% or better
Grid 2/Anode	2,000V at a few µA, variable and controlled
Heater	6.3V at 0.7A, stabilised to 2.5% or better

A new tube should take much less helix current than this – around 0.7mA. As discussed in the last section, tubes tend to be withdrawn from professional service and become

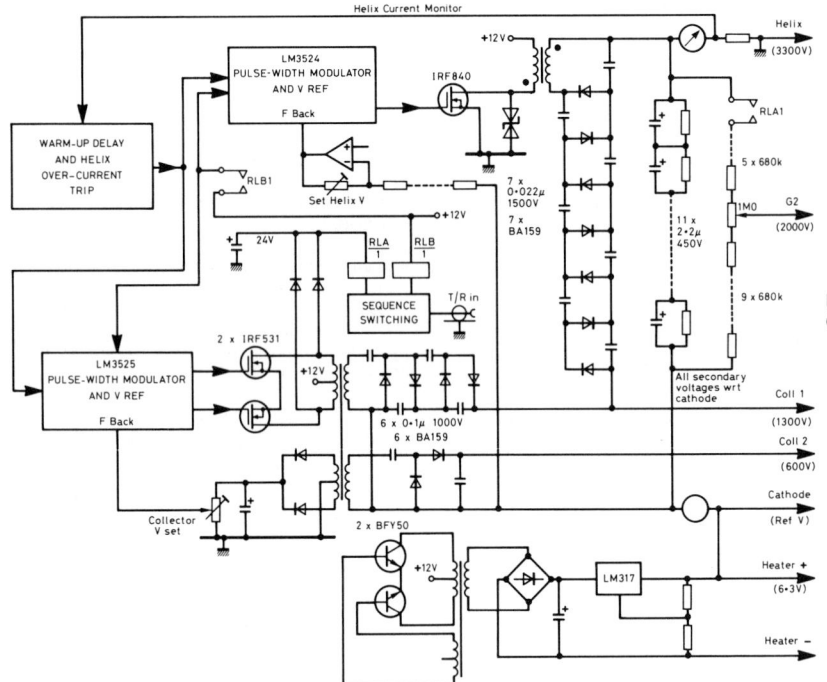

Fig 18.195. Block diagram of the G4JNT switch mode twt power supply unit

available to amateurs because the helix current during operation has risen above a particular threshold, often set at around 2mA. The helix supply, in particular, must be very voltage stable and "clean", i.e. free from hum or noise which will readily modulate the electron beam and therefore the output.

The main problem with powering twts is that although all the voltages are referred to the cathode, it is the helix which is connected to the system ground. Therefore all other supplies, including the heater, hang at very unpleasant voltages *below* ground, connected to a stringently stabilised rail. Transformers which are insulated to 5kV or more must be used, even if the output voltage required from them is much lower.

So that the twta may be used for portable operation, it was appropriate to design a switch-mode power supply (smps) for nominal 12V operation, rather than to search out high-insulation mains transformers with high voltage windings. Double-insulated transformers have sufficiently good insulation but they are not commonly available with outputs of hundreds of volts.

Fig 18.195 is a block diagram showing the approach taken. Three separate smps are used as follows:

A saturating-type push-pull inverter supplies 10V for the heater. This is stabilised down to 6.3V by an LM317 integrated circuit regulator. The core used for the prototype unit was of uncertain type, but many non-gapped ferrite pot-cores of around 25mm (1in) diameter will do. Winding for 1V per turn gives an oscillation frequency of about 200kHz at full load, with a pre-regulator efficiency of about 80%. The secondary was wound with ptfe insulated wire which was further insulated from the primary winding by 3 turns of pvc tape. The heatsink of the LM317 will be at the same potential as the cathode of the twt and must, therefore, be a separate unit, mounted for 5kV isolation – it can be fixed on the same pcb as the collector supplies.

A push-pull inverter, using a pair of IRF531 power fets, supplies the collector voltage requirements. Two secondary windings are used, each supplying 300V, both wound with approximately 200 turns of thin ptfe insulated wire. Three layers of pvc tape give additional insulation. One winding supplies a voltage doubler to give the collector 2 supply and the other is quadrupled for the collector 1 supply.

An LM3525 switch-mode controller ic is used to drive the power fets. The supplies are stabilised by rectifying the output from a tertiary winding and feeding this back to control the LM3525. A Mullard FX3730 core was used for the transformer, operating with an inverter frequency of 25kHz. Since the collector supplies do not need to be either very stable or smoothed, this approach with no additional smoothing capacitors is adequate. The correct relative voltage between the two supplies was obtained by adjusting the number of turns on the transformer secondary. The two supplies are then set together by one preset potentiometer in the controller. The supply is somewhat under-run, giving about 35W in this application, and could be quite easily used as a 70W supply for a 2C39 amplifier, for instance.

A different approach was used for the helix and grid 2 supplies. One side of the 3 – 4.5kV required is connected to ground (albeit the "wrong" side), so very high insulation is no longer required. A "flyback"-type regulator is employed using an IRF840 power mosfet, giving a negative output from the transformer secondary. This is septupled (×7) to around 2.0 – 3.5kV and connected in series with the collector 1 supply to give the full helix supply voltage. This supply is the most stringent with regard to stability, noise and ripple, so stabilisation is performed directly. With the helix connected to ground, the cathode potential is at minus helix volts and this approach is therefore possible. A 40MΩ resistor chain divides down the cathode voltage directly and controls an LM3524 smps ic via an inverting op-amp. Although the helix supply actually comes from two inverters, this method stabilises the total output voltage.

With careful attention to the feedback-loop stability, a helix voltage stabilisation in the order of 0.1% is readily achieved. Ripple is reduced to less than 1V by the output smoothing capacitor which consists of eleven 2.2µF, 450V working electrolytic capacitors in series. The constructor must not forget to include the voltage equalising resistors across each capacitor!

The transformer used was wound on an RM10 *gapped* core with $A_L = 630nH/t^2$. Primary turns = 20, secondary = 50. This supply is just capable of providing 3mA helix current at 4.5kV, but any more would need a bigger transformer core such as the Mullard FX3721.

The grid 2 voltage is derived from the helix supply by a potential divider. The current required is less than 10µA, so the chain can consist of low power resistors. A potentiometer allows fine adjustment of cathode current, coarse setting being achieved by selecting the divider chain resistors.

Two supervisory ("watch-dog") circuits are also incorporated, both using op-amp comparators. A latching trip shuts down both eht power supplies if the helix current exceeds 4mA for more than 200ms. This is reset by means of a momentary-make push-button switch. In addition, a three minute warm-up timer inhibits eht from being applied until the tube cathode has had time to reach its operating temperature. Both of these watch-dogs operate on the shutdown pins of the smps controllers and should be reasonably fool-proof: two leds indicate their status. A timer override is useful, however, when testing the psu into dummy loads, before connecting it to the twt.

Fig 18.196 is the full circuit schematic for the twta psu. Transformer winding details are given in Table 18.14. The pcb layouts, shown in Figs 18.197 to 18.202, were designed for this project by Barry Chambers, G8AGN.

In use, the helix voltage needs to be accurately set to within a few volts. A multiturn preset potentiometer should be used in the feedback path to provide the resolution needed. The helix operating voltage is usually printed on the individual tube, along with the G2 voltage and optimum cathode current. These values should be set initially, adjustments being made for maximum output once rf drive is applied.

When powering-up the tube it is *essential* to delay G2 voltage build-up until all other voltages have reached their correct values. At low helix and/or low collector voltage, excessive helix current is drawn which will operate the trip and the voltages will, therefore, never be applied! A further effect is that the helix supply is momentarily overloaded and the IRF840 can be damaged. A small high-voltage relay, such as the ITT type RF1E which is often available at rallies, is used to switch the supply into the divider chain. A 24V supply, suitable for operating these relays, is obtained by rectifying the voltage at the drains of the fets in the collector power supply inverter; this voltage is only available when the eht is on and could also be used for operating other relays, such as the 18GHz SMA coaxial changeover relays often available as surplus.

By employing a sequence-switching arrangement, it is possible to "key" the eht supplies via the push-to-talk (ptt) line. The ptt line switches the power to the fet driver circuits directly via a relay and, after a 0.2 to 0.5 second delay, the eht relay is enabled. On releasing the ptt line, both relays drop out immediately. This arrangement means that the psu only operates when required and so conserves battery power. The heater supply must, of course, operate continuously, even though the twta is not operative on receive.

Virtually all of the components needed are available from many major UK suppliers such as Verospeed, RS Components or Farnell Electronic Components and, apart from the potential hazards mentioned in the introduction, there should be no problems in reliably reproducing this design.

Table 18.14. Transformer winding details for the G4JNT twt power supply unit

Transformer T1	RM10 gapped core, or Mullard FX3721
	A_L = 630nH
	Primary 20t
	Secondary 50t (this may be adjusted, according to helix voltage requirements)
	Insulate for 1kV
	Observe winding sense as indicated on the circuit schematic
Transformer T2	Mullard FX3730 core (* see note)
	Primary 3 + 3t
	Secondary 1 Approx 100t; insulate for 4kV wkg, adjust turns
	Secondary 2 Approx 100t for Vc1 and Vc2 required
	Tertiary 2 + 2t
Transformer T3	Ferrite pot-core, *no gap*, approx 25mm diameter
	Primary 12 + 12t
	Secondary 10t, insulate for 4kV wkg
	Feedback 3 + 3t
	Observe winding sense as indicated on the circuit schematic

Wire size uncritical: 1mm suggested for all windings
* Note: The smaller FX3720 core can be used, but the turns must be increased to Primary 5 + 5t
    Secondaries 160t each
    Tertiary 3 + 3t

18.122 MICROWAVE HANDBOOK

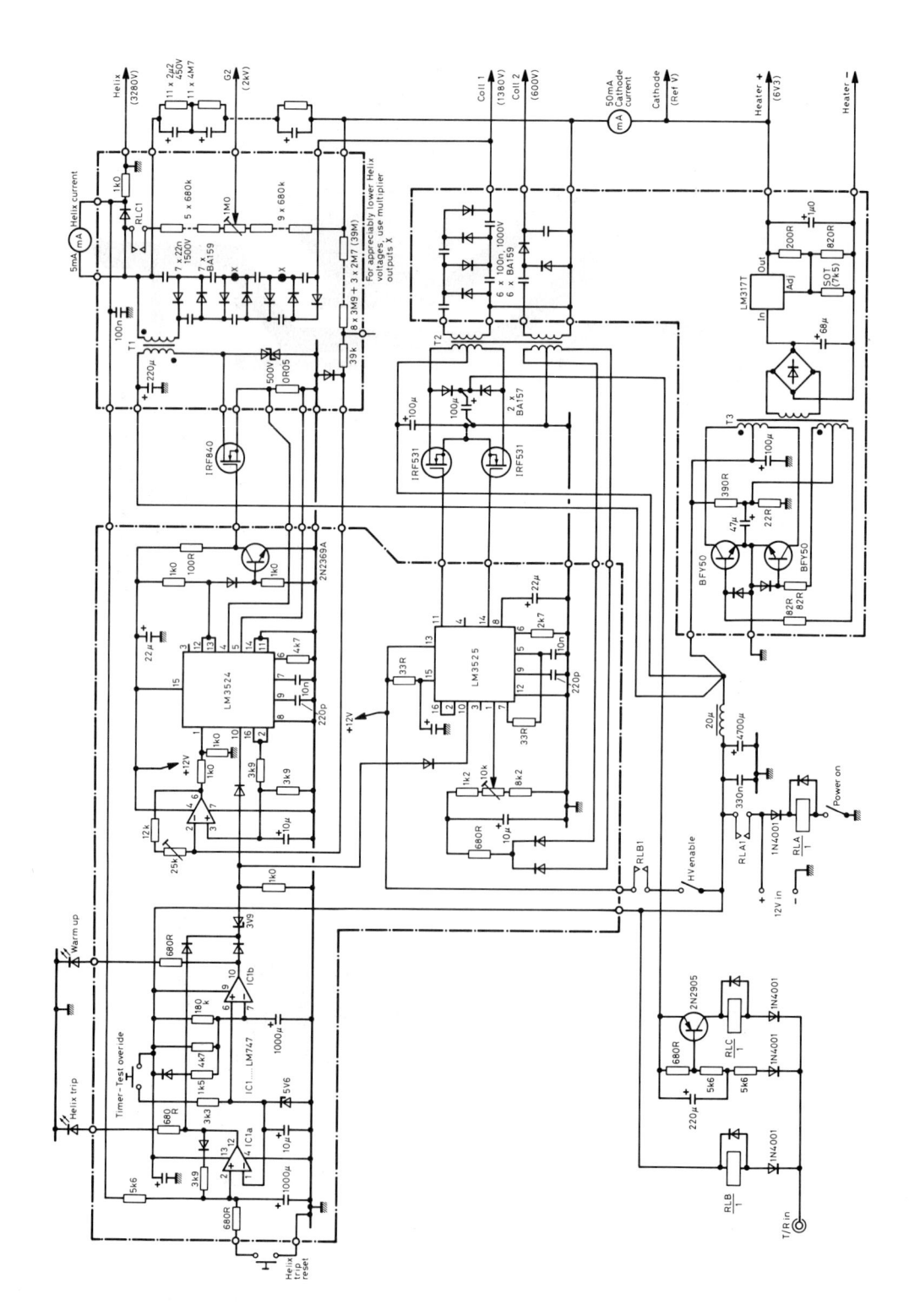

Fig 18.196. Circuit schematic for the G4JNT twt psu. There are three main pcb circuit modules: (1) the heater, cathode and anode supply module, (2) the eht supply for collectors and helix and (3) the "watch-dogs", timers, delay and smps controller module. The remaining circuits and components are "off-board". All unmarked diodes are 1N914, 1N4148, etc. Transformer winding details are given in Table 18.13

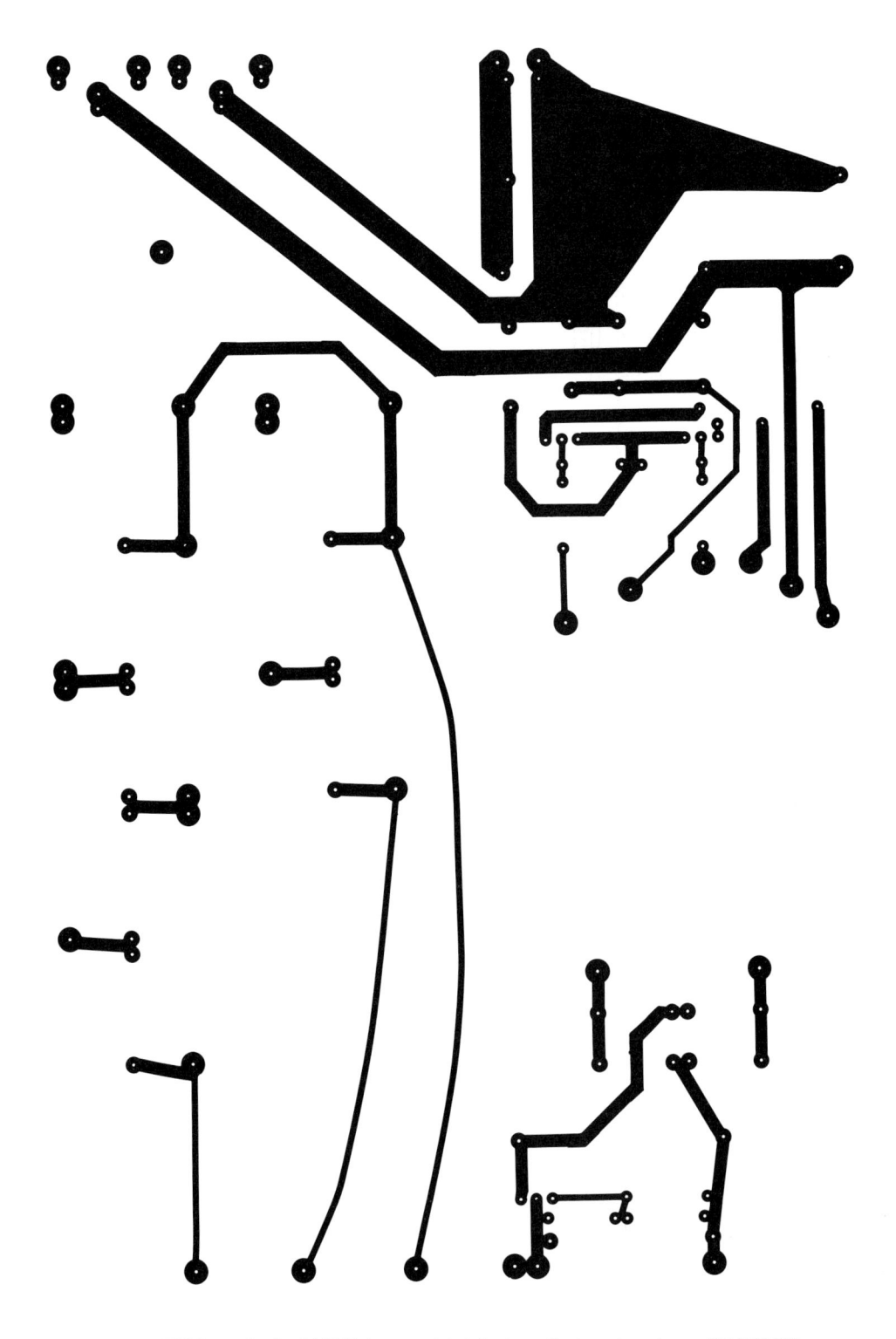

Fig 18.197. PCB layout for the G4JNT twt psu module 1: heater, cathode and anode supplies (G8AGN)

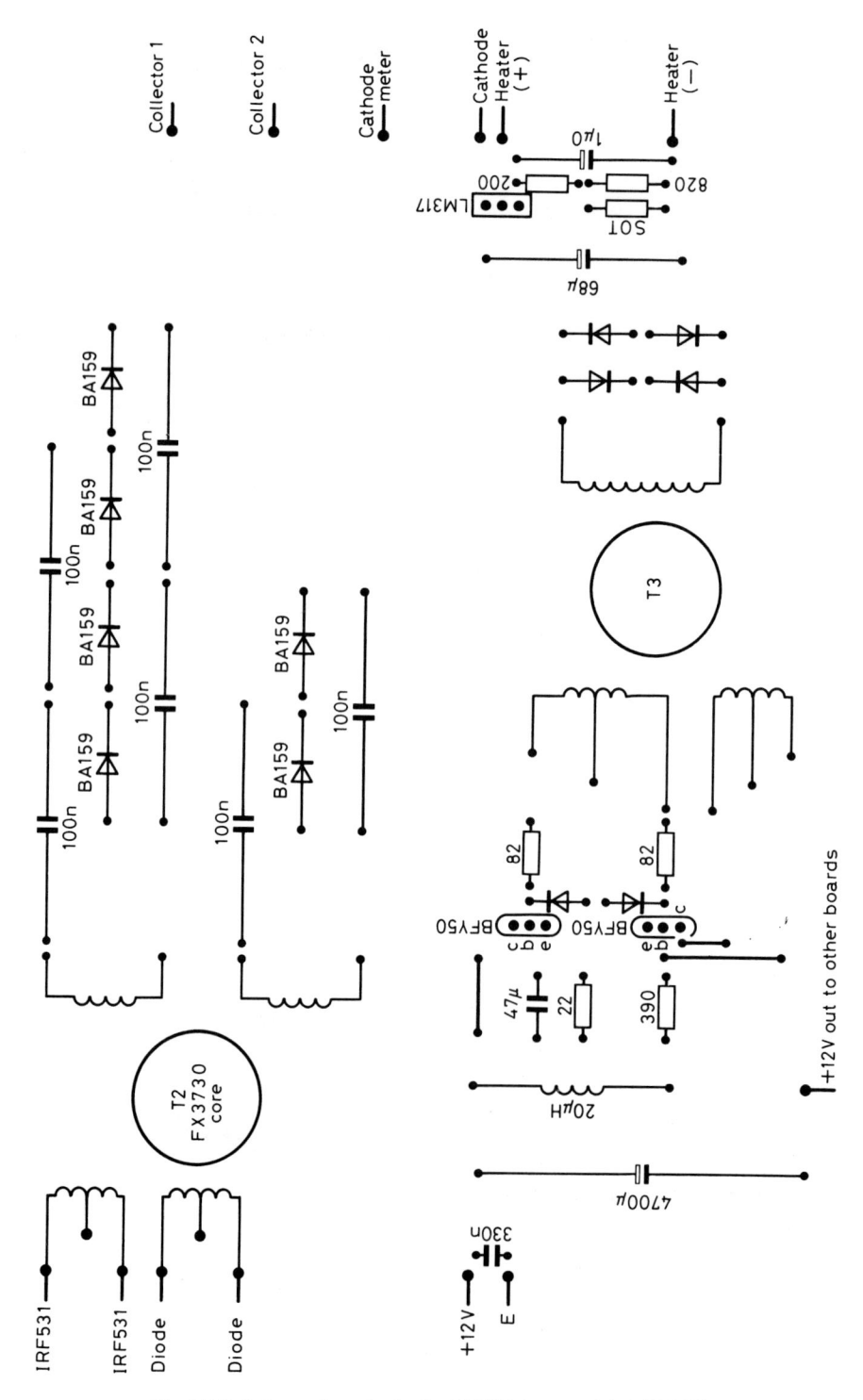

Fig 18.198. Component overlay for the G4JNT twt psu module 1 (G8AGN)

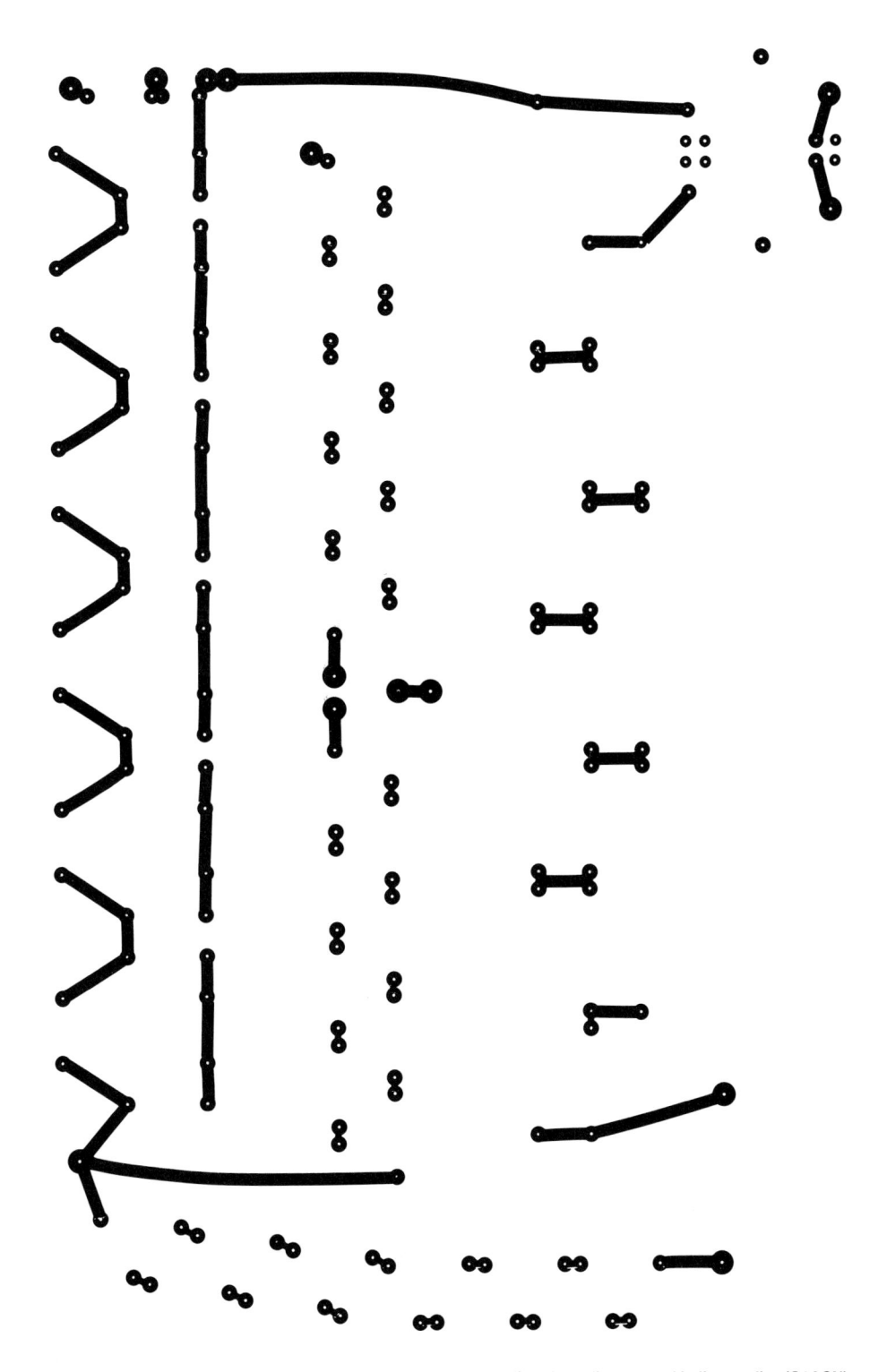

Fig 18.199. PCB layout for the G4JNT twt psu module 2: EHT multipliers for collectors and helix supplies (G8AGN)

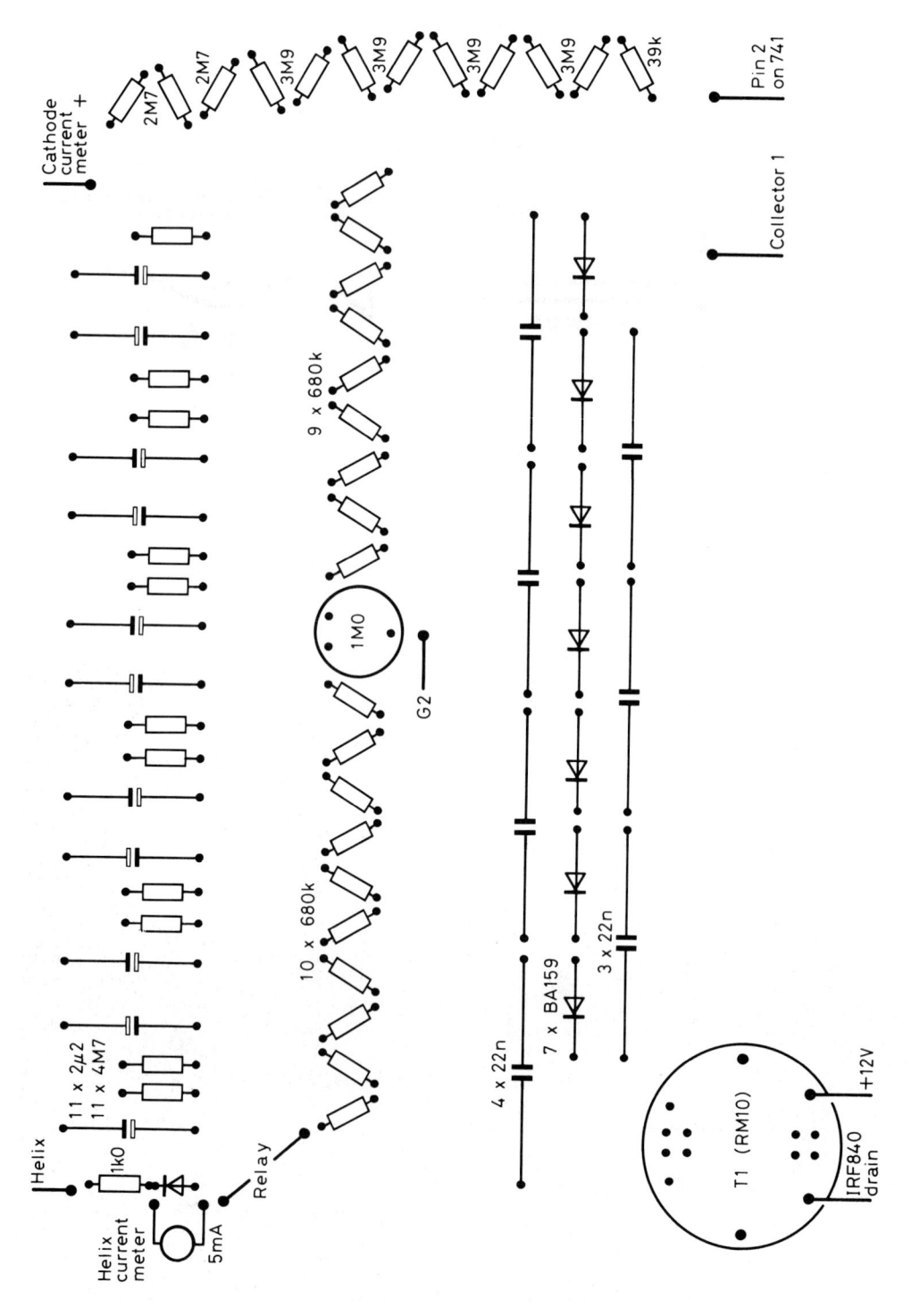

Fig 18.200. Component overlay for the G4JNT twt psu module 2 (G8AGN)

THE 10GHZ ("3CM") BAND 18.127

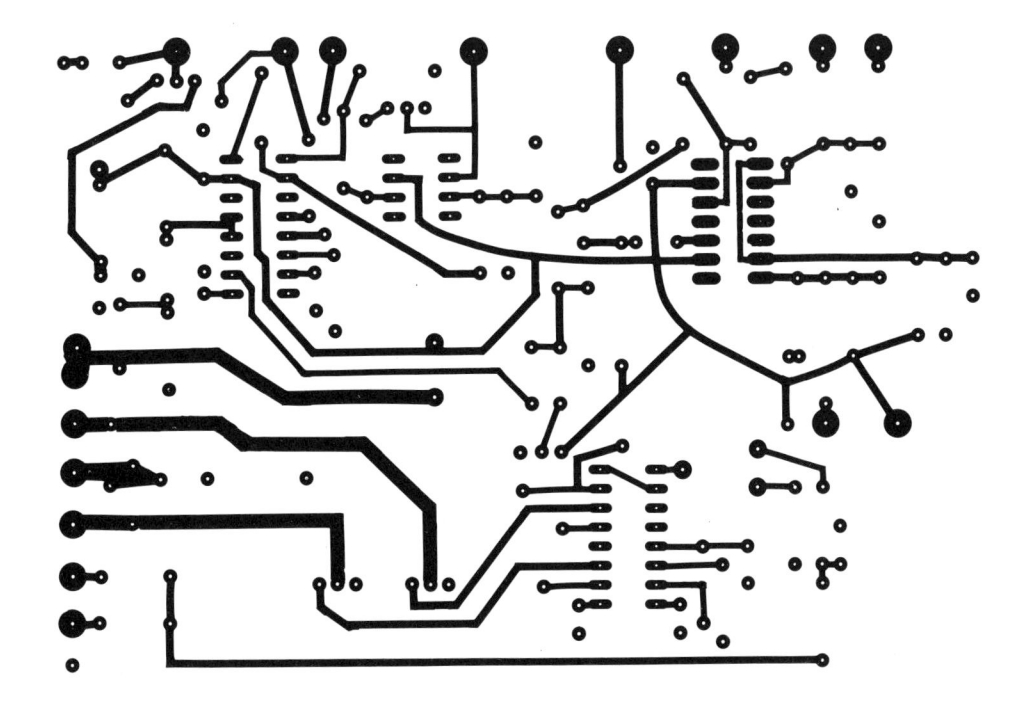

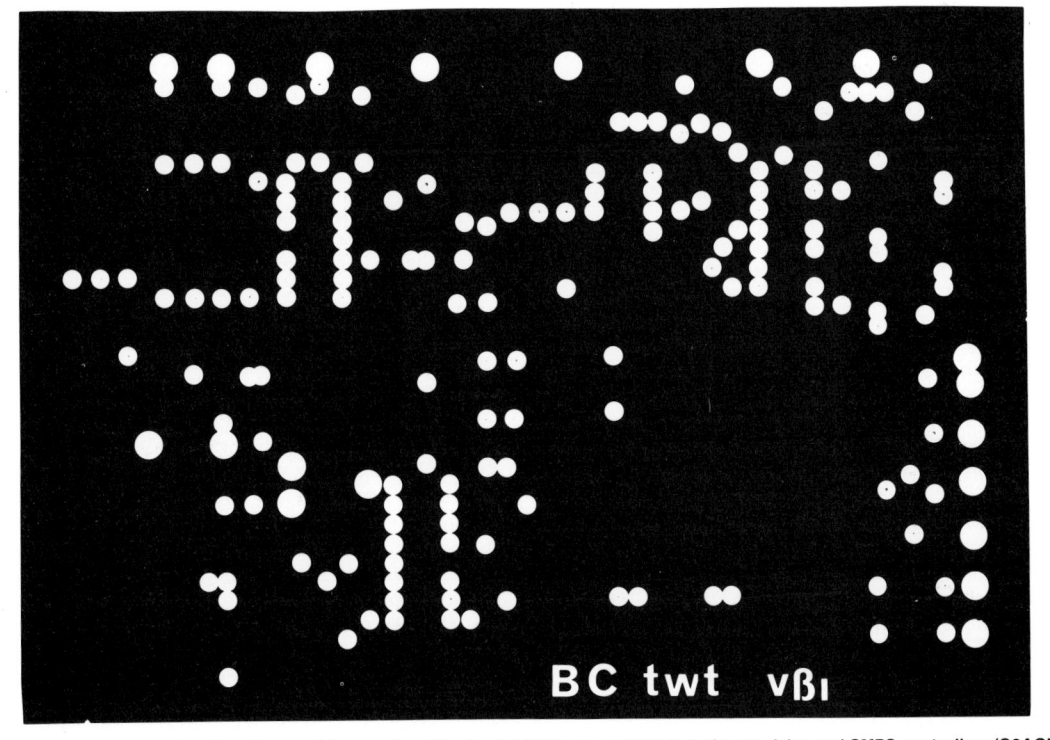

Fig 18.201. PCB layout (track and groundplane pattern) for the G4JNT twt psu module 3: timers, delay and SMPS controllers (G8AGN)

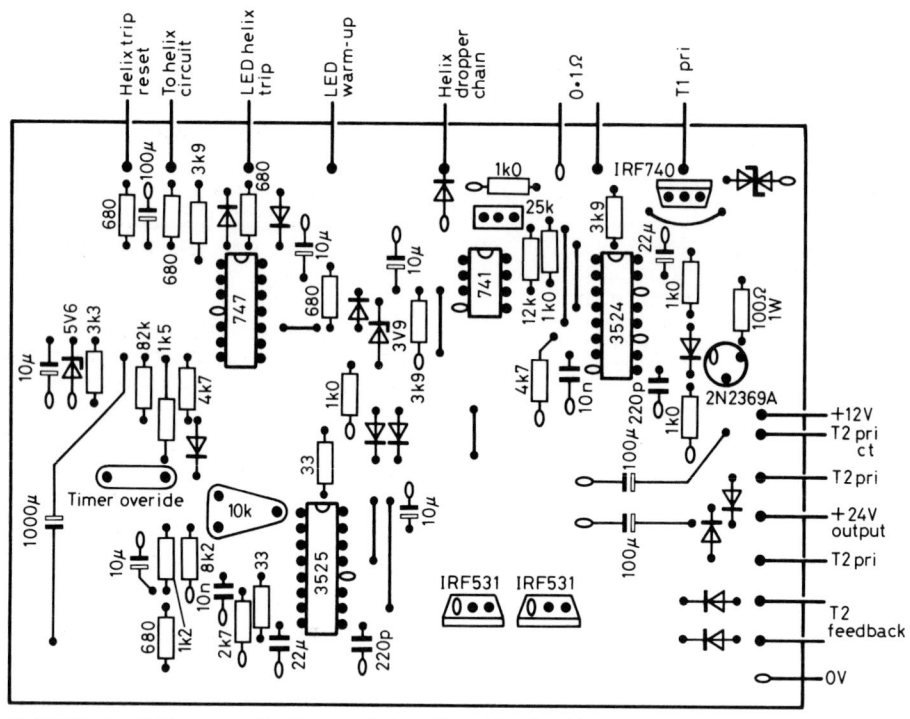

Fig 18.202. Component overlay for the G4JNT twt psu module 3 (G8AGN)

18.12 CONCLUSIONS

The development of the 10GHz amateur band has had a long and varied history. The 10GHz band is "different" from the lower microwave bands in so far as it is the first band where waveguide components are of such a size as to make compact, portable equipment feasible.

Amateurs have easy and ready access to many surplus professional components and to well designed, ready made oscillators and mixers, thereby simplifying the task of assembling wideband equipment. Arising from the simplicity and low cost of such equipment, so the band has proved popular with the beginner and the "simple" concept has been amply illustrated within this chapter. Wide use of simple equipment has led directly to the first generation of more sophisticated and, therefore, more "potent" narrowband equipment, allowing longer and obstructed paths to be worked.

Concurrent with these developments, advances in gallium arsenide (GaAs), microstrip and surface mount device (smd) technology have reached the stage where mass produced consumer devices for use in the 10 to 12GHz region are becoming available at economic prices and this has, in its turn, led to second generation equipment for both the wideband and narrowband modes. A number of practical designs for both wide and narrowband equipments have been described, together with possible alternatives and suggestions for development using the newer technologies. Operating techniques were dealt with elsewhere. Some test equipment and test methods peculiar to 10GHz have been discussed. Other suggestions may be found in chapter 10, "Test equipment".

The 10GHz band, therefore, is expected to grow in use and popularity as more amateurs become involved in mastering the techniques used in pcb-based transmitters and receivers. There has been relatively little work carried out on the setting up and use of 10GHz as a medium for local networks carrying really wideband traffic, such as high speed data links and for fast scan atv. In-band repeaters and the use of the band for linking and "slaving" other repeaters were mentioned in the chapter 9, "Beacons and repeaters", although again, little practical work has been done. Propagation research over difficult paths and by troposcatter are two further areas well worthy of more exploration by amateurs.

As a band where self-training in electronics and communications, construction and improvisation, operating skills and original experimentation are all possible, 10GHz is unsurpassed! The size and content of the chapter should emphasise this. Most of the information presented has come from the practical finding and experiences of many UK amateurs (and others) and has been "distilled" down as far as possible without, hopefully, becoming either too terse or too long.

18.13 ACKNOWLEDGEMENTS

The authors wish to thank the following individuals for their ideas or designs: G3JVL, G3SEK, G3WDG, G4DDK, G4JNT, G4KNZ, G6WWM, G8AGN and G8DEK. We would also like to thank the following organisations who assisted, either by permitting information to be reproduced, or whose products have been described or used to evolve the designs: Plessey Semiconductors Ltd, Microwave Associates and Mitsubishi Corporation. It should not be inferred that other manufacturers' products are unsuitable for the purposes described; simply that those products used were most easily accessible to amateurs.

Our thanks also go to those other individuals, too numerous to name, who have unwittingly contributed in some way, over the years, via the *RSGB Microwave Newsletter*.

18.14 REFERENCES

[1] D Clift, G3BAK, *RSGB "Bulletin"*, March, April and May, 1953.
[2] *VHF/UHF Manual*, 2nd Edition (1971), pp5.31 to 5.34, RSGB.
[3] *VHF/UHF Manual*, 3rd Edition (1976), pp8.23 to 8.26 and 8.33 to 8.35, RSGB.
[4] *VHF/UHF Manual*, 4th Edition (1983), pp9.39 to 9.41, RSGB.
[5] J N Gannaway, G3YGF, *Radio Communication*, August, 1981, pp710 to 714 and 717.
[6] S J Davies, G4KNZ, *RSGB Microwave Newsletter*, 09/82, October 1982.
[7] C Scrase, G8SHF, *RSGB Microwave Newsletter*, 10/81, November 1981.
[8] H Fleckner, DC8UG and G Bors, DB1PM, *VHF Communications*, 1/81, Verlag UKW Berichte.
[9] B Chambers, G8AGN, *RSGB Microwave Newsletter*, 06/85, July 1985.
[10] *Microwave transmission circuits*, G L Ragan, Vol.9, MIT Radiation Labs.
[11] *VHF/UHF Manual*, 4th Edition (1983), p9.27, RSGB.
[12] G Burt, GM3OXX, *Radio Communication*, June 1975.
[13] *VHF/UHF Manual*, 3rd Edition (1976), pp8.31 to 8.33, RSGB.
[14] M T Aylward, G8APP, and D S Evans, G3RPE, *Radio Communication*, February 1976.
[15] D S Evans, G3RPE, and C W Suckling, G3WDG, *Radio Communication*, June 1978.
[16] S Jewell, G4DDK, *RSGB Microwave Newsletter* 5/85, June 1985.
[17] G J Reason, G4EBF, *RSGB Microwave Newsletter* 07/81, August 1981.
[18] D Clift, VK5ZO, *RSGB Microwave Newsletter* 5/85, June 1985.
[19] P Schorah, GW3PPF, *RSGB Microwave Newsletter* 06/83, July 1983.
[20] S J Davies, G4KNZ, *RSGB Microwave Newsletter* 10/81, November 1981.
[21] P Tonbridge, G8DEK and D S Evans, G3RPE, *Radio Communication*, March 1976.
[22] C Elliot, G4MBS, *RSGB Microwave Newsletter* 02/83, March 1983.
[23] R Griek, DK2VF and M Munich, DJ1CR, *VHF Communications*, 2/79, Verlag UKW Berichte.
[24] M C Parkin, G8NDJ, *RSGB Microwave Newsletter*, 06/85, July 1985.
[25] "10GHz Gunnplexer transceivers – construction and practice", J R Fisk, W1HR, *Ham Radio*, January 1979.
[26] J Wood, G3YQC, *CQ-TV*, Issue 122, British Amateur Television Club (BATC).
[27] J Wood, G3YQC, *CQ-TV*, Issue 126, BATC.
[28] M Wooding, G6IQM, *CQ-TV*, Issue 132, BATC.
[29] D Crump, G8GKQ, *CQ-TV*, Issue 136, BATC.
[30] R Humphreys, G4WTV, *CQ-TV*, Issue 136, BATC.
[31] B Chambers, G8AGN, *RSGB Microwave Newsletter*, 06/82.
[32] *VHF/UHF Manual*, 4th Edition (1983), pp9.68 to 9.70, RSGB.
[33] R Evans, G3LQC, *RSGB Microwave Newsletter*, 09/82.
[34] "10GHz transverter in microstripline technique", P Vogl, DL1RQ, *DUBUS Technical Reports*, 2/86.
[35] "Redesigned 10GHz SSB transverter", P Vogl, DL1RQ, *DUBUS Technical Reports*, 4/86.
[36] "SHF microstrip circuits for amplifiers and filters", L Hansen, LA6LCA, *Report to Norwegian VHF Convention*, Geilo, 1986.
[37] "Microwave dielectric resonators", S Jerry Fiedziuszko, *Microwave Journal*, September 1986, pp 189 – 200.
[38] "Resonator filters for 9, 6 and 3cm bands", DK2AB, *DUBUS Technical Reports* 1/86.
[39] "Further information on the G3JVL 10GHz transverter", C W Suckling, G3WDG, *RSGB Microwave Newsletter Technical Collection*.
[40] *Modern High Performance Equipment For 10GHz*, Part 1, "The G3WDG–001 Multiplier/Amplifier" by C W Suckling, G3WDG and G B Beech. Published by the RSGB Microwave Committee, August 1990.
[41] *Modern High Performance Equipment For 10GHz*, Part 2, "The G3WDG–002 Receive Converter" by C W Suckling, G3WDG and G B Beech. Published by the RSGB Microwave Committee, February 1991.
[42] *Modern High Performance Equipment For 10GHz*, Part 3, "The G3WDG–003 Transmit Converter" by C W Suckling, G3WDG. Published by the RSGB Microwave Committee, July 1991.
[43] "144MHz interface and 'Universal' TX-RX Sequencer", I F White, G3SEK, *RSGB Microwave Newsletter*, 10/90, July 1991.
[44] A C Talbot, G4JNT, *RSGB Microwave Newsletter*, 08/90, February 1991.

APPENDIX A1 – PERFORMANCE CHECKING USING NATURAL NOISE SOURCES

Receiver sensitivity (the ability to detect and resolve a weak signal) is ultimately limited by random electrical fluctuations, mainly arising from thermal effects, commonly known as "noise". This noise may be generated within the receiver itself, as a result of random electrical fluctuations through such devices as resistors, diodes and transistors, or it may originate from external natural sources such as galactic sources, the sun, the earth or man-made sources. At microwave frequencies, man-made electrical noise is usually relatively insignificant, but noise from natural sources may be highly significant.

It is within our capability to minimise (but never totally eliminate) a receive system's self-generated noise. In the most demanding professional applications, for instance deep space communications and radio astronomy, the first-stage microwave amplifying devices and circuits may be cryogenically cooled and placed directly at the antenna in order to reduce the receiver noise to an absolute practical minimum. Such extreme measures as cryogenic cooling are unlikely to be realistic for amateurs, although placing the first amplifying stages close to, or at, the antenna is often practised. However, careful choice of devices, careful design of the circuits, proper matching and signal transfer, the use of the minimum practicable receiver bandwidths, the use of well-matched directional antennas and minimum feeder loss all contribute to enhanced performance. Control of many of these are within the capabilities and budgets of many amateurs.

These factors, being variable and interactive, infer that adjustment or alignment are also important in the quest for ever-better performance. What is important in this context is the ultimate signal to noise (S/N) or signal plus noise to noise (S+N/N) ratio at the output of a receiver: no matter what degree of amplification is applied to a weak signal, it will be impossible to detect that signal if it is below the level of the noise generated in the receiver system – including the noise contribution of the antenna and feeder!

Some simple, inexpensive, amateur noise-generating and measuring equipment has already been described in chapter 10, "Test equipment", including a receiver alignment aid containing a noise source and an audio power meter. Whilst equipment such as this may be regarded as essential by the avid microwave constructor/operator, it may not be immediately available to many operators, especially if only used occasionally. After optimising the receiver by instrument (if available), or ear (which is notoriously unreliable!), it is very desirable to be able to realistically estimate the noise performance in order to predict the system potential by the methods described in chapter 3, "System analysis and propagation".

Thermal noise and "black body" radiation

All matter whose temperature is above "absolute zero", i.e. −273°C or 0 kelvin, radiates wideband or "white" thermal noise which arises from the random movements of the molecular, atomic and sub-atomic particles which make up the matter. This noise is often known as "black body radiation" because it emanates from bodies which may not necessarily radiate energy in the visible part of the electromagnetic spectrum: for instance, the moon, the earth and other planets may appear to radiate light but that light is the reflected light of the sun.

The amount of noise radiated is proportional to the absolute temperature of the matter and the total noise power generated in a given receiver bandwidth is defined by the expression:

$$P = kTB$$

where P = noise power (in watts), k = Boltzmann's constant (1.38×10^{-23} joule/K), B = bandwidth (in hertz) and T = temperature (in kelvin). It can be seen that reducing either the receiver bandwidth or the temperature of the critical first amplifying stage(s), or both, will reduce the noise power at the receiver output.

"Cold" objects, for example the sky, radiate low levels of noise whilst "hot" bodies such as the sun, moon, earth, buildings, etc. radiate much higher levels. The higher the temperature, the more energetic the particles of matter in the object and the higher the noise radiated.

These natural sources of thermal noise can be used to check the performance of microwave receivers. It is possible to do this using a directional antenna as a variable noise source: depending where the antenna is pointed, so the receiver "hears" varying amounts of noise – larger amounts when pointed at a "hot" body, such as the sun, moon or ground and smaller amounts when pointed at a "cold" body, such as the sky. By measuring the "hot-cold" difference, it is possible to arrive at quite an accurate estimate of the receive system performance – sufficiently accurate to avoid seriously wrong answers when estimating overall system potential.

Comparing sun and sky noise

Fortunately, there is a very convenient way of checking overall performance: the measurement of sun noise. As already mentioned, the sun is a "hot" body, whilst the sky is a convenient reference "cold" body. The sun has the advantage that it can be located, even on a cloudy day when not directly visible, by approximation – in the northern hemisphere we all know the sun to be in the south at local noon, even if we don't know its exact elevation!

The procedure is simple. The receiving equipment is set up as it would normally be used, but with an audio noise measuring device attached to the output. The antenna is pointed towards the sun and moved in elevation and azimuth for maximum receiver noise output. The sun should be in a clear region of the sky, well clear of buildings, trees, etc. The receiver noise output power is then noted and the antenna moved off the sun to point at a clear region of the sky well away from the sun. The noise level is then noted again, and the ratio of noise power with the antenna on and off the sun determined.

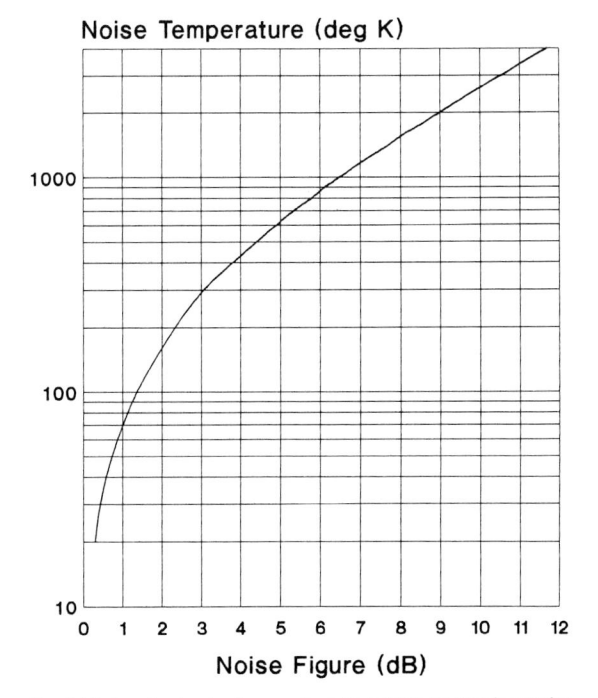

Fig A1.1. Graph of noise temperature (in kelvin) vs. receiver noise figure in decibels

The receiver noise output power when the antenna is pointing at the sky is proportional to the system noise temperature, T_{sys}, which is the sum of the receiver's own noise (proportional to its noise temperature, T_{rx}) and the noise picked up by the antenna (proportional to the effective antenna temperature, T_{ant}). If you are not yet familiar with the concept of noise temperature, the main point to grasp is that noise temperature is just a way of expressing noise powers although, if you have read chapter 3, you should be familiar with the concept! Fig 3.4 in that chapter plotted the relationship between noise temperature and noise figure and is reproduced here for convenience (Fig A1.1).

The receiver noise temperature, T_{rx}, is related to its noise figure by the expression:

Noise Figure (dB) = $10 \log(1 + T_{rx}/290)$,

where 290K, or 17°C, is the "standard" temperature. The effective antenna noise temperature, T_{ant}, is just a way of expressing the noise power coming out of the antenna. The effective noise temperature of a perfect antenna (with a lossless feed) is determined by the noise temperature of the object at which the antenna is pointed.

If the antenna is pointed at the sky, T_{ant} will depend on the noise temperature of the region of the sky at which it is pointed – for instance, at 10GHz, the background radiation from the sky has a noise temperature of 2.7K. However, the antenna will be less than perfect and will receive noise coming from elsewhere. The main contributions to this are pick-up of noise from the "hot" ground behind the antenna, due to "spill-over" from the feed and sidelobes, and noise coming from outer space, from various sources.

It is not possible, unfortunately, to estimate exact values for T_{ant}, since this depends on the type of antenna in use and on the particular region of the sky being observed. However, if the antenna is moved around to find the quietest region of the sky, it will probably appear to be in the range 20 to 40K and typically 30K. Note that, in most cases, T_{ant} is a very small quantity compared with the noise temperature of the receiver (T_{rx}), which means that the receiver sensitivity is governed by the receiver itself, external noise not being an important consideration. On many of the microwave bands, however, where receiver noise figures of 1 to 2dB (or, indeed, significantly less, depending on frequency) are now possible, the effective antenna temperature may well exceed the receiver's own noise, particularly in terrestrial systems where antennas, pointing at the horizon, pick up a lot of ground noise.

When the antenna is pointed at the sun, the receiver noise output is then proportional to the sum of two factors, T_{sys} (the total system noise, i.e. the sum of T_{rx} and T_{ant}) and T_{sun} which represents the extra noise due to the presence of the sun in the antenna beam. The quantity T_{sun} can be determined from the expression:

$$T_{sun} = \frac{F \times G \times \lambda^2}{3.462}$$

where F is the solar flux, G is the gain of the antenna (in real numbers, e.g. 3dB = 2, 6dB = 4, 9dB = 8, etc.) and λ is the wavelength in metres. It should be noted that sun noise exhibits different values at different frequencies, since its noise output is not purely thermal in nature.

F, therefore, varies with frequency and approximate values for the amateur microwave bands up to 24GHz are given in Table A1.1. The equation is also shown solved graphically in Fig A1.2. This shows T_{sun} as a function of dish diameter for these bands. The gain of the dish was calculated using the usual expression:

$$G = \frac{4\pi \times A \times n}{\lambda^2}$$

Table A1.1. Solar and lunar flux

Frequency	Solar Flux*			Lunar Flux**
	(typ)	(min)	(max)	(typ)
1.3GHz	100	35	120	0.08
2.3GHz	110	50	210	0.25
3.4GHz	120	75	270	0.50
5.6GHz	150	130	290	1.50
10GHz	300	270	460	4.50
24GHz	1100	1050	1200	26.00

* 1 Solar flux unit (SFU) = 10^{-22} W/m^2/Hz^{-1}
** Lunar flux expressed in same units (SFU)
Minimum flux represents the noise level of the "quiet" sun. Typical flux represents the noise level of the "normal" sun. Maximum flux represents the noise level of the "active" sun.

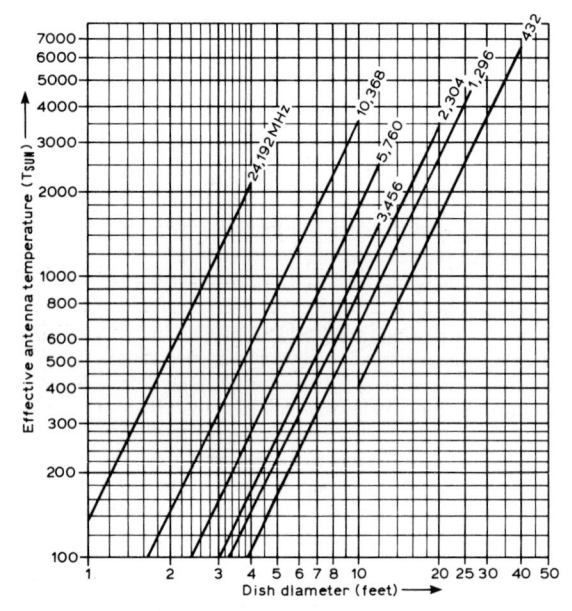

Fig A1.2. Graph of dish diameter (in feet) vs. effective antenna temperature (T_{sun} – see text) for the amateur bands from 432MHz to 24GHz

where A is the area of the aperture of the dish and n the efficiency, which may be taken as typically 0.5 (50%).

As an example, we may calculate the amount of sun noise which one might expect to "hear" at 10GHz, using a 10dB noise figure receiver and a 4ft (1.2m) dish. First we need to determine T_{rx}. Using the first expression, this comes out to be 2,611K. Add 30K as an estimate of T_{ant} to this and we obtain a value of 2,641K for T_{sys}.

Using Fig A1.2, T_{sun} for a 4ft dish at 10GHz is 570K. Thus the noise power from the receiver, when pointing at the sky, is proportional to 2,641K and, when pointing at the sun, to 2,641 + 570 = 3,211K. Thus the ratio of the two powers is given by:

$$\text{Power ratio} = \frac{3,211}{2,641} = 1.215 \text{ or } 0.85\text{dB}$$

The same calculation, assuming the receiver noise figure is reduced to 2.5dB with no other changes, gives an answer indicating that the system should now "hear" a sun/sky noise ratio of 2.5 (or 4dB).

Factors affecting accuracy

The first point to bear in mind is that this figure is only correct for a single channel receiver, i.e. one with a response only at the signal frequency. In simpler microwave receivers there may be no image protection, so that there will be responses at both the signal and image frequencies and twice as much power will be detected from the sun, as it emits broadband noise. In other words, T_{sun} will be twice as great as with a single channel receiver. In the above case, T_{sun} would equal 2×570 = 1,140K for a "wide-open", 10dB nf receiver, giving a sun/sky ratio of:

$$\frac{2,641 + 1,140}{2,641} = 1.43 \text{ or } 1.55\text{dB}$$

As is usually the case, the performance of the equipment will not be as good as expected and a lower figure than that calculated would be observed. The fault could lie in the receiver or antenna or, probably, both! This method of measurement will not give an indication of what is wrong, so both parts of the system will have to be investigated separately. An estimate of the receiver noise can be made using ground noise (see later), leaving the antenna gain as the only variable. Nevertheless, the factor of interest is the overall performance, since it is this which determines how well signals will be received. Any improvements made to the system will immediately show up in increased (observed) sun noise.

The second important point to be considered concerns the accuracy of the measurements. The value of the sun/sky noise ratio is very sensitive, at low values, to the noise figure and antenna gain. Small errors in measurement could lead to large discrepancies in the measured performance of the equipment. Small ratios are best measured using an audio noise voltmeter, such as that described in chapter 10, "Test equipment". The power ratio is then:

$$\text{Power ratio} = 20 \log \left(\frac{\text{noise voltage with antenna at sun}}{\text{noise voltage with antenna at sky}} \right)$$

Fig A1.3 plots the percentage increase in measured noise voltage against the receiver noise figure in dB and also shows an alternative, very simple and unsophisticated audio voltmeter which could be used instead of the meter mentioned above. One difficulty with this type of circuit, in this day and age, is getting hold of a suitable miniature audio transformer!

Larger values are better measured by inserting known values of attenuation at i.f or af when pointing at the sun, to bring the noise level down to the sky value, as monitored on the S-meter or, since S-meters are notoriously unreliable and often agc operated, on an audio noise meter. The value of the attenuator needed is then the required ratio (in decibels). The attenuator should ideally be a calibrated variable type, but either a switched-step attenuator or a range of fixed attenuators could be made according to the information given in chapter 10, "Test equipment" and chapter 13, "Data".

If an audio noise meter is used ensure that, for a.m modes (ssb, cw), the receiver agc is not acting and that, for fm, the receiver is well below limiting. In other words, it is essential that all the receiver stages are linear.

In this discussion, has been assumed that the sun emits a constant level of noise. In practice this is not the case and, often, a higher level is observed than quoted. Except on very rare occasions, the error will not exceed more than a few dB. The likelihood of error being large is less at the higher frequencies. Ideally, several measurements should

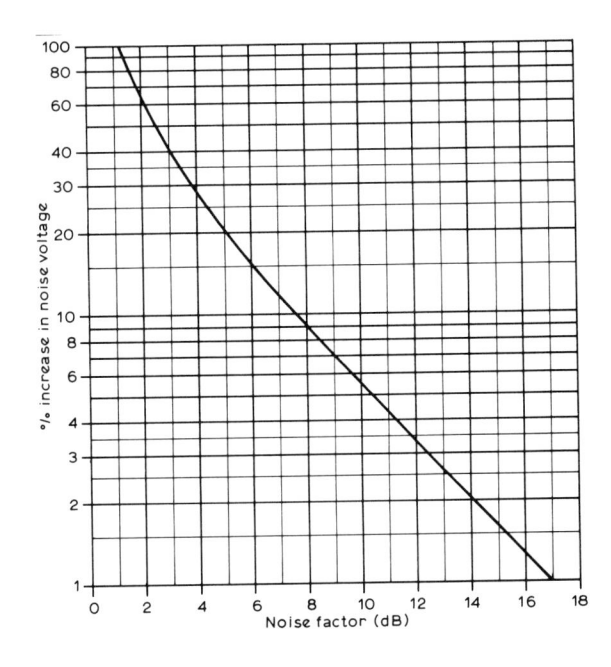

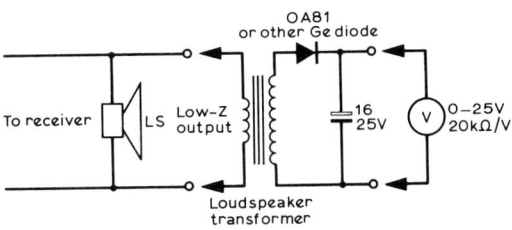

Fig A1.3. (Top) Graph showing percentage increase in receiver noise voltage as a function of noise figure. (Bottom) A simple audio voltmeter circuit which can be used in the absence of more sophisticated equipment

be made at different times on different days to eliminate the chance of a spurious result. The level is unlikely to change significantly over the period of a few minutes needed to make the measurements, allowing relative measurements to be made with a high degree of confidence.

Moon and ground noise

In view of the potential large variability of sun noise with solar activity or time of the solar cycle, two alternative natural sources of noise may be used in a similar manner: moon noise and ground noise, and again the "cold" reference is the sky.

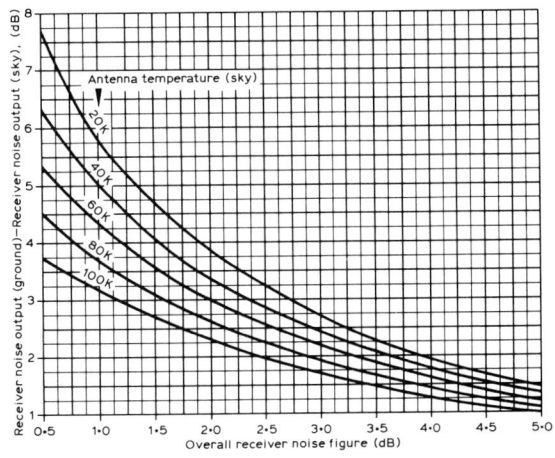

Fig A1.4. Graph of ground/sky noise differential (in dB) vs. overall receiver noise figure for different antenna temperatures (T_{ant} – see text)

Unlike the sun, the moon has an almost constant noise temperature, especially at lower microwave frequencies, because its radio noise emissions are purely thermal in nature and therefore inherently constant. However noise emission, as far as the earth-bound observer is concerned, is affected (particularly above about 5GHz) by both the phase of the moon (as varying areas of the earth-facing surface are heated by solar energy) and the orbital position of the moon relative to the earth, as its orbit around the earth is elliptical. The moon-phase effect can cause a variation of about 7% at 10GHz and 12% at 24GHz, whilst the orbital eccentricity can cause variations of about ±14%. These effects are almost sinusoidal in time and thus due allowance can be made for them. Maximum radio noise emission occurs about 45° (3.5 days) after full moon, at which time the effective moon temperature is 225K. Table A1.1 shows the mean lunar flux levels, expressed in the same units as solar flux (SFU). It can be seen that the lunar flux is very much smaller than that of the sun. Hence the noise emission of the moon is much smaller too.

Pointing the antenna at the ground can provide a somewhat higher and much more constant source of radio noise. The ground temperature is usually taken as 290K. In general, it can be assumed that the effective temperature of man-made structures, such as large buildings, will be the same as the ground, i.e. 290K. The calculated values of ground-sky noise output (in dB) as a function of receiver noise figure (in dB) for a range of effective antenna temperatures (i.e. sky temperatures) are given in Fig A1.4.

CHAPTER 19

The 24GHz ("12mm") band

19.1 INTRODUCTION

The techniques used at 24GHz are similar to those used at 10GHz and the equipment is similar, though physically only about half the size. Gunn diodes (and, less commonly, varactor diodes) can be used for generating power, while point contact diodes and Schottky diodes can be used for mixing and detecting power. These devices are usually mounted in waveguide, and other components such as directional couplers, attenuators, loads, etc. are also constructed in waveguide.

A feature of the band distinguishing it from 10GHz and lower frequencies is the marked dependence of propagation on the weather. The frequency is close to 22GHz, where signals are significantly absorbed by water vapour, and so high humidity and rain can greatly increase propagation losses. For this reason, this part of the spectrum has been little used professionally in the past and, as a consequence, there is little surplus equipment available. Indeed, the range 18 to 26GHz, known as K-band, was often omitted in microwave test equipment and in a manufacturer's range of devices.

The band is now being increasingly used for doppler radar applications such as intruder alarms and speed measurement systems. This type of equipment has to be low cost to be competitive and so Gunn diodes and mixer diodes for 24GHz are now available at reasonable prices. Complete Gunn oscillator and in-line oscillator/mixer assemblies are also available which considerably simplify getting started on the band.

19.2 PROPAGATION

At 24GHz there is a significant loss due to the presence of water in the atmosphere. This attenuation is due to absorption by water vapour molecules and scattering by raindrops. At higher frequencies (e.g. 48GHz) oxygen also causes absorption, but the effect is small at 24GHz (just over 0.01dB/km) compared to water.

The attenuation due to water vapour molecules depends upon the humidity and typically varies from around 0.05 to 0.20dB/km. More comprehensive information on atmospheric absorption losses (provided by G8AGN) is to be found in chapter 20, "The bands above 24GHz", and this is also applicable at 24GHz. Fig 19.1 shows the attenuation (in dB/km) for different temperatures and humidities; a figure of 0.1dB/km is typical for the United Kingdom.

The attenuation due to rain scattering depends upon the size and shape of the raindrops, and the number intercepted along the path. Fig 19.2 shows the expected loss, in dB/km, for rainfall rates between 0.1 and 100mm/hour. The graph strictly applies to steady, widespread rain and horizontally polarised signals. Thus, for a line of sight path, the attenuation comprises the sum of several components:

1. Free space loss
2. Water vapour absorption loss
3. Rain scattering loss (if present)

If it is only raining over part of the path, a sufficiently accurate estimate of the rain scattering loss can be obtained

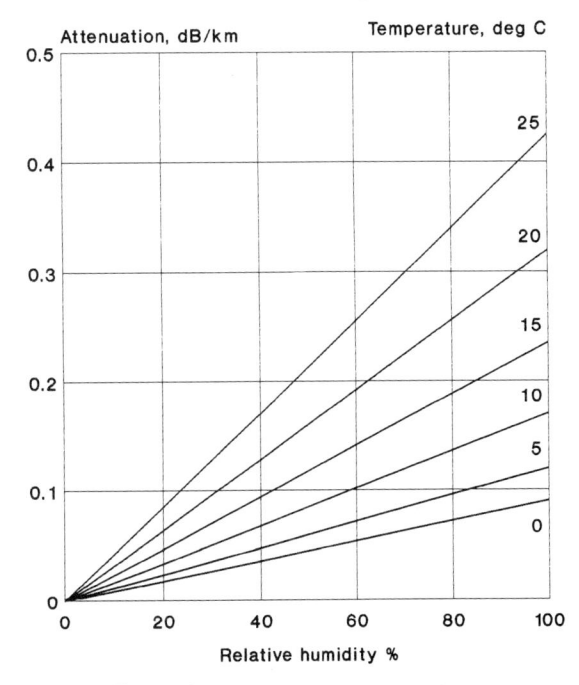

Fig 19.1. Loss due to water vapour at 24GHz

Table 19.1. Path loss capability at 24GHz

Transmitter	Receiver	Path loss capability	Approx. line of sight range
7mW WB 20dB horn	15dB NF 20dB horn	142dB	10km
10mW WB 35dBi dish	10dB NF 35dBi dish	179dB	150km
10mW NB 35dBi dish	10dB NF 35dBi dish	207dB	350km

by multiplying the loss per kilometre (from Fig 19.2) by the length of the path over which it is raining.

The droplets in dense fog also cause scattering and give an attenuation comparable with rainfall of 2.5 to 5.0mm/hour. Snow and hail also cause a significant loss but little data is available on these effects.

Fig 19.3 shows the path losses for various types of path: basic free space, free space plus water vapour, light drizzle and heavy rain. Apart from the free space line, water vapour loss at 0.1dB/km has been included. The light drizzle path loss assumes an additional loss of 0.1dB/km due to 1mm/hour rainfall; the heavy rain path loss assumes a loss of 1.3dB/km due to 10mm/hour rain.

It seems likely that the best time to try a long path may be in winter on a dry, cool, clear day, when the humidity is very low. However, do not discount operation in showery weather; heavy rain is unlikely to occur completely over a long path and it may be possible to take advantage of dense rain clouds by using them as passive reflectors.

Ducting over water should not be discounted either; this phenomenon is well known on 10GHz and should occur on 24GHz. The difference is that the water vapour needed to form ducting conditions will also absorb the signals. Thus, for an enhancement of signals to occur, the gain due to ducting must outweigh the additional absorption.

19.3 EQUIPMENT CAPABILITY

A simple wideband station might consist of a 7mW Gunn oscillator, a 1N26 receive mixer with an overall noise figure of, say, 15dB and a 20dB horn antenna. An advanced wideband station might use a 12inch diameter dish having a gain of 35dBi, an output power of 10mW, and achieve a lower noise figure of 10dB. A bandwidth of 200kHz is used in each case. The path loss capability of these two types of station (in each case, working to a similar station) is shown in Table 19.1. For comparison, a high performance narrow-band station is also included, using a 3kHz bandwidth i.f.

It must be stressed that the ranges shown in Table 19.1 assume an average loss of 0.1dB/km due to water vapour, while in practice this will vary considerably. Figs 19.4 to 19.6 show the capabilities of 24GHz wideband equipment in more detail, with the effect of temperature and humidity taken into account. They are based on equipment using the Plessey GDHM32 module with an assumed (typical) rf

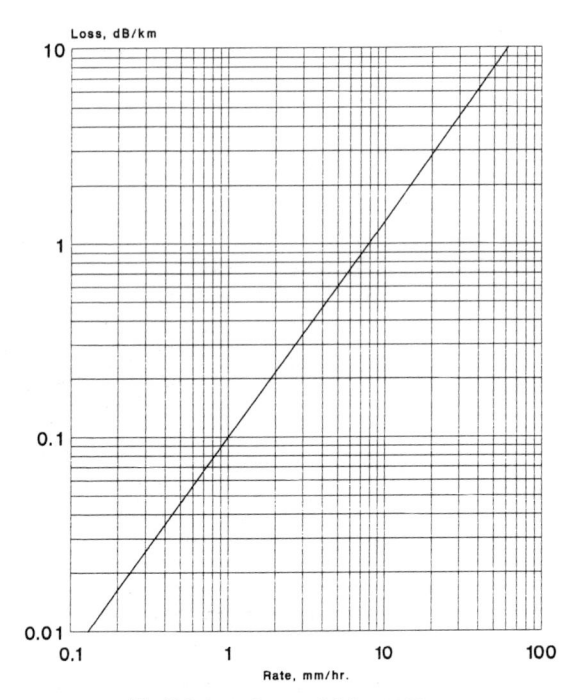

Fig 19.2. Loss due to rainfall at 24GHz

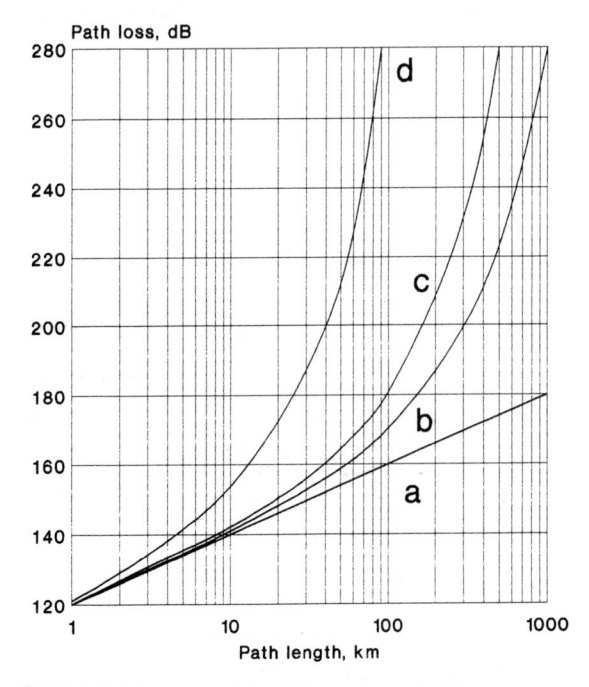

Fig 19.3. Path losses at 24GHz. (a) Free space loss. (b) Free space plus water vapour. (c) Light drizzle. (d) Heavy rain

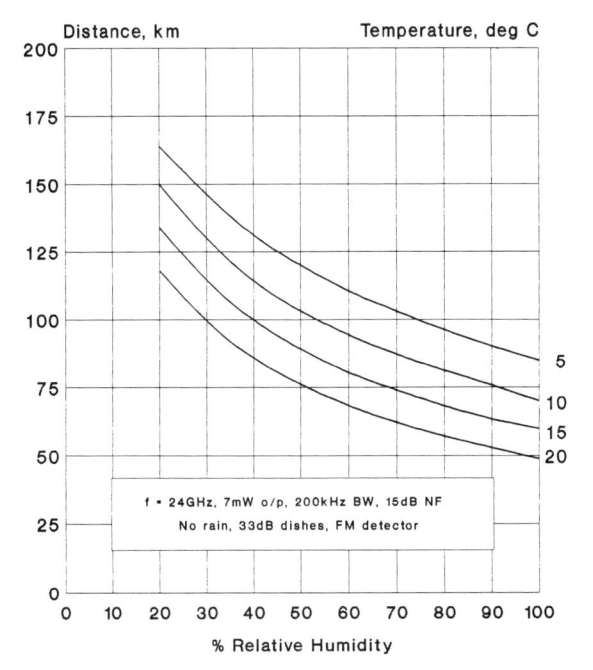

Fig 19.4. Path lengths workable at 24GHz as a function of relative humidity and temperature, for 200kHz bandwidths

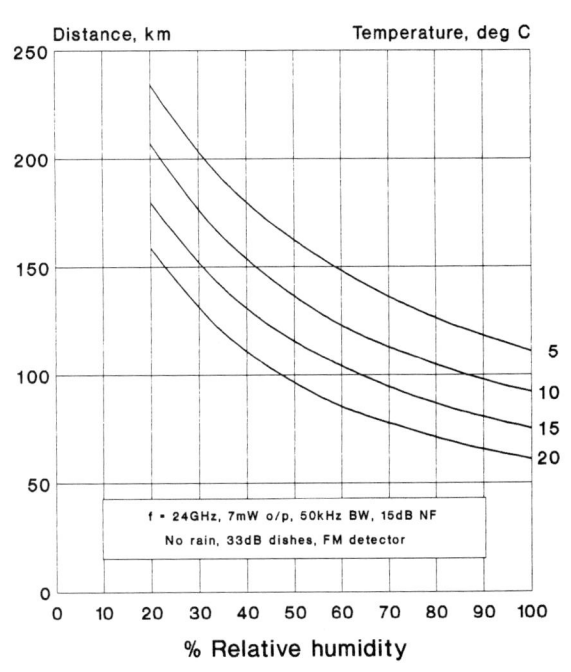

Fig 19.5. Path lengths workable at 24GHz as a function of relative humidity and temperature, for 50kHz bandwidths

output of 7mW and a noise figure of 15dB. Two i.f bandwidths are assumed – 200kHz and 50kHz.

The results suggest that reliable operation in temperate zones such as the UK over paths greater than 150km can be achieved with relatively modest equipment by using narrower bandwidths than commonly used on this band, and coupling this with taking advantage of cold, dry weather conditions. Judging by the success of Italian amateurs in operating over a 450km path on 24GHz in the Mediterranean, it seems likely that long distance contacts may also be assisted by anomalous propagation conditions, e.g. ducting, but this has yet to be investigated properly.

19.4 24GHZ COMPONENTS

19.4.1 Waveguide and flanges

Three standard waveguides can be used – WG20, WG21 and WG22. The alternative names, frequency ranges and cut-off frequencies are shown in Table 19.2.

Avoid WG22, since 24GHz is close to its cut-off frequency. Of the other two types, WG20 is preferred as it is in common use and more readily available. The waveguide wavelength for WG20 is shown in Fig 19.7. The attenuation of copper WG20 is approximately 0.11dB/foot and that of brass WG20 is 0.18dB/foot. There is no overriding reason why a standard guide has to be used. Rectangular sections of a similar size such as used for model making are suitable and are much cheaper. A disadvantage is that

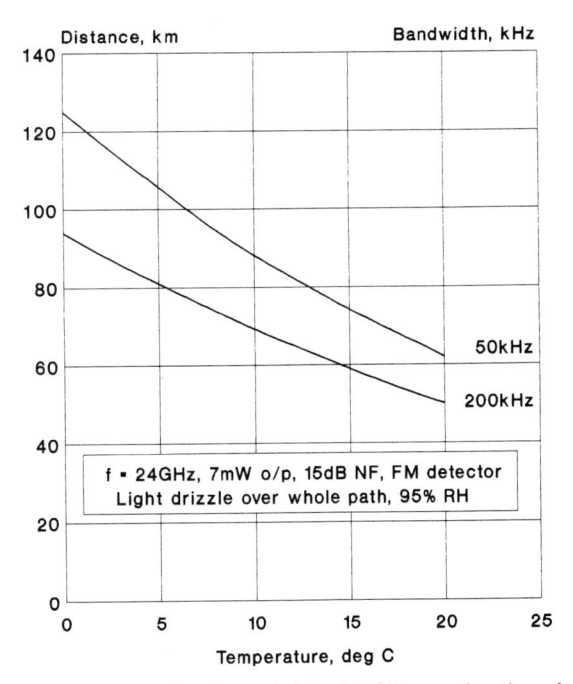

Fig 19.6. Paths lengths workable at 24GHz as a function of temperature, with light drizzle over the wholepath and 95% relative humidity

Table 19.2. Waveguides for 24GHz

WG number	EIA desig	IEC desig	Recommended freq range (GHz)	Cut-off freq (GHz)
WG20	WR42	R220	18.0 to 26.5	14.05
WG21	WR34	R260	22.0 to 33.0	17.36
WG22	WR28	R320	26.5 to 40.0	21.08

The dimensions of these guides are as follows:

WG number	Internal dimensions (mm)	External dimensions (mm)	Wall (mm)	Aspect ratio
WG20	10.67 × 4.318	12.70 × 6.350	1.016	2.47
WG21	8.636 × 4.318	10.67 × 6.350	1.016	2.00
WG22	7.112 × 3.556	9.144 × 5.588	1.016	2.00

Table 19.3. Semi-rigid cable at 24GHz

Diameter	Loss at 24GHz	Overmode frequency
0.085"	1.4dB/foot	45GHz
0.141"	0.9dB/foot	30GHz

diameter semi-rigid, due to the fact that higher modes can propagate down what becomes, in effect, a circular waveguide. Two suitable sizes are shown in Table 19.3.

SMA connectors, whilst normally only rated to 18GHz, are the only common type with acceptable performance. Do not use larger connectors (e.g. N-type) as these will overmode.

19.4.3 Waveguide to coax adapter

To use coax cable, a waveguide to coaxial adapter (or transition) is needed. A design is shown in Fig 19.9, transforming from WG20 to an SMA connector.

The transition consists of a piece of waveguide with one end shorted, and a coupling probe mounted 3/4 λg from this short. Alternatively, the probe can be positioned 1/4 λg away, but this makes construction difficult.

The SMA is a standard socket with a square flange. The inner conductor socket spill is used as the probe and the length trimmed for best matching on 24GHz (between 5 and 6mm); the spill diameter should be 1.27mm.

The easiest method of trimming the spill length is to build two adaptors and connect the coaxial sockets together with an SMA back-to-back adaptor. Then connect a source to one waveguide flange and a detector to the other, preferably with waveguide attenuators, as shown in Fig 19.10. The spills can then be cut down in length until the detected power reaches a maximum, ideally only a dB or so less than the power measured without the two adaptors in circuit.

19.4.4 RF sources and detectors

The easiest way of generating power in the band is to use a Gunn diode mounted in a waveguide cavity and the free-running oscillator obtained is suitable for use as a transmitter or as a local oscillator in a receiver. However, wideband

standard flanges cannot then be used. Also, unless a transition is made up, it will not be easy to connect the system to standard measuring equipment.

As with WG16 at 10GHz, both square and round flanges are used with WG20. To further complicate matters, several different types of round flanges may be found (on older equipment) which are not interchangeable. The square type is preferred and dimensions of a standard flange are shown in Fig 19.8.

19.4.2 Coaxial cable

Equipment for 24GHz is usually built in waveguide as the loss of coaxial cable is excessive. Occasionally, however, it may be necessary to use cable and a semi-rigid type should be used. Braided cables are not suitable; neither is large

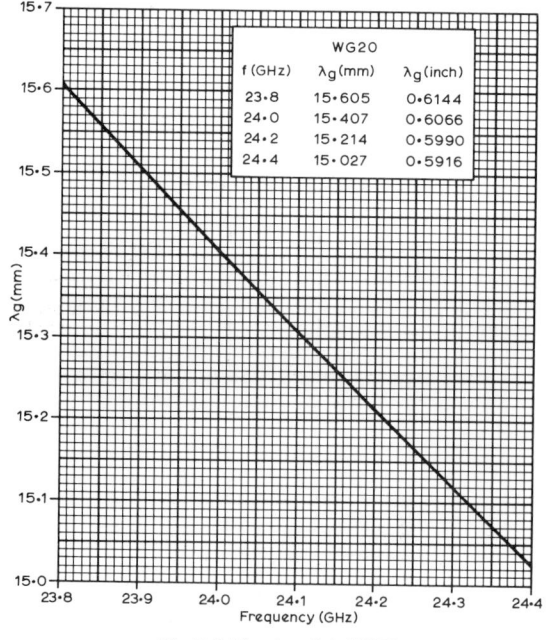

WG20		
f (GHz)	λg (mm)	λg (inch)
23.8	15.605	0.6144
24.0	15.407	0.6066
24.2	15.214	0.5990
24.4	15.027	0.5916

Fig 19.7. Wavelength in WG20

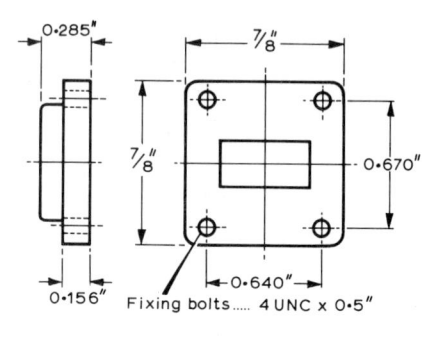

Fig 19.8. Dimensions of a standard WG20 flange

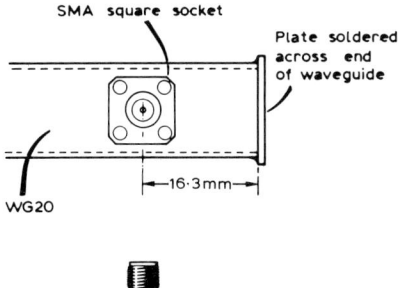

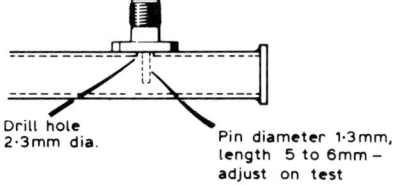

Fig 19.9. Construction of a 24GHz waveguide to coaxial adaptor

fm modulation must be used because of the relatively poor stability, with a minimum receiver bandwidth of one or two hundred kHz. Gunn diodes are available at reasonable cost, and are also available mounted in a waveguide cavity. Purchasing a complete assembly is a more reliable way of getting started, since the device can be obtained ready tuned to the amateur band. See later for details of the Plessey GDHO33, which has a typical power output of 10 to 20mW. An alternative method is to multiply up a lower frequency signal using a varactor diode or step recovery diode. In this case the stability will be determined by the stability of the source, which will usually be crystal controlled. This type of source will be needed for narrowband (cw or ssb) communication.

For mixing and the detection of received signals, two types of diode are suitable – the point contact diode and the silicon Schottky diode. The point contact device commonly used is the 1N26 and derivatives (1N26A, 1N26B, etc.) – the letter suffixes indicate later versions with a lower noise figure. The silicon Schottky devices are more recent devices and capable of better performance, but their cost is usually prohibitive.

An alternative is a complete Gunn/mixer assembly designed for doppler radar applications. An example is the GDHM32 from Plessey, which could be used to form the basis of a wideband rig, and may be successfully used with

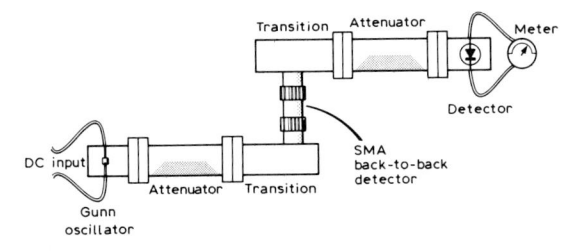

Fig 19.10. Set-up for checking operation of the waveguide to coaxial adaptor

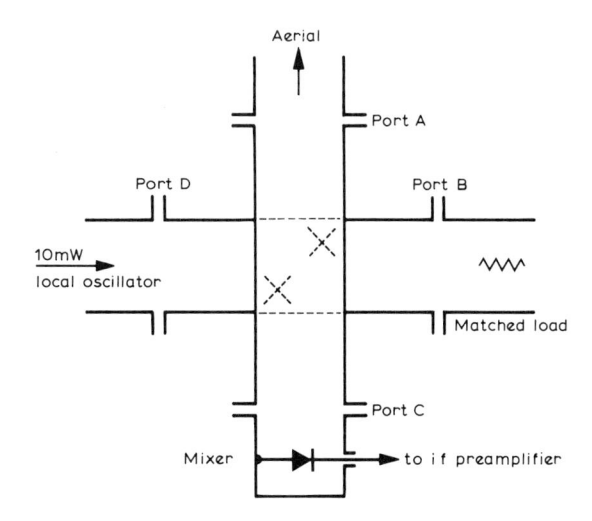

Fig 19.11. Wideband cross-coupler transceiver, set up for receiving

the basic 10.7MHz receiver described in chapter 18, "The 10GHz band", since it is available with negative earth, i.e. bias positive.

19.5 WIDEBAND TRANSCEIVER

The following design for a 24GHz transceiver is based around a directional coupler as shown in Figs 19.11 and 19.12. In the receive mode the Gunn local oscillator is connected to port 1, a matched load to port 2, the mixer and wavemeter to port 3 and the antenna to port 4. The Gunn

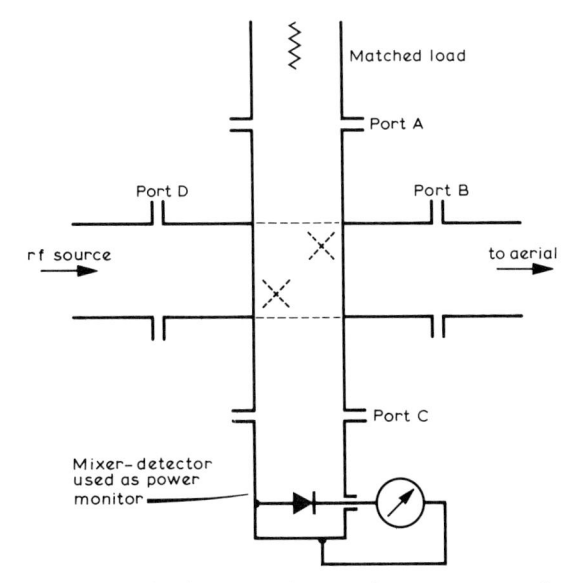

Fig 19.12. Wideband cross-coupler transceiver, set up to transmit

Fig 19.13. Plessey GDHO33 oscillator

oscillator operates into a matched load for best stability, with a small portion of the power being coupled into the mixer diode. The signal from the antenna is also fed into the mixer.

On transmit the connections to ports 1 and 3 are reversed and the Gunn oscillator now operates directly into the antenna. The mixer can be used as a power indicator and to check the frequency using the wavemeter. A local oscillator power of around 0.5 to 1.0mW is usually required by the mixer; for a 10mW Gunn source this means a coupler of 10 to 13dB is required. In the following sections, construction of such a coupler and the other individual components is described.

The directional coupler is also a very useful piece of test equipment. It can be used to monitor the vswr of any component, for example an antenna, while adjustments are made to the matching. In this mode, the Gunn is connected to port 1, a load to port 2, a mixer and wavemeter to port 4 and the antenna to port 3. The diode current then indicates power reflected from the antenna. The reading corresponding to an infinite vswr can be seen by placing a metal plate across the flange on port 3. The accuracy of the measurements will depend on how carefully the coupler is assembled and soldered.

19.5.1 GDO33 and GDHO33 oscillators

The availability of the Plessey GDO33 and GDHO33 Gunn diode oscillator assemblies means that a complete oscillator module can now be purchased working in the amateur band.

The GDHO33 is shown in Fig 19.13. It consists of a Gunn diode mounted in a waveguide cavity, with an iris plate cast into the assembly through which the output is coupled into standard WG20. The power output is around 10 to 20mW, and it is normally supplied set to 24.125GHz.

It requires a −5V supply at approximately 250mA. Tuning is via a small metal screw protruding into the cavity. This oscillator is ideal as a transmitter, where the frequency need not be altered once set. It may also be used as a receive local oscillator where tuning is possible by altering the supply voltage.

The GDO33 is somewhat different in construction and has no iris plate. This means it is much less stable than the GDHO33 – it can be pulled over 100MHz by varying the load impedance, and it is thus necessary to fit an iris plate. The plate can be fitted externally, by extending the cavity length to 4λ with a spacer plate and this allows a nylon or ptfe tuning screw to be fitted also. This means the oscillator is then readily tuneable and it is suitable for use as a receive local oscillator. The modification, designed by G3YGF, is shown in Fig 19.14.

The length of the GDO33 is extended by 9mm using a brass or aluminium plate as a spacer. The plate is cut to the size of a flange, and a flange then used as a template to drill 4 holes through the spacer in the corners. Then several holes are drilled in the centre of the plate, and the holes filed out to WG20 inside dimensions (approx 9mm × 4mm). A 4BA (or M4) hole is then drilled and tapped in the centre of the broad face of the spacer for a tuning screw. The iris plate is made from a 3 to 6 thou (0.08 to 0.15mm) brass or copper sheet with a 13/64" (5.2mm) hole in the middle of it. All the components are held together by four bolts which pass through the spacer and iris plate, the flange on the oscillator module and the WG20 flange to be connected. The GDO33, spacer and iris plate are shown in Fig 19.15.

With the hole diameter as given, the output power will be reduced by about 3dB. The tuning range is about 200MHz with a 4BA nylon screw (100MHz with ptfe screw) which should be adequate for most purposes. Fine tuning can then be provided by adjustment of the supply voltage.

19.5.2 Gunn power supply

This power supply, due to G4CNV, is designed to be used with the GDO33 and GDHO33 Gunn diode oscillators. It

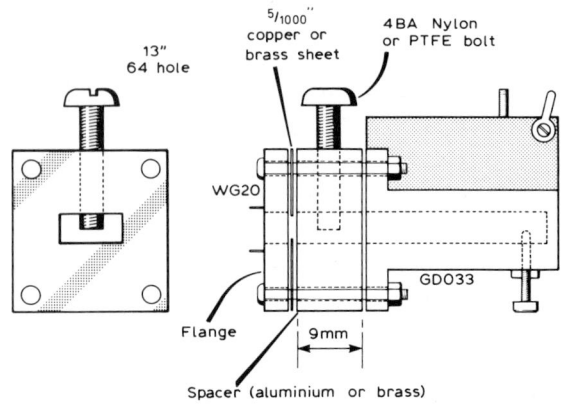

Fig 19.14. Schematic of modification to Plessey GDO33 oscillator

produces up to 250mA at −5V using the circuit shown in Fig 19.16.

One of its features is the inclusion of a current limiter, which makes the power supply short-circuit proof. Tone modulation is provided by a cmos oscillator followed by a simple RC filter. The microphone amplifier was designed for use with a crystal microphone, but is suitable for most types. In the event of more gain being required, the 330Ω resistor in the transistor TR4 should be bypassed with a 10μF capacitor. AFC inputs of either sense are provided, but if a negative earth i.f strip is used, some form of level translator will be necessary. No connections are made to these inputs if the afc facility is not required.

A pcb layout for the power supply/modulator is shown in Fig 19.17, and the component layout in Fig 19.18. Double sided copper clad board should be used, the top surface being left unetched to act as a groundplane. A number of components are mounted externally, these being the tone/audio selector switch, the TIP2955 transistor (mounted on a heatsink, such as the diecast box used to house the power supply), the microphone gain potentiometer and the output voltage setting potentiometer. The latter should be of the 10 turn variety for ease of tuning, since the available range of output voltage (3.5 to 5.5V) will tune the oscillator over almost 100MHz.

If the rest of the equipment has a negative earth, it will be necessary to run the Gunn supply from a separate source, such as another battery. A convenient alternative would be to use a small inverter (with a very well smoothed output), running from the main +12V supply, to produce −12V.

More recently, the use of the Plessey GDHM32 Gunn oscillator/in-line mixer module has proved popular, as it enables those familiar with similar techniques at 10GHz to get going on the band with a minimum of complication. The power output, as a transmitter, is nominally 7mW

Fig 19.15. Plessey GDO33 oscillator with spacer and iris plate

and the unit may be purchased optimised for the amateur band.

One main difference between this module and the others, mentioned earlier, is that the "amateur band" version can be supplied with negative earth, i.e. it requires a more normal positive bias supply. This means that many of the power supply/modulator circuits discussed in chapter 18 can be used to supply and modulate the module, the only proviso being that the voltage range be restricted to that recommended by the manufacturer. This is normally about 3.5 to 6.5V, whereas a 10GHz Gunn may require 6 to 10V. Means of reducing the voltage range of stabilised Gunn supplies were discussed in chapter 18.

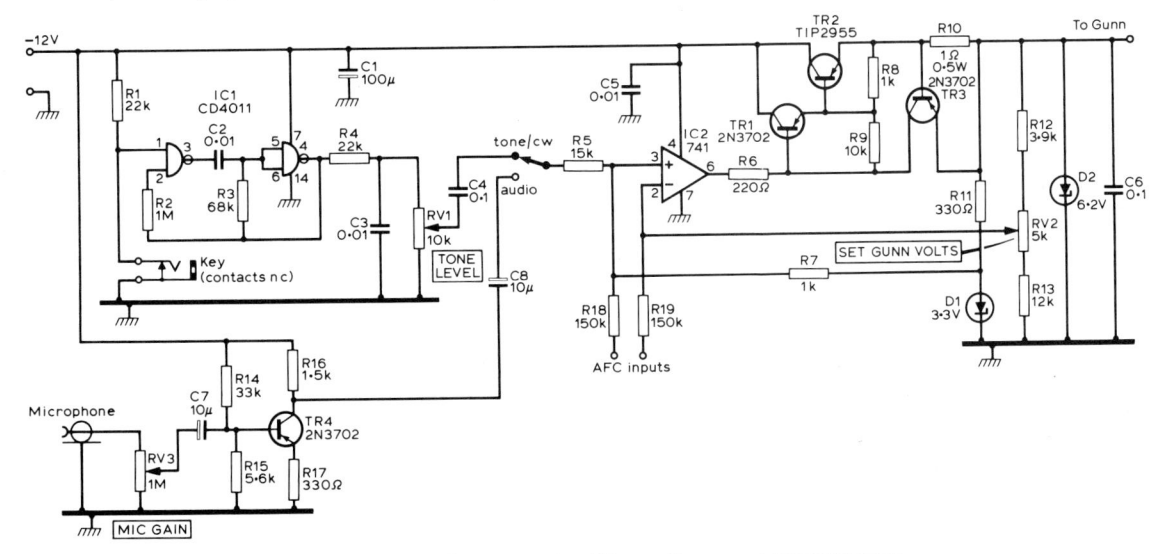

Fig 19.16. Schematic of power supply unit for positive ground GDO33/GDHO33

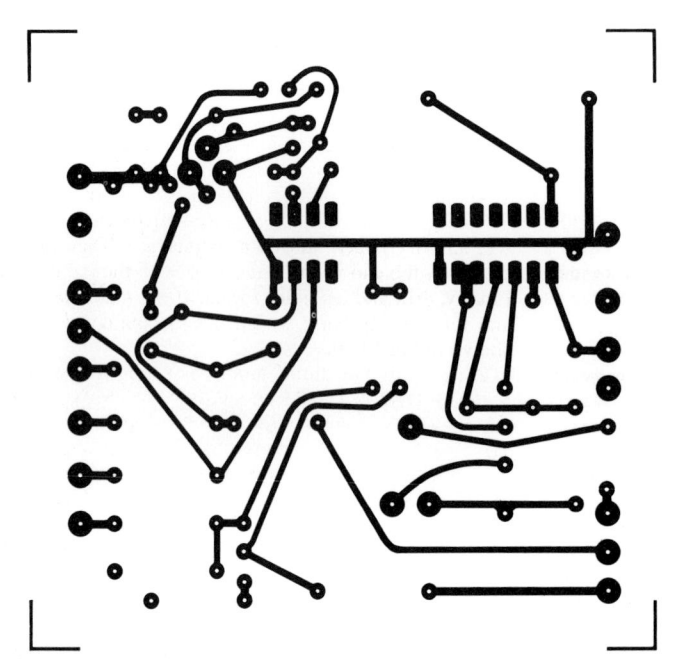

Fig 19.17. PCB track pattern for positive ground power supply unit

19.5.3 1N26 diode mount

The mixer is constructed out of WG20 and can be built using a vertical bench drill and hand tools only. Constructional details are shown in Fig 19.19.

First cut a 30mm length of copper or brass waveguide WG20. Square off both ends of the guide by marking out with a set square and filing to the marks. The length of guide from the flange end to the first matching screw is not critical, but a minimum of 6.5mm is needed for the locking nut and flange. Then drill and tap a 0BA (or M6) hole 7.1mm from the squared-off end. The tapping should be commenced using a tapered tap, changing over to a plug

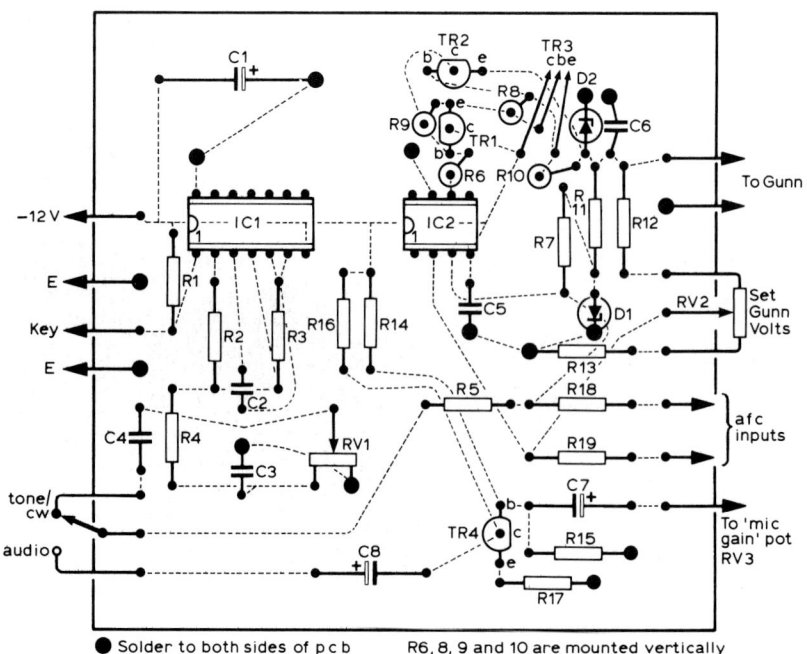

Fig 19.18. Position of components on psu pcb

● Solder to both sides of p c b   R6, 8, 9 and 10 are mounted vertically
• Wire passes through board and soldered to reverse side of p c b

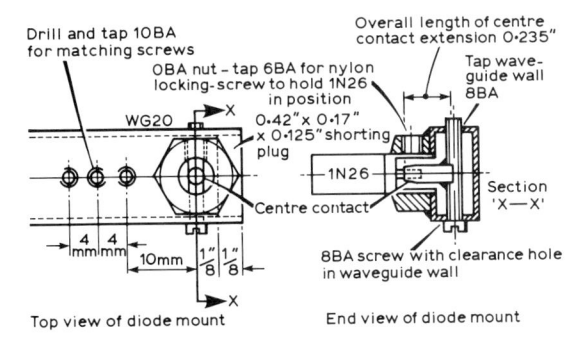

Fig 19.19. Construction of a WG20 1N26 diode mixer/detector mount

tap when the tapered tap comes up against the opposite wall. Next, assemble a 0BA nut on a 0BA screw, the latter preferably cadmium plated or sufficiently dirty so as not to take solder readily. Screw this a short distance into the waveguide and tighten the nut finger-tight against the waveguide wall. Using a hot plate or small flame to provide heat, solder the nut and flange in place. When cool, remove the screw and drill out the nut using a 5.7mm drill. Drill and tap the nut 6BA (or M2.5) through one of its faces, and remove the burr on the inside using the 5.7mm drill.

Drill with a 2.2mm diameter drill through the centre of the narrow wall of the waveguide, 7.1mm from the end, continuing through the opposite wall. Fit a 20mm long piece of 8BA (or M2) brass studding through the holes, fit a nut on each end and tighten up. Using the 5.7mm drill inserted through the nut, make a mark on the studding. Remove the studding and using a small needle file make a 0.090inch (2.29mm) wide slot in the studding, centred on the marked point, to half way through its diameter.

Next prepare the centre conductor for the 1N26 in the following way. Remove the centre connecting piece from a cable mounting BNC socket, saw off the end remote from the metal fingers and file to length (6.0mm). Carefully squeeze the metal fingers, using the wire cutting edges of a pair of pliers, so that an opposite pair of fingers are nearly touching. The connecting piece should then be a good push fit on the centre pin on the 1N26. An alternative connecting piece which does not require such modification can be one of the contacts from the high quality ptfe type of transistor holder.

Fit Sellotape collars onto the ends of the 8BA studding, leaving enough room to fit the nuts on the ends. Refit the studding into the waveguide. Next fit a ptfe washer on each end on the studding to insulate the nuts; these can be cut from a thin sheet. Then fit a washer and nut on one end, and a solder tag and nut on the other. Tighten up the nuts keeping the filed slot on the studding central and facing upwards.

Fit the modified centre conductor onto the 1N26 and insert the diode into the 0BA nut, pushing it in until the end of the centre conductor engages in the slot in the studding. Using as large a soldering iron as will fit, quickly solder the centre conductor to the studding using a minimum of solder. Take care not to leave a large blob of solder on the

joint. As soon as the solder solidifies, remove the diode to reduce the chance of damage. Check that the Sellotape insulation has not been damaged, using a resistance meter between the solder tag and waveguide.

The shorting plug is then made by sawing out a piece of 1/8inch (3.175mm) brass sheet and carefully filing to size. Check the dimensions during the filing to ensure accuracy. The plug is best finished of by rubbing on emery paper or wet and dry paper, laid on a flat surface. Then press the plug into the end of the waveguide with the aid of a vice. Finally, drill and tap holes for the 10BA matching screws, and remove any burrs with a file. A completed mount is shown in Fig 19.20.

Setting up the diode mount is straightforward; the matching screws should be adjusted for maximum diode current (measured between the centre conductor and the waveguide body). The screws should be retained in their optimum positions with locking nuts. In addition to its use as a mixer, this type of mount is very useful as a power indicator. This could consist of a second mount constructed as above with a 50 or 100μA meter mounted on the waveguide, similar to the design for 10GHz in chapter 18.

19.5.4 Directional coupler

The round-hole type of coupler is an attractive proposition, but at the power level of the GDHO33 (10 to 20mW) insufficient coupling can be obtained to give an adequate level of local oscillator injection, and it is necessary to use the Moreno cross-coupler to obtain higher coupling. A design for a 10dB coupler which can be built using only hand tools is shown in Fig 19.21.

Take two pieces of WG20 and file away the broad face of both to leave 0.5inch (12.7mm) long gaps. The two pieces should fit snugly together – if this is not the case further filing will be necessary. The most difficult part of the coupler to make is the septum plate, but if the following procedure is used a minimum of trouble should be encountered.

First mark out the outside edges on the septum plate on a larger piece of brass, but do not cut out at this stage. Next mark out the centre lines of the crosses, and drill holes

Fig 19.20. 1N26 diode mount (G4KNZ)

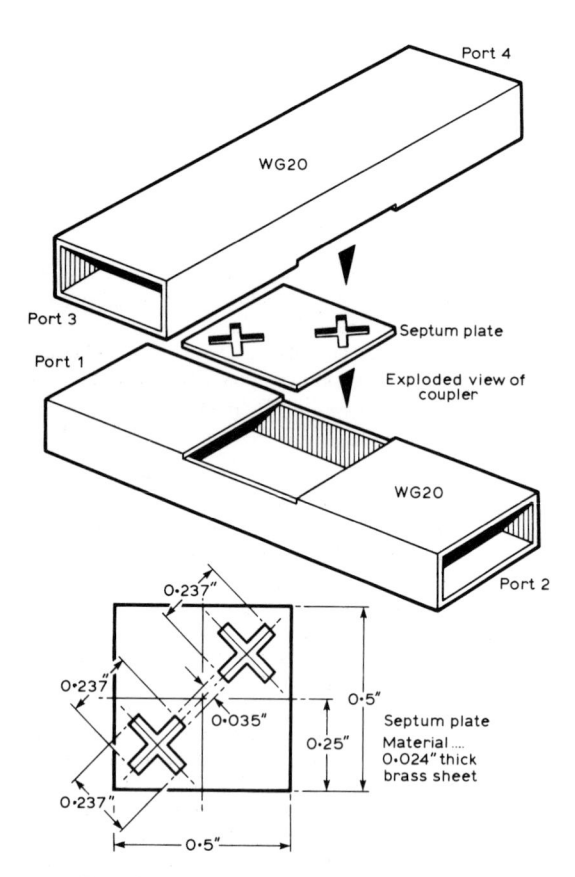

Fig 19.21. Construction of a WG20 directional coupler

Fig 19.22. Directional coupler (G4KNZ)

approximately 0.030inch (0.76mm) in diameter adjacent to each other along these lines so that their edges are just touching. Take a small pointed rectangular-section needle file and reduce its thickness with a grinding wheel to about 0.030inch (0.76mm) over the first quarter inch of its length. The thin edge of the file is then used to remove the metal between the holes and the larger edge to file the edges of the slots to the correct size. The dimensions of the slots should be monitored using vernier calipers. Once the slots are correct, the septum plate can be sawn out from the sheet, filed to size and deburred.

The septum plate is then placed in the cut-out of one of the pieces of waveguide, and the other piece placed over it so that the pieces are jigged together. The waveguides should then be firmly clamped together, heated with a small gas flame and soldered along all joining edges. A completed coupler is shown in Fig 19.22. After cooling, the coupler should be inspected for signs of solder in the coupling slots or inside the waveguides. If this has happened the soldering will have to be done again.

The directivity can be checked by assembling the components as shown in Fig 19.23. The mixer and load on ports 3 and 4 are then interchanged. The diode current reading at port 3 represents the power coupled in the correct direction, that at port 4 the power coupled due to the imperfect directivity. This should be at least 10dB down on the power in the desired direction at port 3.

19.5.5 Matched load

The load is constructed from a piece of wood mounted in the waveguide, as shown in Fig 19.24. A hardwood should be used, as it is easier to work and there is less danger of breaking the tip of the load.

First cut a short length of copper or brass waveguide WG20, long enough to house the wood section and solder a flange to one end. The other end will be left open.

Next cut a piece of wood 32mm long to be an exact fit in the waveguide. It will probably be necessary to use a piece cut from a standard size (e.g. 1/2" × 3/16") and plane it down. Then mark out the taper on the block and plane the end down to a point; this can be done by clamping the block in a vice at the angle of the taper, and planing horizontally. The exact angle is not important, so long as the taper is smooth.

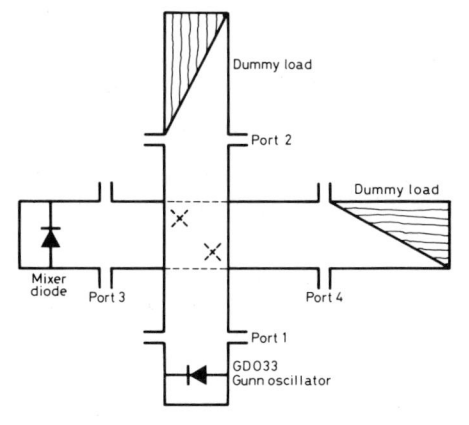

Fig 19.23. Checking performance of the WG20 directional coupler

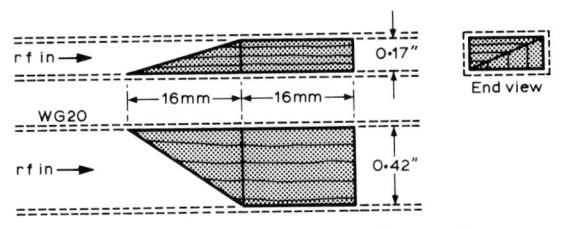

Fig 19.24. WG20 load constructed from wood

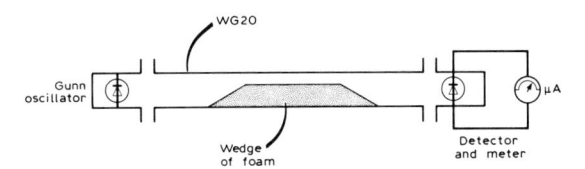

Fig 19.25. Measuring the performance of cmos foam

The finished load should give a vswr of better than 1.1 at 24.1GHz, which is more than adequate.

19.5.6 Setting up the equipment

The components should be connected as shown in Fig 19.11 (the wavemeter may be omitted at this stage if not available). A wideband i.f receiver should be connected to the 1N26 mixer, with a means of monitoring the mixer current (see chapter 8, "Common equipment" and chapter 18, "The 10GHz band").

On applying power to the Gunn oscillator, some mixer current should be observed. The matching screws in the mixer should be adjusted for maximum diode current and then locked into position with lock nuts. A current of approximately 1mA should be observed. Should a much lower figure be obtained, for example less than 200μA, then the components should be checked. A poor mixer diode may be responsible, which should be compared against a known good diode. If this does not appear to be the problem, then the directional coupler should be checked.

The voltage to the Gunn diode should be set at a value which gives good power output and stable operation over the tuning range (–5V should be suitable for the GDO33 and GDHO33). Finally, the deviation in both speech and tone modulation should be set to suit the bandwidth of the i.f employed, typically 200kHz. Note that on receive, tone modulating the Gunn local oscillator will enable cw signals such as that from a crystal calibrator or from narrowband equipment to be heard as an audio tone.

The mixer matching screws can be adjusted to give the best signal to noise ratio using weak signals or a noise generator.

19.6 ATTENUATORS AND LOADS

The design of a waveguide attenuator or load is quite straightforward. An attenuator consists of a piece of lossy material which is tapered at each end to provide a good match, mounted in waveguide. A matched load is similar but need only be tapered at one end.

A load can be built either with the end of the guide open or, more usually, with a short circuit across the end. In the first case, the load should be designed so that the signal is attenuated to a sufficiently low level before it is radiated into space. In the second case, the signal reflected back up the guide must be sufficiently small after passing through the absorbing material twice that it does not represent a significant swr. This attenuation need only be about 15 or 20dB which will give a return loss of 30 to 40dB. This corresponds to an swr of 1.1 to 1.02 and is more than adequate.

A good load can be made from wood and such a load in WG20 has already been described (see Fig 19.24). Alternatively, the conductive foam that is used for packing cmos ics can be used. However, this material comes in a much wider range of resistances than wood. This can sometimes give rise to unexpected results. When measured by sticking the probes of a test meter into the foam a few mm apart, the usual cmos foam has a resistance of megohms to tens of megohms. Some foam has a much lower resistance, perhaps tens of kilohms, while the foam used to line anechoic chambers is in the region of a few kilohms. The lower resistance foam will probably not give a very good match in waveguide, and will reflect a lot of power. The medium and high resistance ones are suitable for loads, and the high resistance ones for attenuators. The best way of measuring the performance of a particular foam is to mount a tapered wedge of it in the guide between an rf source of a few mW and a diode detector, and to measure the amount of power that is transmitted through it. See Fig 19.25. Note that the detected diode current is fairly linear with power below a few hundred microwatts if it is connected to a low resistance load, e.g. a milliammeter. Start out with a fairly small amount of foam in the guide, and gradually increase the width and length of the centre section of the foam until the attenuation is about 10 to 20dB.

Avoid a foam which gives a high attenuation in a relatively short length. It is likely that the discontinuity caused by the tapered foam wedge will reflect a significant amount of power, and degrade the swr.

An attenuator and load built using cmos foam are shown in Figs 19.26 and 19.27. The dimensions are given for guidance only – due to the unknown properties of each foam, you will have to re-measure each type yourself. The foam used in the ones shown had a resistance of tens of megohms between meter probes, even when pressed quite hard into the foam.

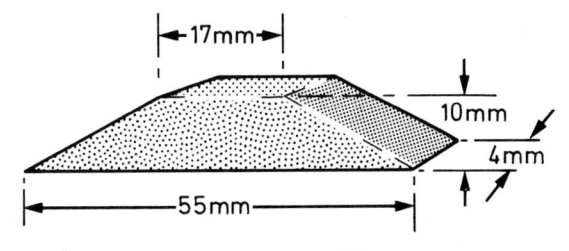

Fig 19.26. Foam wedge for a WG20 attenuator

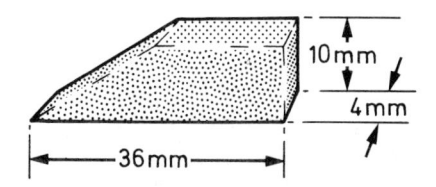

Fig 19.27. Foam wedge for a WG20 load

The foam is tapered at an angle of about 30° at both ends for the attenuator, and at the input end of the load. The angle of 30° is not critical but ensures that the taper occurs over a distance of at least a wavelength. The absorbing section is made a snug fit inside the waveguide and is also at least a wavelength long.

The performance of a load without an end plate can be checked very simply by moving a metal plate to and fro at the rear of the load as shown in Fig 19.28. No change in the diode current indicates that the tapered section is long enough and is absorbing all the power incident on it. Alternatively the diode mount and meter can be used to detect power leaking past the load.

19.7 WAVEMETERS

Most wideband transmitters (and receivers) in use at 24GHz employ a free-running Gunn oscillator for the signal (or local oscillator) source. The stability of such an oscillator can be quite good, say 1 part in 10,000. However, some means of setting the frequency is required unless a commercial Gunn is used unmodified. Also, the frequency of most Gunns can be pulled by a change in load impedance, possibly by 100MHz or more.

Thus, an essential item of test equipment is a wavemeter and two types are described here – a self calibrating wavemeter and a high Q wavemeter. Both are of the absorption type and used in conjunction with a detector diode and meter.

The self calibrating type is a low-Q device and, while very useful for finding the band without recourse to any other test equipment, suffers from the disadvantage that precise frequency setting is difficult. If a good micrometer

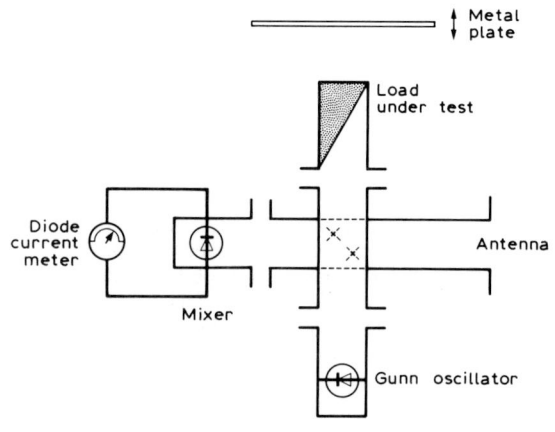

Fig 19.28. Measuring performance of the WG20 matched load

head is used, it should be possible to measure the frequency to within about 50MHz.

The high-Q wavemeter is much more accurate and frequency setting to a few MHz should be possible. However, it requires calibrating against a known instrument.

19.7.1 Self calibrating wavemeter

Details of a self-calibrating wavemeter are shown in Fig 19.29. It consists of a rod of adjustable length mounted coaxially in a cavity. The cavity is loosely coupled to the waveguide through a small hole and absorption of power from the waveguide occurs when the rod resonates at an odd number of quarter wavelengths.

The wavelength at 24GHz is short (12.5mm) and, if the first resonance were used, the tuning rate would be very high. Thus the 5/4 and 7/4 wavelength resonant modes are used which have slower tuning rates. The rate is still high (1,700 and 1,100MHz/mm, respectively) and a micrometer is used to measure the spindle movement.

The main body of the wavemeter is made from brass in which a 3/8" (9.5mm) hole is drilled. This single hole both locates the micrometer stem and forms the cavity, thus ensuring their alignment. The micrometer spindle passes through a quarter wave choke which defines the end of the

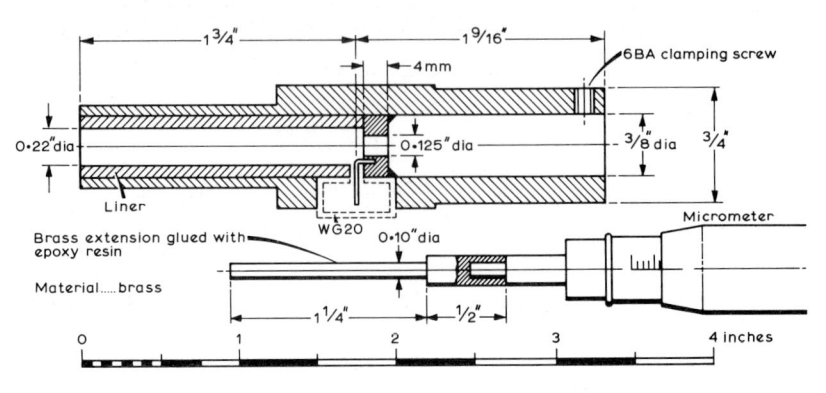

Fig 19.29. Self-calibrating wavemeter for 24GHz

resonator more reliably than a metal contact. The gap between the spindle and choke must be kept small for a high Q, but the two parts must not touch. A short probe from this choke passes through a hole in a thinned part of the cavity wall into the waveguide.

The micrometer is of the 0 to 15mm type with the spindle extended using a length of brass rod, and great care must be taken to ensure these two parts are concentric. The choke, liner and probe are soldered in one operation. See Fig 19.30 for a photograph of a finished wavemeter.

To calibrate the wavemeter, an rf source and detector are needed; these will normally be parts of the transceiver with which the wavemeter is to be used. First, the length of the probe should be trimmed to the minimum length which gives an easily detectable 'suck-out'. This should normally cause a reduction in mixer current of around 10%. Next, for each of a number of (unknown) frequencies the micrometer readings R1 and R2 should be made for both the 5/4 and 7/4 wavelength resonance. The difference between the two readings corresponds to the half wavelength at the frequency measured. Hence:

$$\text{Frequency} = \frac{99.8}{2 \times (R1-R2)}$$

where R1 and R2 are in millimetres and the frequency is given in GHz.

A calibration chart can be made up by making these measurements at several frequencies and plotting the results on a graph. Normally the 7/4 wavelength mode should be chosen for the slower tuning rate.

19.7.2 High Q wavemeter

Details of a high-Q wavemeter covering 24GHz are shown in Fig 19.31. The device is a resonant cavity, tuned by a micrometer-driven plunger, which is coupled to the waveguide through a small hole. The main body of the wavemeter is made from copper, but brass would probably be satisfactory. Other constructional details should be apparent from the drawings. Access to both a lathe and a milling machine will be necessary to duplicate the wavemeter. The micrometer used is a Moore & Wright instrument (0 to 25mm). If this type is not available, modifications could be made to the cavity body to allow the fitting of another type, but care should be taken that the resonant length of the cavity can be reached with the particular micrometer used.

The wavemeter body is clamped to the waveguide after removing the broad wall of the WG20 so that the wavemeter body itself forms the wall of the waveguide.

If the plunger is to remain exactly coaxial with the cavity and not touch the side, it must be firmly held onto the micrometer shaft. The best method is to fix the plunger to the shaft with Araldite. When the glue is fully cured, the micrometer is held in the lathe, and the final turning of the plunger carried out. Only very light cuts should be taken to avoid breaking the epoxy. In this manner exact alignment is assured.

Fig 19.30. Photograph of the self-calibrating wavemeter

The wavemeter has to be calibrated before it is of use, and access must be available to a known instrument. Once a rough calibration is known, accurate calibration points can be obtained by using a crystal calibrator in conjunction with a receiver.

19.8 ANTENNAS

At 24GHz, very high gains can be achieved with quite compact antennas, and either horns or dishes will give useful gains. A reasonable sized horn will give 30dB gain (5° beamwidth) and will be fairly uncritical to build. Dish profiles will need to be accurate to within about 1mm but very high gains can be achieved. A suitable feed is described later. These very high gains will be associated with very small beamwidths, so some careful thought must be given to the construction of a rigid mount so that the antenna can be pointed accurately.

19.8.1 Horns

A horn is a very easy antenna to construct, as its dimensions are not critical and it will provide a good match without any tuning. The dimensions, as per Fig 19.32, for optimum gain horns of various gains (at 24GHz) are shown in Table 19.4. The horn design program given in chapter 4, "Microwave antennas", can be used for designing horns of

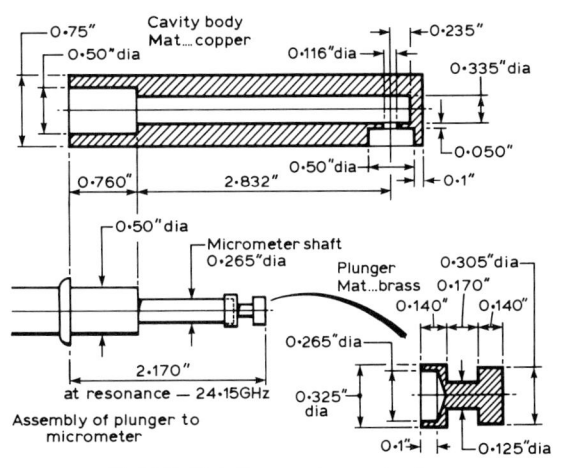

Fig 19.31. High Q wavemeter construction

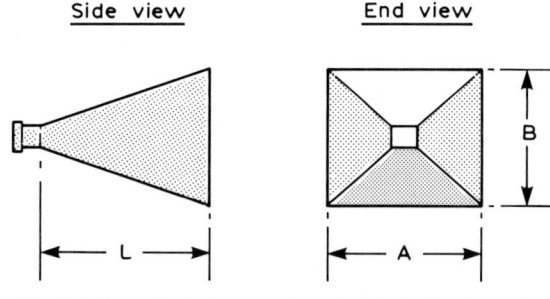

Fig 19.32. Dimensions of a horn antenna for 24GHz. The dimensions A, B and L for optimal gain or sectoral horns can be easily calculated using the BASIC design program given in chapter 4, "Microwave antennas"

Table 19.4. Gain of 24GHz horns

Gain	Beamwidth (degrees)	Length (mm)	A (mm)	B (mm)
15dB	30	26	31	25
20dB	17	81	55	45
25dB	9	270	98	79
30dB	5	810	174	141

other types and gains. In practice, the useful range of horns is 15 to 25dB. Above 25dB the horn becomes very long and unwieldy and it becomes more practical to use a dish.

Suitable materials are pcb material, brass or copper sheet or tin plate. If pcb material is used, the doubled sided type will enable the joins to be soldered both inside and out for extra strength. An ideal source of tin plate is an empty 1 gallon oil can.

The transition from the guide to the horn should be smooth. Use either a butt joint or file the waveguide walls to a sharp edge – see Fig 19.33. The material should be cut to size and then soldered together and onto the end of a piece of waveguide. Alternatively the horn may be coupled directly to the inside of a flange. In this case the material should be the same thickness as the waveguide would be and it is bent as it enters the flange, or the flange may be filed to be part of the taper – see Fig 19.34.

19.8.2 Dish gains

Very high gains can be achieved at 24GHz with relatively compact dishes. However, the resulting narrow beamwidth will make accurate pointing difficult. The most useful size for portable working is probably 12in or 18in. The gains and beamwidths for various sizes of dish are shown in Table 19.5.

As at 10GHz, only solid dishes are suitable and the accuracy of the profile needs to be within $\lambda/10$, i.e. within about 1mm. Further information on dishes was given in the chapter 4, "Microwave antennas".

19.8.3 Dish feeds

A convenient feed for a dish is the so-called "penny feed" due to G4ALN. The dimensions of a version for 24GHz constructed in WG20 are shown in Fig 19.35a. The method used to construct the feed is as follows.

A piece of WG20 of sufficient length to reach the focus of the dish is cut, and its ends squared off by filing. The positions of the slots are marked out using vernier calipers and a right angle, and the slots filed out with a needle file. Repeated checking of the dimensions of the slots during filing will ensure accuracy, paying most attention to the length of the slots.

The end disc is made from 0.036inch (0.9mm) thick brass sheet. A 0.5inch (12.7mm) square piece of this is cut out and soldered to an 0BA brass washer. Using the washer as a guide, the corners are filed off until the brass is the same size as the 0BA washer. A small amount of further filing is then sufficient to reach the final size. The brass disc is then unsoldered from the washer, deburred, and the solder filed off.

The assembly of the disc on the waveguide requires special care to ensure accurate alignment. The clamping arrangement shown in Fig 19.35b is used to hold the disc firmly against the end of the waveguide. The disc is then moved around until it was centrally located, as indicated by measurement with vernier calipers. With the clamp still in place, the disc is soldered to the waveguide above a small gas flame, with the waveguide held vertically. Even though a minimum of solder is used some solder may flow into the slots, and this can be removed after soldering by cutting it away with a scalpel blade, followed by the insertion of the end of a junior hacksaw blade into the slots (after removing one of the pins from the hacksaw blade).

The assembly is completed by sliding a 1/8inch (3.175mm) thick brass plate, with a 0.25 by 0.5inch (6.35 × 12.7mm) slot filed in its centre, on to the waveguide. This plate is for bolting to the dish to hold the feed in place. A

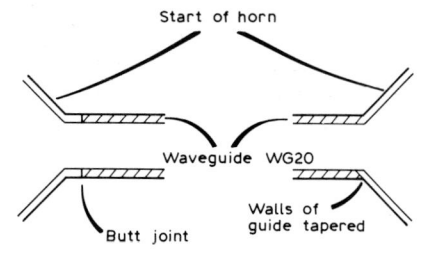

Fig 19.33. Joining a horn to the waveguide

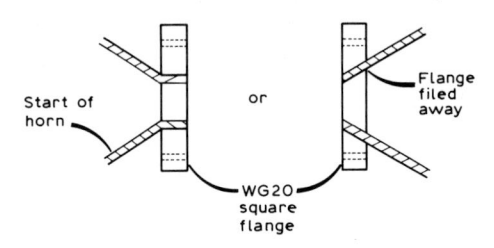

Fig 19.34. Joining a horn directly to a flange

THE 24GHZ ("12MM") BAND

Table 19.5. Gain of 24GHz dishes

Diameter	Gain	Beamwidth (degrees)
12in	35dB	3
18in	38dB	2
24in	41dB	1.5

home-made WG20 flange is then soldered on to the end of the waveguide. The assembly is held in the dish, and the feed slid backwards and forwards to find the point of maximum gain by listening to a remote signal source. The brass plate is then soldered in position, using a set square to ensure that the plate is perpendicular to the waveguide in both planes.

19.9 NARROWBAND TRANSCEIVER

If a narrowband transceiver can be built with the same power output as the wideband equipment already described, the system performance will be enhanced due to both the narrower bandwidth and the use of cw or ssb modulation. The reduction of the bandwidth will typically win 20dB – for a reduction from 300kHz to 3kHz, and a further 10dB will be won on weak signals due to removing the fm threshold effect. Before describing the transceiver, it is useful to compare methods of power generation.

19.9.1 Power generation

There are two main methods available to amateurs to generate power at the higher microwave frequencies:

a) Gunn diodes

Gunn diodes are able to produce from 5 to 250mW power at 24GHz at reasonable cost. However, they suffer from thermal and voltage drift and therefore need a wide bandwidth for reception. The power input required is small and their final mechanical size and assembly can be very compact making a Gunn oscillator very attractive for portable operation.

b) Crystal controlled sources

These can conveniently start with a crystal frequency around 96MHz which is then be multiplied up in successive stages to the final radiated frequency. A low noise oscillator is essential due to the high order of multiplication – in this case 252 times. Sufficient power may be generated at around 384MHz or 1,152MHz (4 or 12 times crystal frequency) where low cost transistors and ics are readily available. This level can vary with the requirement for receive or transmit, and can be 2 to 10W. With a level of 2W, for instance, 1 to 2mW can be available for a receive mixer at 24GHz. This system will be frequency stable but will need critical setting up of the varactor multipliers, preferably with isolation and filtering between stages.

Other methods include impatt diodes and klystrons, both of which are similar to Gunn oscillators, and need a wide receiver bandwidth.

The equipment to be described was designed by G3BNL and uses the advantages of both a and b. The high output power of the Gunn diode is combined with the stability of a crystal controlled source to produce a system suitable for narrowband cw, fm and fsk. The technique is known as 'phase locking'.

19.9.2 Phase locked loop

Using a phase locked loop, a Gunn diode operating with an output power of up to 50mW can be combined with a crystal controlled source having an output of 1 or 2mW at 24GHz. The two sources are separated by the i.f frequency, which in this case will be 144MHz. The local oscillator will start with a crystal frequency of 96.000MHz. The local oscillator frequency is 24,192MHz and the transmit frequency 24,048MHz. Reference should now be made to Fig 19.36.

The waveguide components are constructed in WG20, based around a directional cross coupler (20dB). One arm contains the transmit Gunn diode mount and terminates in the flange to connect to the antenna when used on transmit (port A).

The other arm commences with ×21 multiplier made from a modified GDO33, multiplying from a crystal controlled source at 1,152MHz. The multiplier is followed by a waveguide filter and an in-line mixer, and the other end is terminated in a flange which connects to the antenna on receive (port B).

The mixer diode is a 1N26, with a diode current of about 200µA in the receive mode. With the Gunn diode operating on transmit the current should rise by about 10% dependent on the loading of the waveguide termination and the coupling of the cross-guide coupler (nominally 20dB). The input at 1,152MHz needs to be around 600 to 800mW.

A point to note is the waveguide filter following the ×21 multiplier. This is an important component which should pass only the required frequency of 24,192MHz (21st harmonic) and offer a high rejection to the other harmonics supported in WG20.

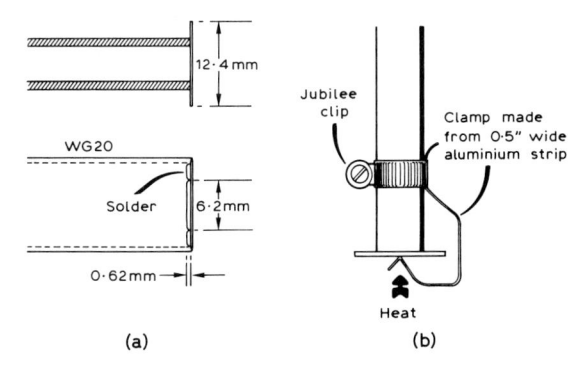

Fig 19.35. Construction of a dish feed for 24GHz

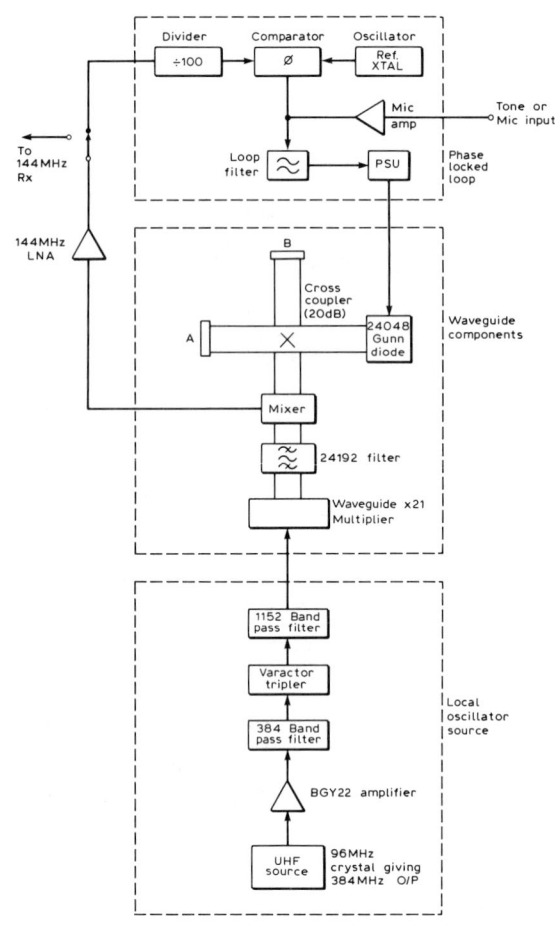

Fig 19.36. 24GHz phase locked loop (pll) system (G3BNL)

19.9.3 Local oscillator source

The original local oscillator source used the RSGB Microwave Committee uhf source, which was described fully in chapter 8, "Common equipment". The pcb uses a crystal of 96.000MHz in a stable low noise Butler oscillator circuit and its 4th harmonic at 384MHz is directly buffered and amplified to a level of about 100mW into 50Ω. The stability is of sufficient order that the frequency at 24GHz varies some 0 to 80kHz over a wide temperature variation of 10 to 40°C; temperature compensation can reduce this by a factor of ten.

Good mechanical stability is essential and this board is best mounted in an STC diecast box, which also helps to stabilise thermal conditions.

The output from this module is followed by a 3dB attenuator (resistive network) and the resulting 50mW fed to a Mullard BGY22 wideband integrated amplifier which takes the output to 2W. The 3dB resistive pad can be formed on the output pins of the 2N4427 and then taken via a short length of miniature 50Ω coax directly to the input tag of the BGY22. Earth the screen of the coax as near to the input of the BGY22 as possible and screw the BGY22 firmly to the diecast lid, treating with heat sink compound first.

The output is coupled to a 384MHz bandpass filter and the fed to a varactor tripler (see Fig 19.37). The tripler operates at about 50% efficiency and the resultant output fed to a bandpass filter on 1,152MHz. Further information on a suitable filter and bandpass filters can be found in the chapter 8, "Common equipment" and chapter 14, "The 1.3GHz band".

This completes the local oscillator source and a short, low-loss coaxial lead is all that is needed to connect it to the waveguide ×21 multiplier. Approximately 800mW should be available at 1,152MHz.

An alternative local oscillator source could consist of the G4DDK–001 and G4DDK–002 modules, also described in chapter 8, "Common equipment". This combination produces a spectrally clean output at 1,152MHz, at the required level, without the need for either a uhf bandpass filter or, indeed, an 1,152MHz filter. For better stability, the crystal can be thermally stabilised by fitting a Murata "Posistor" crystal heater, as specified for the G4DDK–004 2.5GHz source used to drive the various G3WDG 10GHz modules described in the last chapter.

19.9.4 Waveguide multiplier

The multiplier is made from a modified Plessey GDO33 mount (see Fig 19.15), with the Gunn diode removed and replaced with a snap varactor, such as the Microwave Associates MD4901.

The choke assembly is also replaced by an equivalent length of 4mm brass rod, as shown in Fig 19.38. This is drilled at one end to accept the MD4901 diode and the

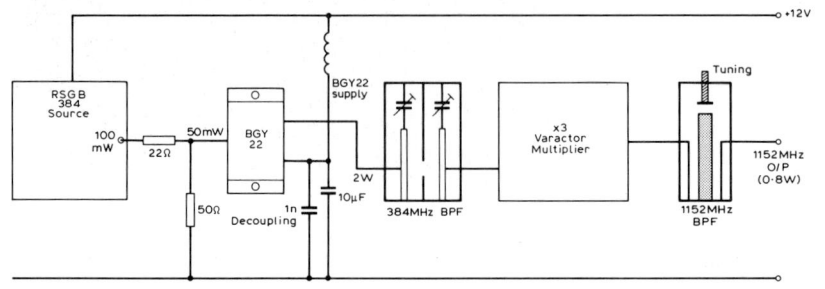

Fig 19.37. 1,152 oscillator source for the 24GHz pll system. That shown is the original source (G3BNL). Other possibilities now exist, such as the G4DDK–001 and G4DDK–002 modules described in chapter 8, "Common equipment"

THE 24GHZ ("12MM") BAND

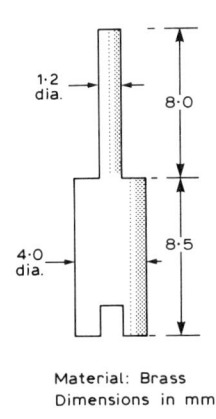

Fig 19.38. Dimension of the choke for the multiplier (G3BNL)

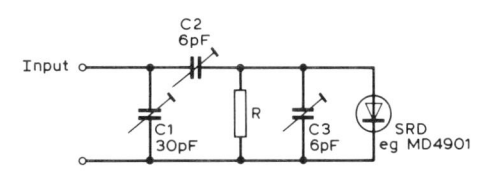

Material: Brass
Dimensions in mm

Fig 19.39. Input matching to the multiplier. Above, layout. Below, schematic. (G3BNL)

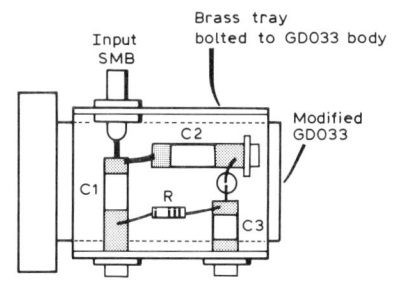

$R \approx 33k$ (S.O.T.), (bias resistor)
C1,2,3 – high Q ceramic tubular trimmers (eg Johansan)

other end turned down to go through the securing nut and a suitable insulating bush recovered from the original choke assembly.

The top of the GDO33 is drilled and tapped with $2 \times 6BA$ holes to hold a small 1.5×1 inch U-shaped tray made of brass sheet for the matching network and SMB input socket (see Fig 19.39). A bias resistor of $33k\Omega$ should be suitable for setting the operating point of the varactor diode. If a different diode is used, the resistor should be adjusted for smooth bias operation, reliable starting and good output. The input matching network is shown in Fig 19.39.

Setting up should initially be done by setting the capacitors in the π-input network for maximum diode volts. Subsequently a detector head may be used on the output of the waveguide section of the GDO33 and the power output optimised. The filter can now be fitted and the multiplier set up for maximum mixer current after the mixer head has been coupled to it. The tuning screw on the GDO33 body should be adjusted at this stage for maximum mixer current. The completed multiplier and filter are shown in Fig 19.40.

19.9.5 Bandpass filter

This filter is constructed from a length of copper WG20 with standard square flanges fitted at either end, as shown in Fig 19.41.

Construction is as follows: first, cut the guide to length, square off the ends and solder flanges to each end. Then carefully drill the holes for the filter elements and slightly countersink the top and bottom of the waveguide to contain the solder spills. Wire size swg 20 should be used together with a suitable drill that allows a tight fit mechanically. The wire can then be soldered in place, preferably with a lower melting point solder than that used for the flanges. Alternatively, the flanges can be wrapped in a damp cloth to prevent the solder round them melting, while the wires are soldered with a small flame. The holes for the tuning screws can then be tapped for 10BA and this size is also used for the three-screw matching sections at either end. Brass screws should be used.

Alignment of the filter is done initially by using the Gunn diode tuned to 24,192MHz as a source, with a suitable detector used at the other end of the guide. Tuning should be sharp with the screws only just protruding into the guide. Final adjustment is done with the filter connected to the output of the $\times 21$ multiplier and the mixer connected to the other end of the filter. The multiplier output is adjusted (as described earlier), together with the filter matching screws, for maximum mixer current.

Fig 19.40. Completed multiplier and filter assembly. In this photograph, the multiplier has been made from a modified Plessey GDO33 Gunn Oscillator cavity (see text). (Photo: G3BNL)

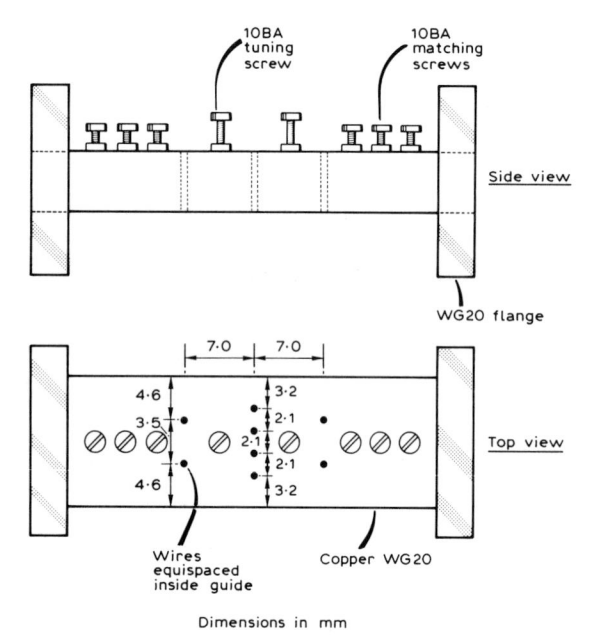

Fig 19.41. Dimensions and construction of waveguide filter (G3BNL)

19.9.6 Mixer

This is probably the most difficult-to-make component, and the original mixer consisted of two halves of machined brass bolted together. An alternative form of construction is shown in Fig 19.42 and 19.43.

As seen the centre bar supporting its 1N26 diode is extended and insulated before entering the feed guide from the filter. The penetration should be set for maximum crystal current and up to 1–2mA should finally be available with all tuning screws matched up. It is essential that either the antenna or a matched load is in place at this time, due to reflection from an open port if it is not.

The hole for the 1N26 should be carefully drilled to allow a comfortable sliding fit. The contact supported from the centre of the horizontal cross bar to the diode contact pin was retrieved from an old ic socket and may be suitably scaled to allow the skirt of the 1N26 to sit just above the upper surface of the waveguide section in the brass block.

The front of the brass block should be drilled with four holes to take a small extension piece of WG20 to interface with the antenna. Behind the diode is a sliding short carefully filed from brass which, when it has been adjusted for best received signal, should be soldered or clamped in place. A similar termination of the local oscillator feed guide behind the diode probe should be adjusted for maximum crystal current before securing. The connections to the mixer diode are shown in Fig 19.44.

19.9.7 Gunn oscillator

This is a standard Plessey GDHO33 (see Fig 19.13) with a suitable diode fitted for the output power needed. Note that high power diodes have the heatsink end negative and are therefore suitable for this equipment. Diodes below about 15mW output are reverse connected (heatsink positive) and are *not* suitable.

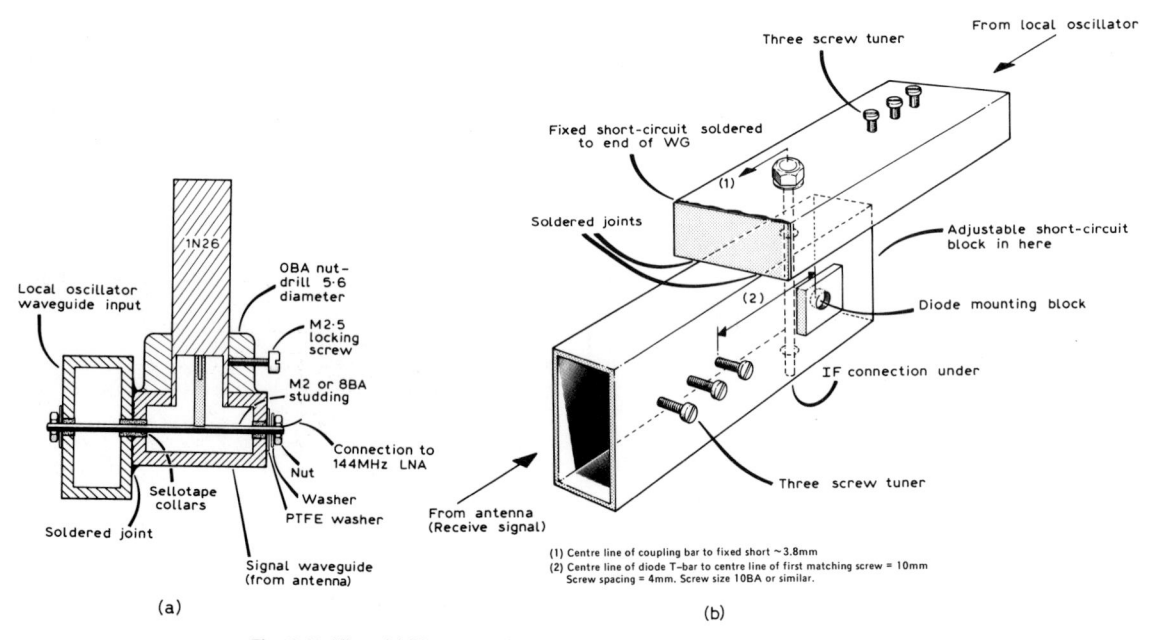

Fig 19.42. Mixer. (a) Diagrammatic section. (b) Mixer constructed from waveguide

THE 24GHZ ("12MM") BAND

Fig 19.43. Photograph of the mixer: brass blocks support the waveguide and the 1N26 mixer diode body is just visible at the top of the block. (Photo: G3BNL)

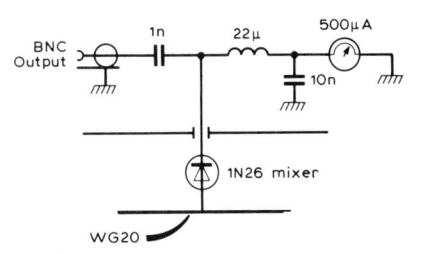

Fig 19.44. Connections to the mixer

19.9.8 Cross-coupler

The cross-coupler is similar to that described earlier, except that round coupling holes are used. Either brass or copper waveguide can be used, but brass may be found easier to work and solder. The coupler is shown in Fig 19.45.

For the coupling holes, a 1/2" square is cut or filed from the crossing surfaces of each waveguide and a square of similar material made to fit the resultant aperture. It is easier to file the holes in this square and then solder the whole lot into position.

19.9.9 PLL circuits

The phase locked loop (pll) circuits are shown in Figs 19.46 and 19.47. On transmit, a free running Gunn oscillator is phase locked to a crystal reference source, giving it a similar stability to the reference. The Gunn is set up to free-run at approximately 24,048MHz by use of a wavemeter. The mixed signal from the Gunn, and that of the

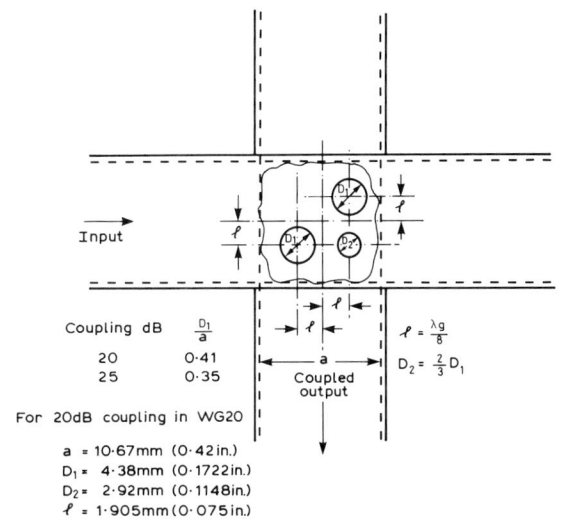

Coupling dB	$\frac{D_1}{a}$
20	0.41
25	0.35

For 20dB coupling in WG20

a = 10·67mm (0·42in.)
D_1 = 4·38mm (0·1722in.)
D_2 = 2·92mm (0·1148in.)
ℓ = 1·905mm (0·075in.)

$\ell = \frac{\lambda_g}{8}$

$D_2 = \frac{2}{3} D_1$

Fig 19.45. Round-hole cross-coupler dimensions

reference at 24,192MHz, results in an output of about 144MHz at approximately 300mV from the low noise amplifier. This is fed to the input of the divide by 100 chip, IC3, the Plessey SP8629. This ecl output is level-shifted by

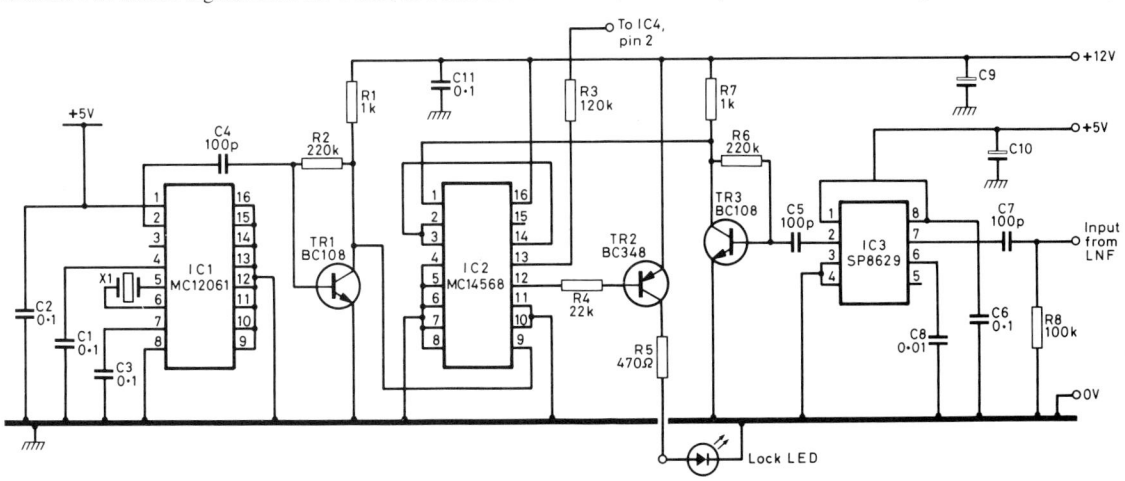

Fig 19.46. PLL divider and comparator circuits (G3BNL)

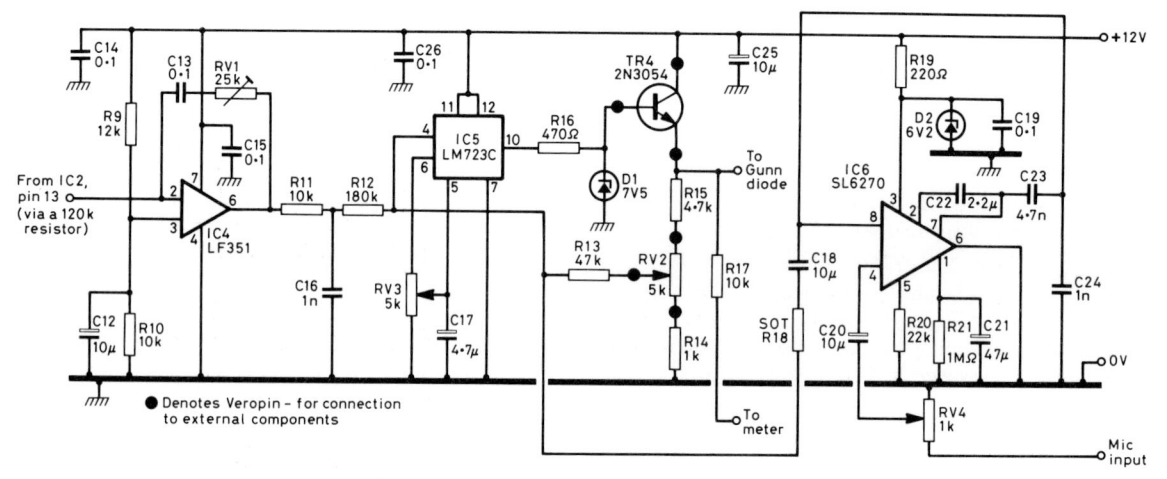

Fig 19.47. Loop filter, Gunn psu and microphone amplifier circuits

TR3, a BC108, to the cmos input level required by the phase comparator, IC2, the Motorola MC14568.

For the reference source a Motorola MC12061 oscillator chip, IC1, is used with a 5.760MHz crystal. This is divided by 4 in IC2 to provide the comparison frequency of 1.440MHz. A slightly higher frequency crystal may be chosen to take the reference frequency slightly away from the edge of the 144MHz band, e.g. to 144.1MHz.

Depending on the phase relationship of the two input ports of the comparator when switched on, pulses whose widths are proportional to the phase displacement will appear at pin 13 of the MC14568. These are fed to a second-order loop filter and integrated, providing a dc voltage for control of the Gunn diode bias supply.

The LF351 active filter circuit (IC4) uses a variable 25kΩ potentiometer in the integration network which can be used to optimise the loop bandwidth. If a spectral display is available, examination of the output of the filter will readily show the required setting. Alternatively, the 144MHz output may be examined and the potentiometer set for minimum noise sidebands. Otherwise, the potentiometer may be adjusted by listening on a suitable 2m receiver and setting for minimum close-in carrier noise. Following the LF351 is a low pass filter, before the signal is fed to pin 4 on IC5.

IC5, an LM723C, is used to buffer the output from the filter and provide sufficient voltage drive to TR4. TR4 is used as an emitter follower to supply the high current to the Gunn diode. RV3 provides a reference voltage for the other input to the LM723 (non inverting).

Feedback is taken from the emitter of TR4 via RV2 through a 47kΩ resistor to pin 4 of IC5, the inverting input. The Gunn diode requires about +6V to operate and the lower and upper limits of this supply may be set by adjustment of RV3 and RV2. A range of between 5 and 6.5V should be chosen. To prevent an excessive voltage being applied to the Gunn diode at switch-on or out-of-lock conditions, a zener diode (7.5V) clamps the base input to TR4. The emitter of TR4 is also taken, via a 10kΩ resistor, to a 1mA meter to monitor the Gunn voltage.

Modulation

For fm modulation, a Plessey SL6270 (IC6) is used to amplify a signal from a dynamic microphone, applied through RV4 to pin 4. This is a vogad device and the input of between 1 and 10mV results in an output on pin 8 levelled to approximately 90mV. This 90mV is fed through a select-on-test (SOT) resistor to provide input summing on pin 4 of IC5. The value of the resistor will probably be in excess of 100kΩ and should be adjusted to provide the maximum deviation required for the particular bandwidth needed. This could be in the order of 5 to 6kHz for nbfm, or much higher for wide band operation. RV4 adjusts the sensitivity of the microphone used.

Construction

The phase locked loop electronics is split into two parts, which should be built on separate pcbs and each housed in a separate screened box. See Fig 19.48. The two parts (as per the two circuit diagrams) comprise:

1. A board containing the pll input digital circuits:

    the divide by 100 chip (SP8629)
    level shifter (BC108)
    reference oscillator chip (MC12021)
    level shifter (BC108)
    phase comparator (MC14568)

This board should preferably by made from double-sided glass fibre to keep the phase noise as low as possible. Layout is not important provided rf connections are kept short and supply points adequately bypassed. Entry points in and out of the box should be via screw-in 1,000pF feed-throughs with the exception of the output on pin 13 of the phase detector. This should be a simple ptfe feed-through and adjacent to the box containing the Gunn psu circuits. The lock led can be mounted on the front panel. The artwork and layout is given in Fig 19.49.

2. A board containing the Gunn psu circuits and loop filter:

> the loop filter (LF351)
> the psu chip (LM723C)
> the psu output transistor (TR4)
> the microphone vogad (SL6270)

Single sided pcb is sufficient for these circuits, and again the board should be fitted in a small diecast box. TR1 is mounted on the wall of the smaller box housing the Gunn psu circuit. The artwork and component layout is given in Fig 19.50.

The 25kΩ potentiometer in the filter can be a miniature horizontal pre-set and so can RV1. RV2 should be a 10 turn helical with control knob and mounted on the front panel for access. The meter is 1mA fsd and scaled 0 to 10. With the 10kΩ series resistor, it will read the voltage supplied to the Gunn diode. Alternatively a 0 to 100 microamp can be used with a 100kΩ resistor. Do not use a 0 to 10V volt-meter because this will cause distortion to the Gunn control line under locked conditions.

Set RV3 to give a minimum scale voltage of, say, 5V. As RV2 is rotated the voltage should then range between 5 and 7V with the components listed.

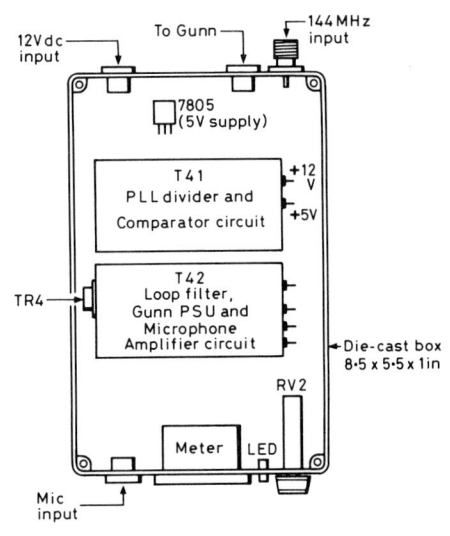

Fig 19.48. PLL transceiver screening-box layout

19.9.10 Operation

The completed equipment is shown in Fig 19.51. Apart from the components already described, a 144MHz low noise preamp (lna) will also be needed. This is connected to the output of the 1N26 mixer to amplify the signal level

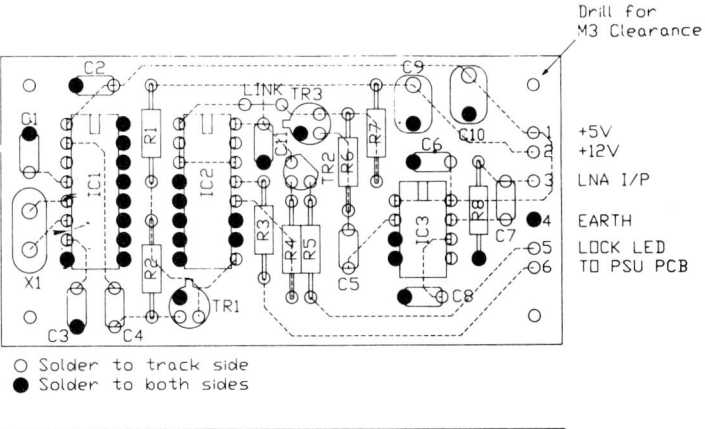

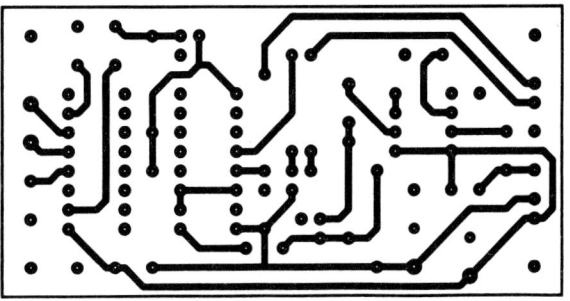

Fig 19.49. PLL divider pcb layout. (a), above, component overlay. (b), below, track pattern (G4KNZ/G8AGN)

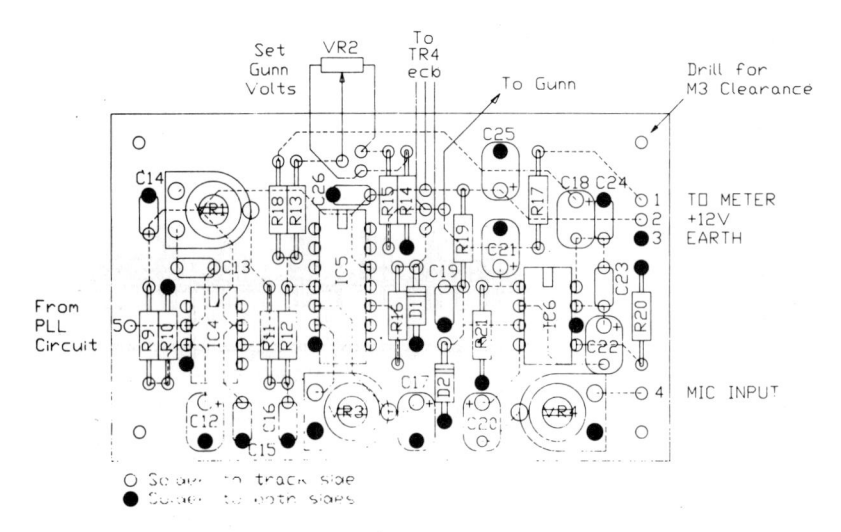

Fig 19.50. Gunn power supply layout. (a), above, component overlay, (b), below, track pattern (G4KNZ/G8AGN)

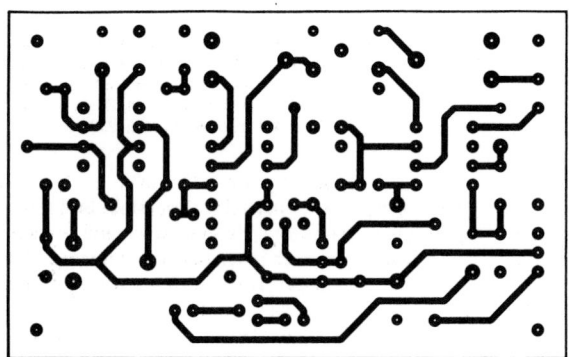

up to that required by the pll circuit. It should have at least 20dB of gain.

If the Gunn diode is now energised, the mixer current should rise by about 10%. This is not too critical, the only requirement being that the 144MHz output from the lna should be over 300mV for correct operation of the detector chain on transmit. To transmit, the antenna should be placed on the waveguide at position 'A' with 'B' left un-terminated (see Fig 19.36).

The output from the low noise amplifier is routed to the input of the divide by 100 chip. The supply is now applied to the phase locked loop unit and the Gunn voltage adjusted by RV2 to the operating voltage of the diode, as monitored on the 1mA meter.

If the mechanical tuning of the Gunn supply has previously been set to within about 20MHz of the transmitting frequency (24,048MHz) then lock should be immediately be obtained and the lock led will go out. Should this not be so then delicate adjustment of RV2, one way or the other, should cause the loop to lock up.

The equipment is now transmitting with the stability of the local oscillator source, which means cw fsk or nbfm qsos are possible. Tone may be introduced by feeding a 1kHz voltage into the microphone circuit. The locking can be checked by monitoring the 144MHz loop input signal to the divider on a 2m receiver. A relatively strong carrier should be audible without direct connection and should be

Fig 19.51. Completed narrowband transceiver mounted on a tripod with dish antenna attached. The pll circuitry is housed in the diecast box at the right-hand side (Photo: G3BNL)

Table 19.6. Waveguide filters

1.
Centre frequency:		24.048GHz
Bandwidth:		250MHz
Passband ripple:		0.1dB
Hole sizes	d1	3.9mm
	d2	2.4mm
	d3	2.4mm
	d4	3.9mm
Cavity lengths	L1	6.6mm
	L2	6.9mm
	L3	6.6mm
Iris plate thickness:		0.5mm
Theoretical insertion loss:		0.8dB

2.
Centre frequency:		24.048GHz
Bandwidth:		100MHz
Passband ripple:		0.1dB
Holes sizes	d1	3.5mm
	d2	1.9mm
	d3	1.9mm
	d4	3.5mm
Cavity lengths	L1	6.6mm
	L2	6.9mm
	L3	6.6mm
Iris plate thickness:		0.5mm
Theoretical insertion loss:		2.0dB

clean and stable. It may also be monitored on a spectrum analyser, or on deviation meter when fm is used.

19.9.11 Summary

The narrowband equipment designed by G3BNL combines both the stability of a crystal controlled source and the higher output power of a Gunn oscillator.

Alternatively it would be possible to build narrowband equipment using simply a multiplier for the transmitter, at the expense of output power. This equipment would still have an advantage over the wideband equipment though, since the system gain due to using cw and the narrow bandwidth outweighs the system loss due to the lower power level.

It should be apparent that if the loop is left unlocked (i.e. disconnect lna to divider), then the equipment can be used in a wideband mode with a suitable wideband receiver.

19.10 FILTERS

As at 10GHz, a bandpass filter can be conveniently made from a section of waveguide partitioned off into resonant halfwave cavities. A simple two cavity design is described earlier in the phase locked loop narrowband system, where the cavities are defined by wires between the broad walls of the waveguide. An alternative design is described here where the cavities are defined by iris plates, and power is coupled into and out of the cavities through the round (or rectangular) holes in the plates – see Fig 19.52.

The unloaded Q of a halfwave cavity built from copper WG20 will be approximately 3,600. Be careful not to leave any excess solder in the cavities. This is particularly a problem at 24GHz since the waveguide is so small. Any solder will drastically lower the unloaded Q and increase the insertion loss – indeed it is possible to double the loss quite easily. Brass (and aluminium) should be avoided for a high Q structure such as a filter. The unloaded Q of a brass cavity will be about half that of copper and the insertion loss of a filter thus about double.

The hole sizes and cavity lengths for a number of filters constructed in WG20 are shown in Table 19.6. The cavity lengths have been reduced by 10% to enable the filter to be tuned precisely onto frequency using tuning screws. Note that there is not much difference in the iris hole sizes between the two designs – so the holes must be drilled quite accurately or the response will not turn out to be that expected. A photograph of a completed filter of this type is shown in Fig 19.53.

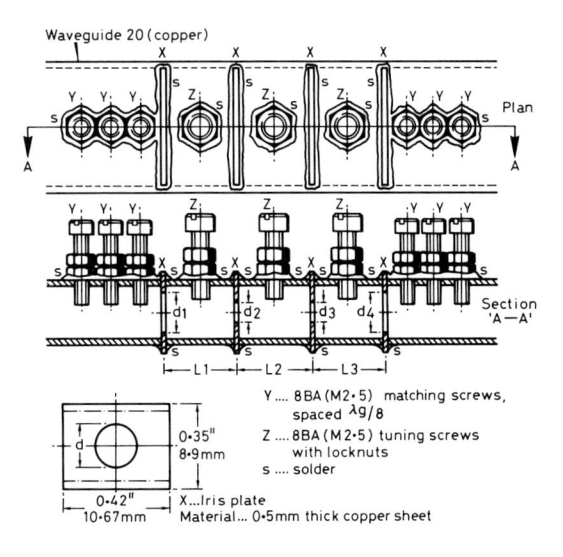

Fig 19.52. Configuration of WG20 iris-coupled bandpass filters. Dimensions for two different filters are given in the text. Matching screws are 10BA or 1.6mm, tuning screws are 8BA or 2.5mm. To preserve the highest Q, the waveguide, iris plates and screws must be made from copper. (G4KNZ)

Fig 19.53. Photo of a WG20 iris-coupled bandpass filter (G4KNZ)

19.11 CONCLUSION

The techniques currently used at 24GHz for wideband closely resemble those described in chapter 18 for the 10GHz band; with the exception of the dimensions of the waveguide and components constructed in waveguide, the techniques employed are very similar. Since the wavelength is a little less than half that at 10GHz, the inference is that constructional accuracy needs to be better by the same factor, i.e. double! This should not deter the determined constructor! A number of professionally produced units (oscillators and oscillator mixers) take much of the onus out of construction of equipment for this band and ensure that its accessibility, to the amateur of limited constructional skills, is not much less than that of the 10GHz band. However, a greater degree of unpredictability of propagation exists and this should add to the impetus to experiment, rather than detract from it!

The recommended frequencies for wideband operation in the UK are now 24.000 to 24.050GHz where it is no longer necessary to seek special written permission to operate, this part of the band being "amateur primary". In the UK, the commonly used narrowband frequency is 24.195GHz (1,152MHz × 21). From June 1990, written permission for frequencies between 24,050 and 24,150MHz is still required, although frequencies above 24,150MHz no longer require such permission. However, at the time of writing, it is probable that narrowband operation will move to the same sub-band at 24,048 to 24,050MHz, which is "common" to the majority of countries in IARU Region 1 (Europe).

CHAPTER 20

The bands above 24GHz

20.1 INTRODUCTION

Until 1979 the highest frequency microwave band available to amateurs was 24GHz. The World Administrative Radio Conference 1979 (WARC 79) allocated a number of new bands above 24GHz to the Amateur and Amateur Satellite Services. They are listed in Table 20.1 and in the text they are referred to as the 47, 76, 120, 142 and 241GHz bands.

Since that time there has been comparatively little work done on these new higher bands, largely because of the non-availability of suitable components to amateurs; components such as mixer, detector, Gunn and multiplier diodes, GaAs fets and the like have been available to professional engineers, but their price has precluded amateur use. As a result, the development of amateur techniques and practical designs have lagged behind developments on other microwave bands. Therefore much of what follows in this chapter is tentative and based on the limited experiences of a few amateurs. We look forward, however, to significant developments (at least at 47 and 76GHz) in the near future.

20.2 THE BANDS

The actual bands available to amateurs at present vary from country to country, but it is to be expected that at least the exclusive bands will be eventually allocated worldwide. Parts of these bands are harmonically related to presently active lower frequency microwave bands and the International Amateur Radio Union (IARU), part of the International Telecommunications Union (ITU), has recommended that initial operation in the new bands should use these harmonically related frequencies. It is thus possible to verify the frequency of operation by listening for harmonics from transmitters in the bands below and, eventually, when the necessary techniques have been developed, it will be possible to generate rf at the higher frequencies by frequency multiplication.

Fig 20.1 illustrates some of these harmonic relationships together with some other useful relationships *not* related to amateur frequency allocations. Fig 20.2 gives some simple relationships based upon harmonics from readily available Gunn sources in the 23 to 25GHz range. Some of the recent successful European work, described later, has been based

Table 20.1. Amateur millimetre bands

Band	Status
47.0 – 47.2GHz	Exclusive
75.5 – 76.0GHz	Eexclusive
76.0 – 81.0GHz	Secondary to radiolocation
119.98 – 120.02GHz	Footnote
142.0 – 144.0GHz	Exclusive
144.0 – 149.0GHz	Secondary to radiolocation
241.0 – 248.0GHz	Exclusive
248.0 – 250.0GHz	Secondary to radiolocation

These bands, with the exception of 120GHz, are allocated to the amateur and amateur satellite services.

upon multipliers from frequencies in this range. The awkward multiplication factor of ×7 from 144MHz is best avoided, for instance by generating 1,008MHz by some other route and providing amplification to a satisfactory power level at 1,008MHz, before further multiplying up into some of the millimetre bands.

It is also practical to start at higher frequencies in the tables; thus an oscillator at 8.064, 16.128GHz or in the range 23 to 25GHz would be a prolific source for the millimetre bands.

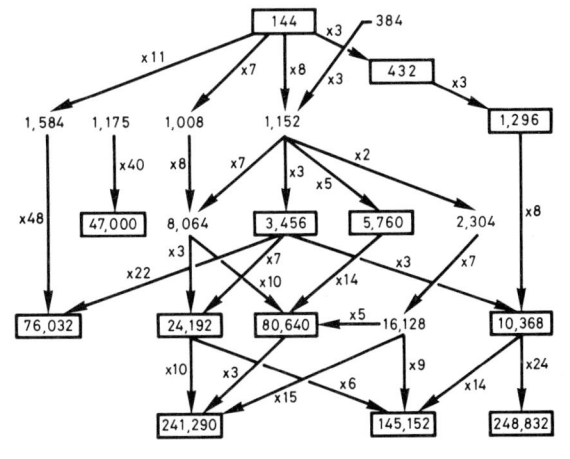

Fig 20.1. Harmonic relationships of the millimetre amateur bands. Frequencies in MHz. Frequencies in "boxes" are in amateur bands, others are not

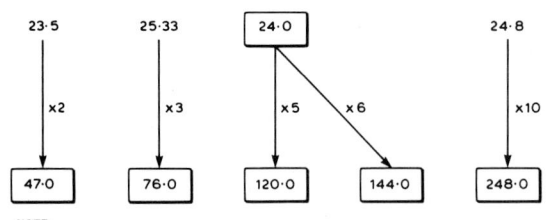

Fig 20.2. Simple harmonic relationships derived from tuneable Gunn oscillators. Frequencies in GHz

NOTE: All frequencies are in GHz
'Open ended' Gunn oscillators may contain significant harmonic output which could be selected by means of a suitable filter.

20.3 PROPAGATION

All of the propagation phenomena known on the lower microwave bands can be expected to manifest themselves in the millimetre bands, but the scale of the phenomena will be smaller. Effective ducts may be only a few metres thick. Smaller objects are effective reflectors and local weather may be of more consequence than is the case on the lower frequency bands.

One phenomenon appears on the millimetric bands which is of much less importance on bands below 20GHz – atmospheric absorption. We are used to referring to the atmosphere as transparent to light, but a moment's thought will remind us that we can see further on a "clear" day than a misty one. The same effect manifests itself with radio frequencies above about 20GHz. Depending on the humidity, there is a variable loss due to water vapour; there is a fixed loss due mainly to oxygen; and finally there is a considerable occasional loss due to rain (or other hydrometeors).

Loss due to absorption follows a totally different law to that associated with radiation; the loss is a *linear* function of distance and is expressed in dB/km, ie. if there is a loss of, say 5dB in the first km, then there is the same loss in each subsequent km. This causes the loss to build up much more rapidly than is due to the inverse square law of radiation and, sooner or later, the loss due to absorption will determine the maximum range that can be worked.

Figure 20.3, taken from CCIR reports, shows the attenuation per km on a horizontal path due to oxygen and water vapour (at $7.5g/m^3$), the two gases which cause the greater part of the losses. *This is in addition to the normal free space loss.*

At a Microwave Round Table [1], G8AGN gave a valuable assessment paper which looked in more detail at the effects of atmospheric absorption and rain attenuation and considered the potential of the amateur millimetre bands. Much reference was made to the 24GHz band as a model, because it may be considered as a "longer mm waveband" and also lies close to the principal atmospheric water absorption band. The remainder of this section is based largely on that paper and it should become apparent that the mm bands above offer more potential than at first sight. A microcomputer prediction program was also given and is included later.

20.3.1 Free space loss

For line of sight paths, the free space path loss (in dB) may be calculated from the classical formula:

$$\text{Loss} = 92.45 + 20\log(f) + 20\log(d)$$

where f is the frequency in GHz and d is the path length in km. Fig 20.4 shows this relationship in graphical form for all the bands above 10GHz, with the exception of 120GHz.

20.3.2 Water vapour and oxygen attenuation

Fig 20.3 has already given a general overview of the additional path losses to be expected due to water vapour and oxygen absorption. The loss due to oxygen is fixed and there is nothing that can be done about it. In particular, at 120GHz there is an absorption band due to oxygen. The maximum loss occurs just below the amateur band at 118.75GHz but, even so, a loss of around 0.5dB/km is to be expected. This is enough to make long-distance contacts unlikely. The loss due to water vapour, however, is not fixed; it is dependent upon the total amount of water held in the air, the absolute humidity. This may be quite high in the summer but very low on a cold night in the winter.

A more useful picture is therefore given by the data derived from a Jet Propulsion Laboratory model which is shown in Fig 20.5. Here absorption predictions for several values of atmospheric water vapour content are shown.

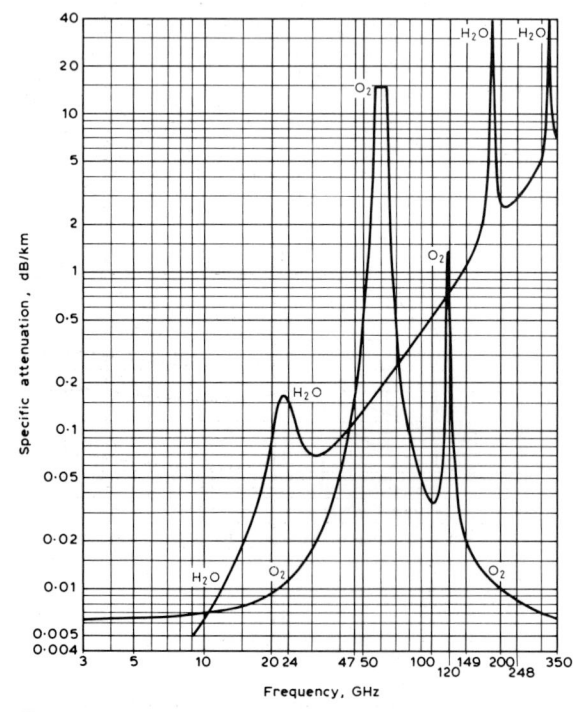

Fig 20.3. Attenuation due to oxygen and water vapour at pressure of 1 atmosphere, temperature 20°C, water vapour $7.5g/m^3$ (source, CCIR)

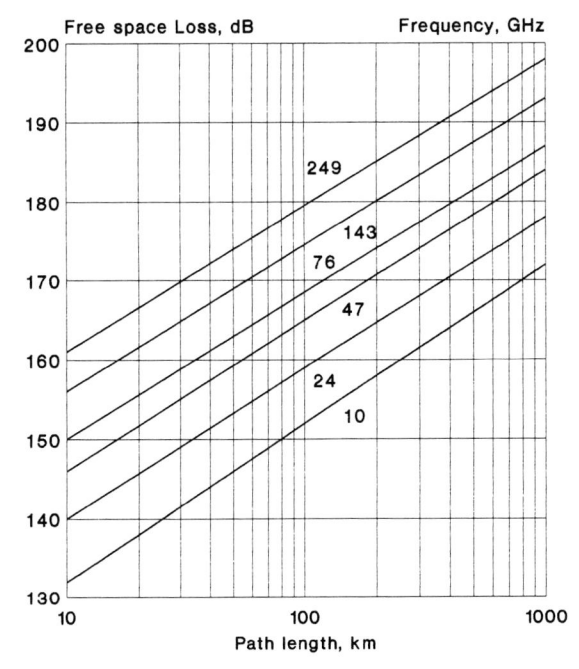

Fig 20.4. "Free space" path loss versus frequency for the bands from 10 to 249GHz (except 120GHz)

Table 20.2. Estimated atmospheric absorption loss

Band	Loss
10GHz	α = 0.0066 + 0.0011W
24GHz	α = 0.012 + 0.0185W
47GHz	α = 0.13 + 0.016W
76GHz	α = 0.2 + 0.034W
142GHz	α = 0.152W
241GHz	α = 0.417W

Loss (α) in dB/km; W in gm water/cubic metre

From these curves, a set of empirical relationships between absorption loss (in dB/km) and atmospheric water content (in gm per cubic metre) have been derived for the amateur bands above 10GHz and these are shown in Table 20.2. These relationships consist of two terms – a constant term, due largely to oxygen, and a term proportional directly to the water vapour content.

For these to be of use, a knowledge of the atmospheric water content over the path of interest is thus required but such information is not directly available and hence must be sought. Fortunately it is at hand in the form of the "relative humidity" (RH). This may be defined as:

$$\frac{\text{Actual water content of air in gm/m}^3 \text{ at temperature T}}{\text{Water content in gm/m}^3 \text{ if air saturated at temperature T}}$$

Note that RH is thus expressed as a percentage.

It is important to note that RH is mainly dependent on the temperature and, although not obvious from the formula given above, it is also slightly dependent on pressure, but this is ignored here. The weight of water, W, held in saturated air at different temperatures is given by the curve shown in Fig 20.6 and an approximate relationship between W and T has been derived from this. Hence, if the air temperature and RH can be measured, then the actual water content and, thus, the additional absorption term in the path loss can be determined.

The most satisfactory way for the radio amateur to measure RH is by using a pair of mercury thermometers, one of which has its bulb cooled by evaporation from a water-wetted wick surrounding it. This records the so-called "wet bulb" temperature. Provided that the air flowing over the bulbs is moving at a velocity of at least 1m/sec, then the RH can be determined from simultaneous reading of both thermometers, as outlined in Table 20.3. Such measurements are facilitated by using an instrument known as a "whirling psychrometer" – two identical thermometers, one "dry" and one "wet", mounted side by side in what looks like an old-fashioned football supporters' rattle. This is whirled rapidly round and the wet and dry bulb temperatures read off the two thermometers without delay. Calculation is eliminated by use of a "psychrometric table" supplied with

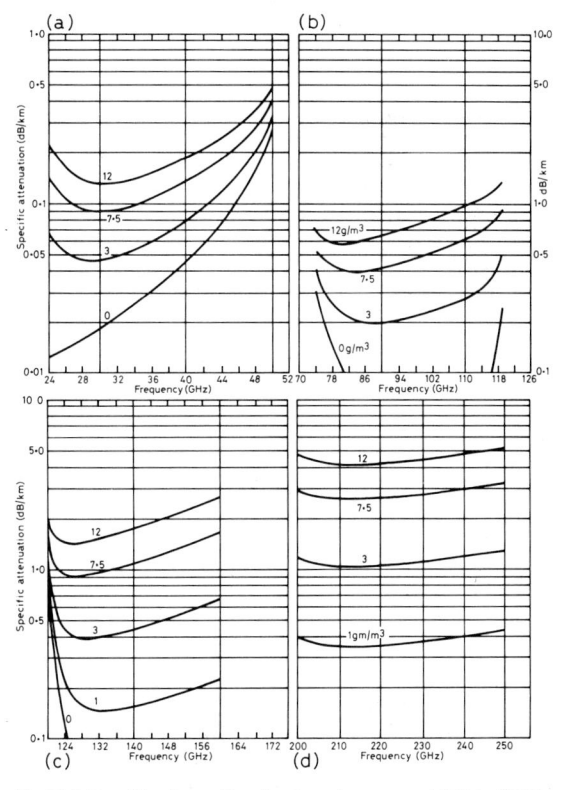

Fig 20.5. Specific attenuation due to water vapour. (a) 24 to 52GHz, (b) 70 to 126GHz, (c) 120 to 176GHz, (d) 200 to 250GHz. The figures on the curves indicate the water vapour content of the atmosphere, expressed in g/m³ (Source JPL model)

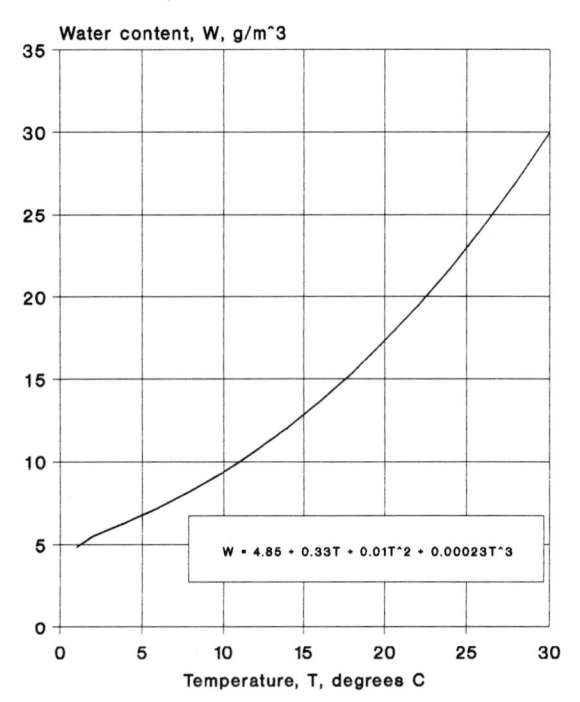

Fig 20.6. Graph of weight of water, W, in saturated air plotted as a function of temperature, T

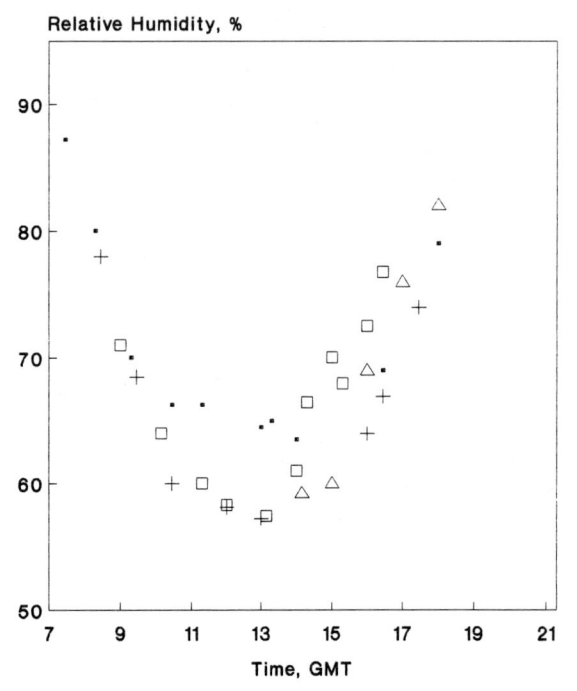

$W = 4.85 + 0.33T + 0.01T^2 + 0.00023T^3$

Fig 20.7. Typical diurnal variation of humidity in northern England as measured over several days by G8AGN

Table 20.3. Determination of relative humidity

$$RH = \frac{100[e_w - AP(T_d - T_w)]}{e_d}$$

P	=	atmospheric pressure in mB (milliBar)
T_d	=	dry bulb temperature in degrees C
T_w	=	wet bulb temperature in degrees C
e_d	=	Saturated water vapour pressure at T_d
e_w	=	Saturated water vapour pressure at T_w
A	=	A constant dependent on the velocity of the air flowing over the thermometer bulbs

For an air velocity of 1 to 1.5 m/sec and T_d 0 deg.C

A = 0.000799

e_d and e_w are related to T_d and T_w by:

$$e = t^{4.9283} \cdot 10^{(23.5518 - (2937.4/t))}$$

where t = T + 273 (ie. degrees K)

Table 20.4. Effect of water vapour at various temperatures

Temperature (°C)	Maximum water content (gm/m³)	Attenuation (dB/km)			
		47GHz	76GHz	142GHz	241GHz
0	4.8	0.20	0.36	0.73	2.0
5	6.8	0.24	0.43	1.03	2.8
10	9.4	0.28	0.52	1.43	3.9
15	12.8	0.33	0.64	1.95	5.3
20	17.3	0.41	0.79	2.63	7.2
25	23.1	0.50	0.99	3.51	9.6
30	30.4	0.62	1.23	4.62	12.7
35	39.6	0.76	1.55	6.02	16.5
40	51.0	0.95	1.93	7.75	21.3

the instrument. It becomes merely a matter of reading off the dry bulb temperature against the difference "dry" minus "wet" (depression) to obtain RH directly.

Fig 20.7 shows the typical variation of RH throughout several days, as measured by G8AGN in the northern UK. At first sight it would appear that the most favourable time for operation on the mm bands would be in the early afternoon. This figure does not, however, tell the whole story since it is the *actual* water content of the air which matters and this is dependent on temperature. Thus, even though the RH may be quite low around the early afternoon, the air temperature will be high, as will be the amount of water that the air could hold if saturated. In practice, therefore, the optimum time for operation may be determined by measuring both RH and air temperature at both ends of the path and estimating the increased path losses due to water absorption.

Table 20.4 gives the maximum water content in the atmosphere at various temperatures and the corresponding attenuation for the four principal millimetre bands. Below 0°C, the absolute humidity falls rapidly so that, at least in the 47 and 76GHz bands, the loss due to water vapour can be ignored. The figures are maxima for the bands when the air is saturated with water, ie 100% relative humidity. The

Table 20.5. Estimated attenuation due to rain

Attenuation, alpha $= AR^B$ — (dB/km)

where R (rainfall) is in mm/hour
A $= cf^d$ (f = frequency in GHz)
B $= ef^g$ (f = frequency in GHz)

For
f < 2.9 GHz	c = 6.39 × 10⁻⁵	d = 2.03
f > 2.9 and < 54 GHz	c = 4.21 × 10⁻⁵	d = 2.42
f > 54 and < 180 GHz	c = 4.09 × 10⁻²	d = 0.699
f > 180 GHz	c = 3.38	d = −0.151

For
f < 8.5 GHz	e = 0.851	g = 0.158
f > 8.5 and < 25 GHz	e = 1.41	g = −0.0779
f > 25 and < 164 GHz	e = 2.63	g = −0.272
f > 164 GHz	e = 0.616	g = 0.0126

actual loss can be taken as proportional to the relative humidity at any temperature. In temperate climates, 100% relative humidity over a large area is uncommon, but 90% occurs sufficiently frequently to limit the range of regular contacts. In arid regions the humidity may be less than 10% for much of the time. It follows that the best way of working very long ranges on the millimetre bands is either to operate in an arid climate or on a freezing winter's night!

20.3.3 Rain attenuation

The other main factor limiting long distance contacts is the increased path loss (α) due to rain attenuation. This may be estimated by using the relationship:

$$\alpha \text{ (dB/km)} = A \times R^B$$

where R is the rainfall rate in mm/hour and the constants A and B are given in Table 20.5. Actual values of rain attenuation may be estimated using the curves of Fig 20.8, given that typical rainfall rates are 0.25, 1, 4 and 16 mm/hour for drizzle, light, moderate and heavy rain respectively. In practice, such rainfall rates do not apply uniformly along the whole path, since rain is often concentrated in localised "cells" which are typically 1 to 10 km in extent.

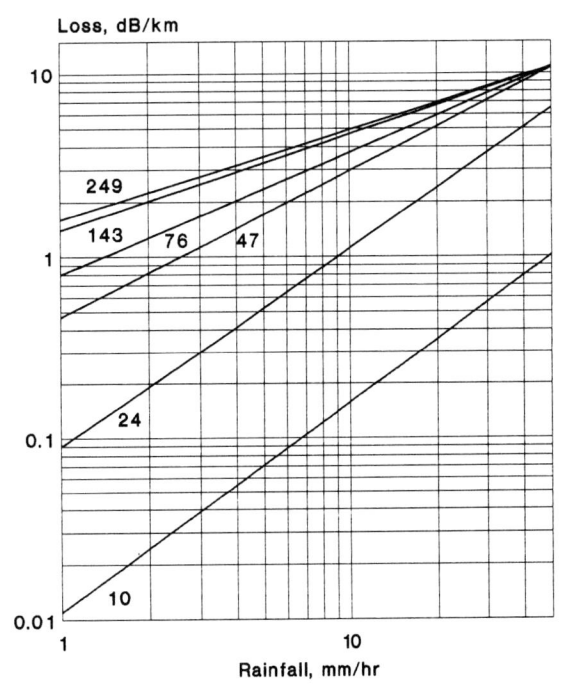

Fig 20.8. Attenuation in the amateur millimetre bands due to rain, plotted as a function of the rate of rainfall

20.4 BAND CAPABILITIES AND USES

20.4.1 The 47 and 76 GHz bands

The 47 and 76 GHz bands can be used for point-to-point operation in the same way as the 10 and 24 GHz bands, with the proviso that, at least in Europe and the coastal regions of America, water vapour will limit ranges to a few tens of kilometres for most of the time. Fig 20.9 shows the expected path losses at 47 GHz with the capability of simple wideband equipment and state-of-the-art low power

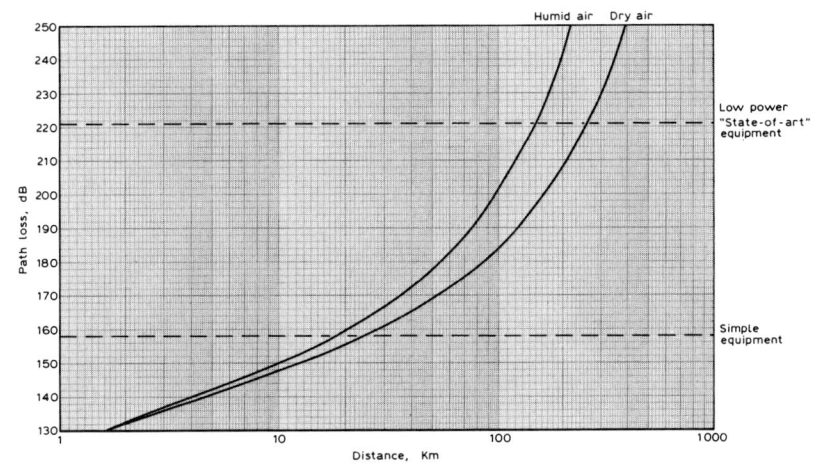

Fig 20.9. Path loss versus frequency for the 47 GHz band, for both dry and humid conditions

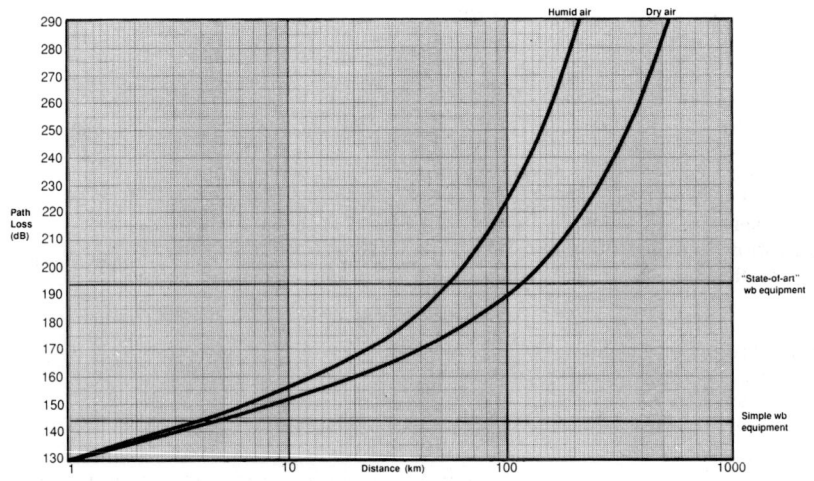

Fig 20.10. Path loss versus frequency for the 76GHz band, for both dry and humid conditions

narrowband equipment marked for reference. Fig 20.10 shows the expected path losses at 76GHz.

The additional path loss terms due to water vapour and oxygen absorption and rain attenuation have been incorporated into a special version of G8AGN's system analysis program given in chapter 2, "Operating techniques", and this is given in section 20.4.3. The modified program has been used to estimate the maximum potential of wideband equipment operating in the 47 and 76GHz bands under a range of climatic conditions. Fig 20.11 shows the path lengths possible at 47GHz for wideband equipment based, perhaps, on generating some second harmonic from a 23.5GHz oscillator and using a harmonic mixer. For comparison, the first world record for this band, a little over 50km, was made using just such wideband equipment (see later). Fig 20.12 shows that paths of 10 to 20km should be possible at 76GHz using very modest equipment, again based on harmonic transmitters and mixers.

It can be seen that, even with very modest equipment, the 47 and 76GHz bands offer some considerable scope for

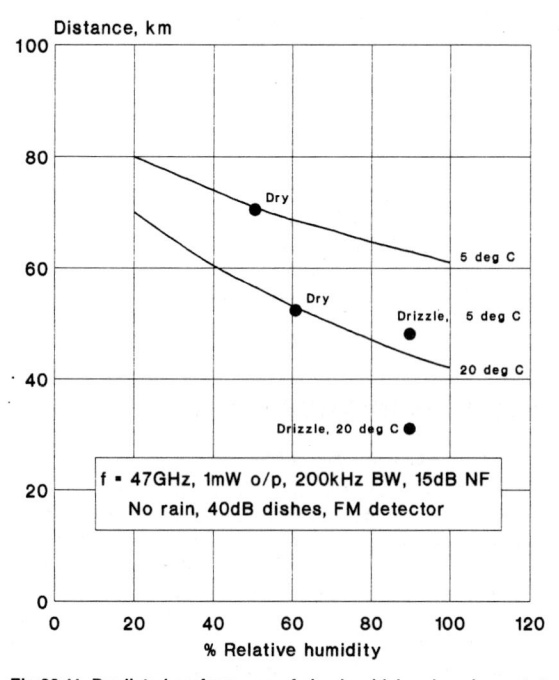

Fig 20.11. Predicted performance of simple wideband equipment at 47GHz

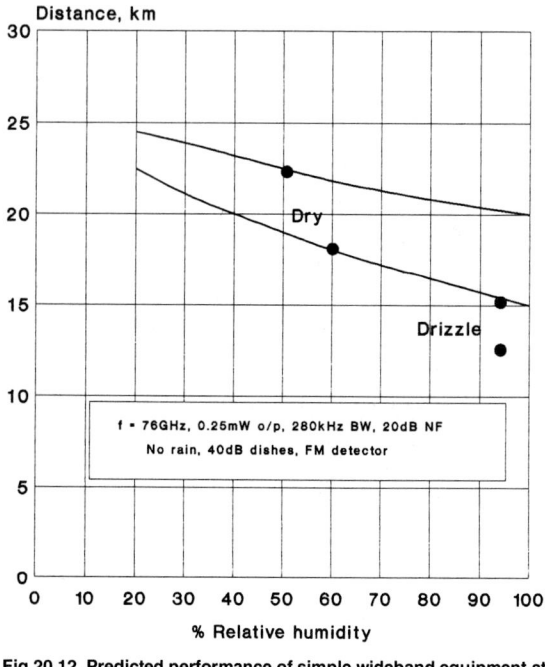

Fig 20.12. Predicted performance of simple wideband equipment at 76GHz

experimentation and dx working. As pointed out later, it is expected that mm-wave devices will be used more widely in both the professional and consumer areas and so will become available to amateurs on the "surplus" market. When this happens, then the prospect of narrowband operation at these frequencies will transform the situation and allow longer distance contacts that those predicted here – certainly on the lower mm wavebands. Indeed the recent establishment (August 1988), by USA amateurs, of a new world record of just over 105km for narrowband operation on the 47GHz band gives support to this thesis.

20.4.2 The higher mm bands

The 120GHz band is on the edge of an absorption band due to oxygen. The loss from this gas of about 0.4dB to 0.5dB per kilometre is more than that to be expected from water vapour and is such that quite potent equipment would be required for communication over any great distance. What this band does offer is privacy, because long distance transmission is practically impossible; so are interference and "eavesdropping".

Possible uses of the 142 and 241GHz bands are more problematical. Consolation can be found in the fact that, apart from the inevitable move to higher frequencies as the lower bands fill, no clear consensus has appeared among professional users as how to best exploit the special characteristics of the higher millimetre bands.

20.4.3 System performance analysis program

This program is a modified version of that given in chapter 2, section 2.7.6, and includes the losses due to water vapour, oxygen and rain attenuation.

```
A MILLIMETRE - BAND RADIO SYSTEM PERFORMANCE PROGRAM
1  REM    RADIO SYSTEM PERFORMANCE ANALYSIS
2  REM
3  REM    v 1.0 (C) 1985 B Chambers G8AGN
4  REM
5  REM    v 1.1 (C) May 1988, 24GHz water losses included
6  REM
7  REM    v 2.0 (C) Oct 1988, Atmospheric and water losses included
8  REM               for the mm bands.
9  REM
10 REM    Program in BBC BASIC
11 REM
70 DATA 0,2.6,10,16,22: REM Detector thresholds
75 DATA 3.41,3.66,3.93,4.22,4.52,4.83,5.19,5.56,6.36,6.80,7.26,7.75,8.27,
8.8,2,9.40
76 DATA 10.01,10.66,11.35,12.07,12.83,13.63,14.84,15.37,16,21,17.30,18.34,
19.43,20.58
77 DATA 21.78,23.05,24.38,25.78,27.24,28.78,30.38: REM Saturated air data
78 @%=&20109: MODE 3: REM Output format F 9.1
80 rfl=-1:afl=-1
90 CLS:PRINT:PRINT"Microwave System Performance Analysis"
100 PRINT:PRINT
110 INPUT "Enter frequency in MHz     "F
120 L=299.8/F
130 INPUT "Enter Rx i.f bandwidth in KHz    "RB
140 LRB=10*LOG(RB)+30
150 INPUT "Enter Rx noise figure in dB    "NF
160 INPUT "Enter Rx antenna gain in dB    "RG
170 BR=SQR(27000/10^(0.1*RG))
180 PRINT:PRINT "Nominal 3dB beamwidth = ";BR;" Degs."
190 PRINT:PRINT "Enter Rx feeder loss in dB    "RF
200 PRINT "Choose type of detector: "
210 PRINT " 1.     SSB "
220 PRINT " 2.     AM  "
230 PRINT " 3.     FM  "
240 PRINT " 4.     FM no limiter "
250 PRINT " 5.     FM slope detectoin"
260 PRINT: INPUT "Enter choice    "DT
270 IF DT<1 OR DT>5 THEN 260
280 RESTORE 70: FOR i= 1 TO DT: READ DTH: NEXT
290 NT=290*(10^(0.1*(NF+RF))-1)
300 PRINT:PRINT"Effective receiver noise temperature = ";NT;"Deg K"
310 NT=10*LOG(NT)
320 ERS=-228.6+LRB+NT+DTH-RG
330 PRINT:PRINT"Effective receiver sensitivity =    ";ERS;" dBW"
340 INPUT"Enter transmitter power in Watts   "TP
350 LTP=10*LOG(TP)
360 INPUT"Enter Tx antenna gain in dB    "TG
370 BT=SQR(27000/10^(0.1*TG))
380 PRINT:PRINT"Nominal 3dB beamwidth = ";BT;" degs"
390 PRINT:PRINT"Enter Tx feeder loss in dB    "TF
400 EI=LTP+TG-TF
410 PRINT:PRINT"Transmitter eirp =    ";EI;" dBW"
420 PLC=EI-ERS
430 PRINT:PRINT"Path loss capability =    ";PLC;" dB"
440 PRINT:PRINT "  Options are:"
450 PRINT"  1.     C/N over a given LOS path"
460 PRINT"  2.     Maximum LOS range of the equipment"
470 PRINT:INPUT"Enter choice    "C$
480 IF C$<>"1" AND C$<>"2" THEN 440
490 IF C$="2" THEN 600
500 PRINT:INPUT"Enter path length in km    "PL
510 LO=32.45+20*LOG(F)+20*LOG(PL)
520 PRINT:PRINT"Free space loss =    ";LO;" dB"
530 ATL=0:PROCatmos
540 RL=0:PROCrain
550 LO=LO+ATL+RL
580 PRINT:PRINT"Nominal path loss =    ";LO;" dB"
590 PRINT:PRINT"Estimated LOS CARRIER TO NOISE =    ";PLC-LO;" dB"
600 PROCpl
605 PRINT:PRINT"Atmospheric absorption loss = ";ATL;" dB"
610 IF RL>0.0 THEN @%=&4020205:PRINT:PRINT"Rain loss =    ";RL;" dB"; @%=&20109
620 PRINT:PRINT"Max. LOS path length = ";PL;" km":@%=10:END
670 DEFPROCpl
680 DL=1:PL=1
700 ATL:PROCatmos:PROCrain
710 X=PLC-20*LOG(4000*PI*PL/L)-ATL-RL
720 IF X>0 THEN PL=PL+DL:GOTO 700
730 IF X=0 THEN 770
740 IF X<0 AND DL>0.01 THEN PL=PL-DL:DL=0.1*DL:PL=PL+DL:GOTO700 ELSE PL=PL-
0.05:GOTO 770
760 PRINT:PRINT"No solution found for path length"
770 ENDPROC
2000 DEFPROCrain
2004 IF rfl=0 THEN ENDPROC
2005 IF rfl=1 THEN 2120
2010 PRINT:INPUT"Is there rain along the path ?    "RP$
2020 rfl=INSTR(RP$,"Y"):   IF rfl=0 THEN ENDPROC
2030 PRINT:PRINT"Enter type of rain: ":PRINT: PRINT"1.    Drizzle over whole path"
2040 PRINT"2.    Light rain over a 10km cell"
2050 PRINT"3.    Moderate"
2060 PRINT"4.    Heavy"
2070 PRINT:INPUT"Enter choice    "CH
2080 IF CH<1 OR CH>4 THEN 2070
2090 IF CH=1 THEN R=0.25
2100 IF CH=2 THEN R=1
2110 IF CH=3 THEN R=4
2112 IF CH=4 THEN R=16
2120 f=0.001*F
2130 IF f<2.9 THEN c=6.39E-5:d=2.03
2140 IF f<54 AND f>=2.9 THEN c=4.21E-5:d=2.42
2150 IF f<180 AND f>=54 THEN c=4.09E-2:d=0.699
2160 IF f>=180 THEN c=3.38:d=-0.151
2170 A=c*f^d
2180 IF f<8.5 THEN e=0.851:g=0.158
2190 IF f<25 AND f>8.5 THEN e=1.41:g=-0.0779
2200 IF f<164 AND f>=25 THEN e=2.63:g=-0.272
2210 IF f>164 THEN e=0.616:g=0.0126
2220 B=e*f^g
2230 alpha=A*R^B
2240 IF CH=1 THEN RL=alpha*PL ELSE RL=alpha*10
2245 IF rfl=0 THEN @%=&4020205:PRINT:PRINT"Rain loss = ";RL;" dB":@%=&20109
2250 ENDPROC
3000 DEFPROCatmos
3005 IF afl=1 THEN 3100
3010 REM Determine atmospheric water content
3020 RESTORE 75
3030 PRINT:INPUT"Enter air temperature in deg C    "T
3040 PRINT:INPUT"Enter relative humidity in %RH    "RH:RH=0.01*RH
3050 FOR IT=-5 TO T
3060 READ W
3070 NEXT
3075 w=W*RH
3080 PRINT:PRINT"Atmospheric water content =    ";w;" gm/m^3"
3100 f=0.001*F
3110 IF f>=10 AND f<=10.5 THEN al=0.0066*0.0011*w
3120 IF f>=24 AND f<=24.25 THEN al=0+0.0185*w
3130 IF f>=47 AND f<=47.2 THEN al=0.13+0.016*w
3140 IF f>=75.5 AND f<=76.0 THEN al=0.40.024*w
3150 IF f>=142 AND f<=144 THEN al=0.152*w
3160 IF f>=248 AND f<=250 THEN al=0.417*w
3170 ATL=al*PL
3175 IF afl=0 THEN PRINT:PRINT"Atmospheric absorption loss = ";ATL;" dB"
3180 IF ATL=0 THEN PRINT:PRINT"Atmospheric data not available"
3185 afl=1
3190 ENDPROC
```

20.5 TRANSMISSION LINES

20.5.1 Waveguide and flanges

Table 20.6 lists the standard waveguide sizes suitable for the amateur millimetre bands together with a range of rectangular brass tubing available from model shops and which can be used for the 47 and 76GHz bands. The quite high losses to be expected from standard rectangular waveguide should be noted. The values given in Table 20.6 are for copper; brass waveguide will give figures some 30% worse and silver about the same amount better.

Two flange types are available for millimetre waveguide: square flanges with four bolt holes and round flanges with

Table 20.6. Waveguide for the millimetre bands

Type no.	Outside dimensions (mm)	Inside dimensions (mm)	Cut-off (GHz)	Band (GHz)	λ_0 (mm)	λ_g (mm)	λ_g / λ_0	Loss dB/m
WR42 WG20	12.7/6.35	10.67/4.32	14.047	24.0	12.49	15.41	1.234	0.45
K&S 268	9.53/4.76	8.74/4.00	17.1	24.0	12.49	18.18	1.454	0.60
WR28 WG22	9.14/5.59	7.11/3.56	21.07	24.0	12.49	26.11	2.090	0.75
K&S 266	7.93/3.97	7.14/3.18	21.0	24.0	2.49	25.72	2.060	0.75
WR22 WG24	7.72/4.88	4.78/2.39	26.34	47.0	6.38	7.70	1.208	0.78
K&S 264	6.35/3.18	5.55/2.38	26.98	47.0	6.38	7.79	1.221	0.90
K&S 262	4.76/2.38	3.96/1.60	37.8	47.0	6.38	18.06	2.830	1.8
WR12 WG26	5.13/3.58	3.10/1.55	48.35	76.0 / 81.0	3.94 / 3.70	5.11 / 4.61	1.296 / 1.247	2.5
WR8 WG28	3.96 dia	2.03/1.02	73.84	120.0	2.50	3.17	1.268	5.0
WR7 WG29	3.96 dia	1.65/0.825	90.84	142.0 / 149.0	2.11 / 2.01	2.75 / 2.54	1.301 / 1.261	6.0
WR5 WG30	3.96 dia	1.30/0.648	115.75	142.0 / 149.0	2.11 / 2.01	3.64 / 3.19	1.725 / 1.587	9.0
WR4 WG31	3.96 dia	1.09/0.546	137.52	241.0 / 250.0	1.24 / 1.20	1.51 / 1.44	1.215 / 1.196	12.0

removable, screwed rings. The latter are more popular among professional users but are more expensive. Fig 20.13 gives the dimensions of a general purpose waveguide flange much used professionally for test gear. The same-sized flange is used with a range of waveguide sizes and it can therefore be used as a standard connector between non-standard waveguides.

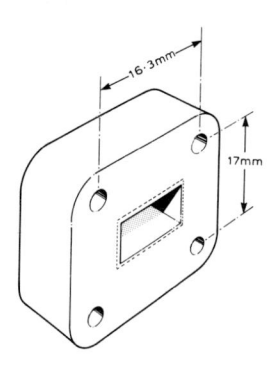

Fig 20.13. General purpose waveguide flange. The holes are 0.144in for either 4BA or M3 bolts. The waveguide is central with respect to bolt holes. Overall dimensions are not critical

As an alternative to the use of flanges, it is possible to adapt popular coaxial connectors for use with waveguide. Fig 20.14 shows the use of a BNC or TNC connector in this way. Since these two connectors have the same basic dimensions and differ only in the use of a bayonet or screwed clamp, it is possible to use the same arrangement with BNC fittings for easy connection and disconnection in the workshop and TNC fittings for a more reliable connection in the field. The internal coaxial components of either type of plug are discarded and for the larger waveguides the connector will need to be drilled through to allow the waveguide to fit.

20.5.2 Overmoded waveguide

As can be seen from Table 20.6, the loss of conventional waveguide is prohibitive, except for very short runs inside the equipment. If we wish to operate with an elevated antenna which involves a long transmission line, then the losses associated with a waveguide run are too high and other techniques are required. If a waveguide is larger than necessary to propagate the fundamental waveguide mode then a variety of other modes are possible in the guide. If not controlled, this "overmoding" is undesirable but, if controlled, particular modes may propagate with much lower loss than the fundamental TE10 mode used in conventional rectangular waveguide.

The most useful low-loss modes occur in a circular tube, and are the TE01 and TM01 modes. They are symmetrical; the TM01 mode has a radial electric field and a circumferential magnetic field, while the TE01 mode has a radial magnetic field and a circumferential electric field. These are shown in Fig 20.15. In each case the fields are low at the circumference of the metal guide; this means that the currents flowing in the waveguide wall are minimal, so the losses are correspondingly small.

The TE01 mode shows the lowest loss; a 15mm diameter copper tube operating at 47 or 76GHz can be expected to have a loss of not more than 0.01dB/m. The mode can be launched from a rectangular guide by the arrangement shown in Fig 20.16. The transition is quite badly mismatched and matching screws are required in the rectangular guide as shown. The TM01 mode has a higher loss; a 15mm copper tube, as above, would have a loss of 0.1dB/m, but the mode is much easier to launch by the arrangement shown in Fig 20.17 and, for the short feeder runs likely to be used by amateur stations, the additional loss is not important. Since these waveguide modes are radially symmetrical, the connections at the two ends need not be in the same plane, and if the circular guide is run up a mast, the antenna can be arranged to rotate about the centre line of the circular guide and a simple choke joint is all that is required to provide for rotation.

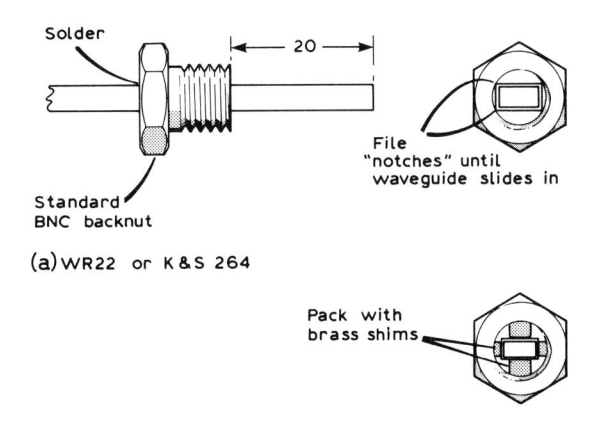

It is important that the overmoded waveguide runs are kept straight; the effect of bends is to couple between

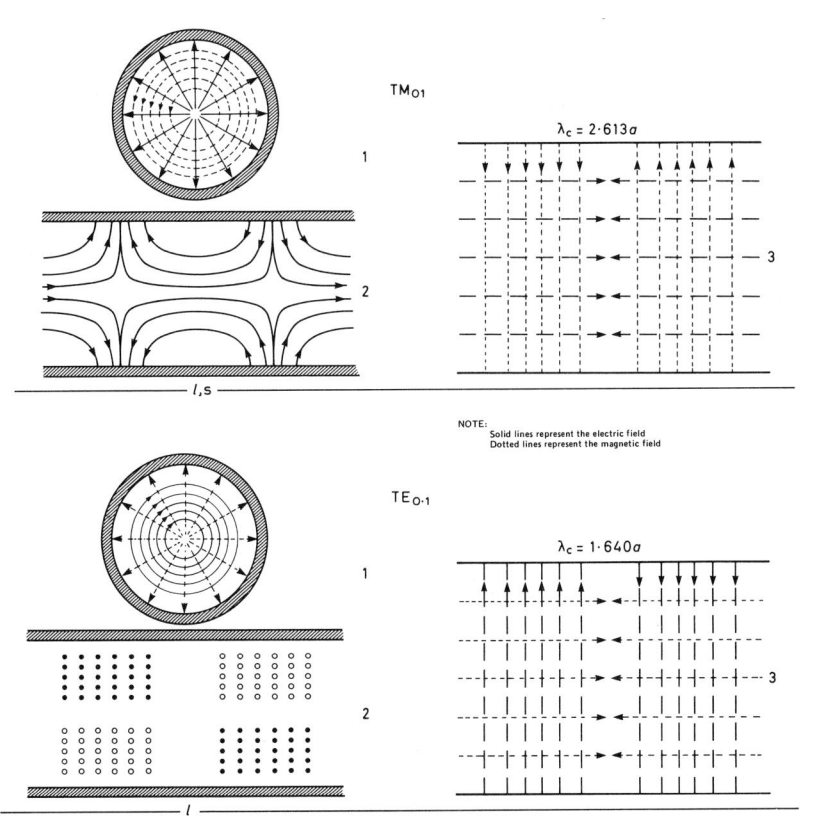

Fig 20.14. Use of BNC or TNC connector instead of a flange

Fig 20.15. TM and TE modes in circular waveguide. Solid lines represent E-field and dotted lines the H-field. © Artech House, Inc., from Theodore S. Saad, *Microwave Engineers Handbook*, Vol 2, 1971. Reprinted by permission

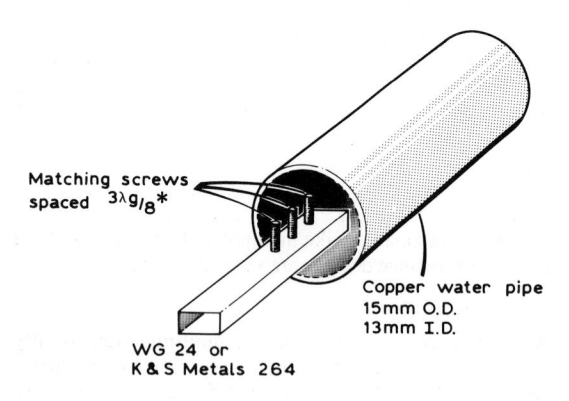

Fig 20.16. Mode transition for rectangular to circular waveguide

$3\lambda g/8$ (WG24) = 2·89mm

$3\lambda g/8$ (K & S 264) = 2·92mm

modes, and energy is transferred into other higher loss modes. These modes are also mismatched at the terminations, so that very high ratio standing waves are set up and a variety of unacceptable resonances will be produced in the guide, leading to further losses.

20.5.3 Other transmission lines

Whilst amateurs are most likely to use waveguide, it has several disadvantages for the professional user, including: close tolerances are needed in building components, it is difficult to use mass production techniques and waveguide does not lend itself to integration techniques. There are alternative transmission lines can be used that overcome these problems; one is microstrip, which is popular at centimetre wavelengths, but at millimetre wavelengths the losses start to become excessive. An alternative is dielectric waveguide.

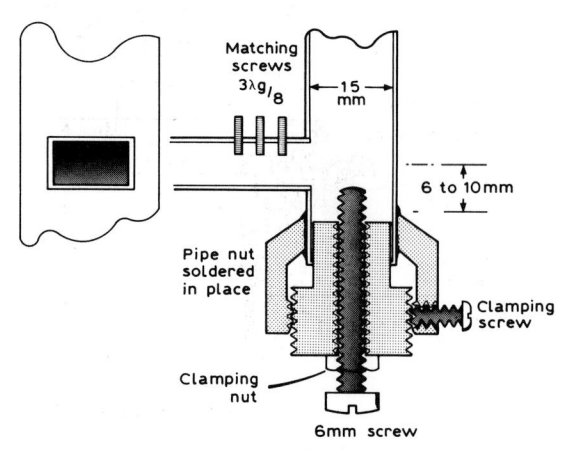

20.5.4 Dielectric waveguide

This consists of a rectangular strip of dielectric and is normally used spaced from a ground-plane by a separate, thick dielectric layer, as shown in Fig 20.18. The electromagnetic field is confined in the rectangular dielectric due to its refractive properties, much as optical fibres behave with light. It can provide fairly good performance, much better than that of microstrip, as shown in Table 20.7. Most of the familiar waveguide components can be made from dielectric waveguide – directional couplers, attenuators, phase-shifters, filters and isolators are all used professionally.

20.5.5 Coaxial lines

Coaxial cables are almost impractical, in the 47GHz band and above, because of overmoding which occurs when the circumference of the cable approaches one wavelength in the dielectric. For example, at 47GHz the largest coaxial line with polythene dielectric (dielectric constant of 2) that can be used would have a circumference of:

$$\frac{300}{47 \times \sqrt{2}} = 4.5\text{mm}$$

giving a diameter of 1.4mm. At 76GHz, the maximum safe diameter is 0.9mm. Cables of such diameter are available (small diameter semi-rigid cable), but are generally too lossy to be useful.

20.5.6 "Optical transmission lines"

In the higher frequency bands it is possible to dispense with the metal walls of the waveguide by focusing radiation into a sufficiently "tight" beam and "catching" all of

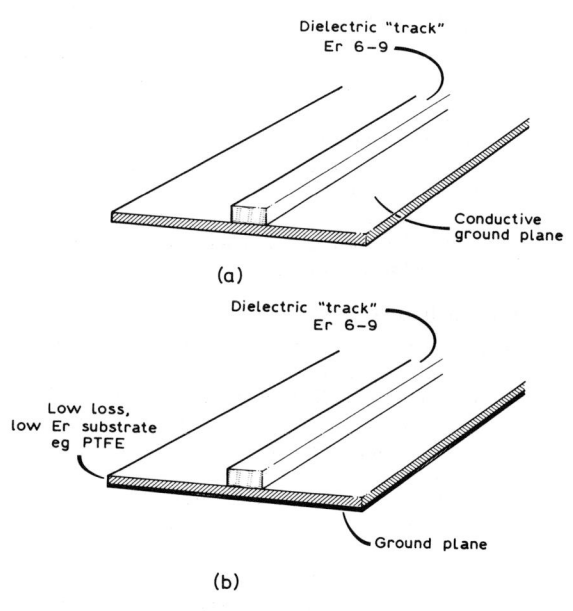

Fig 20.17. Launching the TM01 mode: rectangular to circular guide transition

Fig 20.18. Dielectric waveguide

Table 20.7. Comparison of transmission line losses

Transmission line	Loss at 47GHz (dB/cm)	Loss at 76GHz (dB/cm)
Rectangular guide (copper)	0.011	0.022
50Ω microstrip (on quartz)	0.112	0.225
Dielectric guide (alumina)	0.039	0.078

the radiated power in a similar collector. By using a high-gain antenna which produces an almost parallel beam for a short distance and collecting the power in another antenna of similar size, it is possible to transmit power over tens of metres with relatively low loss. The losses due to sidelobes or other defects of the antennas are likely to be less than those from the same length of conventional waveguide. The antennas are necessarily located within each others' near fields and the best results will be obtained with identical antennas at the two ends of the link.

20.6 SOLID STATE DEVICES

Devices for the millimetre bands are not yet commonplace. There are only a few manufacturers of both oscillator and detector/mixer diodes up to about 100GHz, notably Alpha Industries and M/A-COM (Microwave Associates). Devices at these frequencies are often listed in a catalogue by band (as in X-band for 10GHz), but confusion can arise since there are several letter designation systems and some frequency bands are different in each system. Common designations are:

- Ka-band, covering 18 to 26.5GHz. This stands for "K-above", since K-band is 12.4 to 26.5GHz, and Ku (for "K-under") is 12.4 to 18GHz.
- Q-band, covering 40 to 75GHz. However, the range 40 to 60GHz is sometimes referred to as U-band (eg by Alpha Industries).
- W-band, covering 75 to 110GHz.

See also chapter 13, "Data". All in all, it is considered better to avoid using these lettering systems and to use frequency ranges instead. GaAs fets will soon be available for the millimetre bands, but their cost is likely to be too high for amateur use in the near future.

20.6.1 Oscillators

Klystrons may be found and semiconductor oscillators, Gunn devices and impatt diodes are becoming available for the lower frequency millimetre bands. Low power Gunns will give around 10mW of output power, while high power devices giving some 100mW are available but are very expensive. They are usually operated with 4.5 to 5.5V bias, similar to 24GHz devices. A typical package for a device operating in the 40 to 75GHz range is shown in Fig 20.19.

20.6.2 Mixers and detectors

Probably most diodes in the pill package (SOD31, SOD45, SOD46 and similar outlines) will function as useful rectifiers even if they are not too good as mixers. In general, the chips used in modern microwave diodes are usable up to at least 100GHz; it is the mount which sets the limit to the frequency of use and the problem is getting the rf into the chip. Techniques to overcome this are described later.

Quite low noise figure diodes are available using Schottky barrier construction. Noise figures as low as 8dB at 60GHz are attainable. Like high power Gunn diodes, these better performance devices are very expensive.

20.6.3 Multipliers

At millimetre-wave frequencies, multiplication usually means high-order multiplication and step recovery diodes are available, suitable for output frequencies of up to 100GHz. They, too, are expensive and the amateur is probably better off trying to use lower frequency devices. These will function with reduced efficiency, but should give enough output power for a receiver local oscillator or a harmonic generator/calibrator.

20.7 TECHNIQUES

20.7.1 Mounting components

Most of the readily available pre-packaged devices are physically large compared with a wavelength and it is quite impracticable to arrange "connections" to "terminals" in the sense appropriate to lower frequencies. Usually the actual semiconductor chip is physically small and, in devices designed to operate efficiently at frequencies of 20 or 30GHz, it is usual for the chip to perform quite well at several times that frequency. The problem is, then, to get the rf into or out of the chip.

Fig 20.20 illustrates the internal construction of a typical mixer diode in a pill (SOD31) package, showing the device

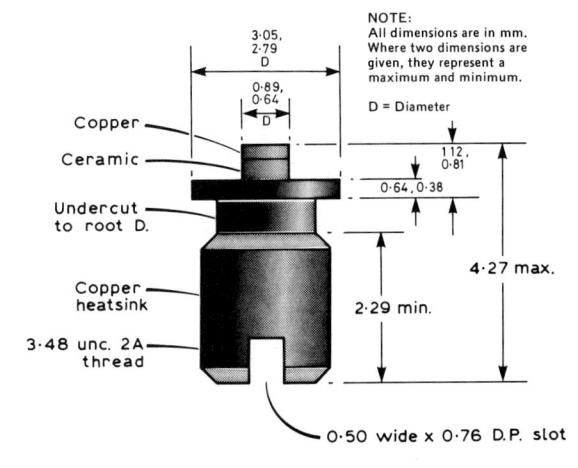

$C_p = 0.10pF$ $L_i = 0.10nH$

Fig 20.19. Typical Gunn diode package

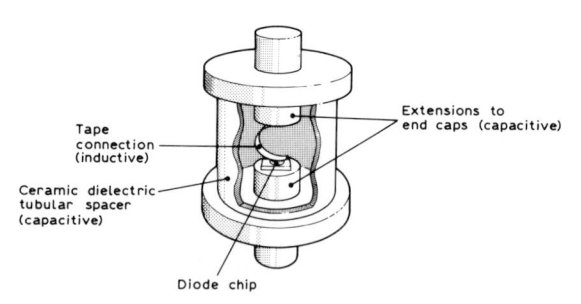

Fig 20.20. Internal construction of a 'pill' mixer diode

chip mounted on one cap of the package and tape connection made from the other cap. Gunn devices, varactors and pin diodes usually have similar arrangements. Since the physical size of the device package is an appreciable part of a wavelength even at 47GHz, we need to find some way to bypass the terminals. The technique used in waveguide mixers is, therefore, to immerse the diode package in the rf field and to compensate as best we can for the reactance introduced by the packaging.

The ceramic package is equivalent to bulk capacitive susceptance across the waveguide and is tuned out by a sliding short behind the diode. A series of tuning screws then match the resistive term and tune out the internal inductance of the mount. Since we are interested only in relatively narrow bandwidths – the whole 47GHz band is only 0.43% wide – the restricted bandwidth introduced by these matching methods is of no consequence. The method is illustrated in the description of a mixer mount for 47GHz which follows later.

20.7.2 Self-oscillating mixers

A receiver can be built using a Gunn diode as both the local oscillator and a mixer – it is referred to as a "self-oscillating mixer". This technique means that a complete receiver (and transmitter, of course) can be built which requires only one expensive semiconductor device. Construction is also greatly simplified – only one diode mount need be built and, of course, there is no need for a cross-coupler.

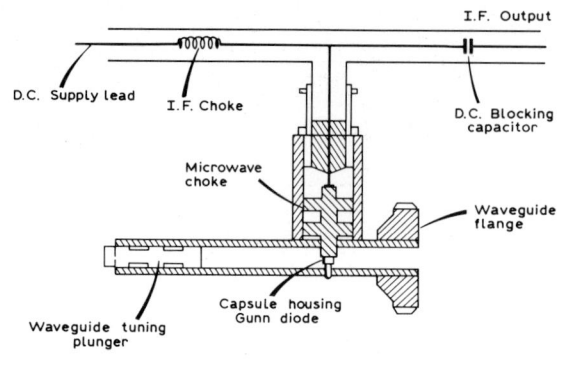

Fig 20.21. Self-oscillating mixer assembly

Fig 20.21 shows the cross section of a 47GHz self-oscillating mixer assembly, with separate dc and i.f feeds. The oscillator uses a plunger tuning arrangement. One problem to watch out for is the occurrence of low frequency parasitic oscillations which are normally suppressed in Gunn oscillators by a shunt capacitor across the choke. A capacitor cannot be used here because it would shunt the i.f signal to ground; in the example given, the Gunn was operated at the highest permissible voltage and this cured the problem. At these frequencies the receive performance can approach that of a separate oscillator and mixer and a system built in this way is an easy way of getting started on the band.

20.8 ANTENNAS

20.8.1 Horns

The simplest antenna to make and use is a horn; quite large gains, eg 30dB, can easily be achieved in a compact size. For higher gains, a parabolic dish can be used. There are other alternatives, eg lenses and microstrip antennas, but these are really only of practical interest to professional users.

Small horn antennas are easy to design, using the design methods described in chapter 4, "Microwave antennas". Being so small, they are not particularly easy to make but are, at least, not as difficult as dish antennas of the same gain!

Methods described for 10 and 24GHz horns can be adapted to make horns for these higher frequencies. A satisfactory method is to make a former from hardwood and build up the horn soldering the four sides on a cut-and-try basis, and finishing off with a wrapping of copper wire or some reinforcing plates at the joint of the horn with the waveguide. The same former can be used for several horns, producing transmit and receive antennas for two or more experimental stations. Some examples of calculated horn dimensions for 47 and 76GHz are given in Fig 20.22.

20.8.2 Dishes

Small dishes are useful as directional antennas but large dishes are hopelessly impracticable. For instance, a gain of 46dB implies a beamwidth of less than one degree in both azimuth and elevation, an accurate knowledge of the location of the station to be worked and a very rigid mounting to enable the antenna to be aligned and kept pointed correctly. Exceptionally precise setting-up would be required to point the dish at the other station, probably involving optical sighting tubes or telescopes. Fig 20.23 extends the range of dish size versus gain to the millimetre bands. Note that the diameter is now given in centimetres, not feet!

Parabolic antennas for the millimetre bands can be cut from solid on a lathe. If a block of metal is used, then the reflector can be used immediately. If a block of hardwood is used then it should be covered in kitchen aluminium foil. The foil should be laid in parallel strips with an overlap of at least $\lambda/4$ at the lowest frequency of operation and parallel to the electric plane of polarisation. Make strips wide enough so that nowhere are there more than two layers and do not attempt to lay the strips radially as it is too difficult to avoid a "muddle" at the centre of the dish.

The required accuracy is not very high, $\lambda/10$, which is 0.6mm at 47GHz, falling to 0.1mm at 240GHz. This should present no difficulty on a metal-working lathe. If a front feed is used then an f/D ratio of 0.5 is recommended; this is the ratio correctly fed with a sectoral horn that tapers in the narrow face only. Initially, a satisfactory feed can be made for experimental work by just radiating from an open guide. Such dishes for the millimetre bands are so small that it is quite good practice to make them oversized and then "underfeed" them. This produces a highly efficient design in the sense that no rf is "wasted" and the best gain is obtained for a given beamwidth. For example, for an f/D ratio of 0.5, the complete dish is turned and the feed is in the plane of the aperture where it is very easy to support.

20.8.3 Lenses

At millimetre wavelengths, techniques similar to optics become practical. Components are much larger than the wavelength and dielectrics exhibit properties which refract the waves in the same way as light is refracted ("bent") in a lens. Thus, not only will a parabolic dish focus microwaves as a concave mirror does light, so a dielectric "lens" will focus millimetre waves. It might be worthwhile for the experimenter to produce a plano-convex lens by turning a disc of a dielectric such as polythene or perspex and trying this as an alternative to a dish or horn!

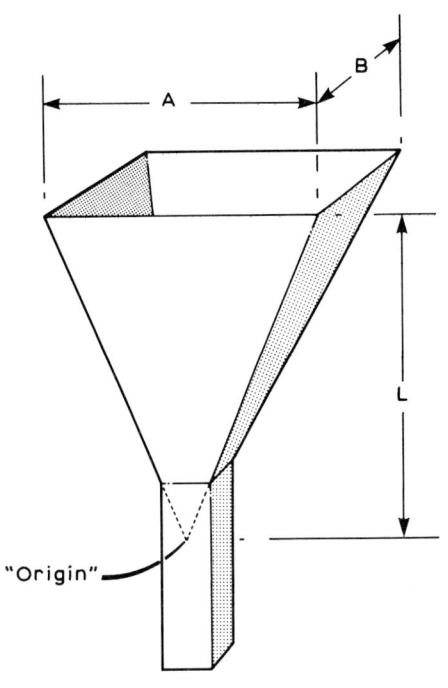

Note:
Dimensions A & B are internal

Fig 20.22. Typical horn dimensions for 47 and 76GHz. For 47GHz WG24 is used. 20dB horn: A=28.5, B=22.3, L=33.2; 30dB horn: A=88.6, B=71.5, L=385.7; For 76GHz, WG26 is used. 20dB horn: A=17.6, B=13.8, L=20.4; 30dB horn: A=54.8, B=44.2, L=238. All dimensions in mm. Optimum gain and sectoral horns can be easily designed using the program in Chapter 4, "Microwave antennas"

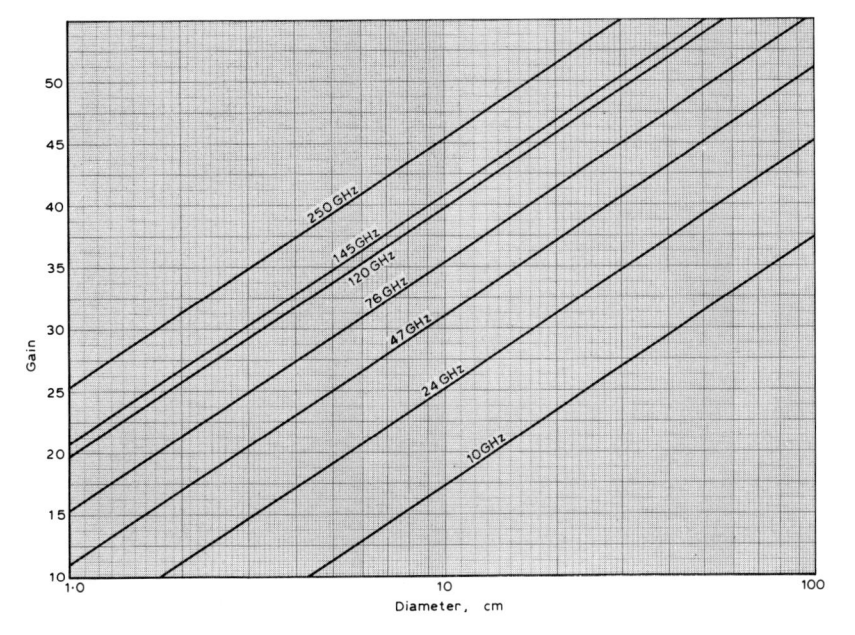

Fig 20.23. Gain of various sized dish antennas

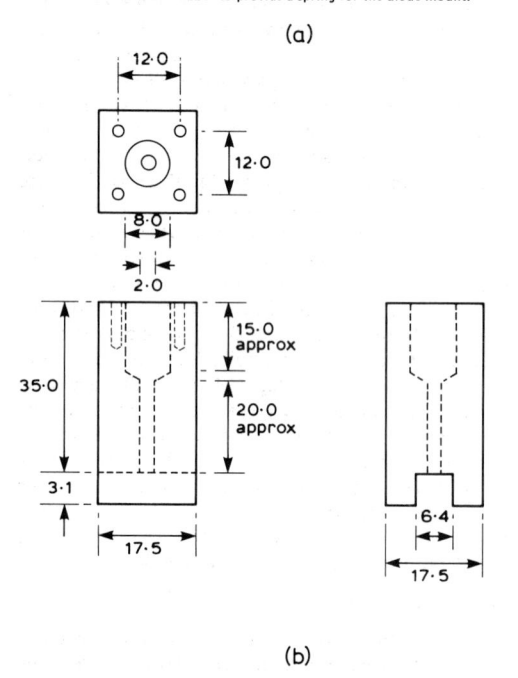

(a)

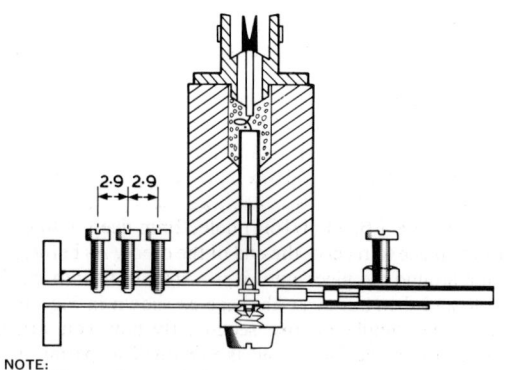

NOTE:
The space between the output choke and the co-axial connector is filled with foam rubber to provide a spring for the diode mount.

(b)

Fig 20.24. (a) and (b): A mixer/detector for 47GHz

(a) Sliding short tuning plunger for waveguide mixer

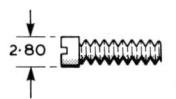

(b) Matching screw with turned down head

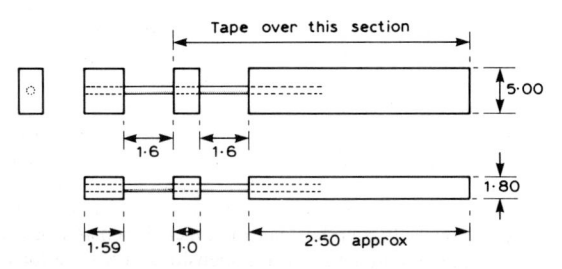

Fig 20.25. (a) and (b): Detail of tuning plunger for the 47GHz mixer/detector

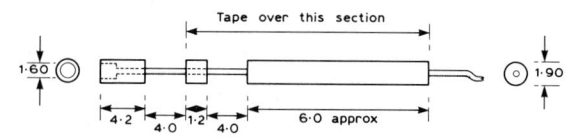

Fig 20.26. Choke for the 47GHz mixer/detector

Diode holding screw (use standard 4BA cheese head screw)

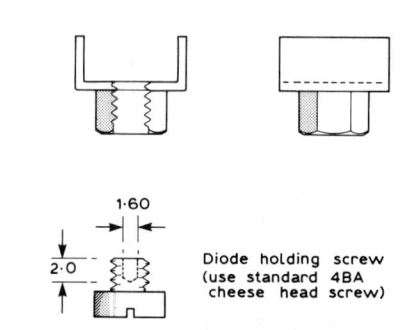

Fig 20.27. Screw to hold the diode for the 47GHz mixer/detector

20.9 EXAMPLES OF EXPERIMENTAL EQUIPMENT FOR 47GHz

20.9.1 Availability of components

The bands immediately below the four principal amateur bands are assigned to the fixed and mobile services while the upper sections of the bands are shared with radiolocation. It is thus to be expected that components will be developed that can be adapted to conventional station-to-station QSOs in the lower parts of the bands and that it will be worth watching the "surplus" market for components suitable for wideband operation in the higher parts of the bands.

20.9.2 A mixer/detector for 47GHz.

This design is based on old designs for 10GHz, but using a "pill" diode instead of a 1N23/1N415 type diode. Fig 20.24 shows the general assembly and the loose components are shown in Figs 20.25, 20.26 and 20.27. The waveguide is soldered to the end of a squared-up brass block. The drawing shows the waveguide set into a slot cut into the block, but this is not really necessary. The size of the block is not important; the sizes shown are suitable for a BNC connector and can be adjusted to suit a different connector. The minimum height is 25mm to allow for the coaxial choke in the output lead and the spill on the output connector. A piece of brass sheet is also soldered to the top of the waveguide at the input to provide thickness for the threads of the matching screws. A flange, or other connector, should be soldered in place at the same time.

The sliding short comprises two sections of high/low transformer, the first of which has an air dielectric and the second plastic. It is backed up by a larger piece of metal for convenience of handling. The prototype was made of brass. The block is first filed up in the solid to be about 0.1mm (say 5 thou inch) smaller than the inner dimension of the waveguide you are using. The dimensions given are for K & S Metals size 264 tubing. The changes for "professional" waveguide are obvious. Do not be too concerned if these dimensions are not precise; what is important is that the sides of the block are parallel.

Next, the block is drilled lengthways with a number 74 drill (for 22swg wire) or 0.5mm drill (for 0.5mm wire). First select your wire and then drill to suit. It does not matter if the hole is off-centre (in fact the block is easier to assemble if the hole is not exactly on the centre line) but do try to make the hole parallel with the long axis. Cut off a small piece of the block and file it down to 1.59mm long. It is worth taking time and care over this dimension, making sure that the block is not tapered. This is one of the reasons the original block was made too long; you can afford to throw away your failures! Make two of this length. Cut off another small piece and file down to 1mm. This component is less important electrically, its main job is for mechanical support.

Now re-assemble on a piece of wire as shown in Fig 20.25. The best method is to apply a touch of solder paste to the parts to be joined and to apply a hot iron to the remote end of the big block and allow the heat to work its way through. Do not underestimate the time it takes for the block to heat and cool. Clean up the block, eg with a nylon burnishing pencil available from model shops.

Next, bend and twist the end sections gently until they look straight and wind on a few layers of thin, self-adhesive plastic tape. The tape should preferably be ptfe, but polyester or even Sellotape will do. Here, accuracy of the end section tells; the better it is, the less important is the dielectric used in the back section. Don't use double sided sticky tape. Now peel off the tape, one face of the block at a time, until the main block is a nice, smooth fit in the waveguide and make final adjustments to the alignment of the end sections so the whole block moves smoothly. The point here is to test the size using the large block and "tweak" the smaller blocks to line up. Do not cut extra tape from the small blocks to make them slide. Now cut away the tape from the end block so that this block is left with a 0.05mm air gap between it and the waveguide.

Next, we tackle the output choke. Because a coaxial transmission line of too large diameter will permit a waveguide mode so that rf bypasses the filter, we are limited to a maximum diameter of 2.0mm. We first drill through the whole block and the waveguide with a 2mm drill. Next open out the top of the block with a larger drill to clear the connector you are using and to provide for the compression pad. The choke inner is turned from brass to a diameter of 1.9mm and while the piece is still in the lathe, a hole (drill 71 or 0.5mm as before) is drilled the full length. Next, pieces are cut off and faced off to lengths of 4.18 and 1.0mm. The three pieces are then assembled on to a straight piece of wire, which at this stage can protrude at both ends, and soldered up. The wire extended at the connector end becomes the output terminal and is cut to length. The piece is then returned to the lathe and the copper wire faced off flush with the brass end and touched with a centre drill. Now drill a 1.6mm (1/16in) hole, 2mm deep, into the end of the piece to receive the end of the diode. As before, wrap tape around the piece until it is a smooth fit into the 2mm hole and then cut away the tape as shown in Fig 20.26 to form the complete choke section. By now you will have noticed that the 2mm hole is too small to permit assembling this mixer by dropping the diode down the output tube, as we do with a 1N23/1N415 diode mixer on 10GHz. Thus a detachable diode mount is needed. This is a 4BA screw fitted to a nut soldered to the face of the waveguide. First measure a few nuts and pick half a dozen or so with the same thickness; these will be our stock of such mounts. Cut off a short (10mm) piece of waveguide, drill a 3.1mm hole (size 32) in one face and cut away the other face, taking care not to distort the piece. Tap the hole 4BA and using a chrome plated or rusty steel screw to align the nut, solder this to the outer face of the waveguide. Ease out the end of the nut with a 4.0mm drill (size 26) so that a 4BA screw will screw down hard and now you will have the jig shown in Fig 20.27(a). This will enable you to file down a 4BA screw to the required length. Do this and put the screw in the lathe and drill the 1.6mm hole, 2mm deep, as shown in Fig 20.27(b). Drill and tap 10BA for the matching screws and fit the nut for the clamping screw for the sliding short, once again using an untinnable screw to locate it.

Finally make some matching screws. The heads of the screws must be turned down so that they can be spaced 2.9mm ($3\lambda/8$) apart without fouling each other. 10BA screws are preferred, but 8BA are just possible if the head is turned down to the same diameter as the thread.

Testing and setting up

The instructions for setting up assume you have no source of 47GHz, but do have a source at 24GHz. There is no need, yet, to re-tune it to 23.5GHz. Set the matching screws to be flush on the inside of the guide. Connect the most sensitive meter you have to the output of the diode unit and offer it up to the open end of the source waveguide. Adjust the sliding short for maximum reading; if you are lucky you will find a series of settings about 4mm apart corresponding to half wavelengths in the guide. Choose the setting nearest to the diode and clamp the short in place. Finally set the matching screws for maximum meter reading.

If your source has a high harmonic output, you may be confused by the presence of 72GHz, and the setting of the short may not be too obvious. Move the mount as far from the source as possible while still getting a reasonable meter reading, and draw up a graph of the meter reading against position of the short. You should then see a pattern and be able to select the best working position.

20.16 MICROWAVE HANDBOOK

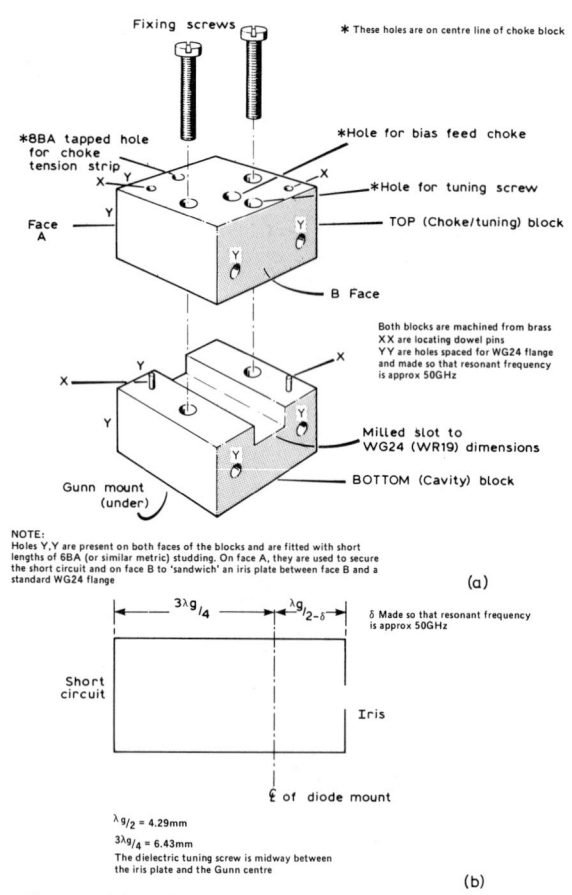

Fig 20.28. (a) and (b): 47GHz Gunn oscillator construction. General view of oscillator block

Labels on figure (a):
- Fixing screws
- *These holes are on centre line of choke block
- *8BA tapped hole for choke tension strip
- *Hole for bias feed choke
- *Hole for tuning screw
- Face A
- TOP (Choke/tuning) block
- B Face
- Both blocks are machined from brass
- XX are locating dowel pins
- YY are holes spaced for WG24 flange and made so that resonant frequency is approx 50GHz
- Milled slot to WG24 (WR19) dimensions
- Gunn mount (under)
- BOTTOM (Cavity) block

NOTE:
Holes Y,Y are present on both faces of the blocks and are fitted with short lengths of 6BA (or similar metric) studding. On face A, they are used to secure the short circuit and on face B to 'sandwich' an iris plate between face B and a standard WG24 flange

Labels on figure (b):
- $3\lambda g/4$
- $\lambda g/2 - \delta$
- δ Made so that resonant frequency is approx 50GHz
- Short circuit
- Iris
- ℄ of diode mount
- $\lambda g/2 = 4.29$mm
- $3\lambda g/4 = 6.43$mm
- The dielectric tuning screw is midway between the iris plate and the Gunn centre

Finally, retune the oscillator to 23.5GHz and repeat all the optimisations described above for the new frequency, which should now lie in the 47GHz band.

20.9.3 An experimental Gunn oscillator for 47GHz

The following design ideas (due to G3WDG and G8AGN) are mainly unproven in practical operation. The description is given to indicate the type of approach which may be feasible for the more advanced microwave constructor but the design cannot, at the moment, be considered as a really practical one. First, the cost of the Gunn devices is very high, their availability low and their robustness suspect. Second, the dimensions and form of construction are critical, as was indicated by the findings on a prototype oscillator of somewhat different construction. The first prototype used a milled cavity with internal dimensions similar to WG22, but with rounded corners. It was found to be off frequency, despite careful construction, and unstable. The tests carried out suggested the dimensioning of both the choke feed and the profile of the cavity to be very critical and, while some rf power was generated in the vicinity of 42GHz, there was considerably more power being developed over a range of undefinable (parasitic) frequencies. Without spectrum analysis available, it was assumed that the parasitic output was similar to that occasionally seen in 10GHz oscillators of early design, where at some bias and tuning settings the device output breaks up into what is best described as rf "white noise" centred roughly around the cavity resonant frequency but extending for perhaps several hundred MHz in a 10GHz oscillator.

The main features of the second design is shown in Fig 20.28(a) and (b). A completely different method of construction has been adopted to ensure *exact* cavity dimensions, with the cavity reduced in size to that of WG24. The choke bias feed arrangement of the first prototype was retained and the general format of the oscillator is similar to that commonly used in the recent iris coupled designs for 10GHz Gunn oscillators described in chapter 18.

It must be stressed that at the time of writing the second prototype oscillator has not been tested (another difficulty in millimetre band construction!), so the design is not offered as a practical solution, but solely to indicate methods of construction which might produce the right results. An accurate milling machine, a model maker's lathe, accurate measuring instruments and considerable machining skills are *essential* before tackling a job of this nature.

First, a piece of suitable brass stock is machined into a right-angled block with dimensions 10.43mm by 25.4mm by, say, 29mm long, with the first two dimensions accurate and flat, as these are the finished dimensions of the oscillator block. Next, four holes, D and E, are drilled to the sizes in Fig 20.29, to a depth of about 20mm. Initially, holes D should be 6BA tapping (or a comparable metric size, not critical in size or placing) and holes E of a size to suit any dowel pins available. The centre line of the long dimension is now scribed around the block and small punch marks made on the block, one each side of the scribe line. These will identify which way round the blocks will fit after the next operation. Now carefully saw the block into two smaller, equal sized blocks. The newly cut faces of each block are milled, flat and square, to the final dimensions ie so that the height of each block is 12.7mm. The two blocks should now measure 25.4mm by 12.7mm by 10.43mm. The two blocks should fit closely together if the job has been done correctly. The block which has holes right through it is the choke block (A) and the other, with blind-ended holes will be the cavity block (B). Holes D in the choke block should now be opened up to clearance size for the bolts to be used and then three more through-holes of the sizes and centres shown in Fig 20.29 drilled in this block. Note that the sizes and placing of the holes A and B are critical, hole C is not.

The cavity block (B) now needs more work. First, holes D are tapped to the appropriate size, taking care to avoid the "bell mouth" effect which was described in chapter 7, "Constructional techniques". The block is returned to the milling machine and a channel 4.775mm wide and 2.387mm deep (critical) is milled centrally across the face which mates with the choke block, as shown in Fig 20.30. Finally, the diode mount collet hole, detailed in Fig

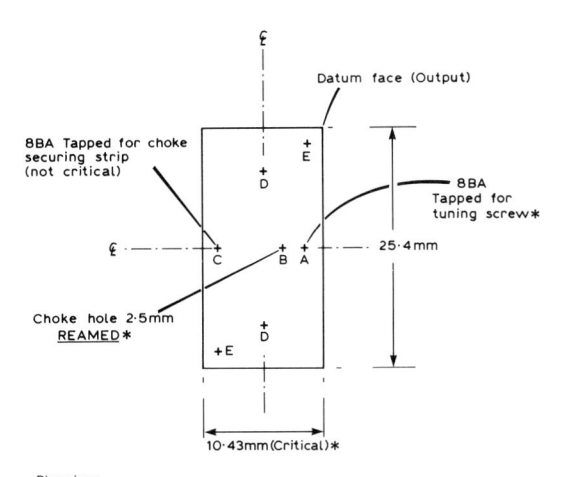

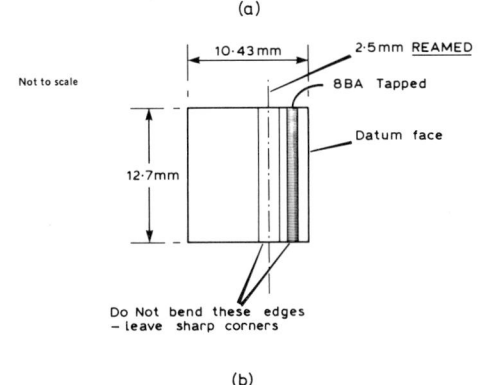

Fig 20.29. (a) and (b): Machining of brass block for the 47GHz Gunn oscillator - choke block

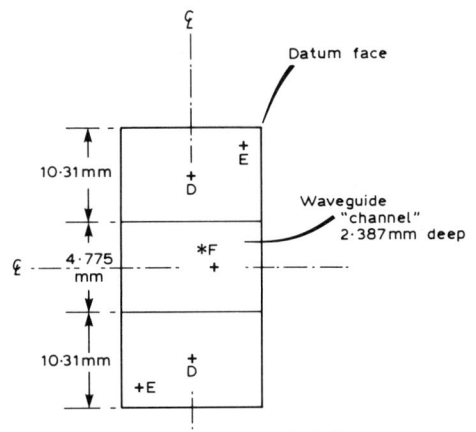

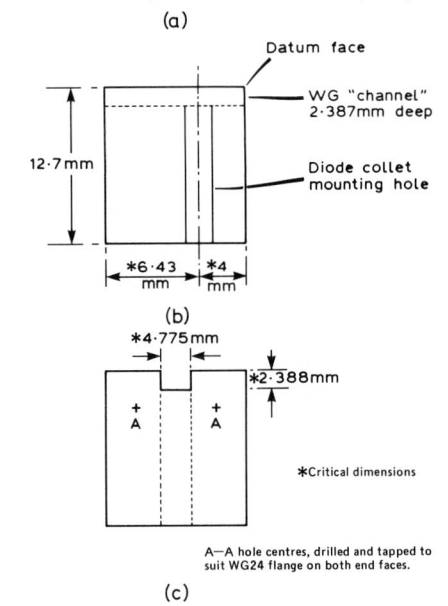

Fig 20.30. (a), (b), and (c): Detail of the cavity block

20.31(b) is drilled and tapped in the position shown in Fig 20.29(b). During milling of the channel it is *essential* that sharp, right-angled corners are achieved.

The dowel pins are gently driven into one of the blocks; the corresponding holes in the other block should be enlarged very slightly to avoid the pins binding when the two blocks are fitted together. Now try the blocks for fit and temporarily bolt them together; they should be a very neat fit! Whilst fitted together, mark the centres for the waveguide flange bolts on each of the two outer faces of the waveguide channel in the assembled block. The holes should be drilled 6BA tapping (or similar metric size) to a depth sufficient to take a short length of studding, which can now be cut, fitted and carefully soldered in place.

A cavity back-plate or short circuit is made from thick (eg 4mm) brass or copper plate, drilled with 6BA clearance holes to centres to correspond to the studs. In the final assembly it is held in place with nuts and shake-proof washers. A diode mounting adapter collet is made from *copper* (heat conduction is important) to the dimensions of Fig 20.31(a) and, on completion, is checked for diode fit and fit into the cavity block mounting hole. At each stage of all the above operations care must be taken to ensure clean, flat, unblemished surfaces on the blocks and to removal of swarf and burrs from surfaces and holes.

The next operation is to fabricate several choke "pins" to the dimensions of Fig 20.30(c). It is suggested that a series of pins be made with the diameter of the diode contacting pin varying between 2.6 and 1.2mm in 0.1mm steps, because it is believed that this dimension is critical and best selected on test. The length of the contacting pin should be

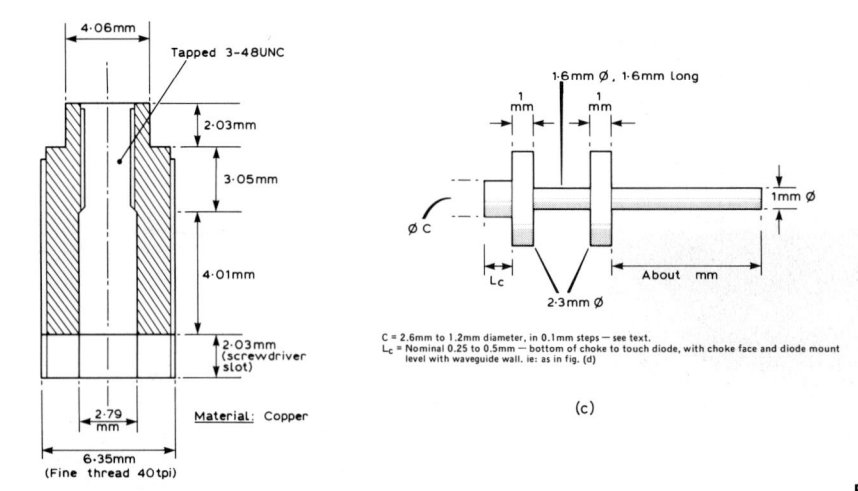

Fig 20.31. (a) Detail of the Gunn diode mounting screw, (b) detail of the diode mounting hole in the oscillator block, (c) detail of the choke, (d) Gunn diode and choke mounting positions

C = 2.6mm to 1.2mm diameter, in 0.1mm steps — see text.
L_c = Nominal 0.25 to 0.5mm — bottom of choke to touch diode, with choke face and diode mount level with waveguide wall. ie: as in fig. (d)

NOTE: Both the mount and the choke ring are flush with the cavity.

adjusted so that the face of the diode is contacted with the diode and choke in the positions shown in Fig 20.30(d).

Insulation of the choke was provided by means of 0.1mm thick shrink sleeving fitted over the choke after completion. The overall diameter should be a sliding fit into the choke block hole and the choke is held in place by a thin plastic strip attached to a screw fitted into the rear tapped hole in the choke block; contact pressure must be small otherwise the diode may be damaged.

An 8BA tuning screw is fabricated either from ptfe or from a brass screw with a ptfe insert fitted into a small centre-drilled hole in the end of an 8BA screw of appropriate length. The iris plate, which completes the assembly, is made from thin brass shim the same size as the standard flange, drilled accordingly and sandwiched between the output end of the cavity and the output flange. The size of the iris hole might start at 0.5mm diameter and be enlarged, on test, to provide the best compromise between power output and stability, as with the corresponding 10GHz designs. A photograph of the completed prototype module is given in Fig 20.32. This should give some idea of the size and general form of construction of the oscillator.

20.9.4 Alternative approaches to 47GHz

The following is an abstract of the ideas used by HB9MIN and HB9AMH [2] which led to the establishment of the first world record for amateur communication at 47GHz, of slightly more than 50km, which stood from 1984 until August 1988. One of the main problems for the amateur (and, it seems, to many professionals too!) is access to suitable test equipment; some home-made test equipment is described in outline in this section.

Three possibilities for the generation of 47GHz signals were considered and examined:

Fig 20.32. Photograph of a prototype 47GHz Gunn oscillator (G8AGN)

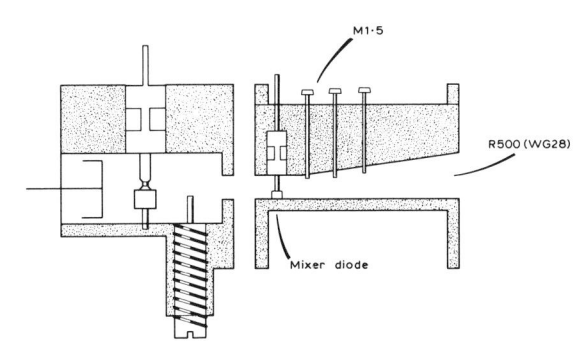

Fig 20.33. Use of 23.5GHz oscillator followed by a doubler (HB9MIN)

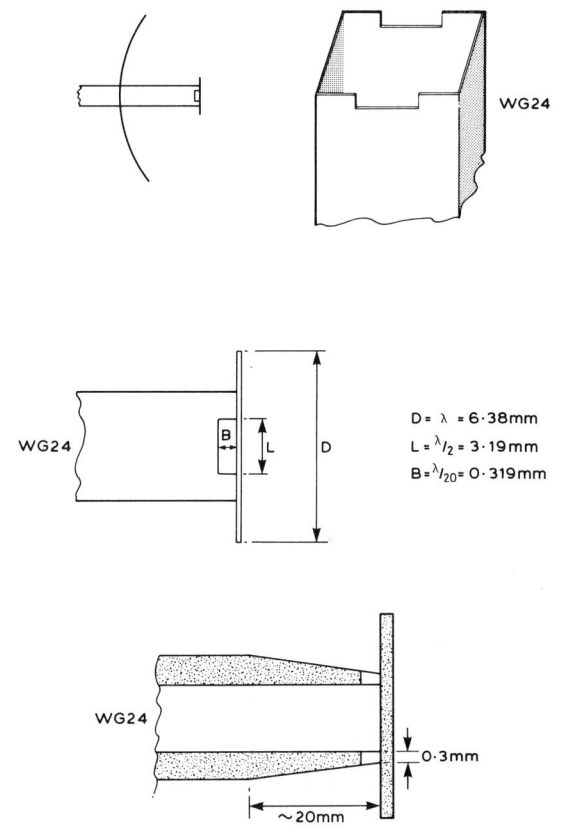

$D = \lambda = 6.38\text{mm}$
$L = \lambda/2 = 3.19\text{mm}$
$B = \lambda/20 = 0.319\text{mm}$

Fig 20.34. Dish feed for 47GHz (HB9MIN)

1. Use of the second harmonic from a 23.5GHz Gunn oscillator, with the fundamental suppressed by filtering.

2. The use of a 23.5GHz Gunn oscillator, followed by a Schottky diode doubler.

3. The use of a 47GHz fundamental Gunn oscillator.

The third alternative was immediately dismissed because of the very high cost of Gunn diodes and professional oscillator assemblies. Such oscillators are very temperature sensitive (about 3MHz/°C) and are pulled by at least 100MHz by slight load variation. Exact design frequency is difficult for the amateur to attain because of the unpredictable nature of the effect of the package parasitic elements on the cavity frequency.

Alternative 1 offers perhaps 1 or 2mW output from a commercial 24GHz oscillator; the temperature effect is typically 1.5MHz/°C whilst frequency pulling will be around 10MHz, both figures being much better than with a 47GHz oscillator.

Although alternative 2 is possible, the increase in power output did not seem to be worth the effort of mounting an expensive diode in a special mount which tapers from WG20 to a reduced height diode section and then back to WG23, especially when the 23.5GHz oscillator could be used effectively as either a sub-harmonic self-oscillating mixer or with a low-barrier or zero-bias Schottky diode in an in-line configuration. Ideally the mixer device should be GaAs because silicon diodes are near their limit, although still just usable.

The equipment used consisted of an M/A-COM 24GHz oscillator retuned for 23.5GHz and a low barrier Schottky diode mounted in WG23 waveguide and used in an in-line configuration as shown in Fig 20.33. The antenna used was a 70cm dish (lamp-shade) with an f/D ratio of 0.3, measured gain of 40dB and fed with a scaled-down version of the "penny feed", shown in Fig 20.34. Great accuracy is needed in the construction of the feed and it may be more practical to use a large horn. The pointing accuracy needed for such a dish is better than 0.5°!

To test and align such transceivers, some form of measuring equipment is essential; professional equipment at these frequencies is not common! Several pieces of home-made equipment are described briefly. The first is an up/down converter based on the self-oscillating mixer principle, the i.f output of which is routed to a counter or spectrum analyser at lower

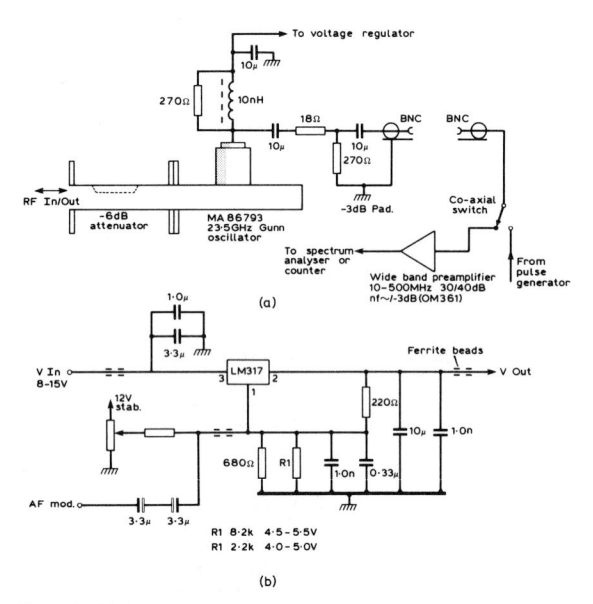

Fig 20.35. (a) Schematic for an up-down converter, and (b) stabilised power supply for the Gunn oscillator. (HB9MIN)

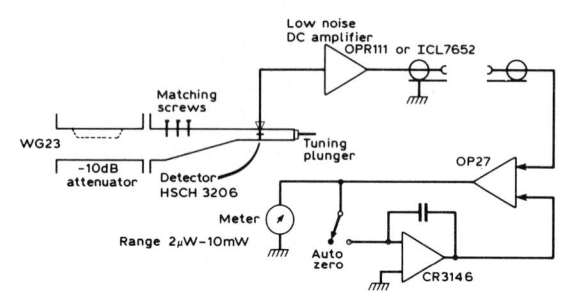

Fig 20.36. Schematic for a power meter (HB9MIN)

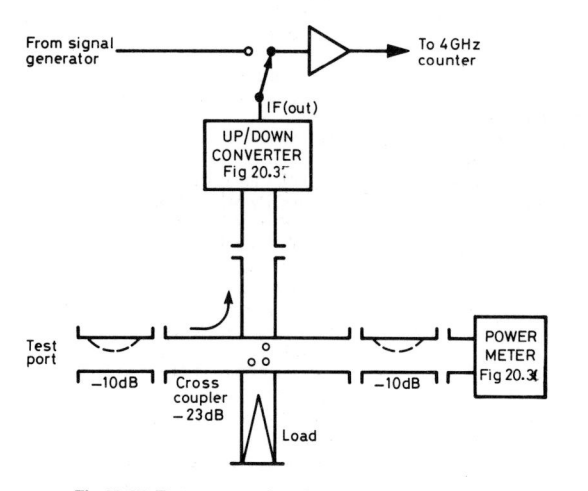

Fig 20.37. Test set up using the power meter (HB9MIN)

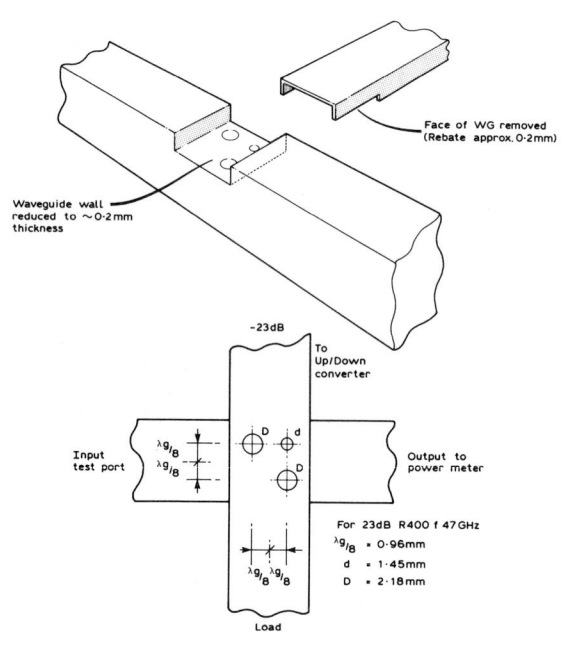

Fig 20.38. Cross coupler dimensions for 47GHz (HB9MIN)

frequency. The block diagram of the converter is given in Fig 20.35. The same converter, when fed from a pulse generator at i.f, will generate test signals in the band.

A power meter was also constructed using a zero bias Schottky diode type HSCH3206, mounted in a reduced height guide, taper matched into WG23 as shown in Fig 20.36. This is normally used with the test set-up shown in Fig 20.37. Fig 20.38 gives some detail of the −23dB cross coupler. This, too, is difficult to make and great accuracy is needed in its construction. It must be pointed out that these ideas, whilst practical, do need a great deal of care and attention in their construction and the degree of inaccuracy which might be tolerable at, say 10GHz, certainly is *not* at 47GHz!

20.10 ACKNOWLEDGEMENTS

The authors would like to acknowledge the assistance of E Zimmermann, HB9MIN, in preparing this chapter.

20.11 REFERENCES

[1] "Propagation above 10GHz – the potential of the amateur mm-wave bands", B Chambers, G8AGN. *Proceedings of the Martlesham Microwave Round Table,* November 1988, published by the Martlesham Radio Society, with acknowledgements to the British Telecom (BT) Research Laboratory, at whose establishment the meeting was held.

[2] "47GHz Praktischer Teil", E Zimmermann, HB9MIN.

Index

1

1,152MHz source, 19.16
1.0 to 1.4GHz source, 14.32
10.7MHz i.f receiver, 18.59
10.7MHz preamplifier, 18.57

2

2.0 to 2.6GHz source, 15.5
2C39 amplifiers
   1.3GHz, 14.18
   2.3GHz, 15.14, 15.18
   using, 14.21
2C39 mixers, 14.16, 15.18
2C39 bias circuits, 14.22
2C39 heater supplies, 14.22

3

384MHz source, 19.16

A

Alford slot
   1.3GHz, 14.27
   2.3GHz, 15.25
Alignment
   G3JVL, 17.6, 18.73,
   multipliers, 18.38, 18.43
   waveguide filters, 18.93
Amateur bands
   1.3GHz, 14.2
   10GHz, 18.1
   2.3GHz, 15.1
   24GHz, 19.24
   3.4GHz, 16.1
   5.7GHz, 17.1
   mm bands, 20.1
Amplifiers
   1.3GHz, 14.18
   10GHz, 18.46
   10GHz GaAs fet, 18.96
   2.3GHz, 14.14, 15.18
   2C39, 14.18, 15.14, 15.18
   modular, 18.99
Antennas
   1.3GHz, 14.22
   10GHz, 18.80
   2.3GHz, 15.23
   24GHz, 19.13
   3.4GHz, 16.10
   5.7GHz, 17.8

Alford slot, 14.27, 15.25
   combiners, 14.24, 15.27
   dish, 14.24, 15.23, 16.10,
   17.8, 18.83, 19.14, 20.12
   helix, 14.27
   matching, 18.93
   mm bands, 20.12
   omni, 14.27, 15.25
   stacking, 14.23
Attenuators
   waveguide, 18.10, 19.11
ATV, 18.63

B

Band capabilities
   1.3GHz, 14.3
   10GHz, 18.2
   120GHz, 20.7
   2.3GHz, 15.2
   24GHz, 19.3
   3.4GHz, 16.2
   5.7GHz, 17.2
   mm bands, 20.5
Band planning
   1.3GHz, 14.2
   10GHz, 18.1
   24GHz, 19.24
Beacons
   1.3GHz, 14.2
   2.3GHz, 15.1
   3.4GHz, 16.1
   5.7GHz, 17.1
Butler oscillator, 14.32, 15.5

C

Changeover system
   G3JVL transverter, 18.74
Circulators, 18.19
Coaxial cable
   use at 24GHz, 19.4
   use on mm bands, 20.10
Combiners
   1.3GHz, 14.24
   2.3GHz, 15.27
Converters
   1.3GHz, 14.4, 14.6
   10GHz, 18.49
   10GHz–144MHz, 18.107
   144MHz–10GHz, 18.112
   2.3GHz, 15.2
   3.4GHz, 16.2

30–10.7MHz, 18.58
5.7GHz, 17.2
Cross-couplers
   24GHz, 19.19
   47GHz, 20.20
Crystal controlled sources
   24GHz, 19.15

D

Detectors
   10GHz, 18.8
   240GHz, 19.4
   for mm bands, 20.11
Dielectric resonators, 18.99
Dielectric waveguide, 20.10
Dipole dish feed, 14.25,
   15.27, 16.12, 18.87
Directional couplers
   10GHz, 18.10
   24GHz, 19.9
   47GHz, 20.20
Dish feeds
   1.3GHz, 14.25
   10GHz, 18.83
   2.3GHz, 15.25
   24GHz, 19.14
   3.4GHz, 16.12
   5.7GHz, 17.8
   dipole, 14.25, 15.27,
     16.12, 18.87
   indirect, 18.87
Dishes
   1.3GHz, 14.24
   10GHz, 18.83
   2.3GHz, 15.23
   24GHz, 19.14
   3.4GHz, 16.10
   dustbin lid, 18.84
   mm bands, 20.12
Doppler radar, 18.49
DRO mixers, 18.54
Droitwich
   frequency standard, 18.79
DROs, 18.29, 18.97
   power supplies, 18.31

E

Equipment capability
   1.3GHz, 14.3
   10GHz, 18.3
   2.3GHz, 15.2

24GHz, 19.2
3.4GHz, 16.1
5.7GHz, 17.2
mm bands, 20.5, 20.7

F

Fast scan atv, 18.63
Feed horns
   1.3GHz, 14.26
   10GHz, 18.84
   2.3GHz, 15.25
   3.4GHz, 16.12
   5.7GHz, 17.9
   circular polarised, 14.26
Filters
   1.3GHz, 14.28
   10GHz, 18.16
   2.3GHz, 15.28
   24GHz, 19.17, 19.23
   3.4GHz, 16.13
   alignment, 18.93, 19.17
   interdigital, 14.30,
     15.28, 16.13
   waveguide, 18.16, 19.23
Flanges
   for mm bands, 20.7
   WG16, 18.4
   WG20, 19.4
Flux
   lunar, 18.131
   solar, 18.131
Frequency
   measurement, 18.98
   off air standards, 18.79

G

G3JVL transverters
   10GHz Mk1, 18.68
   10GHz Mk2, 18.71
   10GHz Mk3, 18.71
   5.7GHz, 17.3
   alignment, 17.6
   performance, 18.76
G3WDG-001, 18.100
G3WDG-002, 18.107
G3WDG-003, 18.112
G4DDK-001 source, 14.32
G4DDK-002 source, 15.5
G4DDK-004 source, 18.101
GaAs fet amplifiers, 14.8,
     15.9, 18.96

Ground noise, 18.133
Gunn diodes
    24GHz, 19.15
    measurements, 18.89
Gunn oscillators
    10GHz, 18.22–28
    24GHz, 19.4, 19.18
    47GHz, 20.16
    tuning, 18.21–22
    GDHM32, 19.7
    GDO33/GDHO33, 19.6
    GDO33, 19.18
    measurements, 18.89
    operating voltage, 18.35
    psus, 18.31, 18.59, 19.6
    tv modulator, 18.65

H

Helix aerials, 14.27
High-Q wavemeter, 19.13
Horns
    10GHz, 18.80
    24GHz, 19.13
    from oil cans, 18.82
    mm bands, 20.12
    sectoral, 18.83
Hybrid-T, 18.20, 18.54

I

Injection locking, 18.44
Interdigital converters
    1.3GHz, 14.4
    2.3GHz, 15.2
    3.4GHz, 16.2
Interdigital filters
    1.3GHz, 14.30
    2.3GHz, 15.28
    3.4GHz, 16.13
Isolators, 18.18

L

Lenses, 20.13
Loads
    waveguide, 18.10, 19.10
Local oscillators
    1.3GHz, 14.32
    2.3GHz, 15.5
Loop Yagis
    1.3GHz, 14.22
    2.3GHz, 15.24
    3.4GHz, 16.11
Lunar flux, 18.131

M

Matched loads, 18.10, 19.10
Matching
    amplifiers, 18.94
    antennas, 18.93
Measurements
    10GHz test bench, 18.88
    frequency, 18.98
Mixers
    1.3GHz, 14.4, 14.6
    10GHz, 18.8, 18.50, 18.68
    2.3GHz, 15.2, 15.13
    24GHz, 19.5, 19.8, 19.18
    3.4GHz, 16.2, 16.10
    47GHz, 20.12, 20.14
    5.7GHz, 17.2, 17.3
    dro, 18.54
    for mm bands, 20.11
    offset mixer, 18.9
    G3JVL, 17.3, 18.6
    self-oscillating, 18.50, 20.12
Modulators,
    pin diode, 18.16
Moonbounce,
    1.3GHz, 14.1, 14.26
Multipliers
    1,152–10,368MHz, 18.41
    2.3GHz varactor, 15.11
    2.5 to 10GHz, 18.100
    24GHz waveguide, 19.16
    3.4GHz varactor, 16.7
    384–1,152MHz, 18.38
    384–10,368MHz, 18.38
    5.7GHz, 17.5, 17.7
    for mm bands, 20.11

N

Narrowband equipment
    10GHz, 18.67
    24GHz, 19.15
Noise figure, 18.131
Noise
    ground, 18.133
    moon, 18.130
    oscillator, 18.77
    sun, 18.130
    thermal, 18.130

O

Omnidirectional antennas
    2.3GHz, 15.25
    1.3GHz, 14.27
Oscillators
    10GHz, 18.20
    dielectric resonator, 18.29
    for mm bands, 20.11
    Gunn, 18.22, 18.24, 18.27, 19.4, 19.6, 19.18, 20.16
    narrowband, 18.36
    noise, 18.46, 18.77
Overmoded waveguide, 20.8

P

Path loss capability
    1.3GHz, 14.4
    10GHz, 18.3
    2.3GHz, 15.2
    24GHz, 19.2
    3.4GHz, 16.2
    5.7GHz, 17.2
    mm bands, 20.6
PCBs at 10GHz, 18.94
Penny feed, 18.86, 19.14

Performance
    checking, 18.130
PIN diode switches, 18.16
PLL, 19.15, 19.19
Power amplifiers
    1.3GHz, 14.18
    uhf, 18.37
Power measurement, 14.28
    47GHz, 20.20
    high power, 14.32
Power meter, 14.31
Power supplies
    dro, 18.31, 18.32
    Gunn, 18.31, 18.59, 19.6
    twt, 18.119
Preamplifiers
    1.3GHz, 14.7, 14.8
    10.7MHz, 18.57
    144MHz, 18.74
    2.3GHz, 15.9
    3.4GHz, 16.5
    i.f, 18.55
    SL560c ic, 18.56
    vhf, 18.55
Propagation
    1.3GHz, 14.3
    2.3GHz, 15.1
    24GHz, 19.1
    3.4GHz, 16.1
    5.7GHz, 17.1
    free space loss, 20.2
    mm bands, 20.2
    oxygen attenuation, 20.2
    rainfall, 20.4–5
    water vapour, 19.1, 20.2

R

Receivers
    1.3GHz, 14.4
    10GHz, 18.45, 18.107
    2.3GHz, 15.2
    3.4GHz, 16.2
    5.7GHz, 17.2
    choice of i.f, 18.46
    image rejection, 18.46
    oscillator noise, 18.46
    television, 18.63
Relative humidity, 20.3
Repeaters, 14.1, 14.3

S

Sectoral horns, 18.83
Slug tuner, 14.32, 15.28
Solar flux, 18.131
Spectrum analysis, 18.91
Stacking antennas, 14.23
Sun noise, 18.130
Switches
    pin diode, 18.16
    waveguide, 18.13

T

Television
    10GHz, 18.63

Gunn modulator, 18.65
Thermal noise, 18.130
Transmitters
    1.3GHz, 14.11–12, 14.16
    10GHz, 18.48, 18.112
    2.3GHz, 15.11
    3.4GHz, 16.6
    5.7GHz, 17.7
Triplers
    1.3GHz, 14.11, 18.38
    3.4GHz, 16.7
Tuners
    slug, 14.32, 15.28
    three-screw, 18.7
TWTs, 18.118
    power supply, 18.119
    surplus, 18.119
TX-RX sequencer, 18.116

V

Varactor
    mixers, 15.13, 16.10
    multipliers, 15.11, 16.7, 17.7
VSWR measurement, 18.7

W

Water attenuation, 20.2
Waveguide
    to coax, 17.9, 18.11, 19.4
    cross-coupler, 19.5
    dielectric, 20.10
    diode mounts, 18.8, 19.8
    filters, 18.16, 19.23
    flanges, 18.4, 19.3, 20.7
    matched loads, 19.10
    matching, 18.6
    mm bands, 20.7
    Old English, 18.6
    overmoded, 20.8
    round, 18.5
    shorts, 18.10
    switches, 18.13
    WG16, 18.4
    WG20, 19.4
Wavemeters
    high-Q, 19.13
    self-calibrating, 19.12
    waveguide, 18.11, 19.12
Wideband equipment
    10GHz, 18.48, 18.62
    24GHz, 19.5
    setting up, 19.11

Y

Yagis
    loop, 14.22, 15.24, 16.11
    stacking, 14.23

Some other RSGB publications...

❏ AMATEUR RADIO CALL BOOK
As well as a list of all UK and Republic of Ireland radio amateurs, this essential reference work also includes an information directory giving useful addresses, EMC advice, lists of amateur radio clubs, operating data, and much more.

❏ PACKET RADIO PRIMER
A light-hearted introduction to the exciting new world of packet radio which will help any beginner to get started with the minimum of fuss. Detailed practical advice on connecting up equipment is followed by a guide through the maze of configurations possible. Then sample logs of contacts with the various forms of 'mailbox' are featured to help you get the best out of the network. Much reference information is also included to supplement your equipment manuals.

❏ SPACE RADIO HANDBOOK
Space exploration by radio is exciting and it is open to anyone! This book shows you how it is done, and the equipment you will need. It covers the whole field of space radio communication and experimentation, including meteor scatter, moonbounce, satellites and simple radio astronomy. A particularly valuable feature is a collection of experiments which will be of interest to schools wishing to explore the many educational possibilities. If you are ready to use radio to explore beyond the atmosphere, let this book be your guide.

❏ VHF/UHF MANUAL
This standard UK textbook on the theory and practice of amateur radio reception and transmission between 30MHz and 24GHz includes full constructional details of many items of equipment. While the contents are intended primarily for the amateur radio enthusiast, there is much information of value to the professional engineer.

❏ LOCATOR MAP OF EUROPE
This map shows IARU locator squares down to the secondary level (IO91 etc). The map covers a geographical area from Iceland to Turkey, and from arctic Russia to Morocco. It measures approximately 900mm by 625mm. An A4 size version of this map (less the instructions) is also available printed onto card for desk-top reference.

❏ LOCATOR MAP OF WESTERN EUROPE
This map shows IARU locator squares down to the tertiary level and covers an area from the Faroe Islands to central Italy, and from southern Sweden to northern Spain. The map measures approximately 1100mm by 900mm.

RADIO SOCIETY OF GREAT BRITAIN
Lambda House, Cranborne Road,
Potters Bar, Herts EN6 3JE

RSGB – representing amateur radio ...
... representing you!

Radio Communication

A magazine which covers a wide range of interests and which features the best and latest amateur radio news. The Society's journal has acquired a world-wide reputation for its content. It strives to maintain its reputation as the best available and is now circulated, free of charge, to members in over 150 countries.

The regular columns in the magazine cater for HF, VHF/UHF, microwaves, SWL, clubs, satellites, data and contests. In addition to technical articles, the highly regarded 'Technical Topics' feature caters for those wishing to keep themselves briefed on recent developments in technical matters. There is also a special column for Novice licensees.

The 'Last Word' is a lively feature in which members can put forward their views and opinions and be sure of receiving a wide audience. To keep members in touch with what's going on in the hobby, events diaries are published each month.

Subsidised advertisements for the equipment you wish to sell can be placed in the magazine, with the advantages of short deadlines and large circulation.

QSL Bureau

Members enjoy the use of the QSL Bureau free of charge for both outgoing and incoming cards. This can save you a good deal of postage.

Special Event Callsigns

Special Event Callsigns in the GB series are handled by RSGB. They give amateurs special facilities for displaying amateur radio to the general public.

Specialised News Sheets

The Society publishes the weekly *DX News-sheet* for HF enthusiasts and the *Microwave Newsletter* for those operating above 1GHz.

Specialised Equipment Insurance

Insurance for your valuable equipment which has been arranged specially for members. The rates are very advantageous.

Audio Visual Library

Films, audio and video tapes are available through one of the Society's Honorary Officers for all affiliated groups and clubs.

Reciprocal Licensing Information

Details are available for most countries on the RSGB computer database.

Government Liaison

One of the most vital features of the work of the RSGB is the ongoing liaison with the UK Licensing Authority – presently the Radiocommunications Agency of the Department of Trade and Industry. Setting and maintaining the proper framework in which amateur radio can thrive and develop is essential to the well-being of amateur radio. The Society spares no effort in defence of amateur radio's most precious assets – the amateur bands.

Beacons and Repeaters

The RSGB supports financially all repeaters and beacons which are looked after by the appropriate committee of the Society, ie, 1.8-30MHz by the HF Committee, 30-1000MHz (1GHz) by the VHF Committee and frequencies above 1GHz by the Microwave Committee. For repeaters, the Society's Repeater Management Group has played a major role. Society books such as the *Amateur Radio Call Book* give further details, and computer-based lists giving up-to-date operational status can be obtained by post from HQ.

Operating Awards

A wide range of operating awards are available via the responsible officers: their names can be found in the front pages of *Radio Communication* and in the Society's *Amateur Radio Call Book*. The RSGB also publishes a book which gives details of most major awards.

Contests (HF/VHF/Microwave)

The Society has two contest committees which carry out all work associated with the running of contests. The HF Contests Committee deals with contests below 30MHz, whilst events on frequencies above 30MHz are dealt with by the VHF Contests Committee.

Morse Testing

In April 1986 the Society took over responsibility for morse testing of radio amateurs in the UK. If you wish to take a morse test, write direct to RSGB HQ (Morse tests) for an application form.

Slow Morse

Many volunteers all over the country give up their time to send slow morse over the air to those who are preparing for the 5 and 12 words per minute morse tests. The Society also produces morse instruction tapes.

RSGB Books

The Society publishes a range of books for the radio amateur and imports many others. RSGB members are entitled to a discount on all books purchased from the Society. This discount can offset the cost of membership.

Propagation

The Society's Propagation Studies Committee is highly respected – both within the amateur community and professionally – for its work. Predictions are given in the weekly GB2RS news bulletins and the Society's monthly magazine *Radio Communication*.

Technical and EMC Advice

Although the role of the Society's Technical and Publications Advisory Committee is largely to vet material intended for publication, its members and HQ staff are always willing to help with any technical matters.

Breakthrough in domestic entertainment equipment can be a difficult problem to solve as well as having licensing implications. The Society's EMC Committee is able to offer practical assistance in many cases. The Society also publishes a special book to assist you. Additional advice can be obtained from the EMC Committee Chairman via RSGB HQ.

Planning Permission

There is a special booklet and expert help available to members seeking assistance with planning matters.

GB2RS

A special radio news bulletin transmitted each week and aimed especially at the UK radio amateur and short wave listener. The script is prepared each week by the Society's HQ staff. The transmission schedule for GB2RS is printed regularly in *Radio Communication*, or it can be obtained via the Membership Services Department at HQ. It also appears in the *Amateur Radio Call Book*. The GB2RS bulletin is also sent out over the packet radio network.

Raynet (Radio Amateur Emergency Network)

Several thousand radio amateurs give up their free time to help with local, national and sometimes international emergencies. There is also ample opportunity to practise communication and liaison skills at non-emergency events, such as county shows and charity walks, as a service to the people. For more information or full details of how to join, contact the Membership Services Department at RSGB HQ.

RSGB Exhibitions and Mobile Rallies

The Society's Exhibition and Rally Committee organizes an annual exhibition and an annual mobile rally. Full details and rally calendar can be found in *Radio Communication*.

RSGB Conventions

The Society's diary in *Radio Communication* contains details of all special conventions which are open to all radio amateurs. The Society holds several major conventions each year.

Observation Service

A number of leading national radio societies have volunteers who monitor the amateur bands as a service to the amateur community. Their task is to spot licence infringements and defective transmissions, and report them in a friendly way to the originating station.

Intruder Watch

This helps to protect the exclusive amateur bands by monitoring for stations not authorised to use them.

Send for our Membership Information Pack today and discover how you too can benefit from these services. Write to:

RADIO SOCIETY OF GREAT BRITAIN, Lambda House, Cranborne Road, Potters Bar, Herts EN6 3JE

Notes